Lecture Notes in Computer Science

Commenced Publication in 1973
Founding and Former Series Editors:
Gerhard Goos, Juris Hartmanis, and Jan van Leeuwen

T0237878

Torben Bach Pedersen   Mukesh K. Mohania
A Min Tjoa (Eds.)

Data Warehousing and Knowledge Discovery

11th International Conference, DaWaK 2009
Linz, Austria, August 31–September 2, 2009
Proceedings

 Springer

Volume Editors

Torben Bach Pedersen
Aalborg University
Department of Computer Science
Selma Lagerlöfsvej 300, 9220 Aalborg Ø, Denmark
E-mail: tbp@cs.aau.dk

Mukesh K. Mohania
IBM India Research Lab
Plot No. 4, Block C, Institutional Area
Vasant Kunj, New Delhi 110 070, India
E-mail: mkmukesh@in.ibm.com

A Min Tjoa
Vienna University of Technology
Institute of Software Technology and Interactive Systems
Favoritenstr. 9-11/188, 1040 Wien, Austria
E-mail: amin@ifs.tuwien.ac.at

Library of Congress Control Number: 2009932136

CR Subject Classification (1998): H.2, H.4, H.3, J.1, H.2.8, H.3.3, I.5.3

LNCS Sublibrary: SL 3 – Information Systems and Application, incl. Internet/Web and HCI

ISSN   0302-9743
ISBN-10   3-642-03729-1 Springer Berlin Heidelberg New York
ISBN-13   978-3-642-03729-0 Springer Berlin Heidelberg New York

springer.com

© Springer-Verlag Berlin Heidelberg 2009
Printed in Germany

Typesetting: Camera-ready by author, data conversion by Scientific Publishing Services, Chennai, India
Printed on acid-free paper   SPIN: 12737444   06/3180   5 4 3 2 1 0

Preface

Data warehousing and knowledge discovery are increasingly becoming mission-critical technologies for most organizations, both commercial and public, as it becomes increasingly important to derive important knowledge from both internal and external data sources. With the ever growing amount and complexity of the data and information available for decision making, the process of data integration, analysis, and knowledge discovery continues to meet new challenges, leading to a wealth of new and exciting research challenges within the area.

Over the last decade, the International Conference on Data Warehousing and Knowledge Discovery (DaWaK) has established itself as one of the most important international scientific events within data warehousing and knowledge discovery. DaWaK brings together a wide range of researchers and practitioners working on these topics. The DaWaK conference series thus serves as a leading forum for discussing novel research results and experiences within data warehousing and knowledge discovery. This year's conference, the 11th International Conference on Data Warehousing and Knowledge Discovery (DaWaK 2009), continued the tradition by disseminating and discussing innovative models, methods, algorithms, and solutions to the challenges faced by data warehousing and knowledge discovery technologies.

The papers presented at DaWaK 2009 covered a wide range of aspects within data warehousing and knowledge discovery. Within data warehousing and analytical processing, the topics covered data warehouse modeling including advanced issues such as spatio-temporal warehouses and DW security, OLAP on data streams, physical design of data warehouses, storage and query processing for data cubes, advanced analytics functionality, and OLAP recommendation. Within knowledge discovery and data mining, the topics included stream mining, pattern mining for advanced types of patterns, advanced rule mining issues, advanced clustering techniques, spatio-temporal data mining, data mining applications, as well as a number of advanced data mining techniques. It was encouraging to see that many papers covered emerging important issues such as spatio-temporal data, streaming data, non-standard pattern types, advanced types of data cubes, complex analytical functionality including recommendations, multimedia data, mssing and noisy data, as well as real-world applications within genetics and within the clothing and telecom industries. The wide range of topics bears witness to the fact that the data warehousing and knowledge discovery field is dynamically responding to the new challenges posed by novel types of data and applications.

From 124 submitted abstracts, we received 100 papers from 17 countries in Europe, North America and Asia. The Program Committee finally selected 36 papers, yielding an acceptance rate of 36%.

We would like to express our most sincere gratitude to the members of the Program Committee and the external reviewers, who made a huge effort to review the papers in a timely and thorough manner. Due to the tight timing constraints and the high number of submissions, the reviewing and discussion process was a very challenging task, but the commitment of the reviewers ensured that a very satisfactory result was achieved.

We would also like to thank all authors who submitted papers to DaWaK 2009, for their contribution to making the technical program so excellent.

Finally, we extend our warmest thanks to Gabriela Wagner for delivering an outstanding level of support within all aspects of the practical organization of DaWaK 2009. We also thank Amin Anjomshoaa for his support with the conference management software.

August 2009   Torben Bach Pedersen
                                                  Mukesh Mohania
                                                      A Min Tjoa

Organization

Program Chairs

Torben Bach Pedersen   Aalborg University, Denmark
Mukesh Mohania   IBM India Research Lab, India
A Min Tjoa   Vienna University of Technology, Austria

Publicity Chair

Alfredo Cuzzocrea   ICAR-CNR and University of Calabria, Italy

Program Committee

Alberto Abello Gamazo   Universitat Politecnica de Catalunya, Spain
Elena Baralis   Politecnico di Torino, Italy
Ladjel Bellatreche   Poitiers University, France
Petr Berka   University of Economics, Prague, Czech Republic
Jorge Bernardino   Instituto Superior de Engenharia de Coimbra, Portugal
Elisa Bertino   Purdue University, USA
Mokrane Bouzeghoub   CNRS - Université de Versailles SQY, France
Stephane Bressan   National University of Singapore, Singapore
Peter Brezany   University of Vienna, Austria
Robert Bruckner   Microsoft, USA
Erik Buchmann   Universität Karlsruhe, Germany
Jesús Cerquides   Universitat de Barcelona, Spain
Zhiyuan Chen   University of Maryland Baltimore County, USA
Sunil Choenni   The Netherlands Ministry of Justice, The Netherlands
Frans Coenen   University of Liverpool, UK
Bruno Cremilleux   Université de Caen, France
Alfredo Cuzzocrea   ICAR-CNR and University of Calabria, Italy
Agnieszka Dardzińska   University of North Carolina at Chapel Hill, Poland
Karen C. Davis   University of Cincinnati, USA
Kevin Desouza   University of Washington, USA
Curtis Dyreson   Utah State University, USA
Todd Eavis   Concordia University, USA
Johann Eder   University of Klagenfurt, Austria
Tapio Elomaa   Tampere University of Technology, Finland
Roberto Esposito   Università di Torino, Italy

Vladimir Estivill-Castro	Griffith University, Australia
Christie Ezeife	University of Windsor, Canada
Jianping Fan	UNC-Charlotte, USA
Ling Feng	Tsinghua University, China
Eduardo Fernandez-Medina	Universidad de Castilla-La Mancha, Spain
Ada Fu	Chinese University of Hong Kong, Hong Kong
Dragan Gamberger	Ruder Boškovic Institute, Croatia
Chris Giannella	Information Systems Security Operation of Sparta, Inc., USA
Matteo Golfarelli	University of Bologna, Italy
Eui-Hong (Sam) Han	iXmatch Inc., USA
Wook-Shin Han	Kyungpook National University, Korea
Jaakko Hollmén	Helsinki University of Technology, Finland
Xiaohua (Tony) Hu	Drexel University, USA
Jimmy Huang	York University, Canada
Farookh Khadeer Hussain	Curtin University of Technology, Australia
Ryutaro Ichise	Japan National Institute of Informatics, Japan
Mizuho Iwaihara	Kyoto University, Japan
Alípio Mário Jorge	University of Porto, Portugal
Murat Kantarcioglu	University of Texas at Dallas, USA
Jinho Kim	Kangwon National University, Korea
Sang-Wook Kim	Hanyang University , Korea
Jörg Kindermann	Fraunhofer Institute, Germany
Jens Lechtenboerger	Westfälische Wilhelms-Universität Münster, Germany
Wolfgang Lehner	Dresden University of Technology, Germany
Sanjay Madria	University of Missouri-Rolla, USA
Jose Norberto Mazón López	University of Alicante, Spain
Anirban Mondal	University of Tokyo, Japan
Ullas Nambiar	IBM Research, India
Jian Pei	Simon Fraser University, Canada
Evaggelia Pitoura	University of Ioannina, Greece
Stefano Rizzi	University of Bologna, Italy
Monica Scannapieco	University of Rome"La Sapienza", Italy
Alkis Simitsis	HP Labs, USA
Il-Yeol Song	Drexel University, USA
Koichi Takeda	Tokyo Research Laboratory, IBM Research, Japan
Dimitri Theodoratos	New Jersey Institute of Technology, USA
Christian Thomsen	Aalborg University, Denmark
Igor Timko	Free University of Bozen-Bolzano, Italy
Juan-Carlos Trujillo Mondéjar	University of Alicante, Spain
Panos Vassiliadis	University of Ioannina, Greece
Millist Vincent	University of South Australia, Australia
Wolfram Wöß	Johannes Kepler Universität Linz, Austria
Robert Wrembel	Poznan University of Technology, Poland
Xiaofang Zhou	University of Queensland, Australia
Esteban Zimanyi	Université Libre de Bruxelles, Belgium

External Reviewers

Timo Aho
Jussi Kujala
Ryan Bissell-Siders
Marc Plantevit
Francois Rioult
Ke Wang
Jinsoo Lee
Julius Köpke
Marcos Aurelio Domingues
Nuno Escudeiro
Tania Cerquitelli
Paolo Garza
Ibrahim Elsayed
Fakhri Alam Khan
Yuzhang Han
Xiaoying Wu

Table of Contents

Pattern Mining

Data Cubes

Data Mining Applications

Analytics

Data Mining

Clustering

Spatio-Temporal Mining

Rule Mining

Olap Recommendation

New Challenges in Information Integration

Laura M. Haas[1] and Aya Soffer[2]

[1] IBM Almaden Research Center, 650 Harry Road, San Jose, CA 95120, USA
[2] IBM Haifa Research Lab, Haifa University Campus, Mount Carmel, Haifa, 31905 Israel
laura@almaden.ibm.com, ayas@il.ibm.com

Abstract. Information integration is the cornerstone of modern business informatics. It is a pervasive problem; rarely is a new application built without an initial phase of gathering and integrating information. Information integration comes in a wide variety of forms. Historically, two major approaches were recognized: data federation and data warehousing. Today, we need new approaches, as information integration becomes more dynamic, while coping with growing volumes of increasingly dirty and diverse data. At the same time, information integration must be coupled more tightly with the applications and the analytics that will leverage the integrated results, to make the integration process more tractable and the results more consumable.

Keywords: Information integration, analytics, data federation, data warehousing, business intelligence solutions.

1 Introduction

Information integration is the cornerstone of modern business informatics. Every business, organization, and today, every individual, routinely deals with a broad range of data sources. Almost any professional or business task we undertake causes us to integrate information from some subset of those sources. A company needing a new customer management application may start by building a warehouse with an integrated and clean record of all information about its customers from legacy data stores and newer databases supporting web applications. A healthcare organization needs to integrate data on its patients from many siloed laboratory systems and potentially other hospitals or doctors' offices. Individuals planning their trip to Austria may integrate information from several different web sites and databases.

There are many information integration problems [1]. Different environments, data sources, and goals have led to a proliferation of information integration technologies and tools [2], each addressing a different piece of the information integration process, for a particular context. There are tools to help explore data on the web, tools to track metadata in an enterprise, and tools to help identify common objects in different data sources. Other technologies focus on information transformation, specifying what data should be transformed and how to transform it, or actually doing the transformation to create the needed data set.

Two major technologies for information integration are data warehousing and data federation. Data warehousing materializes the integrated information, typically leveraging Extract/Transform/Load (ETL) tools to do scalable processing of complex

T.B. Pedersen, M.K. Mohania, and A M. Tjoa (Eds.): DaWaK 2009, LNCS 5691, pp. 1–8, 2009.

transforms. Data federation, on the other hand, is a form of virtual integration; the data is brought together as it is needed. Both techniques have their strengths; materialization allows for efficient access, and, because data is typically processed offline, enables multiple passes so that more complex analyses and transforms can be handled. However, because data warehouses are built ahead of actual use, they have to process all data that *might* be needed. Data federation, by contrast, only transfers and transforms data that is actually needed by the user. Data warehouses typically take months of planning and effort to create; a data federation can be set up in a few days or weeks. Both techniques focus on creating the necessary data sets, and leave any analysis of the data to the application. Both require the use of additional tools to clean data, discover common entities across data sets, etc.

Today's world calls for different technologies. The goal of information integration is increasingly to bring together information in a way that gives the user some new insights. In fact, in the current business climate, leveraging information to enhance the business can be a key competitive differentiator. So users need to get not only data, but also understanding; they need information in a form they can use, and they need it faster than ever before. There are five key challenges to meeting these needs. First, there are many integration problems, with many different characteristics, each needing different tools and techniques. Second, today's data is not static: we must deal with the dynamics of changing data values, changing data formats, and changing user needs. Third, there are not only more data sources than ever that may need to be connected, but also many types of analytics that may need to be coordinated and applied to the correct data sets as part of the integration. Fourth, there is an opportunity to leverage powerful emerging infrastructures to speed the integration and analysis process, but new algorithms or abstractions are required. Finally, in order to really get insight out of the information, the results must be made "consumable" by users – returned in a form in which they are easy to understand, and in which the insights jump out at you.

The rest of this paper is organized as follows. Section 2 explores in more depth the challenges involved in getting insight out of data, especially the many different integration challenges. Section 3 describes some of the research approaches that try to deal with our dynamic world. Section 4 focuses on research into connecting more data sources and analytic engines, and discusses the need to intertwine analytics and traditional integration steps more closely. Section 5 looks at how modern infrastructures and information architectures may allow us to integrate information more rapidly. Section 6 considers briefly work in consumability. We conclude in Section 7.

2 Integration for Insight

The ability to pull value from data is a crucial competitive differentiator for businesses and individuals. Further, substantial value can be realized by making smart data-driven decisions. Due to advances in information technology, vast amounts of heterogeneous data are available online. To derive value from it, however, we increasingly need to analyze and correlate information from diverse sources; the information used must be clean, accurate, consistent, and generally trustworthy.

But information consumers may not know which of these diverse sources are relevant, how to correlate them, what analytics to use, or how to interpret results. Even if the original sources are trustworthy, errors can be easily introduced during the integration process, especially if the consumer is not familiar with the sources. Thus current research in the area of information integration can be viewed as enabling the end-consumer to easily find the relevant data and the appropriate analytics to apply, and ensuring that they get the results in a form that they can understand and interact with.

To achieve this goal, we need to step back and understand what the consumer wants to do with the information. Do they want to have a clean reference dataset, so that they can look up any customer and instantly know how much they've spent this year? Or are they in a very different scenario, trying to correlate information from video, still image, text and databases? Does that correlation need to be done in real-time, to stop a terrorist from getting on a plane, or can it be done overnight, to prepare information on a patient from multiple tests for a doctor who will be seeing the patient the next day? The task at hand, the types of data needed, how quickly it is needed, how accurate it must be – these *desiderata* among others [1] must be considered in creating or choosing tools and technologies to integrate information.

As in the scenarios above, it is often important to be able to identify when multiple sources refer to the same objects. The general problem of finding the data that refers to the same objects or entities is often known as entity resolution. In some scenarios this is harder than in others, and different degrees of precision are needed for different applications. For example, in a hospital there will likely be records from radiology, the lab, and administration all referring to the same patient. In this simple case, there will almost certainly be a patient number of some sort to connect them. But if we add information from other hospitals that might have treated the same patient, finding the common patients becomes harder, as different ids may have been assigned. In this scenario, correctly matching patients may literally be a matter of life and death, so precision is important. Entity resolution may be harder yet in the security scenario in which information from emails, video feeds and criminal record databases must be correlated to find suspects for a recent crime. However, in this application, "recall" (finding all possible candidates) is more important than precision, as further investigations will normally be needed before the information is acted on.

Of course, if multiple data sources refer to the same entity, there is the potential for redundancy and contradiction (actually, these occur within a source, as well). These may be found (and potentially, eliminated) during entity resolution, although they are often looked for earlier, and resolved later. But again, the application scenario will play a major role in how this is done, and to what extent. Some scenarios require a reliable, fully "clean" dataset at the end, as in traditional warehousing scenarios. Others may be focused on the needs of the moment, for example, getting all the data related to the patient that I am about to see. The warehouse will typically exclude redundant information, and force resolution of contradictions. The doctor doesn't want to see three identical records of the same visit, but if those records have some contradictions, she may want to know and see all possibilities. In video surveillance, it may be important to keep all records, no matter how similar they seem.

At all phases of an integration process, the end goal can and should influence the choice of technologies. Information integration design tools – by analogy to physical database design tools – are a worthwhile topic of research. Such tools would be immensely valuable for organizations embarking on any substantial integration project.

3 Dealing with Dynamic Data Environments

To deal with applications that have less rigorous needs for completeness and accuracy, and where an end-user is more directly involved in the process, a more dynamic form of integration is emerging. It was originally inspired by desktop scenarios, in which a user wants to bring together data from email, files, and applications, for example. However, the user only wants a subset of the data for a particular purpose (writing a memo, creating a chart), and, because she is actively involved in the process, is willing to tolerate some errors or inconsistencies. In this environment, integration is done incrementally, as more data is requested. This approach has been labeled "Pay-as-you-go", and the personal work environments that inspired it have been called "Dataspaces" [3].

In these emerging applications, there is no fixed schema that is to be produced, as traditionally existed for warehousing and federation. In fact, in important scenarios such as healthcare patient records or financial transactions there are often hundreds or even thousands of schemas, each of which may be the preferred schema of some user, organization or application. Mapping rules [4] are typically used to relate different schemas, but hundreds of schemas may require thousands of mapping rules, probably incomplete, and evolving over time. Answering queries with that many mapping rules quickly becomes intractable. New techniques are needed to process queries in an almost schema-less world.

With multiple sources of information, overlaps and inconsistencies are the norm. In these dynamic environments, a new challenge is to create methods for on-the-fly data cleansing and entity resolution – when new data is constantly arriving, schemas are not fixed, and there is a specific need. We mentioned earlier the need to find all records relating to a particular patient from many doctors, labs and hospitals. In the desktop environment, a more localized example might be dynamically linking paper authors with emails from them, or finding all information related to an upcoming trip to a conference. These are not requests that today's data cleaning and entity resolution technology can handle.

Another characteristic of these emerging integration scenarios is the availability of additional sources of information that may help – or hinder – integration. For example, the semantic web and linked open data provide a growing volume of semi-structured information that encapsulates additional human knowledge; certain web sites may be seen as authoritative sources for particular domains, e.g., the site of the American Heart Association for heart disease. This knowledge can potentially help with tough integration problems such as entity resolution, or conflict resolution. On the other hand, many such sources are themselves inconsistent, incomplete, or inaccurate, making it difficult to exploit them. Still, ontologies are increasingly being tried as an aid to integration; a recent proposal even leverages people as a source of knowledge directly [5]. In the dynamic world of web data sources, keeping up with changing schemas and interfaces may require the use of metadata and data maintained by the millions of web users, rather than relying on clean, accurate – but instantly stale – individually maintained glossaries or other sources.

4   Converging Integration and Analytics

In the past, information integration typically meant connecting a set of well-structured data sources, but today, the sources are likely to include data that is unstructured (at least, by traditional data standards). Many application scenarios rely on multiple data modalities, such as text, video and structured database data. Typically some analytics must be run on this data before it can be integrated with any other sources. In general, each type of information to be integrated in one of these scenarios may require analysis or other processing in order to extract enough information to permit integration. For example, speech may be turned into text and annotated with keywords or concepts for indexing. Image features, shape recognition, and other analytics may be done on image and video data. The type of analysis may be tuned to the application need. We will extract very different features from text data if it is being used to check for regulatory compliance in banking than we will if the text belongs to patient records and is being used in disease management or diagnosis.

In the past, analytics and information integration were treated in isolation from each other. However, with the incorporation of these complex data types it has become clear that they need, at the least, to be interwoven. Analytics on the underlying data is needed to provide information to the integration process. Integration, in turn, feeds advanced analytics. In some cases, these analytics will drive the need to integrate more data, to enable further analysis.

With this new world view, there are many research challenges to explore. Some research teams are looking at how to analyze particular types of data, discovering the right features to extract, how to correlate them to other types of information, and so on. Others have focused on connecting particular types of information, for example, text with relational data, or speech with video. One newly emerging research challenge is connecting and orchestrating diverse analytic engines. For example, a particular healthcare decision support application might require pipelining patient records through a set of analyses, with the result of one analysis possibly determining other types of analytics to apply. As another example, an application to do content-based targeted advertising for video clips requires several kinds of analytics on the video (object recognition, speech analysis, topic extraction, and so on), plus analytics on the user (profile analysis, sites recently visited, social network analysis, etc). Some of these must be done in sequence; others can be done in parallel. Workflow and streaming engines have been proposed as ways to handle the interconnection and coordination needed to federate these specialized analytics and extract the desired insight while integrating the information flowing through them.

5   Leveraging New Platforms

Many types of analytics require extensive resources, as they involve processing large volumes of data and running complex algorithms over them. For example, large telephone providers (Telcos) in India and China are accumulating call data records at an unprecedented pace (around 1TB of data a day). These records are typically stored for a 3 month period and queried often as part of the Telco's business intelligence and operational functions. Simple business intelligence functions can be performed on the warehouse or data mart where the data is stored. However, running more complex analytics such as social graph analysis for campaign management or chain prediction on these call data records requires different architectures. Hence, massive scale analytic platforms are being explored. The much ballyhooed "cloud computing" has the potential to provide massive parallelism across low-cost processors. However, today the cloud, even with the addition of Hadoop [6], is painful to use for such analytics. A higher level of abstraction is needed, and several groups are working to provide a data model and query language for the cloud [7], [8]. An important area for further work is adapting analytics to these environments. Interesting explorations include using the cloud as a basis for ETL-style structured analytics and integration, for content analytics for text or multimedia data, and for sophisticated mathematical analyses and machine learning algorithms.

While such cloud platforms are built on commodity processors, it is far from trivial to install, run and maintain such systems. In some cases, hosted platforms such as Amazon's EC2 (Elastic Compute Cloud) are adequate. However, this can be an inappropriate solution in the enterprise, when dealing with sensitive company and personal information. Some vendors are therefore offering "private clouds" and services that support these systems and supply tooling for data gathering, integration, and analytics. Again, creating appropriate algorithms and tools that take advantage of the hardware resources and work well in such massive data environments is a significant research challenge.

When the data to be analyzed is public, not private, "Analytics as a Service" models may be explored. These would allow companies to make use of the wealth of information on the Web whether it is factual data (such as companies' financial profiles) or social data (such as blogs) that may have implications for marketing applications (such as brand management or campaign effectiveness). Interesting challenges include how to supply packaged analytics as a service over a public cloud and in particular, how to integrate the results with private information from inside a firewall.

6 Making It Useful – and Usable

Getting answers in real-time is important if the results of an information integration exercise are to be useful. However, much more is needed. It is also important that the information returned be understood by the users and be trustworthy. Information consumers are not typically experts in the data, or in the many analytic tools available. They typically will not understand the process that produced data for their consumption. They may apply additional analytics without really understanding the impact, for example, the compounding of errors that may occur. Consequently, for the end user to feel comfortable with the data and base their decision-making on it, the user needs to understand the rationale behind the displayed results and any specific recommendations,

To help save these users from themselves, researchers are taking a number of different approaches. *Embedded* or *invisible analytics* are an encapsulated set of simple analytic building blocks that can be embedded directly into applications or composed to solve more complicated tasks. The individual building blocks are simple enough to be well understood, and thus they provide a solid foundation for making information integration usable and less error-prone. Examples of invisible analytics include highlighting, entity extraction, and filtering.

Visualizations let users explore different aspects of the result for better understanding. In their simplest form, visualizations are relatively static, and portray a certain aspect of the underlying data. Visualization, however, can be used to go further in the analytics process by serving as a front end that allows interaction with the underlying data. Some examples include testing hypotheses, drilling down into various facets of information, tweaking or refining the analytics process, and linking back to the data source. Delivering on the power of such visualizations once again requires a good link between the application (here the visualization tool) and the integration design.

Provenance of data and analytics helps establish trust by letting the user understand where the results came from, explaining the rationale, promoting reuse and hence, raising confidence. Provenance information has to include information about the base data sets, the way they were integrated (what mappings were used, what entity resolution algorithms, and so on), and the actual analysis applied. Tracking this information, querying it, and relating it to the base data and the end results is an active area of research.

Finally, *situational analytics* make data more usable by taking into account the user's context (task, data, social network, location) and because they are adapted for different environments (secured, mobile, collaborative). Examples of situational analytics may include providing information about products automatically when approaching objects in a store, providing a salesperson with the information about the customer when they arrive on site, or providing advertisers with information about prospective clients in the vicinity. Each of these examples can require information integration, which should further take the context into account.

In general, these approaches aim to make information easier to consume and to help the user derive value or insight from the data more quickly. They leverage integrated information and drive additional information integration needs. But these technologies could also be of value in the integration process itself, as a way to create more effective tools for parts of the integration process that are data-intensive (for example, schema matching or entity resolution, both of which may require looking at large graphs of data). Embedded or situational analytics might help with data cleansing or with "debugging" information integration, when the results are not what was expected.

7 Conclusions

Information integration remains a pervasive and difficult task. There is an increasing diversity of integration problems, technologies, data, sources, and analytics, and all are changing rapidly. Meanwhile, the pressure to deliver more value from integration projects more quickly is mounting. These environmental trends have led to a new generation of research challenges. We identified and described several of these challenges and opportunities. The diversity of integration tasks leads to opportunities for specialized integration solutions and perhaps for an integration design tool to help match the right combination of technologies to particular problems. A dynamic environment is creating interest in iterative approaches to integration, and in postponing processing until the actual need is known. Automatically leveraging community-provided metadata is another approach to a constantly changing world. Where integration and analytics used to be separate, a synergistic and coordinated approach is being explored – which adds further planning options for how to integrate data (what integration steps get done when, with respect to which analytics). Meanwhile, the need for speed drives investigation into the use of new platforms for information integration and analysis, while the need to understand the end result of all these complex logistics is leading to new investigations of how best to provide analytics, how to visualize data and integration processes, and how to record and present their provenance. We foresee no dearth of research topics in this area, and recommend it highly as an area for young researchers.

References

1. Haas, L.: Beauty and the Beast: The Theory and Practice of Information Integration. In: Schwentick, T., Suciu, D. (eds.) ICDT 2007. LNCS, vol. 4353, pp. 28–43. Springer, Heidelberg (2006)
2. Bernstein, P.A., Haas, L.M.: Information integration in the enterprise. Commun. ACM 51(9), 72–79 (2008)
3. Halevy, A.Y., Franklin, M.J., Maier, D.: Principles of dataspace systems. In: PODS, pp. 1–9 (2006)
4. Miller, R.J., Haas, L.M., Hernández, M.A.: Schema Mapping as Query Discovery. In: VLDB, pp. 77–88 (2000)
5. McCann, R., Shen, W., Doan, A.: Matching Schemas in Online Communities: A Web 2.0 Approach. In: IEEE ICDE, pp. 110–119 (2008)
6. Apache Hadoop, http://hadoop.apache.org/
7. Olston, C., Reed, B., Srivastava, U., Kumar, R., Tomkins, A.: Pig latin: a not-so-foreign language for data processing. In: ACM SIGMOD, pp. 1099–1110 (2008)
8. Jaql, http://code.google.com/p/jaql/

What Is Spatio-Temporal Data Warehousing?

Alejandro Vaisman[1] and Esteban Zimányi[2]

[1] Universidad de Buenos Aires, University of Hasselt and
Transnational University of Limburg
alejandro.vaisman@uhasselt.be
[2] Université Libre de Bruxelles
ezimanyi@ulb.ac.be

Abstract. In the last years, extending OLAP (On-Line Analytical Processing) systems with spatial and temporal features has attracted the attention of the GIS (Geographic Information Systems) and database communities. However, there is no a commonly agreed definition of what is a spatio-temporal data warehouse and what functionality such a data warehouse should support. Further, the solutions proposed in the literature vary considerably in the kind of data that can be represented as well as the kind of queries that can be expressed. In this paper we present a conceptual framework for defining spatio-temporal data warehouses using an extensible data type system. We also define a taxonomy of different classes of queries of increasing expressive power, and show how to express such queries using an extension of the tuple relational calculus with aggregated functions.

1 Introduction

Geographic Information Systems (GIS) have been extensively used in various application domains, ranging from economical, ecological, and demographic analysis, to city and route planning [21]. Spatial information in a GIS is typically stored in different so-called *thematic layers* (or *themes*). Information in themes consists of spatial data (i.e., geometric objects) associated to thematic (alphanumeric) information.

OLAP (On-Line Analytical Processing) [7] comprises a set of tools and algorithms that allow efficiently querying *data warehouses*, which are multidimensional databases containing large amounts of data. In this multidimensional model, data are organized as a set of *dimensions* and *fact tables*. Thus, data can be perceived as a *data cube*, where each cell contains measures of interest. Dimensions are further organized in hierarchies that favor the data aggregation process [1]. Several techniques have been developed for query processing, most of them involving some kind of aggregate precomputation.

In spite of the wide corpus of existing work claiming to solve the problem of spatial and spatio-temporal data warehousing and OLAP, there is no clear definition of the meaning of these terms. Moreover, there is no formal notion of "SOLAP query", or "Spatio-temporal OLAP query". Further, existing efforts

T.B. Pedersen, M.K. Mohania, and A M. Tjoa (Eds.): DaWaK 2009, LNCS 5691, pp. 9–23, 2009.

do not clearly specify the kinds of queries addressed. As a consequence, comparing proposals or assessing the capabilities of different approaches is difficult. This paper aims at closing that gap in the following way: (a) first, we define a taxonomy of models that integrate OLAP, spatial data, and moving data types; (b) for each of the classes in this taxonomy, we define the queries that they must support; (c) in order to define these classes of queries, starting from the tuple relational calculus extended with aggregate functions, we propose a spatio-temporal calculus supporting moving data types. We show that each extension defines the kinds of queries in each class of the taxonomy.

The remainder of the paper is organized as follows. Section 2 provides a comprehensive background on existing work in GIS-OLAP integration. Section 3 introduces the calculus and conceptual model that we use throughout the paper, as well as presents our running example. Section 4 introduces the taxonomy of data models. Section 5 defines the queries associated to each class in the taxonomy. We conclude and describe future work in Section 6.

2   Related Work

In the last years, the topic of extending OLAP with spatial and temporal features has attracted the attention of the database and GIS communities. In this section we review relevant efforts in this area.

Rivest et al. [15] introduced the notion of SOLAP (standing for Spatial OLAP), a paradigm aimed at exploring spatial data by drilling on maps, as it is performed in OLAP with tables and charts. They describe the desirable features and operators a SOLAP system should have. Although they do not present a formal model for this, SOLAP concepts and operators have been implemented in a commercial tool called JMAP[1]. Related to the concept of SOLAP, Shekhar et al. [16] introduced MapCube, a visualization tool for spatial data cubes. Given a so-called base map, cartographic preferences, and an aggregation hierarchy, the MapCube operator produces an album of maps that can be navigated via roll-up and drill-down operations.

Several conceptual models have been proposed for spatio-temporal data warehouses. Stefanovic et al. [19] classify spatial dimension hierarchies according to their spatial references in: (a) non-geometric; (b) geometric to non-geometric; and (c) fully geometric. Dimensions of type (a) can be treated as any descriptive dimension. In dimensions of types (b) and (c) a geometry is associated to the hierarchy members. Malinowski and Zimányi [9] defined a multidimensional conceptual model, called MultiDim, that copes with spatial and temporal features. The MultiDim model extends the above classification by considering a dimension level as spatial if it is represented as a spatial data type (e.g., point, region), and where spatial levels may be related through topological relationships (e.g., contains, overlaps). In the models above, spatial measures are characterized in two ways, namely: (a) measures representing a geometry, which can be aggregated along the dimensions; (b) numerical measures, calculated using a topological or

---

[1] http://www.kheops-tech.com/en/jmap/solap.jsp

metric operator. Most proposals support option (a), either as a set of coordinates [15,9], or a set of pointers to geometric objects [19]. Da Silva *et al.* [17] introduced GeoDWFrame, a framework for spatial OLAP, which classifies dimensions in geographic and hybrid, depending on whether they represent only geographic data, or geographic and non-spatial data, respectively. Over this framework, da Silva *et al.* [18] propose GeoMDQL, a query language for spatial data cubes based on MDX and OGC^2 simple features.

It is worth noting that all the above conceptual models follow a *tightly-coupled* approach between the GIS and OLAP components, where the spatial objects are included in the data warehouse. On the contrary, the Piet data model, introduced by Gómez *et al.* [4], follows a *loosely-coupled* approach, where GIS data and data warehouse data are maintained separately, a matching function bounding the two components. Piet supports the notion of *geometric aggregation*, that characterizes a wide range of aggregate queries over regions defined as semi-algebraic sets, addressing four kinds of queries: (a) standard GIS queries; (b) standard OLAP queries; (c) geometric aggregation queries; and (d) integrated GIS-OLAP queries. OLAP-style navigation is also supported in the latter case. Recently, an SQL-like query language was proposed for Piet, denoted Piet-QL. This language, in addition to query types (a) to (d), allows expressing GIS queries filtered by a data cube (i.e., filtered by aggregated data)^3.

Pourabas [13] introduced a conceptual model that uses binding attributes to bridge the gap between spatial databases and a data cube. No implementation of the proposal is discussed. Besides, this approach relies on the assumption that all the cells in the cube contain a value, which is not the usual case in practice. Moreover, the approach also requires modifying the structure of the spatial data.

Traditional data warehouses and OLAP system do not support the evolution of dimension data. *Temporal data warehouses* cope with this issue. Mendelzon and Vaisman [10] proposed a model, denoted TOLAP, and developed a prototype and a Datalog-like query language, based on a temporal star schema. In this model, changes to the structure and/or the instances of the dimension tables are supported, using the concept of transaction and valid time, respectively. Some structural changes also yield different fact table versions. Also, Eder *et al.* [2] propose a data model for temporal OLAP supporting structural changes.

In order to support spatio-temporal data, a data model and associated query language is needed for supporting moving objects, i.e., objects whose geometry evolves over time. This is achieved in Hermes, a system introduced by Pelekis *et al.* [12], and SECONDO [5], a system supporting the model of Güting *et al.* [6]. In spite of their ability to handle spatio-temporal data, neither SECONDO, nor Hermes, are oriented toward addressing the problem of integrating GIS, OLAP, and moving objects. However, in this paper we use many concepts underlying SECONDO to present our approach. Vega López *et al.* [20] present a comprehensive survey on spatio-temporal aggregation.

---

^{2} Open Geospatial Consortium http://www.opengeospatial.org
^{3} A Piet-QL demo can be found at http://piet.exp.dc.uba.ar/pietql

The work by Orlando *et al.* [11] introduces the concept of *trajectory data warehouses*, aimed at providing the infrastructure needed to deliver advanced reporting capabilities and facilitating the use of mining algorithms on aggregate data. This work is also based on the Hermes system. A relevant feature of this proposal is the treatment given to the ETL (Extraction, Transformation and Loading) process, which transforms the raw location data and loads it to the trajectory data warehouse.

3 Preliminaries

3.1 Extending the Conceptual Model

Throughout the paper we use the following real-world example. The Environmental Control Agency of a country has a collection of water stations measuring the value of polluting substances at regular time intervals. The application has maps describing rivers, water stations, and the political division of the country into provinces and districts.

Figure 1 shows the conceptual schema depicting the above scenario using the MultiDim model [9]. There is one fact relationship, WaterPollution, to which several dimensions are related. The fact relationship WaterPollution has two measures, commonArea and load, and is related to five dimensions: Time, District,

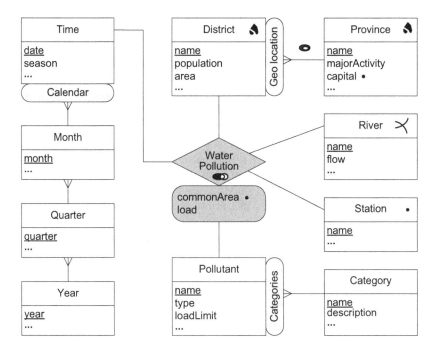

Fig. 1. An example of a spatial data warehouse

River, Station, and Pollutant. Dimensions are composed of levels and hierarchies. For example, while the Station dimension has only one level, the District dimension is composed of two levels, District and Province, with a one-to-many parent-child relationship defined between them.

In the MultiDim model the spatiality of elements is indicated by pictograms. For example, Station, River, and District are spatial levels; they have a geometry represented by a point, a line, and a region, respectively. Similarly, the attribute capital in Province, as well as the measure commonArea in the fact relationship are spatial. Finally, topological relationships may be represented in fact relationships and in parent-child relationships. For example, the topological relationship in WaterPollution indicates that whenever a district, a river, and a water station are related in an instance of the relationship, they must overlap. Similarly, the topological relationship in the hierarchy of dimension District indicates that a district is covered by its parent province.

To address spatio-temporal scenarios we borrow the data types defined by Güting *et al.* [6]. We refer to this work for the complete definition of the type system and the corresponding operations. There is a set of *base types* which are int, real, bool, string, and an *identifier type* id, which is used for the identifiers of level members. There are also *time types* which are instant and periods, the latter being a set of time intervals. There are four *spatial data types*, point, points, line, and region. A value of type point represents a point in the Euclidean plane. A points value is a finite set of points. A line value is a finite set of continuous curves in the plane. A region is a finite set of disjoint parts called *faces*, each of which may have holes. It is allowed that a face lies whith a hole of another face.

Moving types capture the evolution over time of base and spatial types. Moving types are obtained by applying a constructor moving($\cdot$). Hence, a value of type moving(point) is a continuous function f : instant $\rightarrow$ point. Moving types have associated operations that generalize those of the non-temporal types. This is called *lifting*. For example, a distance function with signature moving(point) $\times$ moving(point) $\rightarrow$ moving(real) calculates the distance between two moving points and gives as result a moving real, i.e., a real-valued function of time. Intuitively, the semantics of such lifted operations is that the result is computed at each time instant using the non-lifted operation. Definition 1 summarizes the concepts discussed above.

Definition 1 (Data types). We denote Γ a set of *nontemporal types*, composed of a set of *base types* β, a set of *time types* τ, and a set of *spatial types* ξ. There is also a set of *temporal types* Φ, composed of two sets of temporal types ϕ_β and ϕ_ξ, obtained by applying the moving constructor to elements of β and ξ, respectively.

3.2 Spatio-temporal Calculus

For addressing the issue of querying data warehouses, we use a relational representation of the MultiDim conceptual model. A dimension level is represented by a relation of the same name, having an implicit identifier attribute denoted id, an implicit geometry attribute (if the level is spatial), in addition to the other explicitly indicated attributes. The id attribute (e.g., River.id) identifies a particular instance of the dimension. Dimension levels involved in hierarchies (e.g., District) have also an additional attribute containing the identifier of the parent level (e.g., District.province), and there is a referential integrity constraint for such attributes and the corresponding parent (e.g., Province.id).

A fact relationship is represented by a relation of the same name having an implicit id attribute, one attribute for each dimension, and one attribute for each measure. There is a referential integrity constraint between the dimension attributes in the fact relationship (e.g., WaterPollution.district) and the identifier of the corresponding dimension (e.g., District.id).

We use a query language based on the tuple relational calculus (e.g., [3]) extended with aggregate functions and variable definitions[4]. We explain this language through an example. Consider the following relations from the data warehouse shown in Fig. 1.

District(<u>id</u>, geometry, districtName, population, area, ... , province)
Province(<u>id</u>, geometry, provinceName, majorActivity, capital, governor, ...).

The following query asks the name and population of districts of the Antwerp province.

$$\{d.\text{districtName}, d.\text{population} \mid \text{District}(d) \wedge \exists p \, (\text{Province}(p) \wedge \\ d.\text{province} = p.\text{id} \wedge p.\text{provinceName} = \text{'Antwerp'})\}$$

Suppose that we want to compute the total population of districts of the Antwerp province. A first attempt to write this query would be:

$$\text{sum}(\{d.\text{population} \mid \text{District}(d) \wedge \exists p \, (\text{Province}(p) \wedge \\ d.\text{province} = p.\text{id} \wedge p.\text{provinceName} = \text{'Antwerp'})\})$$

Notice that however, since the relational calculus is based on sets (i.e., collections with no duplicates), if two districts of the Antwerp province happen to have the same population, they would appear only once in the set to which the sum operator is applied. As in Klug's approach [8], this is solved by using aggregate operators that take as argument a set of tuples (instead of a set of values) and that specify on which column the aggregate operator must be applied. Therefore, the above query is more precisely written as follows.

$$\text{sum}_2(\{d.\text{id}, d.\text{population} \mid \text{District}(d) \wedge \exists p \, (\text{Province}(p) \wedge \\ d.\text{province} = p.\text{id} \wedge p.\text{provinceName} = \text{'Antwerp'})\}$$

In this case, the sum operator is applied to a set of pairs $\langle \text{id}, \text{population} \rangle$ and computes the sum of the second attribute.

---

[4] Even though manipulation languages for OLAP exist (e.g., [14]), the choice of the relational calculus is motivated by the fact that it applies to the classical relational model, thus providing a clean and elegant way of defining different spatio-temporal OLAP models and languages.

Finally, suppose we want to calculate the total population by province provided that it is greater than 100,000. In this case we need a recursive definition of queries and variables that bind the results of inner queries to outer queries. The latter query is written as follows.

$\{p.\mathsf{name}, \mathsf{totalPop} \mid \mathsf{Province}(p) \wedge$
$\qquad \mathsf{totalPop} = \mathsf{sum}_2(\{d.\mathsf{id}, d.\mathsf{population} \mid \mathsf{District}(d) \wedge d.\mathsf{province} = p.\mathsf{id}\}) \wedge$
$\qquad \mathsf{totalPop} > 100,000 \,\}$

Here, the outer query fixes a particular province p and the inner query collects the population of districts for that province. The sum of these populations is then bound to the variable totalPop. Notice that this query corresponds to an SQL query with the GROUP BY and HAVING clauses.

Proposition 1. Let us denote $\mathcal{R}_{agg}$ the relational calculus with aggregate functions defined above, over the sets of basic and time types β and τ, respectively. $\mathcal{R}_{agg}$ has the same expressive power of the relational calculus extended with aggregate functions defined in [8].

The idea of the proof follows from the fact that $\mathcal{R}_{agg}$ is a syntactic variation of Klug's calculus. We show later in the paper how we extend $\mathcal{R}_{agg}$ to support spatial and moving data types in order to define a hierarchy of classes of spatio-temporal queries, starting from the expressive power of $\mathcal{R}_{agg}$.

4   A Taxonomy for Spatio-temporal OLAP

Existing proposals for spatial data warehousing cover different functional requirements, but, with limited exceptions, there is no clear specification of the kinds of queries these proposals address. This is probably due to the fact that no taxonomy for these systems has been defined so far. When we talk about GIS, we often refer to *static* GIS, i.e., GIS where the geometry of objects does not change over time. On the other hand, when we talk about OLAP or data warehousing, we assume *static* data warehousing, i.e., data warehouses where dimensions do not change over time. Thus, the term SOLAP refers to the interaction between static GIS and static data warehouses. The schema in Fig. 1 is an example of this approach. When time gets into play, things become more involved, and only partial solutions have been provided. On the one hand, different models exist for temporal data warehousing, depending on the approach followed to implement the warehouse. In this paper we define a temporal data warehouse as a warehouse that keeps track of the history of the *instances* of the warehouse dimensions, i.e., we assume there is no structural (schema) changes. The reason for this is that, as far as we know, only academic implementations of fully temporal data warehouses exist.

We define a taxonomy for spatio-temporal OLAP as follows (see Fig. 2). We start by considering four basic classes: Temporal dimensions, OLAP, GIS, and moving data types. As a derived basic class, adding moving data types to GIS

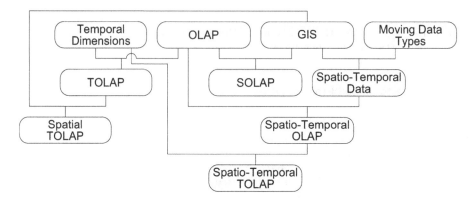

Fig. 2. A taxonomy for spatio-temporal data warehousing

produces *Spatio-Temporal Data*, typically allowing trajectory analysis in a geographic environment. Providing OLAP with the ability of handling temporal dimensions produces the concept of *Temporal OLAP* (TOLAP). The interaction of OLAP and GIS is denoted *Spatial OLAP* (SOLAP). The interaction between GIS and TOLAP is called *Spatial TOLAP* (S-TOLAP). Adding OLAP capabilities to spatio-temporal data results in *Spatio-Temporal OLAP* (ST-OLAP). Finally, if the latter supports temporal dimensions we have *Spatio-Temporal TO-LAP* (ST-TOLAP).

5   Queries

In this section, we define the kinds of queries that should be supported for each one of the classes in the taxonomy of Fig. 2.

5.1   OLAP and Spatial OLAP Queries

We start by showing examples of OLAP queries.

Q1. For water stations located in districts of the Limburg province and polluting agents of organic category give the maximum load by month.

$\{s.\text{name}, p.\text{name}, m.\text{month}, \text{maxLoad} \mid \text{Station}(s) \wedge \text{Pollutant}(p) \wedge$
$\text{Month}(m) \wedge \exists c \, (\text{Category}(c) \wedge p.\text{category} = c.\text{id} \wedge c.\text{name} = \text{'Organic'}) \wedge$
$\text{maxLoad} = \max_1(\{w.\text{load} \mid \text{WaterPollution}(w) \wedge w.\text{station} = s.\text{id} \wedge$
$w.\text{pollutant} = p.\text{id} \wedge \exists d, \exists v, \exists t \, (\text{District}(d) \wedge \text{Province}(v) \wedge$
$\text{Time}(t) \wedge w.\text{district} = d.\text{id} \wedge d.\text{province} = v.\text{id} \wedge$
$v.\text{name} = \text{'Limburg'} \wedge w.\text{time} = t.\text{id} \wedge t.\text{month} = m.\text{id})\})\}$

Q2. For each river, give the total number of stations where, for at least one pollutant, the average load in March 2008 was greater than the load limit for this pollutant.

$\{r.\text{name}, \text{nbStations} \mid \text{River}(r) \land$
$\quad \text{nbStations} = \text{count}(\{s.\text{id} \mid \text{Station}(s) \land \exists p \, (\text{Pollutant}(p) \land$
$\quad \text{avg}_2(\{w.\text{id}, w.\text{load} \mid \text{WaterPollution}(w) \land w.\text{river} = r.\text{id} \land$
$\quad w.\text{station} = s.\text{id} \land w.\text{pollutant} = p.\text{id} \land \exists t(\text{Time}(t) \land w.\text{time} = t.\text{id} \land$
$\quad t.\text{date} \geq 1/3/2008 \land t.\text{date} \leq 31/3/2008)\}) > p.\text{loadLimit}\})\}$

Definition 2 (OLAP queries). Let us call $\mathcal{R}_{agg}$ the relational calculus with aggregate functions defined in Section 3.2. The class of OLAP queries includes all the queries that are expressible by $\mathcal{R}_{agg}$.

We give next some examples of SOLAP queries.

Q3. Total population in the districts within 3 Km from the Ghent district that are crossed by the Schelde river.

$\text{sum}_2(\{d_1.\text{id}, d_1.\text{population} \mid \text{District}(d_1) \land \exists d_2, \exists r (\text{District}(d_2) \land \text{River}(r) \land$
$\quad d_2.\text{name} = \text{'Ghent'} \land \text{distance}(d_1.\text{geometry}, d_2.\text{geometry}) < 3 \land$
$\quad r.\text{name} = \text{'Schelde'} \land \text{intersects}(d_1.\text{geometry}, r.\text{geometry}))\})$

Note that this query do not use a fact relationship. The function distance verifies that the geometries of the two districts are less than 3 Km from each other and the predicate intersects verifies that the district is crossed by the river.

Q4. Stations located over the part of the Schelde river that flows through the Antwerp province, with an average content of nitrates in the last quarter of 2008 above the load limit for that pollutant.

$\{s.\text{name} \mid \text{Station}(s) \land \exists r, \exists p, \exists l, \exists c \, (\text{River}(r) \land \text{Province}(p) \land$
$\quad \text{Pollutant}(l) \land \text{Category}(c) \land r.\text{name} = \text{'Schelde'} \land p.\text{name} = \text{'Antwerp'} \land$
$\quad \text{inside}(s.\text{geometry}, \text{intersection}(r.\text{geometry}, p.\text{geometry})) \land$
$\quad l.\text{category} = c.\text{id} \land c.\text{name} = \text{'Nitrates'} x \land$
$\quad \text{avg}_2(\{w.\text{id}, w.\text{load} \mid \text{WaterPollution}(w) \land w.\text{station} = s.\text{id} \land$
$\quad\quad w.\text{pollutant} = l.\text{id} \land \exists t \, (\text{Time}(t) \land w.\text{time} = t.\text{id} \land$
$\quad\quad t.\text{date} \geq 1/10/2008 \land t.\text{date} \leq 31/12/2008)\}) > l.\text{loadLimit}\}$

Here, the intersection of the river and the district is computed, and then it is verified that the geometry of the station is located inside this intersection.

Definition 3 (SOLAP queries). Let us call $\mathcal{R}_{agg}^{\xi}$ the language $\mathcal{R}_{agg}$ augmented with spatial types in ξ. The class of SOLAP queries is the class composed of all the queries that can be expressed by $\mathcal{R}_{agg}^{\xi}$.

5.2   Temporal OLAP Queries

The notion of Temporal OLAP (TOLAP) arises when evolution of the dimension instances in the data warehouse is supported, a problem also referred to as *slowly-changing dimensions* [7].

This evolution is captured by using temporal types. In other words, when at least one of the dimensions in the data warehouse includes a type in the set ϕ_{β} of Definition 1, we say that the warehouse supports the TOLAP model.

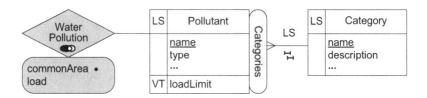

Fig. 3. A temporal dimension Pollutant

To define TOLAP queries, we modify our running example in Fig. 1 making the dimension Pollutant temporal, as shown in Fig. 3. *Temporal levels* are identified by the LS pictogram. In our example, the level Pollutant is temporal, which means that, e.g., a new pollutant may start to be monitored from a particular date. Temporal levels have a predefined attribute called lifespan, of type moving(bool), which keeps track of the validity of a member at each instant. *Temporal attributes* are identified by the VT pictogram. In our example, the attribute loadLimit is temporal, meaning that the load limit varies across time. Temporal attributes are defined over temporal types, for example, moving(real) for the attribute loadLimit.

Finally, *temporal parent-child relationships* are indicated by the LS pictogram. In our example, the relationship between Pollutant and Category is temporal, which means that the association of pollutants to categories varies over time, e.g., at a particular date, a category can be split into two. Temporal relationships are also represented by temporal types. For example, the Pollutant level has an attribute category, of type moving(id), which associates, at each time instant, a category to a pollutant.

The temporal multidimensional model assumes implicit constraints that restrict the lifespan of the instances of temporal levels participating in fact relationships or in temporal parent-child relationships. For example, an instance of the fact relationship relates a time instant t and a pollutant p provided that t is included in the lifespan of p. Further, a pollutant p is related to a pollutant category c at instant t provided that t is included in the lifespan of c.

We consider the following TOLAP queries.

Q5. For each province and pollutant category, give the average load of water pollution by quarter.

$\{p.\text{name}, c.\text{name}, q.\text{quarter}, \text{avgLoad} \mid \text{Province}(p) \land \text{Category}(c) \land$
$\text{Quarter}(q) \land \text{avgLoad} = \text{avg}_2(\{w.\text{id}, w.\text{load} \mid \text{WaterPollution}(w) \land$
$\exists d, \exists t, \exists m, \exists l \ (\text{District}(d) \land \text{Time}(t) \land \text{Month}(m) \land \text{Pollutant}(l) \land$
$w.\text{district} = d.\text{id} \land d.\text{province} = p.\text{id} \land w.\text{time} = t.\text{id} \land$
$t.\text{month} = m.\text{id} \land m.\text{quarter} = q.\text{id} \land w.\text{pollutant} = l.\text{id} \land$
$\text{val}(\text{initial}(\text{atperiods}(l.\text{category}, t))) = c.\text{id}\})\}$

In the last line of the above query, since the parent-child relationship is represented by the temporal attribute category, we need to obtain the value of this attribute at the time defined by the instance t of the Time dimension. As the granularity of the Time dimension is day, function atperiods restricts the

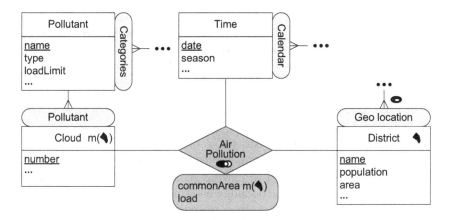

Fig. 4. Examples of a spatio-temporal measure and a spatio-temporal dimension

temporal attribute to that day, function initial takes the first ⟨instant, value⟩ pair of the function, and function val returns the corresponding value. There is no need to verify that the lifespan of the instances of Pollutant or Category include the time t, since these constraints are implicity kept by the model.

Q6. Calculate by day the number of water stations that, for the pollutant lead, had load greater than the maximum value (over its history) of the load limit.

$$\{t.\text{date}, \text{nbStations} \mid \text{Time}(t) \wedge \text{nbStations} = \text{count}_1(\{s.\text{id} \mid \text{Station}(s) \wedge$$
$$\exists w, \exists p \, (\text{WaterPollution}(w) \wedge \text{Pollutant}(p) \wedge w.\text{time} = t.\text{id} \wedge$$
$$w.\text{station} = s.\text{id} \wedge w.\text{pollutant} = p.\text{id} \wedge p.\text{pollutant} = \text{`Lead'} \wedge$$
$$p.\text{load} > \text{val}(\text{initial}(\text{atmax}(p.\text{loadLimit}))))\})\}$$

In the query above, function atmax restricts the temporal attribute to the time instants during which it has its maximum value, function initial takes the ⟨instant, value⟩ pair of the first instant and function val obtains its value.

Definition 4 (TOLAP queries). Let us call $\mathcal{R}_{agg}^{\phi_\beta}$ the language $\mathcal{R}_{agg}$ augmented with the data types in ϕ_β. The class of TOLAP queries is the class composed of all the queries that can be expressed by $\mathcal{R}_{agg}^{\phi_\beta}$.

5.3 Spatio-temporal OLAP Queries

Spatio-temporal OLAP (ST-OLAP) accounts for the case when the spatial objects evolve over time. For this, we need to consider moving types defined by moving(α) where α is a spatial data type in ξ.

In order to define ST-OLAP queries, we add the fact relationship shown in Fig. 4 to our running example in Fig. 1. The Cloud dimension refers to clouds generated by industrial plants. Both the Cloud level and the commonArea measure have a geometry that is a moving region, indicated by the symbol 'm'. Notice that commonArea is a derived measure, i.e., in an instance of the fact relationship that relates a cloud c, a district d, and a date t, the measure keeps the restriction of the trajectory of the cloud at that date and over that district. This is computed by the expression at(atperiods(c.geometry, t), d). Notice also that the Cloud dimension is related to the Pollutant dimension, while in Fig. 1 the Pollutant dimension participates in the fact relationship.

We give next examples of ST-OLAP queries.

Q7. For each district and polluting cloud, give the duration of time when the cloud passed over the district.

$\{d.\text{name}, c.\text{number}, \text{dur} \mid \text{District}(d) \wedge \text{Cloud}(c) \wedge$
$\quad\text{dur} = \text{duration}(\text{deftime}(\text{at}(c.\text{geometry}, d.\text{geometry})))\}$

Function at selects the part of moving geometry that is over the district, function deftime obtains the periods when this happens, and finally function duration calculates the size of the corresponding periods.

Q8. For each district, give by month the total number of persons affected by polluting clouds.

$\{d.\text{name}, m.\text{month}, \text{totalNo} \mid \text{District}(d) \wedge \text{Month}(m) \wedge$
$\quad\text{totalNo} = \text{area}(\text{union}(\{\text{traversed}(p.\text{commonArea}) \mid \text{AirPollution}(p) \wedge$
$\quad p.\text{district} = d.\text{id} \wedge \exists t \, (\text{Time}(t) \wedge t.\text{month} = m.\text{id})\}))/$
$\quad\text{area}(d.\text{geometry}) \times d.\text{population}\}$

The inner query selects all facts relating a given district and a day of a given month; then function traversed projects the moving geometry of the commonArea measure over the plane. A union of all the regions thus obtained yields the part of the district affected by polluting clouds during that month, and the area of this region is then computed. Finally, assuming a uniform distribution of the population, we divide this by the total area of the district and multiply that by its population.

Definition 5 (ST-OLAP queries). Let us call $\mathcal{R}_{agg}^{\phi_\xi}$ the language $\mathcal{R}_{agg}$ augmented with spatial types in ξ and moving spatial types in ϕ_ξ. The class of ST-OLAP queries is the class composed of all the queries that can be expressed by $\mathcal{R}_{agg}^{\phi_\xi}$.

Notice that Definition 5 captures the model of Orlando *et al.* [11] on trajectory data warehousing.

5.4   Spatial TOLAP Queries

Spatial TOLAP (S-TOLAP) covers the case when in addition to having spatial objects and attributes in the data warehouse, the dimensions are also temporal. As we have done in Sect. 5.2, we modify our running example in Fig. 1 so that the dimension Pollutant is temporal, as shown in Fig. 3.

We consider the following S-TOLAP queries.

Q9. For each station located over the Schelde river, give the periods of time during the last quarter of 2008 when the content of nitrates was above the load limit for that pollutant.

$\{s.\text{name}, \text{periods} \mid \text{Station}(s) \land \exists r \, (\text{River}(r) \land r.\text{name} = \text{'Schelde'} \land$
$\quad \text{inside}(s.\text{geometry}, r.\text{geometry}) \land \text{periods} = \text{union}_1(\{t.\text{date} \mid \text{Time}(t) \land$
$\quad\quad \exists w, \exists p, \exists c \, (\text{WaterPollution}(w) \land \text{Pollutant}(p) \land \text{Category}(c) \land$
$\quad\quad w.\text{station} = s.\text{id} \land w.\text{time} = t.\text{id} \land w.\text{pollutant} = p.\text{id} \land$
$\quad\quad p.\text{category} = c.\text{id} \land c.\text{name} = \text{'Nitrates'} \land$
$\quad\quad t.\text{date} \geq 1/10/2008 \land t.\text{date} \leq 31/12/2008 \land$
$\quad\quad w.\text{load} > \text{val}(\text{atinstant}(p.\text{loadLimit}, \text{now})))\})\})\}$

As usual in temporal databases, a distiguished variable, 'now', represents the (moving) current instant. Thus, the above query assumes that the comparison should be made with respect to the *current* value for the limit. Then, the union operator takes a set of dates and construct a minimal set of disjoint periods.

Q10. For each month in 2008, and for each water station in the province of Namur, give the average Biological Oxygen Demand (BOD), if this average is larger than the load limit during the reported month.

$\{m.\text{month}, s.\text{name}, \text{avgBOD} \mid \text{Month}(m) \land \text{Station}(s) \land \exists q, \exists y, \exists p, \exists l \, ($
$\quad \text{Quarter}(q) \land \text{Year}(y) \land \text{Province}(p) \land \text{Pollutant}(l) \land m.\text{quarter} = q.\text{id} \land$
$\quad q.\text{year} = y.\text{id} \land y.\text{year} = 2008 \land p.\text{name} = \text{'Namur'} \land$
$\quad \text{inside}(s.\text{geometry}, p.\text{geometry}) \land l.\text{name} = \text{'BOD'} \land$
$\quad \text{avgBOD} = \text{avg}_2(\{w.\text{id}, w.\text{load} \mid \text{WaterPollution}(w) \land \exists t \, ($
$\quad\quad \text{Time}(t) \land w.\text{station} = s.\text{id} \land w.\text{time} = t.\text{id} \land$
$\quad\quad t.\text{month} = m.\text{id} \land w.\text{pollutant} = l.\text{id})\}) \land$
$\quad \text{avgBOD} > \text{val}(\text{initial}(\text{atperiods}(l.\text{loadLimit}, m.\text{month}))))\}$

In the last term above, function atperiods restrict the load limit to month m, and then functions initial and val obtain the value of this attribute at the first day of the month. This is compared with the average load of that month in avgBOD.

Definition 6 (Spatial TOLAP queries). Let us call $\mathcal{R}_{agg}^{\xi,\phi_\beta}$ the language $\mathcal{R}_{agg}$ augmented with spatial types in ξ and moving types in ϕ_β. We denote S-TOLAP the class of queries composed of all the queries that can be expressed by $\mathcal{R}_{agg}^{\xi,\phi_\beta}$.

5.5 Spatio-Temporal TOLAP Queries

Spatio-Temporal TOLAP (ST-TOLAP) is the most general case where there are moving geometries and the dimensions vary over time. In our running example in Fig. 1 this amounts to replace the temporal dimension Pollutant as in Fig. 3 and to include the AirPollution fact relationship in Fig. 4.

An example of this kind of queries is the following.

Q11. Total number of days when the Gent district has been under at least one cloud of carbon monoxide (CO) such that the average load in the cloud is larger than the load limit at the time when the cloud appeared.

$\text{duration}(\text{union}(\{t.\text{date} \mid \text{Time}(t) \land \exists p, \exists d, \exists c, \exists l \, (\text{AirPollution}(p) \land$
$\quad \text{District}(d) \land \text{Cloud}(c) \land \text{Pollutant}(l) \land p.\text{time} = t.\text{id} \land$
$\quad p.\text{district} = d.\text{id} \land d.\text{name} = \text{'Ghent'} \land p.\text{cloud} = c.\text{id} \land$
$\quad c.\text{pollutant} = l.\text{id} \land l.\text{name} = \text{'CO'} \land p.\text{load} >$
$\quad \text{val}(\text{atinstant}(l.\text{loadLimit}, \text{inst}(\text{initial}(\text{at}(c.\text{lifespan}, \text{true})))))))\}))\}$

To obtain the instant when a cloud appeared we use the functions at (to restrict the lifespan of the cloud), initial, and inst. Functions atinstant and val return the value of the load limit at the instant when the cloud appeared.

Definition 7 (Spatio-Temporal TOLAP queries). Let us call $\mathcal{R}_{agg}^{\phi_\xi,\phi_\beta}$ the language $\mathcal{R}_{agg}$ augmented with spatial types in ξ, moving spatial types in ϕ_ξ, and moving types in ϕ_β. We denote ST-TOLAP the class of queries composed of all the queries that can be expressed by $\mathcal{R}_{agg}^{\xi,\phi_\xi,\phi_\beta}$.

6   Conclusion

In this paper, we have defined a conceptual framework that allows characterizing the functionalities that must be supported by spatio-temporal data warehouses. We have shown that such data warehouses result from the combination of GIS and OLAP technologies, further extended with the support of temporal data types. These data types allow to model both geometries that evolve over time (usually called moving objects) and evolving data warehouse dimensions.

To address the issue of querying spatio-temporal data warehouses, we have defined an extension of the tuple relational calculus with aggregate functions. We defined a taxonomy for spatio-temporal OLAP queries that, starting from the class of traditional OLAP queries, incrementally adds features for defining several classes of queries with increasing expressive power. This is realized by extending the type system underlying the data warehouse and its associated query language. Our taxonomy provides an elegant and uniform way to characterize the features required by spatio-temporal data warehouses and to classify the many different works addressing this issue in the literature.

This work constitutes a first step aiming at defining spatio-temporal data warehouses and therefore many issues remain to be addressed. As we have mentioned above, our framework is defined at a conceptual level and therefore we have omitted any implementation consideration. However, as can be expected, spatio-temporal data warehouses contain huge amounts of data, and therefore optimization issues are of paramount importance. These issues range from appropriate index structures, through pre-aggregation, to efficient query optimization. With respect to the latter issue, our example queries can be expressed in several ways, exploiting either the fact relationship or directly the moving geometries. Although from a formal perspective these alternative queries are equivalent, since they yield the same result, the evaluation time of these queries may vary significatively, depending on the actual population of the data warehouse. Therefore, the translation of our conceptual model into logical and physical models is still another further work.

Acknowledgments. This research has been partially funded by the European Union under the FP6-IST-FET programme, Project n. FP6-14915, GeoP-KDD: Geographic Privacy-Aware Knowledge Discovery and Delivery, and the Argentina Scientific Agency, project PICT 2004 11-21.350.

References

1. Cabibbo, L., Torlone, R.: Querying multidimensional databases. In: Proc. of DBPL, pp. 253–269 (1997)
2. Eder, J., Koncilia, C., Morzy, T.: The COMET metamodel for temporal data warehouses. In: Pidduck, A.B., Mylopoulos, J., Woo, C.C., Ozsu, M.T. (eds.) CAiSE 2002. LNCS, vol. 2348, pp. 83–99. Springer, Heidelberg (2002)
3. Elmasri, R., Navathe, S.: Fundamentals of Database Systems, 5th edn. Addison-Wesley, Reading (2007)
4. Gómez, L., Haesevoets, S., Kuijpers, B., Vaisman, A.: Spatial aggregation: Data model and implementation (2007) CoRR abs/0707.4304
5. Güting, R.H., de Almeida, V.T., Ansorge, D., Behr, T., Ding, Z., Höse, T., Hoffmann, F., Spiekermann, M., Telle, U.: SECONDO: An extensible DBMS platform for research prototyping and teaching. In: Proc. of ICDE, pp. 1115–1116 (2005)
6. Güting, R.H., Schneider, M.: Moving Objects Databases. Morgan Kaufmann, San Francisco (2005)
7. Kimball, R.: The Data Warehouse Toolkit. J. Wiley and Sons, Inc., Chichester (1996)
8. Klug, A.: Equivalence of relational algebra and relational calculus query languages having aggregate functions. Journal of the ACM 29(3), 699–717 (1982)
9. Malinowski, E., Zimányi, E.: Advanced Data Warehouse Design: From Conventional to Spatial and Temporal Applications. Springer, Heidelberg (2008)
10. Mendelzon, A., Vaisman, A.: Temporal queries in OLAP. In: Proc. of VLDB, pp. 242–253 (2000)
11. Orlando, S., Orsini, R., Raffaetà, A., Roncato, A., Silvestri, C.: Spatio-temporal aggregations in trajectory data warehouses. In: Song, I.-Y., Eder, J., Nguyen, T.M. (eds.) DaWaK 2007. LNCS, vol. 4654, pp. 66–77. Springer, Heidelberg (2007)
12. Pelekis, N., Theodoridis, Y., Vosinakis, S., Panayiotopoulos, T.: Hermes: A framework for location-based data management. In: Ioannidis, Y., Scholl, M.H., Schmidt, J.W., Matthes, F., Hatzopoulos, M., Böhm, K., Kemper, A., Grust, T., Böhm, C. (eds.) EDBT 2006. LNCS, vol. 3896, pp. 1130–1134. Springer, Heidelberg (2006)
13. Pourabas, E.: Cooperation with geographic databases. In: Raffanelli, M. (ed.) Multidimensional Databases, pp. 166–199. Idea Group (2003)
14. Ravat, F., Teste, O., Tournier, R., Zurfluh, G.: Algebraic and graphic languages for OLAP manipulations. International Journal of Data Warehousing and Mining 4(1), 17–46 (2008)
15. Rivest, S., Bédard, Y., Marchand, P.: Toward better suppport for spatial decision making: Defining the characteristics of spatial on-line analytical processing (SOLAP). Geomatica 55(4), 539–555 (2001)
16. Shekhar, S., Lu, C., Tan, X., Chawla, S., Vatsavai, R.: MapCube: A visualization tool for spatial data warehouses. In: Miller, H., Han, J. (eds.) Geographic data mining and Knowledge Discovery (GKD), pp. 74–109. Taylor & Francis, Abington (2001)
17. Silva, J., Times, V.C., Salgado, A.C.: An open source and web based framework for geographic and multidimensional processing. In: Proc. of SAC, pp. 63–67 (2006)
18. Silva, J., Castro Vera, A.S., Oliveira, A.G., Fidalgo, R., Salgado, A.C., Times, V.C.: Querying geographical data warehouses with GeoMDQL. In: Proc. of SBBD, pp. 223–237 (2007)
19. Stefanovic, N., Han, J., Koperski, K.: Object-based selective materialization for efficient implementation of spatial data cubes. IEEE Transactions on Knowledge and Data Engineering 12(6), 938–958 (2000)
20. Vega López, I.F., Snodgrass, R.T., Moon, B.: Spatiotemporal aggregate computation: A survey. IEEE Transactions on Knowledge and Data Engineering 17(2), 744–759 (2005)
21. Worboys, M.F., Duckham, M.: GIS: A Computing Perspective, 2nd edn. CRC Press, Boca Raton (2004)

Towards a Modernization Process for Secure Data Warehouses

Carlos Blanco[1], Ricardo Pérez-Castillo[1], Arnulfo Hernández[1],
Eduardo Fernández-Medina[1], and Juan Trujillo[2]

[1] Dep. of Information Technologies and Systems, Escuela Superior de Informática
ALARCOS Research Group - Institute of Information Technologies and Systems
University of Castilla-La Mancha, Paseo de la Universidad, 4, 13071, Ciudad Real, Spain
{Carlos.Blanco,Ricardo.PdelCastillo,Arnulfonapoleon.Hernandez,
Eduardo.Fdezmedina}@uclm.es
[2] Dep. of Information Languages and Systems, Facultad de Informática,
LUCENTIA Research Group, University of Alicante, San Vicente s/n. 03690,
Alicante, Spain
jtrujillo@dlsi.ua.es

Abstract. Data Warehouses (DW) manage crucial enterprise information used for the decision making process which has to be protected from unauthorized accesses. However, security constraints are not properly integrated in the complete DWs' development process, being traditionally considered in the last stages. Furthermore, legacy systems need a reverse engineering process in order to accomplish re-documentation for detecting new security requirements as well as system's design recovery to enable migration and reuse. Thus, we have proposed a model driven architecture (MDA) for secure DWs which takes into account security issues from the early stages of development and provides automatic transformations between models. This paper fulfills this architecture providing an architecture-driven modernization (ADM) process focused on obtaining conceptual security models from legacy OLAP systems.

1 Introduction

Data Warehouses (DWs) manage business' historical information used to take strategic decisions and usually follow a multidimensional approach in which the information is organized in facts classified per subjects called dimensions. In a typical DW architecture, ETL (extraction/transformation/load) processes extract data from heterogeneous Data Sources and then transform and load this information into the DW repository. Finally, this information is analyzed by Data Base Management Systems (DBMS) and On-Line Analytical Processing (OLAP) tools.

Since data in DWs are crucial for enterprises, it is very important to avoid unauthorized accesses to information by considering security constraints in all layers and operations of the DW, from the early stages of development as a strong requirement to the final implementation in DBMS or OLAP tools (Thuraisingham, Kantarcioglu et al. 2007).

In this way, DWs' development can be aligned with the Model Driven Architecture (MDA 2003) approach which proposes a software development focused on models at

T.B. Pedersen, M.K. Mohania, and A M. Tjoa (Eds.): DaWaK 2009, LNCS 5691, pp. 24–35, 2009.

different abstraction levels which separate the specification of the system functionality and its implementation. Firstly, system requirements are included in business models (CIM). Then, conceptual models (PIM) represent the system without including information about specific platforms and technologies which are finally specified in logical models (PSM). Moreover, automatic transformations between models can be defined by using several languages such as Query / Views / Transformations (QVT) (OMG 2005).

Furthermore, MDA architectures support reverse engineering capabilities which consists of analysis of legacy systems to (1) identify the system's elements and their interrelationships and (2) carry out representations of the system at a higher level of abstraction (Chikofsky and Cross 1990). Reverse engineering can be used in the development of DWs to accomplish re-documentation for detecting new security requirements as well as system's design recovery to enable migration and reuse. Nevertheless, reverse engineering takes part in a whole reengineering process (Müller, Jahnke et al. 2000). MDA provides the needed formalization to reengineering process to converge in so-called Architecture-Driven Modernization (ADM), another OMG initiative (OMG 2006). ADM advocates reengineering processes where each artifact involved in these processes is depicted and managed as a model (Khusidman and Ulrich 2007).

We have proposed an MDA architecture to develop secure DWs taking into account security issues in the whole development process (Fernández-Medina, Trujillo et al. 2007). To achieve this goal we have defined an access control and audit model specifically designed for DWs and a set of models which allow the security design of the DW at different abstraction levels (CIM, PIM and PSM). This architecture provides two different paths (a relational path towards DBMS and a multidimensional path towards OLAP tools) and includes rules for the automatic transformation between models and code generation.

This paper improves the architecture by defining an architecture-driven modernization (ADM) process which permits re-documentation and platform migration. Since most of DWs are managed by OLAP tools by using a multidimensional approach, this ADM process is focused on the multidimensional path, obtaining conceptual security models (PIM) from logical multidimensional models (PSM) and legacy OLAP systems.

This paper is organized as follows: Section 2 will present the related work on secure DWs; Section 3 will briefly show our complete MDA architecture for developing secure DWs and will underline the difference between our previous works and the contribution of this paper; Section 4 will present the defined ADM process; Section 5 will use an application example to validate our proposal; Section 6 will finally present our conclusions and future work.

2   Related Work

There are relevant contributions focused on secure information systems development, such as UMLSec (Jürjens 2004) which uses UML to define and evaluate security specifications using formal semantics, or Model Driven Security (MDS) (Basin, Doser et al. 2006) which uses the MDA approach to include security properties in high-level system models and to automatically generate secure system architectures. Within the context of MDS, SecureUML (Lodderstedt, Basin et al. 2002) is proposed as an extension of UML for modeling a generalized role based access control.

However, these proposals do not consider the special characteristics of DWs. In this area, solely Priebe and Pernul propose a complete methodology for develop secure DWs (Priebe and Pernul 2001). This methodology deals with the analysis of security requirements, the conceptual modeling by using ADAPTed UML, and the implementation into commercial tools, but does not establish the connection between levels in order to allow automatic transformations. They use SQL Server Analysis Services (SSAS) creating a Multidimensional Security Constraint Language (MDSCL) by extending multidimensional expressions (MDX) with hide statements for cubes, measures, slices and levels.

Although MDA philosophy has been applied to develop secure DWs (Fernández-Medina, Trujillo et al. 2007) and data reverse engineering field has been widely studied in literature (Aiken 1998; Blaha 2001; Cohen and Feldman 2003; Hainaut, Englebert et al. 2004), there is little research on reengineering of data warehouses following an MDA approach and security concerns are not considered. These reengineering works are performed for: re-documentation, model migration, restructuring, maintenance or improvement, tentative requirements, integration, conversion of legacy data.

3 MDA Architecture for Secure DWs

Our architecture to develop secure DWs proposes several models improved with security capabilities which allow the DW's design considering confidentiality issues in the whole development process, from an early development stage to the final implementation. This proposal has been aligned with an MDA architecture (Fernández-Medina, Trujillo et al. 2007) providing security models at different abstraction levels (CIM, PIM, PSM) and automatic transformations between models (Figure 1).

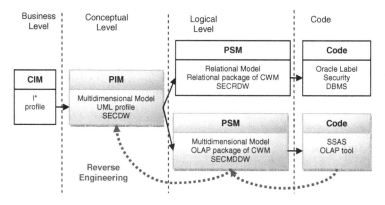

Fig. 1. MDA architecture for Secure DWs

Firstly, security requirements are modeled at business level (CIM) by using a UML profile (Trujillo, Soler et al. 2008) based on the i* framework (Yu 1997), which is an agent oriented approach centered on the agents' intentional characteristics. Then, transformation from secure CIM models to conceptual model (PIM) is achieved applying a methodology described by using the OMG Software Process Engineering Metamodel Specification standard (SPEM) (Trujillo, Soler et al. 2008).

Conceptual models (PIM) are defined according to a UML profile, called SECDW (Fernández-Medina, Trujillo et al. 2007) which has been specifically created for DWs and complemented by an Access Control and Audit (ACA) model focused on DW confidentiality (Fernández-Medina, Trujillo et al. 2006). In this way, SECDW allows the representation of structural aspects of DWs (such as facts, dimensions, base classes, measures or hierarchies) and security constraints which permit the classification of authorization subjects and objects in three ways (into roles (SecurityRole), levels (SecurityLevel) and compartments (SecurityCompartment)) and the definition of several kinds of security rules (Sensitive information assignment rules (SIAR), authorization rules (AUR) and audit rules (AR)).

Multidimensional modeling at the logical level depends of the tool finally used and can be principally classified into online analytical processing by using relational (ROLAP), multidimensional (MOLAP) and hybrid (HOLAP) approaches. Thus, our architecture considers two different paths: a relational path towards DBMS and a multidimensional path towards OLAP tools.

The relational path uses a logical relational metamodel (PSM) called SECRDW (Soler, Trujillo et al. 2008) which is an extension of the relational package of the Common Warehouse Metamodel (CWM 2003) and allows the definition of secure relational elements such as secure tables or columns. Moreover, this relational path is fulfilled with the automatic transformation from conceptual models (Soler, Trujillo et al. 2007) and the eventual implementation into a DBMS, Oracle Label Security.

Furthermore, this MDA architecture was recently improved with a new multidimensional path towards OLAP tools in which a secure multidimensional logical metamodel (PSM), called SECMDDW (Blanco, García-Rodríguez de Guzmán et al. 2008) considers the common structure of OLAP tools and allows to represent a DW model closer to OLAP platforms than conceptual models. SECMDDW is based on a security improvement of the OLAP package from CWM and is composed of: a security configuration metamodel which represents the system's security configuration by using a role-based access control policy (RBAC); a cube metamodel which defines both structural cube aspects such as cubes, measures, related dimensions and hierarchies, and security permissions for cubes and cells; and a dimension metamodel with structural issues of dimensions, bases, attributes and hierarchies, and security permissions which are related to dimensions and attributes.

This path also deals with the automatic transformation from conceptual models by using QVT transformations (Blanco, García-Rodríguez de Guzmán et al. 2008) and the final secure implementation into a specific OLAP platform, SQL Server Analysis Services (SSAS), by using a set of Model-to-Text (M2T) rules.

4 Modernizing Secure DWs

Modernizing DWs provides us several benefits such us to generate diagrams on a high abstraction level in order to identify security lacks in an easy way and to include new security constraints which solve these identified problems. Transformation rules are then applied obtaining an improved logical model and the final implementation. By using the MDA philosophy the system can be also migrate to different technologies (MOLAP, ROLAP, HOLAP, etc.) and different final tools. Since most DWs are managed by OLAP tools using a multidimensional approach (MOLAP), in this section we present a modernization process focused on the multidimensional path obtaining conceptual models from multidimensional logical models (Figure 1).

In a first stage, the multidimensional logical model according to SECMDDW is obtained from the source code of the OLAP tool. To achieve this goal is applied a static analysis (Canfora and Penta 2007) which is a reengineering method based on the generation of lexical and syntactical analyzers for the specific tool. In this way, code files are analyzed and a set of code-to-model transformations create the corresponding elements into the target logical model.

Once logical multidimensional model is obtained several set of QVT rules carry out a model-to-model transformation towards the corresponding conceptual model. Since the source metamodel (SECMDDW) presents three kinds of models (roles configuration, cubes and dimensions) three sets of transformations have been developed (Figure 2). Each transformation is composed of several QVT relations which are focused on transforming structural and security issues.

Role2SECDW transformation creates the security configuration of the system based on a set of security roles. This is an example of a semantic gap between abstractions levels, because conceptual level is richer than logical level and includes support to the definition of security levels, roles and compartment. This transformation presents the relations "RoleFiles2Package" and "Role2SRole" which transform the "RoleFiles" into a "Package" and create security roles "SRole" for each role detected at the logical level. Figure 3 shows the implementation of this transformation and Figure 4 the graphical representation for the "Role2SRole" relation.

Cube2SECDW transformation analyzes cube models and generates at the conceptual level structural aspects and security constraints defined over the multidimensional elements. Table 1 (left column) shows the signatures for the relations included in this transformation.

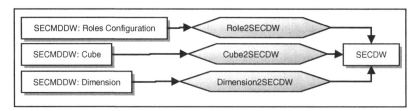

Fig. 2. PSM to PIM transformation overview

```
transformation Role2SECDW (psm:SECMDDW, pim:SECDW) {
    key SECDW::SRole {rootPackage, name};
    top relation RoleFiles2Package {
        xName : String;
        checkonly domain psm rf:SECMDDW::SecurityConfiguration::RoleFiles {
            name = xName };
        enforce domain pim pk:SECDW::Package { name = xName };
        where { rf.ownedRoles->forAll (r:SECMDDW::SecurityConfiguration::Role |
            Role2SRole(r, pk)); }   }
    relation Role2SRole {
        xName : String;
        checkonly domain psm r:SECMDDW::SecurityConfiguration::Role { ID = xName };
        enforce domain pim pk: SECDW::Package{
            ownedMember = sr : SECDW::SRole { name = xName } };   }}
```

Fig. 3. Role2SECDW transformation

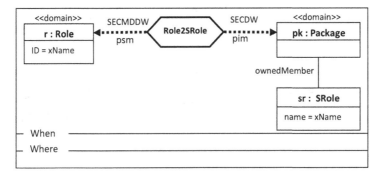

Fig. 4. Graphical representation of Role2SRole relation

Table 1. Relations for Cube2SECDW and Dimension2SECDW transformations

transformation Cube2SECDW	transformation Dimenssion2SECDW
top relation CubeFiles2Package {…}	top relation DimensionFiles2Package {…}
relation Cube2SFact {…}	relation Dimension2SDimension {…}
relation Measures2SFA {…}	relation attribute2SProperty {…}
relation Measure2SProperty {…}	relation hierarchy2SBase {…}
relationDimension2SDimension{…}	realtion attribute2SBaseProperty {…}
relation CubePermission2SClass {…}	relation DimensionPermission2SClass {…}
relation CellPermission2SProperty{…}	relation AttributePermission2SProperty{…}

There are a set of structural rules which transform cubes into secure fact classes ("Cube2SFact" relation) and their related measures and dimensions into secure properties ("Measures2SFA" and "Measure2Property" relations) and secure dimension classes ("Dimension2SDimension" relation). Security permissions related with cubes or cells are transformed into security constraints at the conceptual level ("CubePermission2SClass" and "CellPermission2SProperty" relations).

```
transformation Cube2SECDW (psm:SECMDDW, pim:SECDW) {
    key SECDW::SFact {rootPackage, name};
    top relation CubeFiles2Package {
        xName : String;
        checkonly domain psm cf:SECMDDW::Cubes::CubeFiles { name = xName };
        enforce domain pim pk:SECDW::Package { name = xName };
        where { cf.ownedCubes->forAll (c:SECMDDW::Cubes::Cube I Cube2SFact(c, pk)); } }
    relation Cube2SFact {
        xName : String;
        checkonly domain psm c:SECMDDW::Cubes::Cube { ID = xName };
        enforce domain pim pk: SECDW::Package {
                ownedMember = f : SECDW::SFact { name = xName } };
        where { c.ownedMeasureGroups->forAll (mg:SECMDDW::Cubes::MeasureGroup I
        (mg.ownedMeasures->forAll (m:SECMDDW::Cubes::Measure I Measures2SFA(m, f))));}}
    relation Measures2SFA {
        xName : String;
        checkonly domain psm m:SECMDDW::Cubes::Measure { ID = xName };
        enforce domain pim f:SECDW::SFact {
                attributes = sfa:SECDW::SFA { name = xName } };   }}
```

Fig. 5. Cube2SECDW transformation

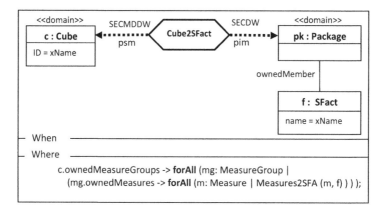

Fig. 6. Graphical representation of Cube2SFact relation

The implementation of some relations is shown in Figure 5 and Figure 6 presents the "Cube2SFact" relation in a graphical way.

Dimension2SECDW transformation focuses on dimension models and creates at the conceptual level structural aspects such as dimension and base classes, properties and hierarchies ("Dimension2SDimension", "attribute2SProperty", "hierarchy2SBase" and "attribute2SBaseProperty" relations) and security constraints related with dimensions, bases and properties ("DimensionPermission2SClass" and "AttributePermission2SProperty" relations). This transformation is composed of several relations which signatures are shown in Table 1 (right column).

The implementation of some relations is shown in Figure 7 and Figure 8 presents the "DimensionPermission2SClass" relation in a graphical way.

```
transformation Dimension2SECDW (psm:SECMDDW, pim:SECDW) {
    key SECDW::SDimension {rootPackage, name};
    key SECDW::SRole {rootPackage, name};
    top relation DimensionFiles2Package {
        xName : String;
        checkonly domain psm df:SECMDDW::Dimensions::DimensionFiles { name = xName };
        enforce domain pim pk:SECDW::Package { name = xName };
        where { df.ownedDimensions->forAll (d:SECMDDW::Dimensions::Dimension |
                Dimension2SDimension(d, pk)); } }
    relation Dimension2SDimension {
        xName : String;
        checkonly domain psm d:SECMDDW::Dimensions::Dimension {ID = xName };
        enforce domain pim pk: SECDW::Package {
                ownedMember = sd : SECDW::SDimension {
                ownedSecInf = si : SECDW::SecureInformation {}, name = xName } };
        where { d.ownedDimensionPermissions->forAll
        (dp:SECMDDW::Dimensions::DimensionPermission |
        (dp.deniedSet.oclIsUndefined()) implies (DimensionPermission2SClass (dp, si, pk)) ); }}
    relation DimensionPermission2SClass {
        xRoleID : String;
        checkonly domain psm dp:SECMDDW::Dimensions::DimensionPermission {
                roleID = xRoleID };
        enforce domain pim  sd :SECDW::SecureInformation {
                securityRoles = sr : SECDW::SRole { name = xRoleID } };
        enforce domain pim pk:SECDW::Package { ownedMember = sr : SECDW::SRole {} };
        when{ dp.deniedSet = ''; }  }}
```

Fig. 7. Dimension2SECDW transformation

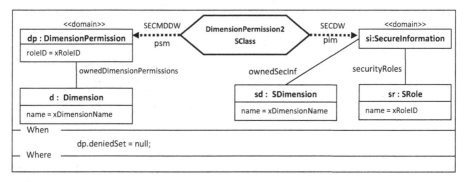

Fig. 8. Graphical representation of DimensionPermission2SClass relation

5 Example

This section shows the defined ADM process by using an example in which the transformation rules are applied into a PSM multidimensional model to obtain the corresponding PIM model. This example uses a DW which manages airport's information about trips involving passengers, baggage, flights, dates and places. This information is analyzed for the airport staff, companies or passengers, and can be used for many purposes, for instance companies can decide to reinforce certain routes with a great number of passengers or can offer to passengers a special price for their top

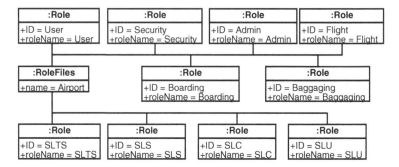

Fig. 9. PSM multidimensional model for security configuration

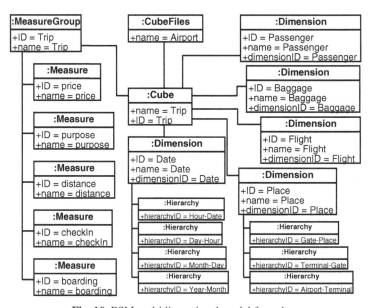

Fig. 10. PSM multidimensional model for cubes

destinations. The source multidimensional PSM model is composed of three parts: security configuration (Figure 9), cubes (Figure 10) and dimensions (Figure 11). Figure 12 finally shows the PIM model obtained after applying the ADM process.

Firstly, **Role2SECDW** transformation analyzes the security configuration model (Figure 9) and creates roles in the PIM model. The conceptual level (PIM) is richer and supports the specification of security levels, compartments and roles, but logical models (PSM) only include information of roles. Thus, transformation rules can only transform each role in the logical model into a role in the conceptual model.

Then, logical cube models (Figure 10) are processed by the **Cube2SECDW** transformation. It creates in the PIM model (Figure 12) the following structural aspects: the secure fact class "Trip", its measures and its related dimensions and hierarchies. Since security permissions related with cubes were not defined, security constraints are not established in the PIM model.

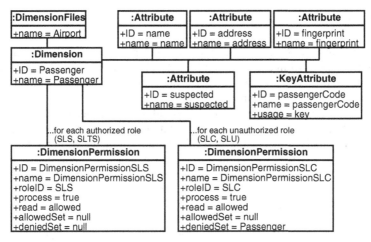

Fig. 11. PSM multidimensional model for dimensions

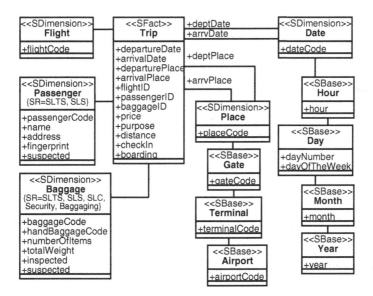

Fig. 12. PIM model

Finally, **Dimension2SECDW** process logical dimension models. Figure 11 shows the PSM model for "Passenger" dimension in which have been defined some attributes and dimension permissions to authorize and deny accesses to certain roles. This structural information is transformed into a secure dimension class "Passenger" with secure properties in the PIM model (Figure 12). Positive security permissions are also transformed by including the authorized roles ("SLTS" and "SLS") as stereotypes of the "Passenger" dimension.

6 Conclusions

We have proposed an MDA architecture for developing secure DWs taking into account security issues from early stages of the development process. We provide security models at different abstraction levels and automatic transformations between models and towards the final implementation.

This work has fulfilled the architecture providing an architecture-driven modernization (ADM) process which allows us to automatically obtain higher abstraction models (PIM). Firstly, code analyzers obtain the logical model from the implementation, and then, QVT rules transform this logical model into a conceptual model. In this way, existing systems can be re-documented and this design at higher abstraction level (PIM) can be easier analyzed in order to include new security constraints. Furthermore, once PIM model is obtained the DW can be migrated to other platforms or final tools.

Our further works will improve this architecture in several aspects: dealing with the inference problem by including dynamic security models which complement the existing models; including new PSM models (such as XOLAP); and giving support to other final platforms (such as Pentaho).

Acknowledgments. This research is part of the ESFINGE (TIN2006-15175-C05-05) Project financed by the Spanish Ministry of Education and Science, the QUASIMO-DO (PAC08-0157-0668) Project financed by the FEDER and the Regional Science and Technology Ministry of Castilla-La Mancha (Spain), the SISTEMAS (PII2109-0150-3135) Project financed by the Regional Science and Technology Ministry of Castilla-La Mancha (Spain) and the MITOS (TC20091098) Project financed by the University of Castilla-La Mancha (Spain).

References

Aiken, P.H.: Reverse engineering of data. IBM Syst. J. 37(2), 246–269 (1998)

Basin, D., Doser, J., et al.: Model Driven Security: from UML Models to Access Control Infrastructures. ACM Transactions on Software Engineering and Methodology 15(1), 39–91 (2006)

Blaha, M.: A Retrospective on Industrial Database Reverse Engineering Projects-Part 1. In: Proceedings of the 8th Working Conference on Reverse Engineering (WCRE 2001), Suttgart, Germany. IEEE Computer Society Press, Los Alamitos (2001)

Blanco, C., García-Rodríguez de Guzmán, I., et al.: Applying QVT in order to implement Secure Data Warehouses in SQL Server Analysis Services. Journal of Research and Practice in Information Technolog (in press) (2008)

Canfora, G., Penta, M.D.: New Frontiers of Reverse Engineering. IEEE Computer Society, Los Alamitos (2007)

Cohen, Y., Feldman, Y.A.: Automatic high-quality reengineering of database programs by abstraction, transformation and reimplementation. ACM Trans. Softw. Eng. Methodol. 12(3), 285–316 (2003)

CWM, OMG: Common Warehouse Metamodel (CWM) (2003)

Chikofsky, E.J., Cross, J.H.: Reverse Engineering and Design Recovery: A Taxonomy. IEEE Softw. 7(1), 13–17 (1990)

Fernández-Medina, E., Trujillo, J., et al.: Model Driven Multidimensional Modeling of Secure Data Warehouses. European Journal of Information Systems 16, 374–389 (2007)

Fernández-Medina, E., Trujillo, J., et al.: Access Control and Audit Model for the Multidimensional Modeling of Data Warehouses. Decision Support Systems 42, 1270–1289 (2006)

Fernández-Medina, E., Trujillo, J., et al.: Developing secure data warehouses with a UML extension. Information Systems 32(6), 826–856 (2007)

Hainaut, J.-L., Englebert, V., et al.: Database reverse engineering: From requirements to CARE tools. Applied Categorical Structures. SpringerLink. 3 (2004)

Jürjens, J.: Secure Systems Development with UML. Springer, Heidelberg (2004)

Khusidman, V., Ulrich, W.: Architecture-Driven Modernization: Transforming the Enterprise. DRAFT V.5, OMG: 7 (2007), http://www.omg.org/docs/admtf/07-12-01.pdf

Lodderstedt, T., Basin, D., Doser, J.: SecureUML: A UML-based modeling language for model-driven security. In: Jézéquel, J.-M., Hussmann, H., Cook, S. (eds.) UML 2002. LNCS, vol. 2460, p. 426. Springer, Heidelberg (2002)

MDA, OMG: Model Driven Architecture Guide (2003)

Müller, H.A., Jahnke, J.H., et al.: Reverse engineering: a roadmap. In: Proceedings of the Conference on The Future of Software Engineering, Limerick, Ireland. ACM Press, New York (2000)

OMG. MOF QVT final adopted specification

OMG, ADM Glossary of Definitions and Terms, OMG: 34 (2006), http://adm.omg.org/ADM_Glossary_Spreadsheet_pdf.pdf

Priebe, T., Pernul, G.: A pragmatic approach to conceptual modeling of OLAP security. In: Kunii, H.S., Jajodia, S., Sølvberg, A. (eds.) ER 2001. LNCS, vol. 2224, p. 311. Springer, Heidelberg (2001)

Soler, E., Trujillo, J., et al.: A Set of QVT relations to Transform PIM to PSM in the Design of Secure Data Warehouses. In: IEEE International Symposium on Frontiers on Availability, Reliability and Security (FARES 2007), Viena, Austria (2007)

Soler, E., Trujillo, J., et al.: Building a secure star schema in data warehouses by an extension of the relational package from CWM. Computer Standard and Interfaces 30(6), 341–350 (2008)

Thuraisingham, B., Kantarcioglu, M., et al.: Extended RBAC-based design and implementation for a secure data warehouse. International Journal of Business Intelligence and Data Mining (IJBIDM) 2(4), 367–382 (2007)

Trujillo, J., Soler, E., et al.: An Engineering Process for Developing Secure Data Warehouses. Information and Software Technology (in Press) (2008)

Trujillo, J., Soler, E., et al.: A UML 2.0 Profile to define Security Requirements for DataWarehouses. Computer Standard and Interfaces (in Press) (2008)

Yu, E.: Towards modelling and reasoning support for early-phase requirements engineering. In: 3rd IEEE International Symposium on Requirements Engineering (RE 1997), Washington, DC (1997)

Visual Modelling of Data Warehousing Flows with UML Profiles*

Jesús Pardillo[1], Matteo Golfarelli[2], Stefano Rizzi[2], and Juan Trujillo[1]

[1] Lucentia Research Group, University of Alicante, Spain
{jesuspv,jtrujillo}@dlsi.ua.es
[2] DEIS, University of Bologna, Italy
{matteo.golfarelli,stefano.rizzi}@unibo.it

Abstract. Data warehousing involves complex processes that transform source data through several stages to deliver suitable information ready to be analysed. Though many techniques for visual modelling of data warehouses from the static point of view have been devised, only few attempts have been made to model the data flows involved in a data warehousing process. Besides, each attempt was mainly aimed at a specific application, such as ETL, OLAP, what-if analysis, data mining. Data flows are typically very complex in this domain; for this reason, we argue, designers would greatly benefit from a technique for uniformly modelling data warehousing flows for all applications. In this paper, we propose an integrated visual modelling technique for data cubes and data flows. This technique is based on UML profiling; its feasibility is evaluated by means of a prototype implementation.

Keywords: OLAP, UML, conceptual modelling, data warehouse.

1 Introduction

Data transformations are the main subject of visual modelling concerning data warehousing dynamics. A data warehouse integrates several data sources and delivers the processed data to many analytical tools to be used by decision makers. Therefore, these data transformations are everywhere: from data sources to the corporate data warehouse by means of the ETL processes, from the corporate repository to the departmental data marts, and finally from data marts to the analytical applications (such as OLAP, data mining, what-if analysis). Data warehousing commonly implies complex data flows, either because of the large number of steps data transformations may consist of, or of the different types of data they carry. These issues rise interesting challenges concerning design-oriented modelling of data warehousing flows. In particular, the thorough visualisation of these models has a deep impact on the current trends for data

* Supported by the Spanish projects: ESPIA (TIN2007-67078), QUASIMODO (PAC08- 0157-0668), and DEMETER (GVPRE/2008/063). Jesús Pardillo is funded by MEC under FPU grant AP2006-00332. Special thanks to the anonymous reviewers for their helpful comments.

T.B. Pedersen, M.K. Mohania, and A M. Tjoa (Eds.): DaWaK 2009, LNCS 5691, pp. 36–47, 2009.
© Springer-Verlag Berlin Heidelberg 2009

warehousing design, where the so-called model-driven technologies [1] promote diagrams as the main, tentatively unique, design artefacts managed by software engineers. Cognitive aspects, such as diagrams readability, are thus related to the productivity of the whole development process.

Nevertheless, the main research efforts made so far have concerned the static modelling of the data warehouse repository [2] and, even when data warehousing flows were considered, it was done within specific business intelligence applications (OLAP, data mining and so on). While these efforts were addressed at designing individual modelling frameworks, all of them characterised nothing but data transformations.

Our contribution in this paper is twofold: firstly, we identify and formally define *data warehousing flows* (f^w's) as the founding concept for every visual modelling technique studied so far, which means that flows can be applied to model any data transformation, from OLAP to data mining (§2). Due to space constraints, this paper is focused on OLAP f^w's that, as a matter of fact, are the backbone of data warehousing. In general, an OLAP session model can be useful in different contexts: (1) it can be a relevant part of more complex data warehousing flows (e.g., it could describe a set of transformations to be applied to multidimensional data in a what-if analysis application); (2) it can be used to design or document a semi-static report, where a limited number of OLAP operators can be applied depending on the data currently visualized; (3) it can represent auditing information showing system administrators the more frequent operators applied to cubes.

Secondly, we present an integrated visual modelling framework for f^w's based on UML [3]. The proposal consists in the diagramming of data cubes, meant as results of multidimensional queries (§3), and data transformations (§4). These diagrams have been implemented in a prototype illustrated in §5. Remarkably, our solution covers the modelling gaps identified in the state-of-the-art as it is shown in §6. Finally, conclusions are drawn in §7.

2   The f^w Framework for Visual Modelling

A data warehousing flow f^w may be functionally characterised as $f : I \rightarrow O$ where f is the (probably complex) data transformation, I is the universe of data objects managed by f and O is the universe of the processed data objects. Moreover, f is defined as the composition of other functions, $f = f_1 \circ \ldots \circ f_n$, that may recursively be defined as compositions.

f^w's may be classified according to the actual data-object type they manipulate, *i.e.*, the domain I and codomain O. Data cubes, we argue, are the most important factor in this classification because they are the building blocks of data warehouses. Let C be the universe of all data cubes involved in an f^w, and X denote any unspecified sort of data objects; we can distinguish the disjoint categories shown in Table 1.

For instance, *mining flows* may be characterised as f^{-c}, because they transform data cubes into other data objects, namely data-mining patterns such as

Table 1. General taxonomy of data flows based on their data-object types

Name	Notation	Definition
OLAP flow	f^c	$f^c : C \rightarrow C$
ETL flow	f^{+c}	$f^{+c} : X \rightarrow C$
mining flow	f^{-c}	$f^{-c} : C \rightarrow X$
object flow	$f^{\emptyset}$	$f^{\emptyset} : X \rightarrow X$

association rules or clusters. Though the names for flow types were chosen according to the field where they mainly appear, in practice they are not bound to a single application domain, such as in the case of what-if analysis, where different types may be involved.

The canonical f^w may be decomposed (regarding Table 1) into: $f^w = f^{\emptyset} \circ f^{-c} \circ f^c \circ f^{+c} \circ f^{\emptyset}$, where $\circ$ is any composition operator and the f^w's from the right to the left respectively characterise: (1) transactional flows for populating data sources; (2) ETL flows for populating the data warehouse; (3) OLAP flows occurring during an OLAP session; (4) mining flows aimed at extracting patterns from the data warehouse; and (5) flows for manipulating and visualising patterns. Though there are in practice many ways of connecting f^w's (*e.g.*, multiple branches are valid structures), it is evident that f^w's involving data cubes either in input or in output are the actual backbone of the data warehousing process. Noticeably, all f^w's that involve data cubes might also be characterised as atomic $f^{\emptyset}$'s at the finest detail level, because cubes can be decomposed into their elements (dimensions, measures, etc.). For this reason, f^{-c}'s and f^{+c}'s will be sometimes visually modelled as a composition of detailed $f^{\emptyset}$'s instead of being considered as atomic. Conversely, visually modelling internal details of f^c's is out of the scope of this paper.

Visual modelling of f^w's should comply with the following wish-list: (i) it should be based on some multidimensional diagrams that model data warehouse facts and dimensions, (ii) it should be easy to understand, (iii) its semantics should have formal foundation, and (iv) it should rely on a standard notation.

The need to manage f^w complexity suggests to create separate diagrams for data cubes. For this reason, our framework provides two kinds of diagrams, namely *data cube* and f^w *diagrams*. Data cube diagrams represent a multidimensional query formulated on the data warehouse. f^w diagrams represent how actions transform data cubes.

For such diagrams to be cognitively effective, their notations should achieve a *conceptual integration* of information from separate diagrams into a coherent user's mental model and a *perceptual integration* by means of perceptuals cues (orienting, contextual and directional information) in order to support navigation between diagrams [4]. In f^w diagrams, data cubes are just rendered as *information scents* [5] (like the scent of food) that encourage readers to look for more detailed diagrams (where more succulent information could be found). Conversely, data cube diagrams render each data cube in detail. In addition, data cubes are visually modelled over the *multidimensional diagrams* [6] of the underlying data warehouse facts, thus providing the required perceptual cues[1]. The following two sections describe data cube and f^w diagrams, respectively.

3   Data Cube Diagrams

Our aim is to propose a visual modelling technique based on standard representations of f^w's. A well-known extension technique such as UML profiling [3] seems then appropriate. Profiling enables to easily but formally extend (in terms of both semantics and notation) the UML language, the *de facto* standard for general-purpose modelling in software engineering. By means of profiling, data cubes can be smoothly hosted in a UML multidimensional diagram describing the data warehouse. To accomplish this goal we will use the UML profile for multidimensional modelling presented in [6], namely the `DataWarehouse` UML profile, which also provides a proper iconography that improves diagram readability.

Fig. 1 illustrates a sample multidimensional diagram (*host* diagram) to which the data cube diagram of Fig. 2 is allocated (*guest diagram*). The host diagram represents the database for analysing the `sales` facts (⊞ in Fig. 1) by `product`, `customer`, `location`, and `date` (⟲). Each dimension allows sales to be aggregated (◉) at different *granularities*. For instance, sales may be aggregated by `month` and `year` (▣). In addition, facts and dimensions can be respectively described by measures (**M**, *e.g.*, `quantity`) and descriptors (**D**, *e.g.*, city `name`).

On the other hand, the guest diagram represents a query that may be easily referred to the host diagram. For instance, Fig. 2 shows a cube of sales `quantity` grouped by `month`, store `city` and product `branch`; in particular, only the branches whose code is food.

3.1   A UML Profile for the Integrated Diagramming of Data Cubes

Our `DataCube` profile is based on the `DataWarehouse` profile by Luján-Mora *et al.* [6]. Both profiles are represented in Fig. 3 with the standard notation for UML profiling [3], *i.e.*, the profile diagrams where stereotypes of UML modelling elements are presented. On the one hand, the `DataWarehouse` profile contains a set of *stereotypes* (*e.g.*, `Fact` and `Dimension`). Each stereotype represents a single multidimensional concept by extending the specific UML *metaclass* considered as the most semantically close to that concept. Fig 3 also shows the proper iconography. On the other hand, the `DataCube` profile (see Fig. 3) introduces five stereotypes[2]:

---

[1] Indeed, visual models contain two kinds of data: statements about the reality that they model and metadata about how they are represented, such as canvas locations.

[2] We chose names for the `DataCube` profile stereotypes according to the terminology of the multidimensional expressions (MDX) from Microsoft, the most spread language for OLAP querying, to emphasise that data cubes are the result of queries.

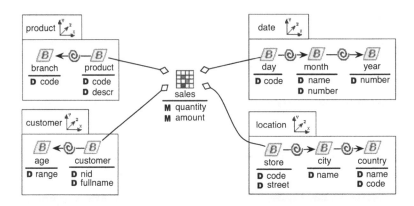

Fig. 1. Multidimensional diagram by using the `DataWarehouse` UML profile

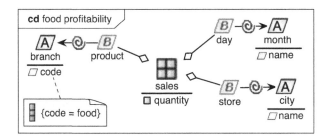

Fig. 2. Data-cube diagram by using the `DataCube` UML profile

Cell, summarising a set of `Facts` of a given `cube`.
Axis, showing each `Base` (component of the group-by clause) of a given `cube`.
CellMember & AxisMember, representing each returned `Measure` and `Descriptor`.
Slice, representing the predicate formulated on `CellMembers` and/or `AxisMembers`.

All these stereotypes specialise the `CubeElement` abstract stereotype. Every `CubeElement` has references to both the extended metaclass and the supporting entity of the multidimensional diagram (shown by a *use* dependency in Fig. 3). The `cube` attribute of the `CubeElement` stereotype is a tag definition referring to all the cubes that contain a cube element. Let F be a class stereotyped as `Fact`. In order to represent a cube c resulting from a query on fact F, you need to additionally stereotype F as `Cell` and annotate class F with tagged value c for the `cube` attribute of `Cell`.

All `CubeElement`s (except `Slices`) have two abstraction levels:

Space Specification, where either `Cell` or `Axis` stereotypes are applied to `Facts` or `Bases`, respectively. Each retrieved cell or axis member are left to the designer as a *variation point* [3] whose options are: (i) all owned properties (measures or descriptors) are retrieved in that cell; or (ii) they still remain unspecified. Unless differently stated, the second option is assumed.

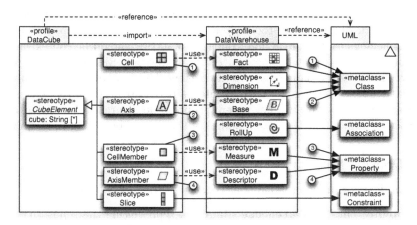

Fig. 3. DataCube and DataWarehouse UML profiles

Member Specification, where CellMember or AxisMember stereotypes must be applied together with the owner Cell or Axis, respectively. The underlying space specification is thus explicitly shown. The application of these stereotypes may be managed by diagramming tools as we shall discuss next.

The rules about how to apply and combine the proposed stereotypes are formally specified in OCL [7] (*i.e.*, a declarative language to query and specify constraints in UML models), and thus, they could be automatically managed by the corresponding checking engine. For instance, the OCL constraint to check the rule of CellMember specifications is

```
context CellMember inv 'Member Specification Rule':
self.base_Property.class.extension_Cell.cube->includesAll(self.cube)
```

3.2   Rendering Data Cube Diagrams

UML profiles allow to adapt the UML notation to include new iconography. In this way, the DataCube profile provides a new version of the DataWarehouse stereotypes whose aim is to emphasise the retrieved data. This marking is accomplished by swapping the DataWarehouse icons, rendered in grayscale, with new coloured versions. The aesthetics decision of colouring is justified by the cognitive studies about *preattentive processing* [8], stating that coloured diagram nodes are distinguished from grayscale ones before conscious attention occurs, thus showing CubeElements "at a glance". In addition, the selected colours are *complementary* (red vs. green), so they can be distinguished very well from each other. Due to black-and-white printing, this iconography also uses shapes (resembling the underlying DataWarehouse elements) for codifying CubeElements. It is worth noting that there is not a DataWarehouse counterpart of the Slice stereotype, because particular predicates regarding specific data instances only concern data cube diagrams, not multidimensional diagrams.

Nevertheless, rendering `DataCube`s over `DataWarehouse`s requires special customisations of diagramming tools. Since every `CubeElement` references all the cubes it belongs to, the `DataCube` iconography setting should be context-aware. The context is the current diagram itself: only the `CubeElement`s, whose `cube` property includes the name given to the current diagram, are actually rendered with `DataCube` icons (though they may internally refer to other cubes). Of course, this assumes that both the `CubeElement::cube`s and diagram names refer to the same set of values, *i.e.*, the actual names of the data cubes being diagrammed.

3.3   Drawing Data Cubes

The workflow for drawing `DataCube`s consists of the following steps:

1. **Copying the `DataWarehouse` diagram**, representing the repository from which the data cube will be retrieved, into a new `DataCube`-to-be diagram.
2. **Renaming this `DataCube` diagram** in order to identify the desired cube.
3. **Specifying `CubeElement`s** by following one of the two specifying conventions discussed above (namely, space or member) and by attaching the proper `Slice`s to the corresponding `CubeElement`s. This step is actually decomposed into (i) application of `CubeElement` stereotypes if they were not already applied for a previous data cube, and (ii) addition of the current diagram name as a `CubeElement::cube` tagged value. Since the second step could be sometimes cumbersome, it can be automatically managed in diagramming tools by implementing the corresponding controllers for context-aware marking.
4. **Hiding undesired `DataWarehouse` elements** (optional) for enhancing the final diagram readability. This step is mainly targeted to visually remove from the diagram (not from the metamodel occurrence) the unused `Measure`s, `Dimension`s, and `Descriptor`s, or to prune aggregations (`Base`s).

4   The Data Warehousing Flows Visual Library

Data cube diagrams visualise the static aspects of f^w's. In this section, we discuss how to manage the dynamic aspects. According to the well-known software-engineering principle of *separation of concerns* [9], the dynamic and static features of f^w's are represented in separate diagrams. By setting aside the complexity of data cubes, f^w's can be visualised in a more readable form. However, we recall that both kinds of diagrams are closely related, and they were devised as artefacts to be used together, as described in §2.

To represent the dynamic aspects of f^w's, we could use any type of diagram aimed at process modelling, such as flow charts, data flows, etc. We selected activity diagrams because they enable designers to model f^w dynamics intuitively by means of UML. In this way, both static and dynamic diagrams discussed in this paper may be smoothly integrated to be managed together. Like for data cubes, the many kinds of activities involved in an f^w require an additional customisation. Therefore, we propose (i) a UML profile for adapting the activity diagram notation to represent data cubes, and (ii) a set of f^w catalogues that capture the f^w diversity. Due to the space constraints, we shall only discuss the OLAP catalogue of the f^w library[3]. However, all of them share the same principles.

4.1 A UML Profile for Diagramming the f^w Library

As for data cube diagrams, we also devised a UML profile to adapt the notation of activity diagrams. The main issues this profile deals with are: (i) notational improvement, emphasising *object flows* ([3], p. 386) for data cubes, and (ii) validation of *action* names that model f^w's ([3], p. 311). As to the first issue, the notation decorates the edges that connect actions by highlighting data cube flows (with the ⊞ icon) as shown in Fig. 4. These flows are related to data cube diagrams. As to the second issue, the naming patterns used for actions express the semantics of actions and of their parameters. In such a way, naming patterns can be checked using regular expressions codified by OCL constraints. Due to space constraints, we overlook the details of the profile definition. All the same, the library that applies this profile for visualising f^w's is described next.

The OLAP catalogue is the main entity of the f^w library due to the relevance of OLAP applications. This catalogue includes the best-known operators of OLAP algebras [10]. With a few exceptions, all operators are f^c. Note that the f^w library contains the best practices in data warehouse modelling, and it is not limited to the presented operators. For instance, some modellers may believe that the *pivoting* operator is relevant enough to be added to the f^w library.

Fig. 4 shows that each OLAP operator is modelled as an action taking cubes in input and output. Naming patterns formalise the vocabulary widely understood among OLAP analysts [10]. Each naming pattern may encode several parameters, represented as `<parameter>`. Optional parameters are enclosed in square brackets. Parameters with multiple occurrences are followed by a "+" mark; in this case, occurrences are separated by commas. Parameters commonly refer to cube elements, and they are instantiated with the same name of the element they refer to. Next, we briefly discuss how the OLAP algebra is rendered; for a deep understanding of this topic, interesting readers may be referred to [10]:

slice by has a criterion (*i.e.*, a constraint) for filtering values of cube members. This constraint may be given in natural language, *e.g.*, this year, or in a formal language such as OCL, *e.g.*, year = now.year.
dice by is similar to slice, but applied to several dimensions at the same time.
roll up & drill down aggregate and disaggregate data cubes. They are parameterised with a dimension and optionally a base when multiple aggregation paths are possible from the current aggregation level (*e.g.*, sales by day could be aggregated into months but also weeks).
md-project selects one or more data cube measures.
drill anywhere groups cells by a set of dimensions. This OLAP operation, also known as *change base*, generalises the add & remove dimension operators, that

---

[3] Herein, the term 'library' refers to the whole metamodel for the f^w dynamic modelling, whereas each part of this is called 'catalogue'.

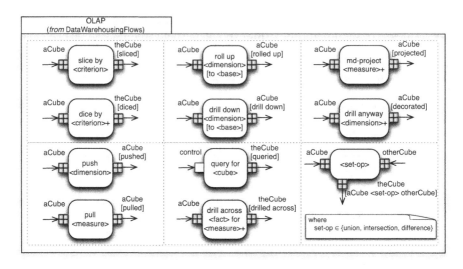

Fig. 4. The f^w catalogue of OLAP actions

can be simulated by using an extended notation: [+-]dimension for adding or removing dimension to/from the current cube base.

push converts a dimension into a measure.

pull converts a measure into a dimension.

query for queries a data warehouse in order to retrieve a data cube. This is the only operation that takes a control flow rather than a cube in input. The cube parameter is bound to a cube diagram.

drill across joins a cube with a fact to change the set of measures.

union, intersection, & difference manipulate data cubes by using set semantics, thus they are the only actions that require two data cubes in input. In addition, other set operations such as symmetric difference may be defined by means of the previous ones.

5   Prototype and Example Application

Our framework has been implemented in the ECLIPSE development platform[4], whose modular, plugin-based architecture enables a proper implementation of the UML extensions proposed. In particular, we have enhanced the plugin for UML modelling, UML2TOOLS, with the functionalities of diagram rendering and event controller for both data cube diagrams and f^w library (see Fig. 5). The proposed models are stored in two kinds of files: those that contain the diagramming metadata and those that contain the modelled elements.

We consider as a case study an OLAP analysis of profitability in the food market domain. The screenshot in Fig. 5 shows in its upper part the whole f^c modelled as an activity diagram: naming patterns are used to name actions, and

---

[4] http://www.eclipse.org (UML2TOOLS are also located here)

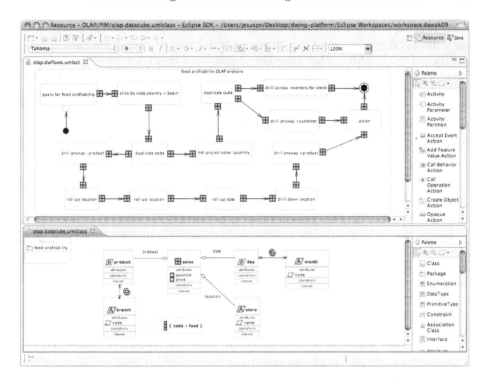

Fig. 5. Prototype for the f^w visual-modelling framework implemented in ECLIPSE

the manipulated data cubes are properly emphasized by the notation introduced, the ⊞ icon. For each data cube involved, a data cube diagram like the one in Fig. 2 may be drawn. For instance, the data cube diagram in the bottom part of the figure shows the input data cube for **food profitability**. Remarkably, it was possible to simply label data cube flows in activity diagrams with a measure name (such as **quantity** or **price**) because data cubes are properly represented in their own diagrams. Furthermore, f^c semantics is clearly stated by naming patterns, so that automated generation of executable code is made possible.

In more detail, the c_0 data cube obtained by action **query for food profitability** is first sliced (flow f_0^c) by selecting Spanish sales only. Then, the resulting cube $f_0^c(c_0)$ is duplicated to fork the analysis process into two branches, one focusing on sales location and time, one on sales quantity. In the first branch, whose overall flow we denote with f_1^c, the product dimension is removed (**drill anyway -product**), then a sequence of **roll up** actions are carried out, twice for **location** (from **store** to **city**, then from **city** to **country**), and once for **date** (from **month** to **year**). After these operations, sales are analysed by location and date at country and year granularity. The analyst then **drills down location** to show details on **cities** by **years**. Finally, products are added to the data cube axes (**drill anyway +product**) to prepare to a later **union**. As to the second branch, it first applies a multidimensional projection to focus only on sales quantity (**md-project sales::quantity**, denoted with f_2^c).

Then it is further decomposed into two sub-branches, the first (f_{2a}^c) performing a `drill across inventory for stock` to traverse interesting data cubes, the second (f_{2b}^c) adding the customer dimension (`drill anyway +customer`). Finally, the two output data cubes c_1 and c_2 for this f^c may be denoted as $c_1 = (f_1^c \circ f_0^c(c_0)) \cup (f_{2b}^c \circ f_2^c \circ f_0^c(c_0)); c_2 = f_{2a}^c \circ f_2^c \circ f_0^c(c_0)$. Note that, though the f^w library is extensively used in this example, additional utility actions–such as `duplicate cube`–have also been modelled.

6   Literature Review

There are many modelling frameworks presented in the scientific literature regarding the particular f^w's. Specifically, concerning OLAP, there are a few works proposing visual modelling techniques for OLAP. [11] presents a specific approach where OLAP sessions are represented by UML state diagrams. Regarding OLAP query modelling, [12] employs a graph-based representation to highlight the f^w dependencies from analytical tools to the data sources. While [12] is oriented to visual modelling of dynamic aspects, [13] achieves static representation of queries by annotating structural conceptual models of a data warehouse, similarly to the data cube diagrams we propose in §3. A compact representation of OLAP queries is achieved by means of UML class-like structures also in [11].

Overall, even OLAP works point out the dichotomy between visualising f^w's as data states *vs.* data transformations. This is related to the classical debate on *state vs. flow charts*: they are complementary and emphasising one aspect rather than another. It is also related to the dilemma of visualising *structural vs. dynamic* aspects of f^w's. Citing [14], "every notation highlights some kinds of information at the expense of obscuring other kinds".

7   Conclusions

The state-of-the-art for visual modelling of f^w's comprises a wide range of techniques, each taking into account specific aspects of application domains, but overlooking their common foundational concepts. In this work we identified two challenging issues concerning design-oriented f^w visual modelling: how to handle complex data structures such as data cubes, and how to specify the semantics of the involved data transformations in a formal and straightforward mode. For this reason, we devised an f^w visual modelling framework where two kinds of diagrams are provided by using UML as scaffolding. Their suitability to visually manage the complexity involved in f^w's is shown by applying them to an example scenario relying on the `Eclipse` platform.

The results of this work have interesting implications for data warehouse practitioners. Regarding the integrated vision of f^w's, the current modelling tools, that were conceived for a specific kind of f^w, may be reused for the others. This fact sets a bridge between current visual modelling techniques. Thanks to the unifying definition of f^w's, we presented a general framework for their modelling. This proposal is aligned with model-driven technologies [1] such as those presented in [15] for designing data warehouses.

Some challenging new research topics appear next. Two of them are specially encouraged: automatic code generation from these diagrams by applying model transformations (according to [15]), and the empirical validation of their enhancement in cognitive issues such as readability.

References

1. Favre, J.: Towards a Basic Theory to Model Model Driven Engineering. In: 3rd Workshop in Software Model Engineering, WiSME (2004)
2. Romero, O., Abelló, A.: A Survey of Multidimensional Modeling Methodologies. International Journal of Data Warehousing & Mining 5(2), 1–23 (2009)
3. OMG: Unified Modeling Language (UML) Superstructure, version 2.1.2. (November 2007), http://www.omg.org/technology/documents/formal/uml.htm
4. Moody, D.: What Makes a Good Diagram? Improving the Cognitive Effectiveness of Diagrams in IS Development. In: Adv. in Inform. Syst. Dev.; New Methods and Pract. for the Networked Soc., pp. 481–492 (2007)
5. Pirolli, P.: Information Foraging Theory: Adaptive Interaction with Information. Oxford University Press, USA (2007)
6. Luján-Mora, S., Trujillo, J., Song, I.Y.: A UML profile for multidimensional modeling in data warehouses. Data Knowl. Eng. 59(3), 725–769 (2006)
7. Object Management Group: Object Constraint Language (OCL), version 2.0. (October 2003), http://www.omg.org/technology/documents/formal/ocl.htm
8. Ware, C.: Information Visualization: Perception for Design, 2nd edn. Morgan Kaufmann, San Francisco (2004)
9. Tarr, P.L., Ossher, H., Harrison, W.H., Sutton Jr., S.M.: N Degrees of Separation: Multi-Dimensional Separation of Concerns. In: Proc. ICSE, pp. 107–119 (1999)
10. Romero, O., Abelló, A.: On the need of a reference algebra for OLAP. In: Song, I.-Y., Eder, J., Nguyen, T.M. (eds.) DaWaK 2007. LNCS, vol. 4654, pp. 99–110. Springer, Heidelberg (2007)
11. Trujillo, J., Luján-Mora, S., Song, I.-Y.: Applying UML For Designing Multidimensional Databases And OLAP Applications. In: Advanced Topics in Database Research, vol. 2, pp. 13–36 (2003)
12. Papastefanatos, G., Vassiliadis, P., Simitsis, A., Vassiliou, Y.: Design Metrics for Data Warehouse Evolution. In: Li, Q., Spaccapietra, S., Yu, E., Olivé, A. (eds.) ER 2008. LNCS, vol. 5231, pp. 440–454. Springer, Heidelberg (2008)
13. Cabibbo, L., Torlone, R.: From a Procedural to a Visual Query Language for OLAP. In: Proc. SSDBM, pp. 74–83 (1998)
14. Green, T.R.G., Petre, M.: Usability Analysis of Visual Programming Environments: A 'Cognitive Dimensions' Framework. J. Vis. Lang. Comput. 7(2), 131–174 (1996)
15. Mazón, J.-N., Trujillo, J.: An MDA approach for the development of data warehouses. Decis. Support Syst. 45(1), 41–58 (2008)

CAMS: OLAPing Multidimensional Data Streams Efficiently

Alfredo Cuzzocrea

ICAR-CNR and University of Calabria, Italy
cuzzocrea@si.deis.unical.it

Abstract. In the context of data stream research, taming the multi-dimensionality of real-life data streams in order to efficiently support *OLAP analysis/mining tasks* is a critical challenge. Inspired by this fundamental motivation, in this paper we introduce CAMS (*Cube-based Acquisition model for Multidimensional Streams*), *a model for efficiently OLAPing multidimensional data streams*. CAMS combines a set of data stream processing methodologies, namely (*i*) the *OLAP dimension flattening process*, which allows us to obtain dimensionality reduction of multidimensional data streams, and (*ii*) the *OLAP stream aggregation scheme*, which aggregates data stream readings according to an OLAP-hierarchy-based *membership* approach. We complete our analytical contribution by means of experimental assessment and analysis of both the *efficiency* and the *scalability* of OLAPing capabilities of CAMS on synthetic multidimensional data streams. Both analytical and experimental results clearly connote CAMS as an *enabling component* for next-generation *Data Stream Management Systems*.

1   Introduction

A critical issue in representing, querying and mining *data streams* [3] consists of the fact that *they are intrinsically multi-level and multidimensional in nature* [6,18], hence they *require to be analyzed by means of a multi-level and a multi-resolution (analysis) model accordingly*. Furthermore, it is a matter of fact to note that enormous data flows generated by a collection of stream sources *naturally* require to be processed by means of advanced analysis/mining models, beyond traditional solutions provided by primitive SQL-based DBMS interfaces. Conventional analysis/mining tools (e.g., DBMS-inspired) cannot carefully take into consideration these kinds of multidimensionality and correlation of real-life data streams, as stated in [6,18]. From this, it follows that, if one tries to process multidimensional and correlated data streams by means of such tools, rough errors are obtained in practice, thus seriously affecting the quality of decision making processes that found on analytical results mined from streaming data. Contrary to conventional tools, *multidimensional analysis* provided by *OnLine Analytical Processing* (OLAP) technology [7,17], which has already reached an high-level of maturity, allows us to efficiently exploit and take advantages from multidimensionality and correlation of data streams, with the final aim of improving

T.B. Pedersen, M.K. Mohania, and A M. Tjoa (Eds.): DaWaK 2009, LNCS 5691, pp. 48–62, 2009.

the quality of both analysis/mining tasks and decision making in streaming environments. OLAP allows us to aggregate data according to (*i*) a fixed logical schema that can be a *star* or a *snowflake schema*, and (*ii*) a given SQL aggregate operator, such as SUM, COUNT, AVG etc. The resulting data structures, called *data cubes* [17], which are usually materialized within *multidimensional arrays* [2], allow us to meaningfully take advantages from the amenity of querying and mining data according to a multidimensional and a multi-resolution vision of the target domain, and from the rich availability of a wide set of OLAP operators, such as *roll-up*, *drill-down*, *slice-&-dice*, *pivot* etc, and OLAP queries.

According to the OLAP stream model, being practically very hard to support OLAPing of multidimensional data streams in an online manner (i.e., *on the fly*), as even clearly stated in [6,18], a popular solution to this end consists in *computing summarized representations of data stream readings in an off-line manner*, and then *making use of these summarized representations in order to efficiently answer aggregate queries that represent the baseline operations for OLAP analysis/mining tasks over multidimensional data streams*. It should be noted that (*i*) aggregate queries are a "natural" way of extracting useful knowledge from summarized data repositories (like data cubes), thus they can easily support OLAPing of multidimensional data streams, and (*ii*) the above-illustrated OLAP analysis paradigm is perfectly compliant with the idea of supporting even complex decision making processes over multidimensional data streams, as highlighted by recent studies [6,14,16,18]. In past research experiences [9,10], we have proposed an approach for supporting efficient *approximate aggregate query answering over data streams* that makes use of (*i*) the off-line processing approach above, and (*ii*), in order to further enhance the performance of query evaluation, the *data compression paradigm* [4], which, initially proposed in the context of massive database and data cube compression, has been then even applied to the context of voluminous data stream compression (e.g., see [8]). In more detail, in [9,10] we propose compressing summarized data stream readings in order to obtain faster approximate answers to aggregate queries over data streams. This approach is fully motivated by the well-known assertion stating that *approximation is completely tolerable in OLAP analysis*, where decimal precision is not necessary, and, rather, decision makers are more interested in performing "qualitative" and trend analysis. With respect to research results provided in [9,10], where simple raw data stream readings are considered, without any particular OLAP undertone, in this research effort we focus on (*i*) *multidimensional* data streams, i.e. data streams whose readings embed a certain multidimensionality, and (*ii*) data stream sources characterized by a set of OLAP hierarchies associated to the dimensions of their proper *multidimensional (data) models*. Both these two aspects, which are fundamental constructs of the OLAP stream model, have not been investigated in [9,10].

Although more or less sophisticated instances of the OLAP stream model for analyzing and mining multidimensional data streams can be devised (e.g., [6,18]), depending on particular application requirements, a leading research challenge in this context is represented by *the issue of effectively and efficiently collecting and representing multidimensional data streams*, being the latter two critical phases of OLAPing multidimensional data streams. In fact, it is easy to understand how both these phases heavily affect the quality of later OLAP analysis/mining tasks over multidimensional data streams, not only from a performance-oriented point of view but also with regard to the proper *semantics* of these tasks. A feasible solution to face-off this leading OLAP stream research challenge is represented by the so-called *acquisition models* for data streams (e.g., [12,13,21]). Starting from these considerations, in this paper we introduce and experimental assess an innovative *acquisition model for OLAPing multidimensional data streams efficiently*, called CAMS (*Cube-based Acquisition model for Multidimensional Streams*), which combines a set of methodologies in order to tame the multidimensionality kept in real-life data streams, and efficiently support OLAP analysis/mining tasks over multidimensional data streams.

2   CAMS Overview

Fig. 1 shows an overview of CAMS and, at the same, the OLAPing multidimensional data streams technique underlying our proposed (acquisition) model. As shown in Fig. 1, we assume a reference application scenario according to which multidimensional data streams are analyzed and mined on the basis of an off-line OLAP approach (see Sect. 1). In such a scenario, a repository of summarized (multidimensional) data stream readings, namely MDSR (*Multidimensional Data Stream Repository*), is collected and stored with the aim of executing OLAP analysis/mining tasks in an off-line manner. As highlighted in Sect. 1, CAMS combines a set of data stream processing methodologies in order to tame the multidimensionality of data streams, thus efficiently supporting OLAPing of such streams. Basically, in the global workflow defined by CAMS, two main stages can be identified (see Fig. 1).

In the first stage of CAMS, the N-dimensional data stream produced by the target data stream source is "flattened" by means of our innovative *OLAP dimension flattening process*, which is a fundamental component of our OLAP stream model (instance). This stage originates a *flattened data stream, whose dimensionality M is lower than the dimensionality of the original data stream (i.e., $M << N$)*. As we describe in Sect. 4, the OLAP dimension flattening process basically consists in flattening a multidimensional data cube model (the one associated to the original data stream) onto a lower data cube model (the one

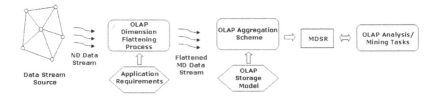

Fig. 1. CAMS overview

associated to the flattened data stream), whose dimensions, called *flattening dimensions*, are selected from the original dimension set of the data stream source in dependence on specific application requirements. The process is performed via systematically merging original hierarchies defined on dimensions of the multidimensional data stream source model. The final flattening dimensions are thus equipped with *specialized hierarchies* generated in dependence on application-oriented requirements. We first introduced the OLAP dimension flattening process in [11], with the goal of efficiently supporting *semantics-aware advanced OLAP visualization of multidimensional data cubes*. In the actual research, this process is specialized to the more interesting and challenging case of dealing with OLAPing multidimensional data streams efficiently. It should be noted that the data stream application scenario poses more problematic issues than the corresponding case experienced in OLAP over conventional data sources. In fact, the so-called *curse of dimensionality* problem (e.g., [5,20]), which, briefly, consists in the fact that when the number of dimensions and size of the target data cube increase, multidimensional data cannot be accessed and queried efficiently, gets worse in data stream environments. As a consequence, the usage of our proposed OLAP dimension flattening process makes sense perfectly towards ensuring effective and efficient OLAPing of multidimensional data streams.

In the second stage of CAMS, readings of the flattened data stream are used to *populate* the summarized data stream repository MDSR, according to a proper *OLAP stream aggregation scheme*. Basically, this scheme determines how readings of the flattened data stream participate to the aggregations defined by the OLAP storage model of MDSR. This, in turn, finally determines the way of populating MDSR. In more detail, in this stage CAMS makes use of efficient *search algorithms* that inspect the OLAP hierarchies associated to dimensions of MDSR in order to determine the *membership* of *dimensional attribute values* of data stream readings to *OLAP classes* defined by members of these hierarchies, thus finally determining the way of simultaneously aggregating data stream readings within MDSR cells along multiple (flattened) dimensions.

3   The Multidimensional Data Stream Model

Consider a set S of N data stream sources denoted by $S = \{s_0, s_1, \ldots, s_{N-1}\}$. Let $M_S = \langle \mathcal{D}(M_S), \mathcal{H}(M_S), \mathcal{M}(M_S) \rangle$ be the N-dimensional (OLAP) model of S, such that: (i) $\mathcal{D}(M_S) = \{d_0, d_1, \ldots, d_{N-1}\}$ denotes the set of N dimensions of M_S; (ii) $\mathcal{H}(M_S) = \{h_0, h_1, \ldots, h_{N-1}\}$ denotes the set of N hierarchies of M_S, where $h_k \in \mathcal{H}(M_S)$ denotes the hierarchy associated to the dimension $d_k \in \mathcal{D}(M_S)$; (iii) $\mathcal{M}(M_S)$ denotes the set of measures of M_S. For the sake of simplicity, in the following we will assume to deal with single-measure OLAP models, i.e. $\mathcal{M}(M_S) = \{m\}$. However, models and algorithms presented in this paper can be straigthforwardly extended to the more challenging case in which multiple-measure OLAP models (i.e., $|\mathcal{M}(M_S)| > 1$) are considered.

For the sake of simplicity, the stream source name $s_i \in S$ will also denote the data stream generated by the source itself. Each stream source $s_i \in S$ produces a multidimensional stream of data, s_i, composed by an unbounded sequence of (data stream) readings of kind: $r_{i,j}$, i.e. $s_i = \langle r_{i,0}, r_{i,1}, r_{i,2}, \ldots \rangle$ with $|s_i| \to \infty$. In more detail, $r_{i,j}$ denotes the j-th reading of the data stream s_i, and it is defined as a tuple $r_{i,j} = \langle id_i, v_{i,j}, ts_{i,j}, a_{i,j,k_0}, a_{i,j,k_1}, \ldots, a_{i,j,k_{P-1}} \rangle$, where: (i) $id_i \in \{0, .., N-1\}$ is the stream source (absolute) identifier; (ii) $v_{i,j}$ is a non-negative integer value representing the measure produced by the stream source s_i identified by id_i, i.e. the reading value; (iii) $ts_{i,j}$ is a *timestamp* that indicates the time when the reading $r_{i,j}$ was produced by the stream source s_i identified by id_i, i.e. the reading timestamp; (iv) a_{i,j,k_p} is the value associated to the dimensional attribute A_{k_p} of the P-dimensional model of the stream source s_i identified by id_i, denoted by $M_{s_i} = \langle \mathcal{D}(M_{s_i}), \mathcal{H}(M_{s_i}), \mathcal{M}(M_{s_i}) \rangle$, being $\mathcal{D}(M_{s_i})$, $\mathcal{H}(M_{s_i})$ and $\mathcal{M}(M_{s_i})$ the set of dimensions, the set of hierarchies and the set of measures of M_{s_i}, respectively.

The definition above adheres to the so-called *multidimensional data stream model*, which is a fundamental component of the OLAP stream model introduced in Sect. 1. According to the multidimensional data stream model, each reading $r_{i,j}$ embeds a *dimensionality*, which is used to meaningfully handle the overall multidimensional stream. This dimensionality is captured by the set of values $DimVal(r_{i,j}) = \{a_{i,j,k_0}, a_{i,j,k_1}, \ldots, a_{i,j,k_{P-1}}\}$ associated to the set of dimensional attributes $DimAtt(r_{i,j}) = \{A_{i,k_0}, A_{i,k_1}, \ldots A_{i,k_{P-1}}\}$ of M_{s_i}. Also, dimensional attribute values in $r_{i,j}$ are logically organized in an (OLAP) hierarchy, denoted by $h_{i,j}$.

For the sake of simplicity, in the following we will refer the set S of stream sources as the "stream source" itself. To give insights, S could identify a sensor network that, in the vest of collection of sensors, is a stream source itself. Another important assertion states that the OLAP stream model assumes that the multidimensional model of S, M_S, is a-priori known, as happens in several real-life scenarios such as sensor networks monitoring environmental parameters (e.g., temperature, pressure, humidity etc).

Given a stream source S and its multidimensional model, M_S, the multidimensional model of each stream source $s_i \in S$, M_{s_i}, can either be *totally* or *partially* mapped onto the multidimensional model of S, M_S. The total or partial mapping relationship only depends on the mutual correspondence between dimensions of the multidimensional models, as the (single) measure is always the same, thus playing the role of invariant for both models. In the first case (i.e., total mapping), M_{s_i} and M_S are equivalent, i.e. $M_{s_i} \equiv M_S$. In the second case (i.e., partial mapping), M_{s_i} is a *multidimensional (proper) sub-model* of M_S, i.e. $M_{s_i} \subset M_S$. Basically, this defines a *containment relationship* between models, and, consequentially, a containment relationship between the multidimensional data models. It should be noted that mapping relationships above are able to capture even complex scenarios occurring in real-life data stream applications and systems. Also, in our research, mapping relationships define the way readings are aggregated during the acquisition phase (see Sect. 5).

4 Flattening Multidimensional Data Streams

The OLAP dimension flattening process allows us to obtain a transformation of the N-dimensional data stream into a flattened M-dimensional data stream, on the basis of application requirements (see Sect. 2). For the sake of simplicity, in order to describe our proposed OLAP dimension flattening process, here we consider as running example the case in which the N-dimensional data stream is flattened into a two-dimensional data stream (i.e., $M = 2$). The more general process for the case $M > 2$ is a straightforward generalization of the one we next describe.

In the running example, the two flattening dimensions, denoted by d_{f_0} and d_{f_1}, respectively, are selected from the set $\mathcal{D}(M_S)$ and then equipped with specialized hierarchies, denoted by h_{f_0} and h_{f_1}, respectively, such that each hierarchy h_{f_i}, with $i \in \{0, 1\}$, is built by meaningfully merging the "original" hierarchy of d_{f_i} with hierarchies of other dimensions in $\mathcal{D}(M_S)$, according to application requirements driven by specific OLAP analysis goals over the target multidimensional data stream. To theoretical consistency purposes, here we assume that $h_{f_0} \in \mathcal{H}(M_S)$ and $h_{f_1} \in \mathcal{H}(H_S)$, respectively.

The final shape of each hierarchy h_{f_i}, with $i \in \{0, 1\}$, depends on the so-called *ordered definition set* $FDef(d_{f_i})$, with $i \in \{0, 1\}$, which constitutes an input parameter for the OLAP dimension flattening process. This set is composed by tuples of kind: $\langle L_j, d_{j+1}, P_{j+1} \rangle$, such that, given two consecutive tuples $\langle L_j, d_{j+1}, P_{j+1} \rangle$ and $\langle L_{j+1}, d_{j+2}, P_{j+2} \rangle$ in $FDef(h_{d_i})$, the sub-tree of h_{j+2} (i.e., the hierarchy of d_{j+2}) rooted at the root node of h_{j+2} and having depth equal to P_{j+2}, said $T_{P_{j+2}}(h_{j+2})$, is merged to h_{j+1} (i.e., the hierarchy of d_{j+1}) via (*i*) appending a *clone* of $T_{P_{j+2}}(h_{j+2})$ to each member $\sigma_{i, L_{j+1}}$ of h_{j+1} at level L_{j+1}, and (*ii*) erasing the original sub-tree rooted at $\sigma_{i, L_{j+1}}$. This process is iterated for each tuple $\langle L_j, d_{j+1}, P_{j+1} \rangle$ in $FDef(d_{f_i})$ until the final hierarchy h_{f_i} is obtained.

To give an example, consider Fig. 2, where the two hierarchies h_{j+1} and h_{j+2} are merged in order to obtain the new hierarchy $h_{f_{j+1}}$ that can also be intended as a "modified" version of the original hierarchy h_{j+1}. Specifically, $h_{f_{j+1}}$ is obtained via setting both L_{j+1} and P_{j+2} equal to 1.

Algorithm `MergeOLAPHierarchies` implements the merging of two OLAP hierarchies. It takes as arguments the two OLAP hierarchies to be merged, h_{j+1} and h_{j+2}, and the parameters needed to perform the merging task (i.e., the level of h_{j+1}, L_{j+1}, to which clones of the sub-tree of h_{j+2} rooted at the root node of h_{j+2} have to be appended, and the depth P_{j+2} of such sub-tree), and returns the modified hierarchy $h'_{j+1} \equiv h_{f_{j+1}}$. Finally, algorithm `FlattenMultidimensionalCubeModel` implements the overall OLAP dimension flattening process by merging the target OLAP hierarchies pair-wise via algorithm `MergeOLAPHierarchies`, which thus plays the role of baseline procedure. Since an N-dimensional data cube model is flattened onto an M-dimensional data cube model, with $M << N$, a *flattening map* FM determines the way groups of OLAP dimensions of the original data cube model must be flattened onto *one new* OLAP dimension of the flattened data cube model. Each entry of

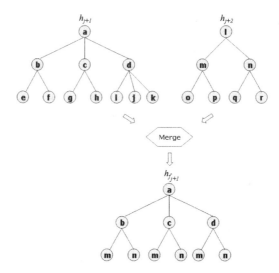

Fig. 2. Merging OLAP hierarchies

FM, denoted by fm_i, stores the definition set $FDef(fm_i)$ of the group of M_i dimensions of the original data cube model to be flattened onto one new dimension of the flattened data cube model. Recall that, for each definition set $FDef(fm_i)$ in FM, the first hierarchy to be processed, h_{f_i}, is the one associated to the first of the M_i flattening dimensions, d_{f_i}. This property univocally determines the flattening dimensions in FM. Algorithm `FlattenMultidimensionalCubeModel` takes as arguments the multidimensional model of the stream source S, M_S, and the flattening map FM, and returns the M flattening dimensions $\mathcal{D}_M = \{d_{f_0}, d_{f_1}, \ldots d_{f_{M-1}}\}$ with modified hierarchies $\mathcal{H}_M = \{h_{f_0}, h_{f_1}, \ldots h_{f_{M-1}}\}$, respectively. It clearly follows that (i) $\bigoplus_{i=0}^{|FM|-1} M_i = M$, and (ii) $|FM| = M$.

5   Computing OLAP Aggregations over Multidimensional Data Streams

Consider a stream source S and its multidimensional model M_S. Given the M-dimensional flattened data stream $s_{MD} \in S$, which is generated by the OLAP dimension flattening process (see Sect. 4), the OLAP stream aggregation scheme deals with the problem of populating the summarized repository of data stream readings MDSR by means of s_{MD} and according to the underlying OLAP stream model (see Sect. 2). As regards storage issues, we assume that MDSR is represented in memory as a multidimensional array [2], like data cubes of conventional OLAP architectures [7]. According to this storage representation model, MDSR can be viewed as a collection of *data cells* that are indexed by means of a certain *multidimensional access method* [15], such that each cell stores an SQL-based aggregation of readings of s_{MD} (e.g., SUM, COUNT, AVG etc). The M dimensions of MDSR are those originated by the OLAP dimension flattening process, and are equipped with specialized hierarchies. The main idea of the OLAP stream aggregation scheme consists in determining how readings in s_{MD} participate to data cell aggregations of MDSR. This, in turn, finally determines the way of populating MDSR.

Similarly to what done with the OLAP dimension flattening process (see Sect. 4), for the sake of simplicity, in order to describe our proposed OLAP stream aggregation scheme, here we consider the case in which MDSR is materialized as a two-dimensional array (i.e., $M = 2$). As a consequence, $s_{MD} \equiv s_{2D}$. The more general scheme for the case $M > 2$ is a straightforward generalization of the one we next describe.

In the running example, we consider the case in which MDSR is characterized by the following two flattening dimensions: the *normal* flattening dimension d_{f_N}, and the *temporal* flattening dimension d_{f_T}. Both hierarchies of d_{f_N} and d_{f_T}, denoted by h_{f_N} and h_{f_T}, respectively, are obtained by means of the OLAP dimension flattening process. Indeed, to further simplify, we assume that h_{f_T} follows the *natural temporal hierarchy* (e.g., $Year \rightarrow Quarter \rightarrow Month \rightarrow Day$), thus, without loss of generality, we assume that h_{f_T} is properly obtained by a *void* flattening process. It should be noted that the temporal dimension allows us to meaningfully capture how data streams evolve over time. However, our OLAP stream model is general enough to handle any kind of dimension arising in real-life data stream applications and systems (e.g., categorial dimensions of retail scenarios).

Given a reading $r_{2D,j}$ embedded in s_{2D} of S, on the basis of traditional OLAP aggregation schemes over conventional data sources like relational data sets stored in DBMS (e.g., [2]), the measure $v_{2D,j}$ of $r_{2D,j}$ has to be aggregated along all the dimensions of the multidimensional model $M_{s_{2D}}$. In our proposed OLAP stream model, this means that the measure $v_{2D,j}$ contributes to a certain (array) cell of MDSR (and updates its value) based on the membership of dimensional attribute values $DimVal(r_{2D,j}) = \{a_{2D,j,k_0}, a_{2D,j,k_1}, \ldots, a_{2D,j,k_{P-1}}\}$ and the timestamp $ts_{2D,j}$ of the reading $r_{2D,j}$ with respect to the normal and temporal hierarchy associated to dimensions of MDSR, respectively. This way, we obtain a *specialized* aggregation scheme for our proposed OLAP stream model able of (*i*) taming the curse of dimensionality problem arising when multidimensional data streams are handled (see Sect. 2), and (*ii*) effectively supporting the simultaneous multidimensional aggregation of data stream readings.

It should be noted that *the OLAP dimension flattening process plays a role in the final way readings are aggregated during the acquisition phase.* Focus the attention on the normal flattening dimension d_{f_N} and the associated hierarchy h_{f_N}. Assume that $\mathcal{D}_N(M_S) = \{d_{k_0}, d_{k_1}, \ldots, d_{k_{F-1}}\}$ is the sub-set of $\mathcal{D}(M_S)$ used to generate d_{f_N} ($\mathcal{D}_N(M_S) \subset \mathcal{D}(M_S)$). Let us now focus on the collection of stream sources of S. Although each stream source $s_i \in S$ could define a total (i.e., $M_{s_i} \equiv M_S$) or partial (i.e., $M_{s_i} \subset M_S$) containment relationship with respect to the multidimensional model of S, M_S, the OLAP dimension flattening process essentially combines dimensions in $\mathcal{D}_N(M_S)$ and, as a consequence, the final multidimensional model of the flattened two-dimensional data stream s_{2D}, $M_{s_{2D}}$, results to be a "combination" of the multidimensional models of data stream sources in S. Intuitively enough, it is easy to observe that, if the multidimensional models $M_{s_{2D}}$ and M_S are coincident (i.e., $M_{s_{2D}} \equiv M_S$), then readings embedded in s_{2D} are simultaneously aggregated along *all* the dimensions in M_S to obtain the final aggregate value in the corresponding MDSR cell. Otherwise, if the multidimensional models $M_{s_{2D}}$ and M_S define a proper containment relationship (i.e., $M_{s_{2D}} \subset M_S$), then readings embedded in s_{2D} are simultaneously aggregated along *a partition* of the dimensions in M_S to obtain the final aggregate value in the corresponding MDSR cell.

Formally, given a reading $r_{2D,j} = \langle id_{2D}, v_{2D,j}, ts_{2D,j}, a_{2D,j,k_0}, a_{2D,j,k_1}, \ldots, a_{2D,j,k_{P-1}} \rangle$ embedded in s_{2D}, on the basis of a top-down approach, starting from the dimensional attribute value at the highest aggregation level of $h_{2D,j}$ (i.e., the hierarchy associated to dimensional attribute values in $r_{2D,j}$ – see Sect. 3), a_{2D,j,k_0}, we first search the hierarchy of the normal flattening dimension d_{f_N}, h_{f_N}, starting from the member at the highest aggregation level, denoted by $\sigma_{0,0}^{\mathcal{N}}$, by means of a *breadth-first tree visiting strategy*, and we check whether a_{2D,j,k_0} belongs to the OLAP class defined by the *current* member of h_{f_N}, $\sigma_{i,L_j}^{\mathcal{N}}$ (when $i = 0$, then $\sigma_{i,L_j}^{\mathcal{N}} \equiv \sigma_{0,0}^{\mathcal{N}}$). When a member of h_{f_N} such that a_{2D,j,k_0} belongs to the class it defines, denoted by $\sigma_{i^*,L_j^*}^{\mathcal{N}}$, is found, then (*i*) the breadth-first search is contextualized to the sub-tree of h_{f_N} rooted at $\sigma_{i^*,L_j^*}^{\mathcal{N}}$, denoted by $T^*(h_{f_N})$, and (*ii*) the current search dimensional attribute value becomes the value that immediately follows a_{2D,j,k_0} in the hierarchy $h_{2D,j}$, i.e. a_{2D,j,k_1}. After that, the whole search is repeated again, and it ends when a leaf node of h_{f_N} is reached, denoted by $\sigma_{i^*,Depth(h_{f_N})}^{\mathcal{N}}$, such that $Depth(h_{f_N})$ denotes the depth of h_{f_N}. Note that the search should end when the *last* dimensional attribute value $a_{2D,j,k_{P-1}}$ is processed accordingly, but, due to the OLAP dimension flattening process and the possible presence of *imprecise or incomplete data*, it could be the case that the search ends before that. For the sake of simplicity, hereafter we assume to deal with hierarchies and readings adhering to the simplest case in which the search ends by reaching a leaf node of h_{f_N} while the last dimensional attribute value $a_{2D,j,k_{P-1}}$ is processed. The described search task allows us to determine an indexer on the first dimension of MDSR, d_{f_N}. Let us denote as $I_{\mathcal{N}}^*$ this indexer. The other indexer on the second dimension of MDSR, d_{f_T}, denoted by $I_{\mathcal{T}}^*$, is determined by means of the same approach exploited for the previous case, with the difference that, in this case, the search term is fixed and represented by the reading timestamp $ts_{2D,j}$. When both indexers $I_{\mathcal{N}}^*$ and $I_{\mathcal{T}}^*$ are determined, an MDSR cell is univocally located, and the reading measure $v_{2D,j}$ is used to finally update the value of this cell.

Let us now focus on a running example showing how our proposed OLAP stream aggregation scheme for multidimensional data stream readings works in practice. Fig. 3 shows the hierarchy h_{f_N} associated to the normal flattening dimension d_{f_N} of the running example, whereas Fig. 4 shows instead the

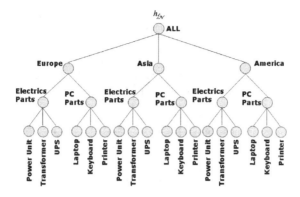

Fig. 3. The hierarchy associated to the normal flattening dimension of the running example

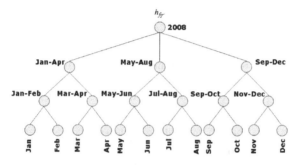

Fig. 4. The hierarchy associated to the temporal flattening dimension of the running example

hierarchy h_{f_T} associated to the temporal flattening dimension d_{f_T}. As suggested by Fig. 3 and Fig. 4, the multidimensional data stream model of the running example describes an application scenario focused on sales of electric and personal computer parts sold in Europe, Asia and America during 2008. The hierarchy h_{f_N} derives from the OLAP dimension flattening process, whereas the hierarchy h_{f_T} follows the natural temporal hierarchy organized by months and groups of months (i.e., a void flattening process). Readings are produced by different locations distributed in Europe, Asia and America, thus defining a proper network of data stream sources. In more detail, the described one is a typical application scenario of modern *Radio Frequency IDentifiers* (RFID) [19] based applications and systems.

Fig. 5 shows the array-based repository MDSR that represents summarized information on readings produced by RFID sources, equipped with the normal and temporal hierarchies. In particular, each MDSR cell stores a SUM-based OLAP aggregation of readings according to both the normal and temporal dimension, simultaneously.

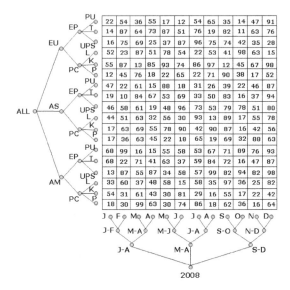

Fig. 5. The array-based repository of summarized data stream readings MDSR of the running example

Now consider the reading $r_{2D,k} = \langle id_{2D}, 5, 6/15/08, India, Delta - Power, Power2500 \rangle$ embedded in s_{2D}, which records the sale of a $Power2500$ transformer, produced by the company $Delta\text{-}Power$ at the price of 5 RP, sold in India on June 15, 2008. Focus the attention on how the value of $r_{2D,k}$ is aggregated within cells of MDSR. On the basis of our proposed OLAP stream aggregation scheme, the final MDSR cell to be updated is finally located by means on two distinct paths on the respective hierarchies h_{f_N} and h_{f_T} determined by the simultaneous membership of dimensional attribute values and timestamp of the reading $r_{2D,k}$ to classes defined by members of these hierarchies, in a top-down manner. Fig. 6 shows the configuration of MDSR after the update. Note that the old value 69 of the target cell has been updated to the new value $69 + 5 = 74$.

Finally, algorithm PopulateRepository implements the proposed OLAP stream aggregation scheme that allows us to populate the target array-based repository of summarized data stream readings MDSR by means of the M-dimensional flattened stream s_{MD}. It takes as arguments the repository MDSR and the input reading $r_{MD,j}$ of s_{MD}, and updates MDSR by the measure value embedded in $r_{MD,j}, v_{MD,j}$, according to the simultaneous membership-based multidimensional aggregation approach described above. Furthermore, since, to efficiency purposes, data streams are usually processed with the aid of a *buffer* B (e.g., [1]) having a certain memory C_B, such that $C_B > 0$, CAMS also implements the buffered version of algorithm PopulateRepository, called BufferedPopulateRepository. Algorithm BufferedPopulateRepository takes as arguments the repository MDSR and the buffer of M-dimensional readings B_{MD}, and updates MDSR via iteratively invoking algorithm PopulateRepository, which thus plays the role of baseline procedure.

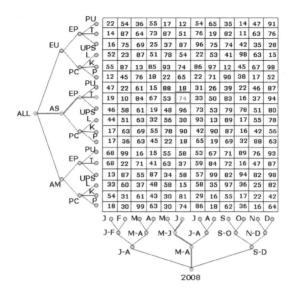

Fig. 6. The array-based repository of summarized data stream readings MDSR of Fig. 5 after the update

6 Experimental Assessment and Analysis

In order to test the OLAPing capabilities of CAMS over multidimensional data streams, we conducted a series of experiments where we stressed both the *efficiency* and the *scalability* of CAMS. The hardware/software platform of our experimental environment was characterized by a 2 GHz *Pentium* equipped with 2 GB RAM and running *Microsoft Windows XP. Microsoft Analysis Services 2000* was the OLAP platform chosen as data layer of the experimental setting. Both the CAMS framework and the experimental environment have been implemented in *Sun Microsystems Java 1.5.* Java software has been interfaced with OLAP data stored in the *Analysis* server by means of a "neutral" XML-based API library. We engineered a synthetic data stream generator producing readings whose values are distributed according to a *Uniform* distribution over a given range $[L_0, L_1]$, with $L_1 > L_0$. Our experiments were focused to stress the overall amount of *memory space* taken by CAMS for computing OLAP aggregations over multidimensional data streams when populating the repository MDSR.

In our experimental environment, we introduced the following (experimental) parameters: (*i*) N, which models the number of dimensions of the original data cube model; (*ii*) M, which models the number of dimensions of the flattened data cube model; (*iii*) $P_M = \{P_{f_0}, P_{f_1}, \ldots, P_{f_{M-1}}\}$, which models the depths of hierarchies associated to dimensions of the flattened data cube model, such that P_{f_i} denotes the depth of the hierarchy $h_{f_i} \in \mathcal{H}_M$; (*iv*) L_0, which models the lower bound of the range of the Uniform distribution generating the data stream reading values; (*v*) L_1, which models the upper bound of the range of the Uniform

distribution generating the data stream reading values; (vi) S, which models
the overall size of data stream readings expressed in terms of K readings. We
combined all these parameters in order to study the variation of memory space,
expressed in MB, needed to compute the final OLAP aggregations of MDSR. In
our experimental environment, the whole experimental setting, denoted by E,
was thus modeled as follows: $E = \langle N, M, \{P_{f_0}, P_{f_1}, \ldots, P_{f_{M-1}}\}, L_0, L_1, S \rangle$. Each
experimental campaign was characterized by an instance of E, denoted by $\widehat{E}$,
which is obtained via setting the values of all the experimental parameters in E.

Fig. 7 (a) shows the variation of memory space with respect to the variation
of the size of data stream readings, S, for several values of the number of dimensions of the flattened data cube model, M, ranging over the interval $[4 : 8]$.
For this experimental campaign, we set the experimental parameters as follows:
$\widehat{E} = \langle 16, \{4, 6, 8\}, P_{f_0} = P_{f_1} = \ldots = P_{f_{M-1}} = 15, 100, 500, [400K, 800K]\rangle$. It
should be noted that this campaign was focused to stress the efficiency of CAMS,
i.e. the capability of CAMS in efficiently computing OLAP aggregations over multidimensional data streams. Fig. 7 (b) shows instead the variation of memory
space with respect to the variation of the number of dimensions of the original data cube model, N, for several values of the size of data stream readings,
S, ranging over the interval $[600K: 800K]$. For this experimental campaign, we
set the experimental parameters as follows: $\widehat{E} = \langle [16, 24], 8, P_{f_0} = P_{f_1} = \ldots =
P_{f_{M-1}} = 15, 100, 500, \{600K, 700K, 800K\}\rangle$. It should be noted that this cam-
paign was focused to stress the scalability of CAMS, i.e. the capability of CAMS
in efficiently scaling-up over data stream sources characterized by increasing-in-
dimensionality data cube models.

From the analysis of the our experimental results shown in Fig. 7, it clearly
follows that CAMS allows us to effectively and efficiently computing OLAP aggregations over multidimensional data streams, while also ensuring a good scalability when data cube models grow in number of dimensions and size. All these

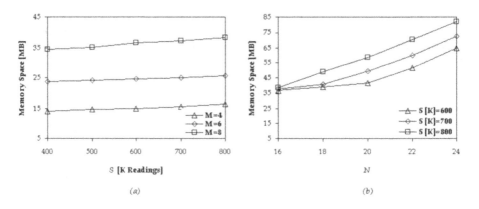

Fig. 7. Variation of memory space needed to compute OLAP aggregations over multidimensional data streams with respect to the variation of the size of data stream
readings (a) and the number of dimensions of the original data cube model (b)

amenities make CAMS an enabling component for next-generation DSMS towards the support of OLAP analysis/mining tasks over multidimensional data streams. In particular, the scalability of CAMS, as demonstrated by our experimental results, plays a critical role in real-life data stream applications and systems, as the latter are very often characterized by massive and high-dimensional flows of data stream readings.

7 Conclusions and Future Work

CAMS, a model for efficiently OLAPing multidimensional data streams has been presented in this paper. CAMS combines a set of data stream processing methodologies in order to tame the multidimensionality of data streams, which is a leading challenge in OLAP stream research. Our experimental assessment and analysis have clearly highlighted the benefits deriving from integrating CAMS within any data-stream-oriented OLAP tool of next-generation DSMS. A further experimental analysis of CAMS will regard stressing other important and critical experimental parameters not considered in this research, such as time needed to compute OLAP aggregations over multidimensional data streams, along with the assessment of CAMS against real-life data stream sets. However, a clear result of the research presented in this paper states that, in the context of OLAP tools for next-generation DSMS, CAMS plays the role of enabling component. Despite this, the new frontier for OLAP stream research is represented by models, techniques, algorithms and architectures for *effective on-the-fly OLAPing of multidimensional data streams*, which is postponed as future work.

References

1. Abadi, D., Carney, D., Cherniack, M., Convey, C., Lee, S., Stonebraker, M., Tatbul, N., Zdonik, S.: Aurora: A New Model and Architecture for Data Stream Management. VLDB Journal 12(2), 120–139 (2003)
2. Agarwal, S., Agrawal, R., Deshpande, P.M., Gupta, A., Naughton, J.F., Ramakrishnan, R., Sarawagi, S.: On the Computation of Multidimensional Aggregates. In: VLDB, pp. 506–521 (1996)
3. Aggarwal, C.: Data Streams: Models and Algorithms. Springer, Heidelberg (2007)
4. Barbarà, D., Du Mouchel, W., Faloutsos, C., Haas, P.J., Hellerstein, J.M., Ioannidis, Y.E., Jagadish, H.V., Johnson, T., Ng, R.T., Poosala, V., Ross, K.A., Sevcik, K.C.: The New Jersey Data Reduction Report. IEEE Data Engineering Bulletin 20(4), 3–45 (1997)
5. Berchtold, S., Böhm, C., Kriegel, H.-P.: The Pyramid-Technique: Towards Breaking the Curse of Dimensionality. In: ACM SIGMOD, pp. 142–153 (1998)
6. Cai, Y.D., Clutterx, D., Papex, G., Han, J., Welgex, M., Auvilx, L.: MAIDS: Mining Alarming Incidents from Data Streams. In: ACM SIGMOD, pp. 919–920 (2004)
7. Chaudhuri, S., Dayal, U.: An Overview of Data Warehousing and OLAP Technology. SIGMOD Record 26(1), 65–74 (1997)

8. Cuzzocrea, A.: Synopsis Data Structures for Representing, Querying, and Mining Data Streams. In: Ferragine, V.E., Doorn, J.H., Rivero, L.C. (eds.) Handbook of Research on Innovations in Database Technologies and Applications: Current and Future Trends, pp. 701–715 (2009)

9. Cuzzocrea, A., Chakravarthy, S.: Event-Based Compression and Mining of Data Streams. In: Lovrek, I., Howlett, R.J., Jain, L.C. (eds.) KES 2008, Part II. LNCS, vol. 5178, pp. 670–681. Springer, Heidelberg (2008)

10. Cuzzocrea, A., Furfaro, F., Masciari, E., Saccà, D., Sirangelo, C.: Approximate Query Answering on Sensor Network Data Streams. In: Stefanidis, A., Nittel, S. (eds.) GeoSensor Networks, pp. 53–63 (2004)

11. Cuzzocrea, A., Saccà, D., Serafino, P.: Semantics-aware Advanced OLAP Visualization of Multidimensional Data Cubes. International Journal of Data Warehousing and Mining 3(4), 1–30 (2007)

12. Deshpande, A., Guestrin, C., Madden, S.: Using Probabilistic Models for Data Management in Acquisitional Environments. In: CIDR, pp. 317–328 (2005)

13. Deshpande, A., Guestrin, C., Madden, S., Hellerstein, J.M., Hong, W.: Model-driven Data Acquisition in Sensor Networks. In: VLDB, pp. 588–599 (2004)

14. Dobra, A., Gehrke, J., Garofalakis, M., Rastogi, R.: Processing Complex Aggregate Queries over Data Streams. In: ACM SIGMOD, pp. 61–72 (2002)

15. Gaede, V., Gunther, O.: Multidimensional Access Methods. ACM Computing Surveys 30(2), 170–231 (1998)

16. Gehrke, J., Korn, F., Srivastava, D.: On Computing Correlated Aggregates over Data Streams. In: ACM SIGMOD, pp. 13–24 (2001)

17. Gray, J., Chaudhuri, S., Bosworth, A., Layman, A., Reichart, D., Venkatrao, M., Pellow, F., Pirahesh, H.: Data Cube: A Relational Aggregation Operator Generalizing Group-By, Cross-Tab, and Sub-Totals. Data Mining and Knowledge Discovery 1(1), 29–54 (1997)

18. Han, J., Chen, Y., Dong, G., Pei, J., Wah, B.W., Wang, J., Cai, Y.D.: Stream Cube: An Architecture for Multi-Dimensional Analysis of Data Streams. Distributed and Parallel Databases 18(2), 173–197 (2005)

19. Han, J., Gonzalez, H., Li, X., Klabjan, D.: Warehousing and Mining Massive RFID Data Sets. In: Li, X., Zaïane, O.R., Li, Z.-h. (eds.) ADMA 2006. LNCS, vol. 4093, pp. 1–18. Springer, Heidelberg (2006)

20. Li, X., Han, J., Gonzalez, H.: High-Dimensional OLAP: A Minimal Cubing Approach. In: VLDB, pp. 528–539 (2004)

21. Madden, S., Franklin, M.J., Hellerstein, J.M., Hong, W.: The Design of an Acquisitional Query Processor for Sensor Networks. In: ACM SIGMOD, pp. 491–502 (2003)

Data Stream Prediction Using Incremental Hidden Markov Models

Kei Wakabayashi and Takao Miura

HOSEI University, Dept.of Elect.& Elect. Engr.,
3-7-2 KajinoCho, Koganei, Tokyo, 184–8584 Japan

Abstract. In this paper, we propose a new technique for time-series prediction. Here we assume that time-series data occur depending on *event* which is unobserved directly, and we estimate future data as output from the most likely event which will happen at the time. In this investigation we model time-series based on event sequence by using *Hidden Markov Model*(HMM), and extract time-series patterns as trained HMM parameters. However, we can't apply HMM approach to *data stream* prediction in a straightforward manner. This is because Baum-Welch algorithm, which is traditional unsupervised HMM training algorithm, requires many stored historical data and scan it many times. Here we apply *incremental Baum-Welch algorithm* which is an on-line HMM training method, and estimate HMM parameters dynamically to adapt new time-series patterns. And we show some experimental results to see the validity of our method.

Keywords: Forecasting, Data Stream, Hidden Markov Model, Incremental Learning.

1 Introduction

Recently there have been a lot of knowledge-based approaches for huge databases, and much attention have been paid on *time-series prediction* techniques [2]. Especially prediction on *data stream*, which is assumed huge amount of data and high speed updating, is important technique in many domains.

There have been many prediction approach proposed so far [4]. One of the traditional approach is *Exponential Smoothing* that is a heuristic method based on weighted mean of past data. In exponential smoothing method, we assume weight of data decrease with time exponentially. Then we can obtain the weighted mean at time t using the weighted mean at time $t-1$ as an aggregated value.

Holt-Winters method is one of the famous prediction approach proposed based on exponential smoothing [3]. In this method they estimate future data based on the weighed mean $\tilde{y}_t$ and mean of $\tilde{y}_t$'s variation F_t. $\tilde{y}_t$ and F_t are updated at each time as follows:

$$\tilde{y}_t = \lambda_1 y_t + (1 - \lambda_1)(\tilde{y}_{t-1} + F_{t-1})$$

$$F_t = \lambda_2(\tilde{y}_t - \tilde{y}_{t-1}) + (1 - \lambda_2)F_{t-1}$$

T.B. Pedersen, M.K. Mohania, and A M. Tjoa (Eds.): DaWaK 2009, LNCS 5691, pp. 63–74, 2009.

λ_1 and λ_2 are called as smoothing parameters. Here $0.0 \leq \lambda \leq 1.0$. If λ is set to bigger value, the weight of past data will decrease more quickly. They estimate future data at time $t + h$ using $\tilde{y}_t$ and F_t as follows:

$$\tilde{y}_{t+h|t} = \tilde{y}_t + hF_t$$

However, by Holt-Winters method it is hard to estimate data depending on event. For example, wind velocity data occur depending on event such as approach of typhoon, does not depend on previous data directly.

In Hassan [5], they discuss how to predict stock market using *Hidden Markov Model* (HMM) which is a stochastic model assumed simple Markov process with hidden state. They estimate hidden state at each observation and estimate future observations based on state transition. We believe we can estimate time-series depending on event sequence effectively by this approach, since the estimated data doesn't depend on previous observation.

However this approach can't estimate a data following new pattern which didn't be appeared in training data, because they train HMM parameters in advance by EM-algorithm which requires large computational time. This is serious problem especially in *data stream* forecasting. Data stream is assumed as infinite time-series and distribution of data changes dynamically [6] [10]. We should train the model incrementally while forecasting data since given training data contain only a part of patterns in whole data stream.

There are some research about incremental learning for HMM [1]. In Stenger [11], they discuss how to update HMM parameters incrementally and propose *incremental Baum-Welch algorithm*. Incremental Baum-Welch algorithm does not require historical data, and we can compute very quickly. They apply this approach to background modeling in real-time video processing.

In this investigation we propose a new forecasting approach using incremental Baum-Welch algorithm. We believe we can achieve adaptive time-series estimation on data stream by using incremental HMM learning. We compare our approach to conventional batch Baum-Welch algorithm and show the effectiveness of proposal method.

In this work, we discuss time-series forecasting issue on data stream in section 2. In section 3 we review incremental Hidden Markov Model and we develop the forecasting technique using incremental HMM. Section 4 contains experimental results. We conclude our discussion in section 5.

2   Time-Series Forecasting Based on Event

First of all, we describe how we predict time-series on data stream. We illustrate our approach in a figure 1. Let us assume a sales history of goods in a store as time-series, such that has a large sale for coffees at day t, beer at day $t + 1$, and fruit juice at day $t + 2$.

In this investigation, we assume that observation in time-series occurs depending on unknown (hidden) state. In the figure 1, we interpret the day t as a busy day for many customer by examining the sales data, because of few beers and

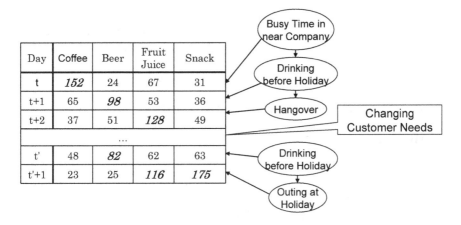

Day	Coffee	Beer	Fruit Juice	Snack
t	*152*	24	67	31
t+1	65	*98*	53	36
t+2	37	51	*128*	49
...				
t'	48	*82*	62	63
t'+1	23	25	*116*	*175*

Fig. 1. Modeling Time-Series based on Event Sequence

many coffees saled. We consider the day $t + 1$ as a day before holiday because many customers want to drink alcohol, and many customers have hangover at day $t + 2$ because of many fruit juices sold.

On this interpretations of data, we see transition from a busy day to a drinking day, and transition from a drinking day to a hangover day. If we see these transition pattern in past data many times, we may estimate sales of goods at $t + 1$ or $t + 2$ when we are at the day t.

In this approach, although we estimate future observation by considering transition of hidden state, we don't have to interpret these unknown state [12]. In this investigation we formalize hidden state explicitly using Hidden Markov Model (HMM), and estimate observation by HMM trained using past observations without interpreting states.

However, during the passage of long time, some change of transition pattern may occur. For example, if a new high school opens near by the store, many students will come to the store and sales of snacks and juices will increase. For such changes of pattern, we should consider adaptation to new pattern even while processing estimation of future observations.

Unfortunately, it is not easy to update HMM parameters *on-line*, that means training model using new observations dynamically after start of process for estimation. This is because we can't obtain the most likely parameters directly from observation sequence, but we must estimate hidden state sequence before parameter estimation. Generally, for unsupervised learning of HMM, we employ Baum-Welch algorithm which requires training observation sequence and scans them many times. However, it is hard to recalculate HMM parameters by Baum-Welch algorithm at each time we get new observation, from the aspect of computational time.

Especially in data stream environment, we can't memorize historical observation sequence because of infinite amount of observation, then we should apply *incremental* parameter updating. In this investigation, we employ *incremental*

Baum-Welch algorithm which doesn't require historical observations for on-line update of HMM parameters. We apply incremental Baum-Welch algorithm to estimating future observation and we discuss how we can achieve adaptive forecasting with on-line training.

3   Forecasting Using Incremental Hidden Markov Model

3.1   Hidden Markov Model

A *Hidden Markov Model* (HMM) is nothing but an automaton with output where both the state transition and the output are defined in a probabilistic manner. The state transition arises according to a simple Markov model but it is assumed that we don't know on which state we are standing now[1], and that we can observe an output symbol at each state. We could estimate the transition sequences through observing output sequence.

A HMM model consists of (Q, Σ, A, B, π) defined below[8]:

(1) $Q = \{q_1, \cdots, q_N\}$ is a finite set of states
(2) $\Sigma = \{o_1, \cdots, o_M\}$ is a set of output symbols.
(3) $A = \{a_{ij}, i, j = 1, ..., N\}$ is a probability matrix of state transition where each a_{ij} means a probability of the transition at q_i to q_j. Note $a_{i1} + ... + a_{iN} = 1.0$.
(4) $B = \{b_i(o_t), i = 1, ..., N, t = 1, ..., M\}$ is a probability of outputs where $b_i(o_t)$ means a probability of an output o_t at a state q_i.
(5) $\pi = \{\pi_i\}$ is an initial probability where π_i means a probability of the initial state q_i. In this work, we assume all states occur in the same probability as initial state, that is $\pi_i = \frac{1}{N}$.

The probability matrix A shows the transition probability within a framework of simple Markov model, which means state change arises in a probabilistic manner depending only on the current state. Thus, for instance, the (i, j)-th component of A^2 describes the transition probability from q_i to q_j with two *hops* of transitions. Similarly the output appears depending only on the current state.

Since we consider multidimensional observation, we define output symbol as numeric vector in this work. We assume each values in vector occurs from normal distribution independently. That is, when $o_t = o_{t1}, o_{t2}, ..., o_{tD}$, we define probability of o_t at state q_i as:

$$b_i(o_{tk}) = \frac{1}{\sqrt{2\pi}\sigma_{ik}} e^{-\frac{(o_{tk} - \mu_{ik})^2}{2\sigma_{ik}^2}}$$

μ_{ik} and σ_{ik} are the mean and variance of normal distribution respectively, and both are parameters of our HMM.

---

[1] This is why we say *hidden*.

We should think about how to obtain parameters, since it is hard to determine the transition probability matrix A and the output probability B definitely. This problem is called a *model calculation* of HMM. Usually we do that by means of some machine learning techniques[9].

One of the typical approach is *supervised learning*. In this approach, we assume *labeled training data* in advance to calculate the model, but the data should be correctly classified by hands since we should extract typical patterns them by examining them. Another approach comes, called *unsupervised learning*. Assume we can't get labeled training data but a mountain of unclassified data except a few. Once we obtain strong similarity between the classified data and unclassified data (such as high correlation), we could extend the training data in a framework of Expectation Maximization (EM) approach[7].

One of the typical approach is known as a *Baum-Welch* algorithm. The algorithm has been proposed based on EM approach. That is, the algorithm adjusts the parameters many times to maximize the likelihood for the generation of the output symbols given as unsupervised data. The process goes just same as EM calculation, i.e., we calculate the expect value of the transition probability and the output probability, then we maximize them. We do that until few change happens.

3.2   Incremental Baum-Welch Algorithm

Let $\gamma_t(q_i)$ be the stay probability on state q_i at time t, and $R_T(q_i)$ be the summation of stay probabilities $\gamma_t(q_i)$, that is:

$$R_T(q_i) = \sum_{t=1}^{T} \gamma_t(q_i) \tag{1}$$

When we obtain a new observation o_T at time T, we update HMM parameters by using *incremental Baum-Welch algorithm*. In first, we calculate $\gamma_T(q_x)$ which is likelihood that we stay state q_x at time T. Here we assume $\gamma_{T-1}(q_i)$ and $R_{T-1}(q_i)$ are memorized in advance for all i. Then, we obtain $\gamma_T(q_i, q_x)$ which is the probability of the transition at time T from state q_i to q_x using current HMM parameters as:

$$\gamma_T(q_i, q_x) = \gamma_{T-1}(q_i) \frac{a_{ix} b_x(o_T)}{\sum_{x=1}^{N} a_{ix} b_x(o_T)}$$

We obtain $\gamma_T(q_x)$ using $\gamma_T(q_i, q_x)$ as:

$$\gamma_T(q_x) = \sum_{i=1}^{N} \gamma_T(q_i, q_x)$$

Then the HMM parameters are updated as [11]:

$$a'_{ij} = \frac{R_{T-1}(q_i) a_{ij} + \gamma_T(q_i, q_j)}{R_T(q_i)}$$

$$\mu_i' = \frac{\sum_{t=1}^T \gamma_t(q_i) o_t}{\sum_{t=1}^T \gamma_t(q_i)} = \frac{R_{T-1}(q_i)\mu_i + \gamma_T(q_i) o_T}{R_T(q_i)}$$

$$\sigma_i' = \frac{\sum_{t=1}^T \gamma_t(q_i)(o_t - \mu_i)^2}{\sum_{t=1}^T \gamma_t(q_i)} = \frac{R_{T-1}(q_i)\sigma_i + \gamma_T(q_i)(o_T - \mu_i)^2}{R_T(q_i)}$$

By using incremental Baum-Welch algorithm, we can update parameters quickly without historical data.

However, change of parameters become smaller and smaller with time because R_T increase directly with time and the weight of new observation become relatively small. Therefore we define *forget function* $w_\lambda(t)$ to reduce the weight of old observations as:

$$w_\lambda(t) = \lambda^t$$

Here let t be the time elapsed since the data observed, λ be constant of forgetting speed and $0.0 \le \lambda \le 1.0$. In this paper, we call $\lambda' = (1.0 - \lambda)$ as *forget coefficient*.

We redefine $R_T(q_i)$ to reduce the weight of old observations as below:

$$R_T(q_i) = \sum_{t=1}^T \gamma_t(q_i) \lambda^{T-t}$$

$$= \sum_{t=1}^{T-1} \gamma_t(q_i) \lambda^{T-1-t} \lambda + \gamma_T(q_i) \lambda^{T-T}$$

$$= \lambda R_{T-1}(q_i) + \gamma_T(q_i)$$

We can derive $R_T(q_i)$ recursively. By applying this new definition, we can update HMM parameters using observations weighted based on forget function by incremental Baum-Welch algorithm.

3.3   Estimating Observation

We develop our theory to estimate future data using incremental Hidden Markov Model. Since incremental Baum-Welch algorithm does not execute iterative parameter estimation such as EM-algorithm, it requires many observations for convergence HMM parameters initialized random values. Here we apply Baum-Welch algorithm as off-line learning to obtain initial model.

First, we generate random numbers for HMM parameters A and μ, and assign 1.0 to all σ. Then we apply Baum-Welch algorithm to given training sequence which contains T observations. After convergence, we calculate $\gamma_t(q_i)$, the stay probabilities on state q_i at time t for $1 \le t \le T$. We obtain its summation $\sum_{t=1}^T \gamma_t(q_i)$ and the stay probabilities at final time T, $\gamma_T(q_i)$.

Then we discuss how to estimate the observation at next time $T + 1$ using HMM. Since observation at time $T + 1$ is unknown, we estimate the likelihood which we stay on state q_i at time $T + 1$ as:

$$\gamma_{T+1}(q_i) = \sum_{j=1}^{N} \gamma_T(q_j) a_{ji}$$

If we are standing at state q_i, the most likely observation at time $T + 1$ is $\mu_i = (\mu_{i1}, \mu_{i2}, ..., \mu_{iD})$. Therefore, the expected value of the observation is estimated as:

$$\sum_{j=1}^{N} \gamma_{T+1}(q_i) \mu_i$$

We generate this value as our estimated observation at time $T + 1$.

When we observe a new data at time $T + 1$, we update HMM parameters by using incremental Baum-Welch algorithm. After updating parameters, we calculate $\gamma_{T+1}(q_i)$ and $R_{T+1}(q_i)$ on each state q_i at time $T + 1$ using updated HMM parameters. That is:

$$\gamma_{T+1}(q_i) = \sum_{j=1}^{N} \gamma_T(q_j) a'_{ji} b'_i(o_{T+1})$$

$$R_{T+1}(q_i) = \lambda R_T(q_i) + \gamma_{T+1}(q_i)$$

We use $\gamma_{T+1}(q_i)$ to estimate next observation.

4   Experimental Results

Here we discuss some experimental results to show the usefulness of our approach. First we show how to obtain the results and to evaluate them, then we show and examine the results.

4.1   Preliminaries

As a test data for our experiments, we take 5 years (1828 days) data in Himawari Weather Data 1996 to 2000 in Tokyo, and select 6 schema, *maximum temperature* (˚C), *minimum temperature* (˚C), *humidity* (%), *maximum wind velocity* (m/s), *duration of sunshine* (hour) and *day rainfall amount* (mm). Table 1 show a part of observation sequence.

We take 10% of observation sequence (182 days) for off-line training data, and we predict the last 90% of observations (1646 days). Here we apply 2 algorithms to estimate observations, HMM method with batch training by conventional Baum-Welch algorithm periodically (batch HMM), and HMM method with incremental training by incremental Baum-Welch algorithm at every observation(Incremental HMM). In batch HMM method, we memorize data observed

Table 1. Test Data

Day	Max. Temp.	Min. Temp.	Humidity	Max. Wind	Sunshine	Rainfall
3/26/1997	17.5	10.4	49	11.3	9.4	0.0
3/27/1997	11.9	8.4	84	16.1	0.1	14.5
3/28/1997	17.2	6.3	47	14.4	9.6	0.0
3/29/1997	17.4	12.0	64	16.9	0.0	10.5
3/30/1997	25.1	10.4	63	17.8	9.9	40.0
3/31/1997	14.6	7.9	47	10.6	7.5	0.0
4/1/1997	15.7	8.0	50	11.5	8.4	0.0

at past w_1 days, and update parameters at every w_2 days. In this experiment we set w_1 (window size) to $100, 200, 400, 800, \infty$, and w_2 (interval) to $1, 10, 20, 40$, and we examine results. In incremental HMM method, we set λ' (forget coefficient) to $0.0, 0.001, 0.002, 0.005, 0.01, 0.02, 0.05, 0.1, 0.2, 0.5$, and we examine results.

We evaluate the prediction accuracy of method using *Mean Square Error* (MSE). Let p_t be the estimated value, x_t be the actual value at time t, then MSE is defined as follows:

$$MSE = \frac{1}{T} \sum_{t=1}^{T} (p_t - x_t)^2$$

We also examine execution time of estimating observation sequence (1646 days) by batch HMM and incremental HMM method. We evaluate MSE and execution time by average of 10 trials. In the following experiments, we give 15 HMM states.

4.2   Results

Table 2 shows the MSE results. In result of maximum temperature and minimum temperature, we got the best MSE 10.17, 5.01 by incremental HMM method in condition of $\lambda' = 0.2, 0.5$, respectively. The best MSE of maximum and minimum temperature prediction by batch HMM method is 18.21, 14.93 respectively, both are given by condition of $w_1 = 200$ and $w_2 = 10$. From this result, we can say that new observations are more important to forecast future temperature data, because the weight of old observations are reduced quickly by setting λ' to 0.2, 0.5 or w_1 to 200. We can see same characteristic in humidity results, that we got the best MSE by incremental HMM method in condition of $\lambda' = 0.2$, and in batch HMM method we got the best MSE in condition of $w_1 = 200$ and $w_2 = 10$.

On the other hand, in result of rainfall amount, we see that better result comes from conditions of lower forget coefficient, larger window size, and small interval of update. We got the best MSE of rainfall prediction, 175.17, by incremental HMM method in condition of $\lambda' = 0.01$. In batch HMM result we got the best MSE, 177.01, in condition of $w_1 = \infty$ and $w_2 = 10$. From this result we can say that both new and old observations are important to forecast future data,

Table 2. Prediction Accuracy (MSE)

Method		MSE					
		Max. Temp.	Min. Temp.	Humidity	Max. Wind	Sunshine	Rainfall
Batch HMM							
window	interval						
100	1	22.42	18.96	191.46	19.16	16.79	186.45
100	10	25.64	22.75	194.54	19.06	16.54	185.43
100	20	23.85	20.87	191.25	19.20	16.61	186.29
100	40	24.31	21.06	195.50	19.23	16.67	186.64
200	1	19.68	17.08	177.19	19.03	15.87	177.34
200	10	**18.21**	**14.93**	**169.79**	18.92	15.62	177.71
200	20	28.87	27.63	185.51	18.86	15.66	180.10
200	40	28.79	26.35	181.91	18.94	15.66	181.16
400	1	33.16	31.83	204.17	**18.81**	15.67	176.75
400	10	25.55	23.21	188.69	18.90	15.70	177.64
400	20	32.87	32.15	196.62	19.14	15.71	179.52
400	40	24.60	22.09	179.01	19.12	15.65	181.05
800	1	57.52	59.54	244.36	18.88	15.87	177.64
800	10	20.80	17.82	179.98	18.94	15.63	177.59
800	20	29.62	27.90	187.00	19.01	**15.60**	180.15
800	40	30.98	29.33	188.29	19.05	15.68	180.81
∞	1	56.68	58.34	246.79	18.94	15.81	177.63
∞	10	20.29	17.32	178.78	18.91	15.61	**177.01**
∞	20	31.82	30.25	190.85	18.94	15.62	178.95
∞	40	31.97	30.36	188.47	18.96	15.67	180.63
Inc. HMM							
λ'							
0.0		59.93	61.58	236.72	18.91	15.62	175.25
0.001		57.35	58.69	232.39	18.86	15.60	175.99
0.002		58.15	59.76	233.40	18.92	15.58	176.21
0.005		53.94	54.99	225.99	18.79	15.40	175.50
0.01		48.09	47.89	217.26	19.01	15.34	**175.17**
0.02		40.58	34.79	255.31	21.10	15.82	176.73
0.05		15.97	12.34	166.98	**18.29**	**15.33**	179.76
0.1		11.88	7.69	158.14	18.50	15.35	183.70
0.2		**10.17**	5.61	**151.93**	19.04	15.64	188.39
0.5		10.52	**5.01**	167.52	22.19	17.80	220.41

because frequent parameter update using many past observations cause better MSE.

Table 3 shows the execution time of estimating observations. In batch HMM method, the execution time increase directly with window size and inversely with interval of update. By incremental HMM method we can predict very quickly compared to batch HMM method. For instance, compared to batch HMM method in condition of $w_1 = 800$ and $w_2 = 1$ that takes 3568244.20ms, we can execute incremental HMM method for about 1/6500 of time.

Table 3. Execution Time

Method		Execution Time(ms)	
		Whole Data	per 1 Data
Batch HMM			
window	interval		
100	1	537782.10	326.72
100	10	53922.80	32.76
100	20	27141.00	16.49
100	40	13535.20	8.22
200	1	1060462.50	644.27
200	10	105598.40	64.15
200	20	52573.70	31.94
200	40	26112.40	15.86
400	1	2042269.70	1240.75
400	10	139620.30	84.82
400	20	59092.60	35.90
400	40	27025.40	16.42
800	1	3568244.20	2167.83
800	10	139652.60	84.84
800	20	58978.80	35.83
800	40	27006.90	16.41
∞	1	5342431.90	3245.71
∞	10	141042.70	85.69
∞	20	59390.80	36.08
∞	40	26963.00	16.38
Inc. HMM			
λ'			
0.0		550.10	0.33
0.001		537.60	0.33
0.002		551.90	0.34
0.005		545.40	0.33
0.01		542.10	0.33
0.02		561.00	0.34
0.05		545.40	0.33
0.1		543.60	0.33
0.2		542.20	0.33
0.5		545.60	0.33

4.3   Discussions

Here we discuss how we can think about our experimental results and especially about our approach.

In result of maximum temperature, minimum temperature and humidity, we got better MSE when we reduce the weight of old observations quickly. This is because these attribute have strong characteristic of *seasonality*, for instance, in summer period it is around 30 ℃ of temperature and 70 % of humidity at most day, and in winter period 10 ℃ of temperature and 35 % of humidity. Because season changes with long period, we can construct specific model for each seasons by reducing weight of old observations quickly, and get good prediction accuracy.

On the other hand, in rainfall amount prediction, we got better MSE by using many past observations. This is because we have to learn the weather patterns based on observation vector sequence to forecast rainfall amount. In other words, rainfall amount strongly depends on weather event (like sunny, rain or typhoon). Frequent update of parameters is also nessesary to forecast rainfall because small w_2 cause better MSE.

In the result of execution time, incremental HMM method requires few execution time compared to batch HMM method. Especially, rainfall amount prediction requires large window size and small update interval, therefore it takes a lot of time for good forecasting by batch HMM method. By using proposal method, we can forecast observations very quickly with about the same accuracy as batch HMM method. In addition, proposal method doesn't require large storage to keep old observations, therefore any storage access doesn't happen even if data amount become huge.

5   Conclusion

In this investigation we have proposed how to estimate future data on data stream by modeling time-series as stochastic process. We have presented adaptive future data estimation using HMM by applying incremental Baum-Welch algorithm. We have discussed some experimental results and shown the usefulness of our approach.

In this work we have given the number of state to HMM in advance. But we need some techniques which can vary number of state concurrently for adapting new patterns more accurately and quickly.

In this case, we have examined numeric vector as observations, but it is possible to apply our approach to nonnumeric time-series. For example, we may have some sort of forecast for accidents by applying our approach to news stream.

References

1. Cavalin, P.R., Sabourin, R., Suen, C.Y., Britto Jr., A.S.: Evaluation of Incremental Learning Algorithms for An HMM-Based Handwritten Isolated Digits Recognizer. In: The 11th International Conference on Frontiers in Handwriting Recognition (ICFHR 2008), Montreal, August 19-21 (2008)
2. Duan, S., Babu, S.: Processing forecasting queries. In: Proc. of the 2007 Intl. Conf. on Very Large Data Bases (2007)
3. Gelper, S., Fried, R., Croux, C.: Robust Forecasting with Exponential and Holt-Winters Smoothing (June 2007)
4. de Gooijer, J.G., Hyndman, R.J.: 25 Years of IIF Time Series Forecasting: A Selective Review, Tinbergen Institute Discussion Papers 05-068/4 (2005)
5. Md. Hassan, R., Nath, B.: StockMarket Forecasting Using Hidden Markov Model: A New Approach. In: Proceedings of the 5th International Conference on Intelligent Systems Design and Applications (2005)

6. Jian, N., Gruenwald, L.: Research Issues in Data Stream Association Rule Mining, SIGMOD Record 35-1, pp.14–19 (2006)
7. Iwasaki, M.: Statistic Analysis for Incomplete Data. EconomistSha, Inc. (2002) (in Japanese)
8. Manning, C.D., Schutze, H.: Foundations of Statistical Natural Language Processing. MIT Press, Cambridge (1999)
9. Mitchell, T.: Machine Learning. McGrawHill Companies (1997)
10. Muthukrishnan, S.: Data streams: algorithms and applications. In: Proceedings of the fourteenth annual ACM-SIAM symposium on discrete algorithms (2003)
11. Stenger, B., Ramesh, V., Paragios, N.: Topology free Hidden Markov Models: Application to background modeling. In: IEEE International Conference on Computer Vision (2001)
12. Wakabayashi, K., Miura, T.: Identifying Event Sequences using Hidden Markov Model. In: Kedad, Z., Lammari, N., Métais, E., Meziane, F., Rezgui, Y. (eds.) NLDB 2007. LNCS, vol. 4592, pp. 84–95. Springer, Heidelberg (2007)

History Guided Low-Cost Change Detection in Streams*

Weiyun Huang, Edward Omiecinski, Leo Mark, and Minh Quoc Nguyen

College of Computing, Georgia Institute of Technology, Atlanta, USA
{wyhuang,edwardo,leomark,quocminh}@cc.gatech.edu

Abstract. Change detection in continuous data streams is very useful in today's computing environment. However, high computation overhead prevents many data mining algorithms from being used for online monitoring. We propose a history-guided low-cost change detection method based on the "s-monitor" approach. The "s-monitor" approach monitors the stream with simple models ("s-monitors") which can reflect changes of complicated models. By interleaving frequent s-monitor checks and infrequent complicated model checks, we can keep a close eye on the stream without heavy computation overhead.

The selection of s-monitors is critical for successful change detection. History can often provide insights to select appropriate s-monitors and monitor the streams. We demonstrate this method using subspace cluster monitoring for log data and frequent item set monitoring for retail data. Our experiments show that this approach can catch more changes in a more timely manner with lower cost than traditional approaches.

The same approach can be applied to different models in various applications, such as monitoring live weather data, stock market fluctuations and network traffic streams.

1 Introduction

With the development of network, data management and ubiquitous computing technology, data streams have become an important type of data source and have attracted much attention [2,7,8,9,12]. A data stream is a sequence of data points which usually can only be read once and does not support random access. Generally the data points are time ordered. Change detection in data streams has become a popular research topic in the data mining community [3,4,10,14,15,18].

Stream data change detection usually involves two steps: model generation and model comparison. We call this a "compute-and-compare" approach, or "C&C" for short. This approach repeatedly generates models from the data stream and then compares them to see if there is any change among these models. Therefore, it incurs very high cost. In some cases, the inherent complexity of the models prevents the algorithms to be executed frequently. Therefore we may fail

---

* This research has been supported by National Science Foundation Grant CCR-0121643.

T.B. Pedersen, M.K. Mohania, and A M. Tjoa (Eds.): DaWaK 2009, LNCS 5691, pp. 75–86, 2009.

to detect any change that only lasts for a short duration, and even if we do discover the change, it may be too late to act on.

Previous work [13] proposes an "s-monitor" approach to tackle this difficult problem. An *s-monitor* with respect to an expensive data model is a simple model that costs much less to compute, and its change reflects the change of the expensive model. By putting "monitors" into the stream of the data, the expensive step of model generation can be avoided as much as possible.

We model a data stream using sliding windows. A window is the portion of data that enters the system within a certain range. Windows may or may not overlap. A checkpoint is the point where the stream is checked for changes, usually at the end of a window. The distance between two consecutive checkpoints is called the "check interval". The s-monitors approach reduces the length of the check interval, therefore can result in better performance for many applications.

How to place the s-monitors is non-trivial and critical to the success of this approach. [13] shows heuristic sampling based methods to allocate s-monitors. In many cases history can provide better selection of monitors. Instead of using heuristcs, this paper divides the whole process into one offline phase and one online phase. The offline phase studies the history and tries to deduce the pattern of changes; and the online phase makes use of the knowledge learned by the offline phase and determines where to put the s-monitors to catch the possible changes in the future.

This paper is organized as follows. Section 2 reviews the related work on change detection. Section 3 describes the "s-monitor" approach and introduces our history-guided s-monitor selection. Section 4 discusses our approach in detail using real and synthetic data as example. Section 5 presents another example using retail data. Section 6 addresses the future work and section 7 concludes the paper.

2   Related Work

Recently more and more attention has been paid on mining the evolution of the data [3,4,13,15,18,19]. Aggarwal [3] uses the velocity density estimation concept to diagnose the changes in an evolving stream. Wang et al. [18] use ensemble methods to detect concept drift. Aggarwal et al. [4] also propose a framework for clustering evolving numerical data streams with the use of both an on-line algorithm and an off-line processing component. Kifer et al [15] lay theoretical foundation by designing statistical tests for one dimensional data. Our approach is different from previous research in that we try to tackle the complexity problem by strategically choosing part of data to process and/or performing low-cost model processing.

Subspace clustering algorithms [5,16,17], although unfortunately not very efficient, are very effective in high dimensional data sets. The MAFIA algorithm is one such algorithm. Like many other subspace clustering algorithms, it divides each dimension into bins and find dense bins, uses them to generate 2-D grids and find dense ones, then goes up to three dimensions, and so forth. This process requires multiple passes of data. Subspace clustering algorithms can capture arbitrary shapes of clusters, and the results do not depend on the initialization of the clusters. The generated model is easy to interpret. Apriori algorithm [6] is widely used to analyze retail data but its complexity is also high. These two algorithms cannot handle high speed streams. We use them to demonstrate how to integrate low-cost s-monitors with expensive model generation algorithms.

Aggarwal et al. [1] propose a framework to maintain "fading cluster structures" for data streams and compute projected clusters for current and historical data. This work examines subsets of dimensions as we do, but it is different from our research. We focus on detecting changes from sliding windows of data, rather than designing a clustering algorithm. Our subspace clustering example uses low-cost s-monitors together with an off-the-shelf algorithm to detect the changes in the data.

Selection of s-monitors is similar to prediction. Using history to predict data values in time series [11] is a common practice. The selection of s-monitors is critical to the success of our approach, and there are multiple ways to allocate the s-monitors. In this paper, we focus on using history to guide the allocation of s-monitors.

3   Proposed Method

3.1   The "S-monitor" Approach

This paper is based on the "s-monitor" approach proposed by [13], therefore we briefly introduce it before we discuss our history-guided s-monitor selection.

A change is the difference between an earlier state and a later state, or rather, an earlier value and a later value of a particular model. "Detection delay" and "detection rate" are two important metrics to measure a change detection algorithm. The former measures the time it takes a detection algorithm to report a change since it happens. The latter measures how much percentage of changes can be captured.

Suppose we need to monitor a stream for the change of a complicated model M. The rationale for "s-monitors" is that in many cases, one does not have to compute the whole model M in order to know that it has changed, and some easy-to-compute models (i.e. "s-monitors") can be a good indicator of the changes. If we can find good s-monitors, then we can check them frequently because of their low cost. Since s-monitors usually cannot reflect all the changes, we may interleave infrequent M checks with those s-monitor checks to improve the detection rate. A similar but more common practice is that to monitor a high-speed stream, one can mainly process a data sample of each window, but occasionally (maybe offline) spend time processing all the data in a window to get a thorough view.

Figure 1 shows two example scenarios in which we can benefit from the s-monitor approach. Suppose we are watching a one-dimensional data stream. The C&C approach generates and compares expensive models at checkpoints $M1$ and $M2$, while our approach can check the stream more frequently using the set of s-monitors (ms). The arrows in figure 1 indicate the checkpoints for both models.

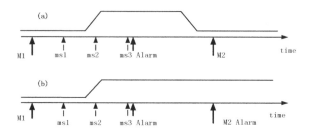

Fig. 1. Change detection examples

In figure 1(b), after getting $M2$, the C&C approach detects the change, while we can do that after $ms3$. In figure 1(a), because the stream changes back to the initial state before $M2$ is obtained, no changes can be detected by C&C, while we can still report a change after $ms3$.

In summary, s-monitors can detect changes with less detection delay. And when the changes are relatively short-lived, the s-monitor approach has a good chance to achieve a higher detection rate than the $C\&C$ approach, since the latter will miss the changes happened (and recovered) between two consecutive M checks (see detailed proof in [13]).

3.2   Defining S-monitors from Historical Knowledge

Selecting monitors for a real data set requires understanding the data set. S-monitor selection is similar to approximate an unknown data distribution with a function: we first decide which function can best describe this kind of distribution, and then decide the parameters of the function. For example, a common practice of statistics is linear regression. We first assume a linear function, then decide the parameters. Whether we should use linear regression to the data, however, depends on the data's characteristics.

We can analyze history using multiple methods, such as time-series analysis [11], data visualization, and aggregation against certain features. Patterns learned from the history can then be used in online stream monitoring. Figure 2 shows the basic steps of low-cost change detection using history.

The whole process is divided into an offline phase and an online phase. The two phases share common knowledge repository ("causal relationship" in the figure) and a processing module (feature selection).

The offline phase processes a history of data, using the expensive algorithm $\mathcal{C}$. The result models (with timestamps on them) are then sent through the feature selection module, to extract some features (denoted by capital letters A, B, $\cdots$, in the figure) that can describe these models. Then, a training module will process the timestamped feature values, find causal rules between earlier values and later values and store these rules in the knowledge repository (knowledge can be of other representation than the rules).

The online phase processes the stream of current data. The rectangles on the stream represent sliding windows. For a window at time t_k we can run the

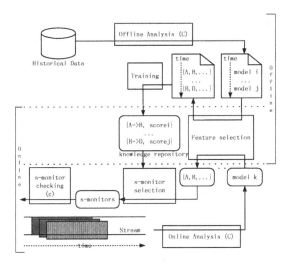

Fig. 2. Using history to guide s-monitor selection

expensive algorithm once and get a model k. Model k will be converted into a set of feature values, which is then fed into the s-monitor selection module. Next, the s-monitors are selected and output to the checking module c.

Knowledge learned from history can degrade or expire over time. Therefore, offline history analysis may need to be repeated periodically.

4   Example: Subspace Cluster Change Detection

In this section, we demonstrate the s-monitor approach using change detection for subspace clusters.

4.1   Web Server Log Data

This data set is the log of a web server which is used to distribute software (mainly linux software). The log contains about 4 months of data, 3414082 entries. We extracted 7 features from the log, namely domain (of the client host), method (GET, POST, etc), server response code, file path, file type, agent (Mozilla, Wget, etc), and file size.

We set every day as a window. We want to find out whether the access pattern in each window changes with respect to the subspace clusters with a given threshold. Such information can be used to maintain the web servers. For example, a subspace cluster $\{(1,1),(3,6),(5,5),(6,0)\}$ can be translated as: "clients are .com hosts, getting a '404 not found' response while trying to access some rpm files under /pub/linux directory". A sudden emergence of such a cluster can imply broken web links for some popular files, especially when some web pages are updated shortly before the time this cluster starts to appear. Subspace clusters can also help the administrator to configure the server so as to provide better performance and screen the users.

We use the MAFIA algorithm [16] as the model generation algorithm $\mathcal{C}$. The subspace clusters make up the high cost model M. The density threshold is set to 5%. All the experiments (including the retail data experiment in section 5) are performed on a Pentium IV 2.80GHz PC with 1GB memory, running Redhat 7.2. Data are in binary format.

4.2   S-monitor Selection Based on Association Rules

Our goal is to discover at least one changed subspace when the model changes. Dimensions and bins essentially divide the whole data spaces into multidimensional grids and subspaces are simply grid cells or a collection of grid cells. With limited system resources (such as memory), we want to choose a subset of the grids/collection of grids that are most likely to change.

By studying the history we find out that the subspace clusters in one window and the changed clusters in next window often show up together. Therefore, we analyze the changes that happened in the past and deduce the "causal" relationship between earlier and later phenomena. For example, we may get a rule like "subspace A is dense in window i and subspace B's density changed either from dense to sparse, or sparse to dense, in window $i + 1$". We keep the number of occurrences in the history as the score for each rule. Such a rule is essentially an association rule [6]. That is, given the set of all subspaces Σ, and a rule $\langle A, B, s_{AB} \rangle$, $A \in \Sigma$ and $B \in \Sigma$, the score s_{AB} is the support of the association rule. The number of times that A appears in a window is called the support of A, or s_A. The confidence of the rule is then s_{AB}/s_A, which is an approximation of the probability $P(B|A)$.

In the online phase, after each M generation, we match the subspace clusters in M with all the rules, and choose the ones with highest confidences to follow. Each matched rule will produce a candidate s-monitor, and one candidate s-monitor can be produced by multiple rules. We associate each candidate with the highest confidence of its rules. For an s-monitor set of size n, the top-n candidates (with the highest confidences) are selected as the s-monitors.

Example. Current window contains subspace clusters A and B, and we have a list of rules $\langle A, C, 30 \rangle$, $\langle B, C, 20 \rangle$, $\langle A, D, 10 \rangle$, $\langle B, D, 20 \rangle$. The challenge is to choose one s-monitor to watch in the next window.

First we compute the confidence of each rule and get $\langle A, C, 0.75 \rangle$, $\langle B, C, 0.50 \rangle$, $\langle A, D, 0.25 \rangle$, $\langle B, D, 0.50 \rangle$. Since all these rules can be matched with A or B, we get candidates C and D, with C associated with confidence 0.75 (since it is higher than 0.50) and D associated with confidence 0.50. Therefore we choose C over D.

4.3   Result

In practice, we interleave one expensive model generation with r s-monitor checks to detect the changes. In this experiment, to show the effectiveness, we run the MAFIA algorithm and our online learning algorithm side by side so we can make more comparisons. We use static data set to simulate a stream so MAFIA

Table 1. Effectiveness of history analysis: predict changes of subspaces with more than 3 dimensions. Total number of changed windows = 67.

$|ms|$	training cost (sec)	$\sum C_{cr}$ (sec)	$\sum C_{ms}$ (sec)	No. missed windows	success rate
10	0.1449	0.3768	0.6555	20	70.2%
20	0.1479	0.4016	1.3515	14	79.1%
40	0.1473	0.4393	2.4985	7	89.6%

can have enough time processing. For every window, our algorithm predicts the s-monitors in the next window, and we use the result of MAFIA for the next window to test if the prediction is accurate. Table 1 shows the result using 10, 20 and 40 s-monitors.

We analyze 68 non-overlapping windows and consider the subspace clusters with 3 or more dimensions. Out of 68 windows, 67 are changed, and with 40 s-monitors, we capture 89.6% changes. The training cost (offline phase) is trivial. The s-monitor creation cost (C_{cr}) and checking cost (C_{ms}) are much less compared with the execution time of the expensive algorithm $\mathcal{C}$, which is about 18 seconds for the same 68 windows.

4.4 Knowledge Update

Accurate prediction depends on the accuracy of the knowledge. As we process the online stream, we need to pay attention to the "freshness" of our knowledge. Our knowledge repository basically stores the causal relationship between subspaces. As time goes by, new causal relationship may be discovered, and the support of old rules may change. In such case, we repeatedly update the knowledge repository by adding new rules and updating the supports. Since this processing mainly involves counting, it can be done in an incremental way. If some knowledge is considered stale, we should remove it. This is more complicated since we need to keep track of timestamps of the rules. In the experiment below we use incremental update only.

4.5 Result for Synthetic Data

Since the web log data set is not very large, we create a synthetic data set to demonstrate the efficiency of our algorithm. We design the data so that the pairwise causal relationship between windows can be used for s-monitor selection.

The synthetic data is a data stream containing 130 non-overlapping windows. Each window can be in one of the three states, each state corresponding to a set of subspace clusters. The first 30 windows are used as "history", in which each window has exactly $\frac{1}{3}$ chances of being in any of the states. In the next 40 windows, the probabilities of being in state 1, 2, 3 are 10%, 70%, 20%, respectively. In the last 60 windows, these probabilities changed to 10%, 20%, 70%. We interleave expensive model generation with 4 s-monitor checkpoints. We report the result for two strategies: the first one without knowledge update, i.e., the

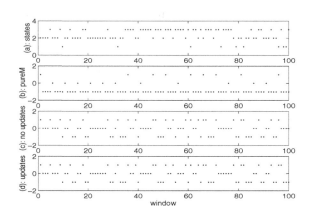

Fig. 3. Effectiveness of change detection algorithms (100 online windows)

s-monitor selection is purely based on the learned knowledge from the first 30 windows. The second strategy updates the knowledge repository after every four times the expensive model is generated.

Figure 3(a) shows the 100 online windows of one such synthetic data set (y values 1,2 and 3 represent the three states each window is in). Figure 3(b) shows the detection output of the expensive model. If $\mathcal{C}$ detects a change, we set a value "1", otherwise we set a value "0". At the position where $\mathcal{C}$ is not run (where the s-monitor checkpoints are), we set a value "-1". We can see that it captured 9 changes (there are 9 dots on the line y=1). Note that some changes are missed since they did not last long enough.

The third plot (figure 3(c)) is about interleaved M and ms checks, without knowledge update. Five s-monitors are checked for each ms checkpoints. A ms check may fail to discover a change if the set of s-monitors are not set to the regions actually changed. If a ms check detects a change, we cancel the following ms checkpoints until the next M-checkpoint. This method captured 27 changes, while ms checkpoints are actually executed 51 times (out of which 5 checks fail to detect changes).

The fourth plot (figure 3(d)) describe the interleaved method with knowledge update. This method captured 28 changes, where ms checkpoints are checked 48 times (out of which 3 checkpoints fail to detect changes).

Table 2 displays the overhead of these methods. The column "$\sum C_{up}$" is the cost to update the knowledge repository.

Table 2. Change detection performance for synthetic data, 3 states (online phase only)

Curve in Fig. 3	No. changes captured	$\sum C_M$ (sec)	$\sum C_{cr}$ (sec)	$\sum C_{ms}$ (sec)	No. ms	Failed ms	$\sum C_{up}$ (sec)
(b)	9	28.0632	-	-	-	-	-
(c)	27	29.1555	0.0339	1.0108	51	5	-
(d)	28	28.2248	0.0343	1.0896	48	3	0.0120

Table 3. Change detection performance for synthetic data, 20 states (online phase only)

Method	No. changes captured	$\sum C_M$ (sec)	$\sum C_{cr}$ (sec)	$\sum C_{ms}$ (sec)	No. ms	Failed ms	$\sum C_{up}$ (sec)
M only	19	35.8403	-	-	-	-	-
M & ms, no updates	43	35.8426	0.1948	1.2602	64	17	-
M & ms, w/ updates	44	35.9972	0.2150	1.1320	57	12	0.1634

We can see that the creation time and average running time of ms is very small compared with the cost of M. By adding $r = 4$ s-monitor checkpoints between every two M-checkpoints, we only increase the processing time by about 4%, while we detect more changes than only using expensive models (28 with updated knowledge vs. 9 with M-checks only).

The 3-state experiment is to make the result easy to visualize and understand. Our algorithm can certainly handle more states. As another example, table 3 shows the result of a similar experiment for a 20-state stream, using 5 s-monitors, over 140 online windows.

5   Retail Data Change Detection

This section shows the effectiveness of our approach with another application. Nowadays many data sources are distributed, homogeneous in nature while a global model is required. A typical example is a retailer chain. Physically distributed stores make transactions every day and large amount of data has to be transferred to a central site for data mining. The data contains sensitive information of each individual customer, such as one's purchasing patterns, therefore the data communication from local stores to processing site has to be secure. More and more research has been focusing on the related security and privacy issues.

Our experiment intends to reduce the data communication by interleaving s-monitor checkpoints between global model generations. To show the effectiveness of our approach, we omit the interleaved expensive model generation. Instead, we put the s-monitor checks and the expensive model checks side by side to see how good the approximation is, with much less data communication needed.

Our global model here is the frequent item sets. To make things easy we only look at item sets containing two items. This method can easily be extended to multiple item case.

Our goal is to use the information from only a few stores to deduce if there is any change in the global model. Usually, with the same support threshold, a local store can generate more frequent item sets than the global model, among those local results, some are common to most of the stores, thus they would make up the global model. If we check all the local frequent item sets, we will be considering many item sets that can never make the global model. We want to limit the sets we need to check. The other knowledge we want to have is that which local stores best reflect the global model. The changes of these stores can better indicate the change of global models than other stores.

Therefore, we use history to deduce two parts of knowledge. One is a collection of item sets that changes most in history (named "feature set"), and the other part is a collection of stores whose local models match the global model the best. In online phase, we check the selected local stores for the frequency of those selected item sets.

Each store maintain a bit vector (called "set vector") for every window – with the ith bit set to 1 if the ith item set is frequent in this window, 0 otherwise. If a set vector is different from the one from previous window, we say there is a change. Also, for each store, we maintain another bit vector for all of its windows, called window vector. Each bit stands for a window, and a "1" means there is a change in this window compared with its predecessor, while a "0" means no change.

Example. Suppose the feature set contains three item sets A, B and C. For store i, A and C are frequent (their support exceeds the threshold) in window 1 but B is not, so a set vector 101 is generated. In window 2 and 3, all A, B and C are frequent, thus the set vectors are both 111. $101 \neq 111$, so we say there is a change in window 2 for store i, and set the second bit of store i's window vector to 1. Window 3 has the same set vector as window 2 so the third bit of store i's window vector is set to 0. The first bit is discarded (denote as "-") as it does not have a predecessor.

By comparing store i's window vector V_i with the global window vector V_g, we can evaluate how well store i reflects the global change. Given a window vector V_a, we call the number of bits V_a and V_g disagree in the online phase the testing error, while the number of disagreeing bits in offline phase the training error. Accuracy is computed by dividing testing error with the number of windows in online phase.

We select stores with the lowest training error as the "voters" for online phase. In online phase, for every window, each store generates its set vector and compare it with the one from last window. If there is a change, this store vote "1", otherwise it votes "0" (no change). We then take the majority of the votes to approximate the change in the global model.

The data set used in this experiment is the Sam's club data[1] provided by Walmart, Inc. It contains the sales data between January 1 and January 31, 2000, from 130 stores. We use overlapped sliding window model. Based on the observation that change of number of transactions has a weekly pattern, the window length is set to seven days and sliding step size is one day. We use Christian Borgelt's implementation[2] of Apriori algorithm to mine the frequent item sets. The support threshold is set to 0.2%.

We put 13 windows in offline phase and 13 for online. Table 4 shows the window vectors and performance for both offline and online phases. The approximation is made with 5, 10, 20 local stores (voters), respectively. The first row

---

[1] http://enterprise.waltoncollege.uark.edu/tun.asp
[2] http://fuzzy.cs.uni-magdeburg.de/~borgelt/apriori.html

Table 4. Change detection accuracy for Sam's club data

	offline windows	online windows	Training Error	Testing Error	online accuracy
Actual	-111000010101	1011011100001	-	-	-
5 voters	-111000010111	1000011000001	1	3	77%
10 voters	-111000010111	1111011100101	1	2	85%
20 voters	-111000010101	1110011100001	0	2	85%

shows the changes of the actual global model. Apparently, with small number of voters (out of the 130 stores), we can have a good approximation of those changes. We only need to check the frequency of the selected item sets from a few voters, which saves a great deal of data communication. Of course the computation cost is also lower, but that is trivial compared with the reduction in data communication cost.

6 Future Work

The idea of deploying low-cost s-monitors into the data space can be applied to various applications. People often think of sampling when large amounts of data need to be processed, similarly, s-monitors can help in many occasions when expensive models are needed for stream change detection. We intend to apply our approach to other types of models and to look for good s-monitors with theoretically guaranteed accuracy.

When history analysis is used to guide the s-monitor selection, the gradual removal of stale knowledge can be necessary for some applications. How to efficiently downgrade the effect of stale data is also an interesting problem.

7 Conclusion

In this paper, we present a new approach for data stream change detection, with respect to the widely used "compute-and-compare" approach. When the stream monitoring task requires the repeated generation of complicated models, we can interleave the model generation with low-cost s-monitor checking, so that we can detect more changes in a more timely manner. We propose a history guided method to select appropriate s-monitors. Then we take subspace cluster monitoring and frequent item set monitoring as examples and demonstrate the effectiveness and performance of our approach. Our analysis and experiments show that in many cases, our low-cost approach can provide much better performance than the traditional approach.

Acknowledgment

We would like to thank Professor Alok Choudhary and his group for providing us the source code of the MAFIA algorithm.

References

1. Aggarwal, C., Han, J., Wang, J., Yu, P.S.: A framework for projected clustering of high dimensional data streams. In: Proc. of VLDB (2004)
2. Aggarwal, C., Han, J., Wang, J., Yu, P.S.: On demand classification of data streams. In: Proc. of the ACM SIGKDD (2004)
3. Aggarwal, C.C.: A framework for diagnosing changes in evolving data streams. In: Proc. of ACM SIGMOD, pp. 575–586 (2003)
4. Aggarwal, C.C., Han, J., Wang, J., Yu, P.S.: A framework for clustering evolving data streams. In: Proc. of VLDB, pp. 81–92 (2003)
5. Agrawal, R., Gehrke, J., Gunopulos, D., Raghavan, P.: Automatic subspace clustering of high dimensional data for data mining applications. In: Proc. of ACM SIGMOD, pp. 94–105 (1998)
6. Agrawal, R., Srikant, R.: Fast algorithms for mining association rules. In: Proc. of VLDB, pp. 487–499 (1994)
7. Babcock, B., Olston, C.: Distributed top-k monitoring. In: Proc. of ACM SIGMOD (2003)
8. Chandrasekaran, S., Franklin, M.J.: Streaming queries over streaming data. In: Proc. of VLDB, pp. 203–214 (2002)
9. Carney, D., et al.: Monitoring streams - a new class of data management applications. In: Proc. of VLDB (2002)
10. Ganti, V., Gehrke, J., Ramakrishnan, R.: Mining data streams under block evolution. SIGKDD Explorations 3(2), 1–10 (2002)
11. Giles, C.L., Lawrence, S., Tsoi, A.C.: Noisy time series prediction using a recurrent neural network and grammatical inference. Machine Learning 44(1/2), 161–183 (2001)
12. Guha, S., Koudas, N., Shim, K.: Data-streams and histograms. In: Proc. of STOC, pp. 471–475 (2001)
13. Huang, W., Omiecinski, E., Mark, L., Zhao, W.: S-monitors: Low-cost change detection in data streams. In: Proc. of AusDM (2005)
14. Hulten, G., Spencer, L., Domingos, P.: Mining time-changing data streams. In: Proc. of ACM SIGKDD (2001)
15. Kifer, D., Ben-David, S., Gehrke, J.: Detecting change in data streams. In: Proc. of VLDB, pp. 180–191 (2004)
16. Nagesh, H., Goil, S., Choudhary, A.: Mafia: Efficient and scalable subspace clustering for very large data sets. Technical Report 9906-010, Northwestern University (1999)
17. Parsons, L., Haque, E., Liu, H.: Subspace clustering for high dimensional data: a review. SIGKDD Explorations 6(1), 90–105 (2004)
18. Wang, H., Fan, W., Yu, P.S., Han, J.: Mining concept-drifting data streams using ensemble classifiers. In: Proc. of ACM SIGKDD (2003)
19. Yang, J., Yan, X., Han, J., Wang, W.: Discovering evolutionary classifier over high speed non-static stream. In: Advanced Methods for Knowledge Discovery from Complex Data. Springer, Heidelberg (2005)

HOBI: Hierarchically Organized Bitmap Index for Indexing Dimensional Data[*]

Jan Chmiel[1], Tadeusz Morzy[2], and Robert Wrembel[2]

[1] QXL Poland (Allegro.pl)
Jan.Chmiel@allegro.pl
[2] Poznań University of Technology, Institute of Computing Science
Poznań, Poland
{tmorzy,rwrembel}@cs.put.poznan.pl

Abstract. In this paper we propose a hierarchically organized bitmap index (HOBI) for optimizing star queries that filter data and compute aggregates along a dimension hierarchy. HOBI is created on a dimension hierarchy. The index is composed of hierarchically organized bitmap indexes, one bitmap index for one dimension level. It supports range predicates on dimensional values as well as roll-up operations along a dimension hierarchy. HOBI was implemented on top on Oracle10g and evaluated experimentally. Its performance was compared to a native Oracle bitmap join index. Experiments were run on a real dataset, coming from the biggest East-European Internet auction platform *Allegro.pl*. The experiments show that HOBI offers better star query performance than the native Oracle bitmap join index.

1 Introduction

A data warehouse architecture has been developed in order to integrate data originally stored in multiple distributed, heterogeneous, and autonomous data sources (DSs) deployed across a company. A core component of this architecture is a database, called a data warehouse (DW), that stores current and historical integrated data from DSs. The content of a DW is analyzed by the so-called On-Line Analytical Processing (OLAP) queries (applications), for the purpose of discovering trends, patterns of behavior, anomalies, and dependencies between data.

In order to support such kinds of analyses, a DW uses a dimensional model. In this model [7], an elementary information being the subject of analysis is called a *fact*. It contains numerical features, called *measures* that quantify the fact and allow to compare different facts. Values of measures depend on a context set up by *dimensions* that often have a hierarchical structure composed of levels. Following a dimension hierarchy from bottom to top, one may obtain more aggregated data in an upper level, based on data computed in a lower level. The

---

[*] This work was supported from the Polish Ministry of Science and Higher Education grant No. N N516 365834.

T.B. Pedersen, M.K. Mohania, and A M. Tjoa (Eds.): DaWaK 2009, LNCS 5691, pp. 87–98, 2009.

dimensional model is often implemented in relational OLAP servers (ROLAP) [3], where fact data are stored in a fact table, and dimension data are stored in dimension level tables. In the paper we will focus on a ROLAP implementation of a DW.

OLAP queries access large volumes of data by means of the so-called star queries. Such queries frequently join fact tables with multiple dimension level tables, filter data by means of query predicates, and compute aggregates along multiple dimension hierarchies. Reducing execution time of star queries is crucial to a DW performance. Different techniques have been developed for reducing execution time of such queries. The techniques include among others: materialized views and query rewriting, partitioning of tables and indexes, parallel processing, and multiple index structures that are discussed below.

1.1   Related Work

The index structures applied to optimizing access to large volumes of data can be categorized as: (1) multi-dimensional indexes, (2) B-tree like indexes, (3) join indexes, (4) bitmap indexes, and (5) multi-level indexes.

For indexing data in multiple dimensions some multi-dimensional indexes have been proposed in the research literature, like for example the family of R-trees [5,18], Quad-tree [4], and K-d-b-tree [16]. Due to their inefficiency in indexing many dimensions and large sizes they have not gained popularity in data warehouses.

The most common indexes applied in traditional relational databases are indexes from the B-tree family [8]. They are efficient only in indexing data of high cardinalities (i.e., wide domains) and they well support queries of high selectivities (i.e., when few records fulfill query criteria). Since OLAP queries are often expressed on attributes of low cardinalities (i.e., narrow domains), B-tree indexes are inappropriate for such applications.

For efficient executions of star queries a B-tree based index, called a join index, was developed [24]. This index stores in its leaves a precomputed join of a fact and a dimension level table. An extension to the join index was made in [14] where the authors represent precomputed joins by means of bitmaps.

Concepts similar to the join index were developed for object databases for the purpose of optimizing the so-called path queries, i.e., queries that follow the chain of references from one object to another ($o_i \rightarrow o_{i+1} \rightarrow \ldots o_{i+n}$). Persistent (precomputed) chains of object references are stored either in the so-called access support relation [9] or in the so-called join index hierarchy [6].

For indexing data of low cardinalities, for efficient filtering large data volumes, and for supporting OLAP queries of low cardinalities the so-called bitmap indexes have been developed [13,15]. A bitmap index is composed of bitmaps, each of which is the vector of bits. Every value from the domain of an indexed attribute $A.T$ (in table T) has associated its own bitmap. The number of bits in each bitmap is equal to the number of rows in T. In a bitmap created for value v of A, bit number n is set to '1' if the value of A of the n-th row is equal to v. Otherwise the bit is set to '0'.

Queries whose predicates involve attributes indexed by bitmap indexes can be answered fast by performing bitwise AND, or OR, or NOT operations on bitmaps. The size of a bitmap index increases when the cardinality of an indexed attribute increases. As a consequence, bitmap indexes defined on attributes of high cardinalities become very large or too large to be efficiently processed in main memory [27]. In order to improve the efficiency of accessing data with the support of bitmap indexes defined on attributes of high cardinalities, either different kinds of bitmap encodings have been proposed, e.g., [2,10,17,23,26] or compression techniques have been developed, e.g., [1,21,22,25].

In [11,12,19,20] indexes of multi-level structures have been proposed. The so-called multi-resolution bitmap index was presented in [19,20] for the purpose of indexing scientific data. The index is composed of multiple levels. Lower levels, are implemented as standard bitmap indexes offering exact data look-ups. Upper levels, are implemented as binned bitmaps, offering data look-ups with false positives. An upper level index (the binned one) is used for retrieving a dataset that totally fulfills query search criteria. A lower level index is used for fetching data from boundary ranges in the case when only some data from bins fulfill query criteria.

In [11,12], the so-called hierarchical bitmap index was proposed for set-valued attributes for the purpose of optimizing subset, superset, and similarity queries. The index, being defined on a given attribute, consists of the set of index keys, where every key represents a single set of values. Every index key comprises signature S. The length of the signature, i.e. the number of bits, equals the size of the domain of the indexed attribute. S is divided into n-bit chunks (called index key leaves) and the set of inner nodes. Index key leaves and inner nodes are organized into a tree structure. Every element from the indexed set is represented once in the signature by assigning value '1' to an appropriate bit in an appropriate index key leaf. The next level of the index key stores information only about these index key leaves that contain '1' on at least one position. A single bit in an inner node represents a single index key leaf. If the bit is set to '1' then the corresponding index key leaf contains at least one bit set to '1'. The i-th index key leaf is represented by j-th position in the k-th inner node, where $k = \lceil i/l \rceil$ and $j = i - (\lceil i/l \rceil - 1) * l$. Every upper level of the inner nodes represents the lower level in an analogous way. This procedure repeats recursively up to the root of the tree.

From the index structures discussed above none was proposed for indexing hierarchical dimensional data in a data warehouse. Moreover, none of the above index structures reflects the hierarchy of dimensions. Such a feature may be useful for computing aggregates in an upper level of a dimension based on data computed for a lower level.

1.2   Paper Contribution and Organization

In this paper we propose a hierarchically organized bitmap index (HOBI) for indexing data in a dimension hierarchy. HOBI is composed of hierarchically organized bitmap indexes. One bitmap index is created for one dimension level.

Thus, an upper level bitmap index can be perceived as the aggregation of lower level bitmap indexes. HOBI supports: (1) query range predicates expressed on attributes from dimension level tables at any level of a dimension hierarchy and (2) roll-up operations along a dimension hierarchy. HOBI was implemented as a software layer located on top of Oracle10g DBMS. The performance of HOBI has been evaluated experimentally and compared to the performance of a native Oracle bitmap join index that is built in the DBMS. Experiments were run on a real dataset, coming from the biggest East-European Internet auction platform *Allegro.pl*. As our experimental results demonstrate, HOBI offers better star query performance than the native Oracle bitmap join index, regardless the data volume fetched by test queries.

This paper is organized as follows. Section 2 presents examples that motivate the need for an index along a dimension hierarchy. Section 3 presents the hierarchically organized bitmap index that we developed. Section 4 discusses the experimental results evaluating the performance of HOBI. Finally, Section 5 summarizes and concludes the paper.

2   Motivating Examples

2.1   Example Data Warehouse

Let us consider as an example a simplified DW schema, cf. Figure 1, built for the purpose of analyzing Internet auction data. The schema includes the *Auctions* fact table storing data about finished Internet auctions. The schema allows to analyze the number of finished auctions, and aggregate purchase costs with respect to time, location of a customer, and product sold. To this end, *Auctions* is connected to three dimensions via foreign keys, namely *Product*, *Location*, and *Time*, all of them composed of hierarchically connected levels [1]. For example, dimension *Location* is composed of three levels, namely *Cities*, *Regions*, and *Countries*, such that *Cities* belong to *Regions* that, in turn, belong to *Countries*.

In the example schema multiple analytical range, set, roll-up, and drill-down queries are executed. Some of them are discussed below.

2.2   Range Query

Let us consider the below query counting the number of auctions that started between April 13th and July 31st, 2007. Typically, the query can be optimized first, by fetching from level table *Days* rows that fulfill the query criteria, and second, by joining the fetched rows with the content of *Auctions*. The query could be optimized by means of a join index on attribute *Days.d_date*. For queries of low selectivities (wide range of dates) a query optimizer will not profit from this index. Moreover, for queries specifying ranges of months or years, the index will not prevent from joining upper level tables, resulting in query performance deterioration.

---

[1] For simplicity reasons table attributes are not shown in Figure 1.

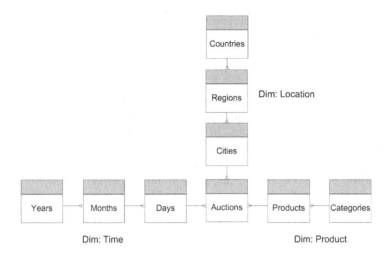

Fig. 1. An example data warehouse schema

```
SELECT COUNT(1)
FROM auctions, days
WHERE auctions.start_day = days.d_date
AND days.d_date BETWEEN to_date('13-04-2007','dd-mm-yyyy')
    AND to_date('31-07-2007','dd-mm-yyyy');
```

2.3   Set Query

Let us consider the below query counting the number of auctions concerning all products from category 'Handheld', namely 'Mio DigiWalker', 'HP iPaq', 'Palm Treo', 'Asus P320', and 'Toshiba Portege', as well as only one product from category 'Mini Notebooks', namely 'Asus Eee PC'. Typically, the query can be optimized by means of a join index on attribute *Products.name*. Notice that all products from category 'Handheld' have been selected by the query. Therefore, if an additional data structure existed at level *Categories* that could point to appropriate auctions, then the query could be executed more efficiently.

```
SELECT COUNT(1)
FROM auctions, products
WHERE auctions.product = products.id
AND products.name IN ('Mio DigiWalker','HP iPaq', 'Palm Treo',
  'Asus P320', 'Toshiba Portege','Asus Eee PC');
```

2.4   Roll-Up/Drill-Down Query

Let us consider the below query computing the sum of sales prices of products, per cities where customers live.

```
SELECT cities.name, SUM(auctions.price)
FROM auctions, cities
WHERE auctions.city = cities.id
GROUP BY cities.name;
```

By rolling up the result of the above query along the hierarchy of dimension *Location*, one can compute sum of product sales prices per regions (cf. the below query) and, further, per countries. In order to perform these roll-ups, a query optimizer has to join level tables *Cities* and *Regions* and join them with fact table *Auctions*. Having an additional data structure at level *Regions*, pointing to appropriate regional auctions, the query could be better optimized.

```
SELECT regions.name, SUM(auctions.price)
FROM auctions, cities, regions
WHERE auctions.city = cities.id AND cities.region = region.id
group by regions.name;
```

3   Hierarchically Organized Bitmap Index

In order to optimize queries joining fact and dimension level tables along a dimension hierarchy, we propose the so-called *hierarchically organized bitmap index* (HOBI). HOBI is composed of bitmap indexes created for every level of a dimension hierarchy. The bitmap indexes are also organized in a hierarchy that reflects the dimension hierarchy, such that a bitmap index at an upper level aggregates bitmap indexes from a lower level.

In order to present the concept of HOBI let us assume that dimension D is composed of level tables $LT_1, \ldots, LT_n$. We assume that every level table includes a key attribute.

Level tables form a hierarchy in D. The hierarchy is denoted as $LT_1 \rightarrow LT_2 \rightarrow \ldots \rightarrow LT_n$, where symbol $\rightarrow$ represents a relationship between a lower level table and its direct upper level table, cf. Figure 2. The relationship also exists between level data (table rows), such that multiple rows from a lower level LT_i are related to one row from a direct upper level LT_{i+1}. Rows from level LT_i are represented by values of the key attribute of LT_i and they are denoted as $key_j^{LT_i}$, where $j = \{1, 2, \ldots\}$.

Let $LT_i \rightarrow LT_{i+1}$. Let LT_i store rows denoted as $key_j^{LT_i}$ ($j = \{1, 2, \ldots, m\}$) and $key_k^{LT_i}$ ($k = \{m + 1, \ldots, q\}$). Let further LT_{i+1} store rows denoted as $key_o^{LT_{i+1}}$ and $key_p^{LT_{i+1}}$. Let further $key_j^{LT_i} \rightarrow key_o^{LT_{i+1}}$ and $key_k^{LT_i} \rightarrow key_p^{LT_{i+1}}$.

In the hierarchy of D, every key attribute of LT_i ($i = \{1, 2, \ldots\}$) has defined its bitmap index that is denoted as BI_i. BI_i is composed of the set of bitmaps $\{b_1^i, b_2^i, \ldots, b_m^i, b_{m+1}^i, \ldots, b_q^i\}$, where q represents the cardinality of the indexed key attribute of LT_i. In a direct upper level LT_{i+1}, bitmap b_o^{i+1} for key value $key_o^{LT_{i+1}}$ is computed as follows: $b_o^{i+1} = b_1^i$ OR b_2^i OR $\ldots$ OR b_m^i. Bitmap b_p^{i+1} for key value $key_p^{LT_{i+1}}$ is computed as follows: $b_p^{i+1} = b_{m+1}^{i+1}$ OR

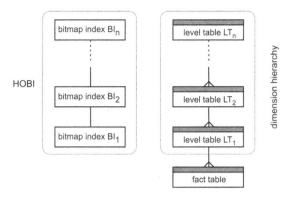

Fig. 2. The concept of HOBI

b_{m+2}^{i+1} OR ... OR b_q^{i+1}. This procedure repeats recursively up to the root of the hierarchy.

In order to illustrate the concept of HOBI let us consider fact table *Auctions* storing 10 auction sales rows, cf. Figure 3. Since *Auctions* is connected to dimension *Product* composed of two levels, namely *Products* and *Categories*, HOBI consists of two levels. At the level of *Products* there are 8 bitmaps, each of which describes auction sales of one product. At the upper level *Categories* there are 2 bitmaps, one bitmap for one category of sold products. The 'Handheld' bitmap describes auction sales of all products from this category, i.e., 'Mio DigiWalker', 'Toshiba Portege', 'Asus P320', 'HP iPaq', and 'Palm Treo'. The 'Handheld' bitmap is computed by OR-ing the bitmaps from level *Products*. Similarly, the 'Mini notebook' bitmap describes auction sales of products from this category and it is constructed by OR-ing bitmaps for 'MSI Wind', 'Asus Eee PC', and 'Macbook Air'.

In order to illustrate the usage of HOBI for range queries (cf. Section 2.2) let us consider a query computing the sum of sales of products 'Mio DigiWalker', 'Toshiba Portege', 'Asus P320', 'HP iPaq', and 'Palm Treo', as well as 'Macbook Air'. Since all products from category 'Handheld' are selected by the query, the 'Handheld' bitmap index, defined for level *Category*, is used for retrieving all auction sales concerning products from this category, rather than using individual bitmaps for products. For retrieving auction sales of 'Macbook Air' the bitmap index on level *Products* will be used.

In order to illustrate the usage of HOBI for roll-up queries (cf. Section 2.4) let us consider a query computing the sum of sales concerning products in category 'Mini notebook'. In order to answer this query, a query optimizer uses bitmap 'Mini notebook' (defined for level *Categories*) for finding appropriate auction sales rows. In this case, fact table *Auctions* need not be joined with the *Product* dimension level table.

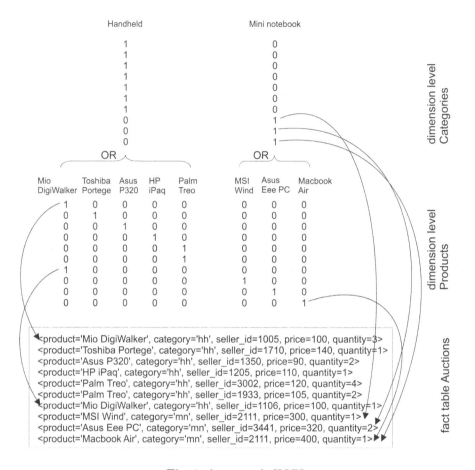

Fig. 3. An example HOBI

4   Experimental Evaluation

The performance of HOBI was evaluated experimentally and compared to the performance of a native Oracle bitmap join index that is built in the DBMS (the usage of a particular index in Oracle can be forced by the INDEX hint). In the experiments we focused on the performance of range queries (cf. Section 2.2) and roll-up queries (cf. Section 2.4). The test queries were executed in a simplified auctions DW, shown in Figure 1. The experiments were run on a real dataset acquired from the Allegro Group, which is the leader of on-line auctions market in Eastern Europe. The test dataset contained 100,000,000 of fact rows describing finished auctions from the period of time between April 2007 and March 2008.

Data were stored in an Oracle 10g DBMS. HOBI was implemented as an application on top of the DBMS. The experiments were run on a PC (Intel Dual

Core 2GHz, 2GB RAM) under Ubuntu Linux 8.04). In order to eliminate the influence of caching, database buffer cache and shared pool were cleared before executing every query. The same query was run 10 times in order to check the repeatability of obtained results.

In the experiments we have measured times of finding rows fulfilling query selection criteria, and we have not measured times of fetching data rows from disk, since for both compared indexes identical number of rows was being fetched. For this purpose, in the test queries we used count as an aggregate function.

4.1   Range Scan

This experiment aimed at measuring the efficiency of HOBI for a range scan query. The test query computed the number of auctions created by sellers in a given time period defined on the *Days* level, as shown below. For this experiment, the bitmap join index was created on attribute *Days.d_ date* and HOBI was created for the *Time* dimension. Time period was parameterized and equaled to 1, 3, 6, and 9 months.

```
SELECT COUNT(1)
FROM auctions, days
WHERE auctions.start_day = days.d_date
AND days.d_date >= date-A AND days.d_date <= date-B;
```

The obtained average query execution times are shown in Figure 4. As we can observe from the chart, HOBI offers better performance than the bitmap

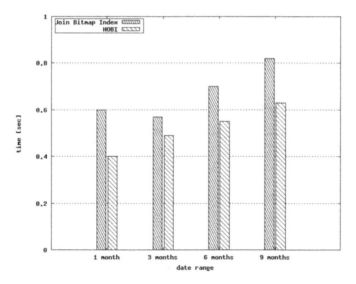

Fig. 4. Different time ranges selected by a test query

join index for every tested time range. For the data visualized in Figure 4 we computed ratio t_{BJI}/t_{HOBI}, where t_{BJI} and t_{HOBI} are average query response times using the bitmap join index and HOBI, respectively. The ratio ranges from 1.16 (for the 3-months query) to 1.48 (for the 1-month and 9-months queries).

4.2 Roll-Up/Drill-Down

This experiment aimed at measuring the efficiency of HOBI for roll-up operations. The query used for the experiment computed the number of auctions created by sellers in a given time period defined on the upper level *Months*. The bitmap join index was created on attribute *Days.d_ date* and HOBI was created for the *Time* dimension, as in the previous experiment. Time period was parameterized and encompassed from 10% to 100% of days stored in the database.

```
SELECT COUNT(1)
FROM auctions, days, months
WHERE auctions.start_day = days.d_date and days.month = months.id
AND months.month >= date-A AND months.month <= date-B;
```

The obtained average query execution times are shown in Figure 5. As we can observe from the chart, HOBI offers better performance than the bitmap join index regardless the selectivity of the test query. For the data visualized in Figure 5 we computed ratio t_{BJI}/t_{HOBI}, similarly as in Section 4.1. The ratio ranges from 1.22 (for the selectivity 10%-20%) to 1.50 (for the selectivity of 75%-85%).

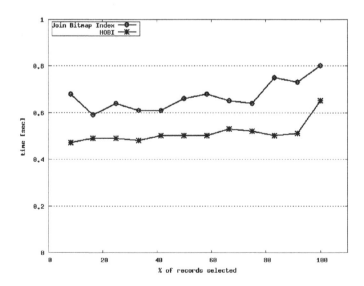

Fig. 5. Variable selectivity of a test query

4.3   Observation

In both of the above experiments HOBI offered better performance than the native Oracle bitmap join index. HOBI profited from the fact that days are grouped to months. For those time periods where queried days formed full months, a bitmap index on *Months* was used instead of an index on *Days*. This way fewer bitmaps were read. Note that HOBI was implemented as an application on top of the DBMS resulting in some additional time overheads. One may expect yet better performance of HOBI while building it in the DBMS so that it is efficiently managed by a system and efficiently used by a query optimizer.

5   Conclusions

In this paper we proposed the HOBI index for indexing hierarchical dimensions. The index is composed of hierarchically organized bitmap indexes, where one bitmap index is created for one dimension level. HOBI supports range predicates expressed on different levels of a dimension hierarchy as well as roll-up operations along a dimension hierarchy. Experimental evaluation of HOBI on a real auction dataset shows its better performance as compared to a native Oracle bitmap join index.

Future work will focus on: (1) developing algorithms for maintaining HOBI after a DW refreshing, (2) evaluating the algorithms experimentally, (3) running the experiments on both real auction data and on synthetic data from the TPC-H benchmark, (4) applying bitmap compression techniques to HOBI for attributes of high cardinalities, (5) developing algorithms for processing compressed HOBI.

References

1. Antoshenkov, G., Ziauddin, M.: Query processing and optimization in Oracle RDB. VLDB Journal 5(4), 229–237 (1996)
2. Chan, C., Ioannidis, Y.: Bitmap index design and evaluation. In: Proc. of ACM SIGMOD Int. Conference on Management of Data, pp. 355–366 (1998)
3. Chaudhuri, S., Dayal, U.: An overview of data warehousing and OLAP technology. SIGMOD Record 26(1), 65–74 (1997)
4. Finkel, R.A., Bentley, J.L.: Quad trees: A data structure for retrieval on composite keys. Acta Informatica 4, 1–9 (1974)
5. Guttman, A.: R-trees: a dynamic index structure for spatial searching. In: SIGMOD, pp. 47–57. ACM Press, New York (1984)
6. Han, J., Xie, Z., Fu, Y.: Join index hierarchy: An indexing structure for efficient navigation in object-oriented databases. IEEE Transactions on Knowledge and Data Engineering 11, 321–337 (1999)
7. Jarke, M., Lenzerini, M., Vassiliou, Y., Vassiliadis, P.: Fundamentals of Data Warehouses. Springer, Heidelberg (2003)
8. Johnson, T., Sasha, D.: The performance of current B-tree algorithms. ACM Transactions on Database Systems (TODS) 18(1), 51–101 (1993)
9. Kemper, A., Moerkotte, G.: Access support in object bases. In: Proc. of ACM SIGMOD Int. Conference on Management of Data, pp. 364–374 (1989)

10. Koudas, N.: Space efficient bitmap indexing. In: Proc. of ACM Conference on Information and Knowledge Management (CIKM), pp. 194–201 (2000)
11. Morzy, M.: Advanced database structure for efficient association rule mining. PhD thesis, Poznań University of Technology, Institute of Computing Science (2004)
12. Morzy, M., Morzy, T., Nanopoulos, A., Manolopoulos, Y.: Hierarchical bitmap index: An efficient and scalable indexing technique for set-valued attributes. In: Kalinichenko, L.A., Manthey, R., Thalheim, B., Wloka, U. (eds.) ADBIS 2003. LNCS, vol. 2798, pp. 236–252. Springer, Heidelberg (2003)
13. O'Neil, P.: Model 204 architecture and performance. In: Gawlick, D., Reuter, A., Haynie, M. (eds.) HPTS 1987. LNCS, vol. 359, pp. 40–59. Springer, Heidelberg (1987)
14. O'Neil, P., Graefe, G.: Multi-table joins through bitmapped join indices. SIGMOD Record 24(3), 8–11 (1995)
15. O'Neil, P., Quass, D.: Improved query performance with variant indexes. In: Proc. of ACM SIGMOD Int. Conference on Management of Data, pp. 38–49 (1997)
16. Robinson, J.T.: The K-D-B-tree: a search structure for large multidimensional dynamic indexes. In: SIGMOD, pp. 10–18. ACM, New York (1981)
17. Rotem, D., Stockinger, K., Wu, K.: Optimizing candidate check costs for bitmap indices. In: Proc. of ACM Conference on Information and Knowledge Management (CIKM), pp. 648–655 (2005)
18. Sellis, T.K., Roussopoulos, N., Faloutsos, C.: The R+-Tree: A dynamic index for multi-dimensional objects. In: VLDB, pp. 507–518 (1987)
19. Sinha, R.R., Mitra, S., Winslett, M.: Bitmap indexes for large scientific data sets: A case study. In: Parallel and Distributed Processing Symposium. IEEE, Los Alamitos (2006)
20. Sinha, R.R., Winslett, M.: Multi-resolution bitmap indexes for scientific data. ACM Transactions on Database Systems (TODS) 32(3), 1–38 (2007)
21. Stabno, M., Wrembel, R.: RLH: Bitmap compression technique based on run-length and Huffman encoding. Information Systems 34(4-5), 400–414 (2009)
22. Stockinger, K., Wu, K.: Bitmap indices for data warehouses. In: Wrembel, R., Koncilia, C. (eds.) Data Warehouses and OLAP: Concepts, Architectures and Solutions, vol. 5, pp. 157–178. Idea Group Inc. (2007)
23. Stockinger, K., Wu, K., Shoshani, A.: Evaluation strategies for bitmap indices with binning. In: Galindo, F., Takizawa, M., Traunmüller, R. (eds.) DEXA 2004. LNCS, vol. 3180, pp. 120–129. Springer, Heidelberg (2004)
24. Valduriez, P.: Join indices. ACM Transactions on Database Systems (TODS) 12(2), 218–246 (1987)
25. Wu, K., Otoo, E.J., Shoshani, A.: Optimizing bitmap indices with efficient compression. ACM Transactions on Database Systems (TODS) 31(1), 1–38 (2006)
26. Wu, K., Yu, P.: Range-based bitmap indexing for high cardinality attributes with skew. In: Int. Computer Software and Applications Conference (COMPSAC), pp. 61–67 (1998)
27. Wu, M., Buchmann, A.: Encoded bitmap indexing for data warehouses. In: Proc. of Int. Conference on Data Engineering (ICDE), pp. 220–230 (1998)

A Joint Design Approach of Partitioning and Allocation in Parallel Data Warehouses

Ladjel Bellatreche[1] and Soumia Benkrid[2]

[1] LISI/ENSMA Poitiers University
Futuroscope, France
bellatreche@ensma.fr
[2] National High School for Computer Science (ESI)
Algiers, Algeria
s_benkrid@esi.dz

Abstract. Traditionally, designing a parallel data warehouse consists first in fragmenting its schema and then allocating the generated fragments over the nodes of the parallel machine. The main drawback of this approach is that interdependency between fragmentation and allocation processes is not taken into account during the design phase. This interdependency is characterized by the fact that generated of fragments are one of the inputs of the allocation problem and both processes optimize the same set of queries. In this paper, we present a new approach for designing parallel relational data warehouses on a shared nothing machine, where the fragmentation and the allocation are done simultaneously. To allocate efficiently query workload over nodes, a load balancing method is given. Finally, a validation of our proposals is presented.

1 Introduction

Over the last decade, the size of most data warehouses has grown by a factor of 10 according to The Data Warehouse Institute (TDWI) [7]. Given this fact, implementing highly performing complex queries becomes a big challenge. The sole use of classical query optimization techniques such as materialized views, indexing, data partitioning, data compression is insufficient [18]. Parallel processing is becoming increasingly important in the world of database computing. More and more organizations are relying on parallel processing technologies to achieve the performance, scalability, and reliability they need [13]. Most of the major commercial database systems support parallelism. Rather than relying on a single monolithic processor, parallel systems exploit fast and inexpensive micro processors to achieve high performance. Designing traditional parallel databases was largely studied [1,6,10,11,14,15,17] compared to the parallel data warehouse design, where it did not get the same attention from the research community which concentrates its efforts on the selection of classical optimization techniques, except in the work done by [9,12,18].

Designing a parallel data warehouse consists of five main steps: (i) choosing the hard architecture, (ii) fragmenting the data warehouse schema, (iii) allocating

T.B. Pedersen, M.K. Mohania, and A M. Tjoa (Eds.): DaWaK 2009, LNCS 5691, pp. 99–110, 2009.

the generated fragments, (iv) balancing the load over the nodes of the parallel machine and (v) processing queries. There are three widely used architectures for parallelizing work: (a) *shared memory* (b) *shared disk* and (c) *shared nothing*. Shared nothing architectures are especially well suited to the star queries running on data warehouses modelled using a star schema, as only a very limited amount of communication bandwidth is required to join one or more (typically small) dimension tables with the (typically much larger) fact table [7]. In this work we adopt this architecture.

Once the architecture chosen, data warehouse designer partitions its schema. Fragmentation is a *pre condition* of parallel data warehouse design. It may be *horizontal*, where table instances are decomposed into disjoint partitions or *vertical*, where tables are split in disjoint sets of attributes. The horizontal partitioning is mainly used for designing parallel data warehouses [18,19,9,12].

The data allocation is the process that places generated fragments over nodes of parallel machine. This allocation may be either *redundant* (with replication) or *non redundant* (without replication). Once fragments are placed, global queries are then rewritten over fragments and executed on the parallel machine. During their execution phase, the load balancing should be verified. Load balancing refers to workload allocation over nodes [17].

By exploring the most important works done on designing traditional parallel databases, in general, and parallel data warehouse, in particular, we identify two mains limitations: (1) The fragmentation and allocation processes are usually done in an isolation way (or iteratively): the designer first partitions his/her data warehouse using his/her favourite fragmentation algorithm and then allocates generated fragments on the parallel machine using his/her favourite allocation algorithm. The data partitioning and fragment allocation are usually done to optimize the same set of queries. The *iterative design* ignores interdependency between fragmentation and allocation processes. This ignorance led to the birth of *two research communities*: one working on the problem of selecting optimal fragmentation schema [3,4] and other one on allocating fragments/tables over various sites (nodes) [1,14,15,10]. Consequently, most of work done on data allocation does not care about of the generation of fragments [15,10]. Note that the fragments are one of the inputs of data allocation problem. (2) There does not exist a complete methodology including the five above steps of parallel data warehouse design, except the work done by [18], where it considers fragmentation and allocation problems in an iterative way on a shared disk machine. In this paper, we propose a complete methodology for designing parallel relational data warehouses overtaking the above limitations. To the best of our knowledge, this work is the first one addressing this issue.

This paper is divided into six sections: Section 2 summarizes existing work done on iterative design of parallel data warehouses. Section 3 details the steps of the iterative approach and identifies its limitations. Section 4 describes our joint approach, where partitioning and allocation are done simultaneously. Section 5 presents the query rewriting phase and proposes an algorithm that balances the load between nodes. Section 6 shows the experimental results obtained using the

APB-1 release II benchmark. Section 7 concludes the paper summarizing the main findings and suggesting future work.

2 Related Work

In this section, we describe the main works done on the fragmentation and allocation in the context of parallel data warehouses [9,12,18,19].

In [18], the authors propose an approach to construct and manage a relational data warehouse on a disk shared parallel machine. This warehouse is modelled using a star schema composed of one fact table and a number of dimension tables. A *Multi-dimensional hierarchical fragmentation* method is used to partition the warehouse. It consists in horizontally fragmenting each dimension table[1] by *Range mode* based on its attributes belonging to lower levels of the dimension hierarchy, then using the fragmentation schemes of dimension tables to partition the fact table. Dimension tables and their mono-table indexes (B-Trees) are duplicated over each disk the parallel machine. Multi-table indexes (bitmap join indexes) are selected using attributes of dimension tables belonging to higher levels of the dimension hierarchy. Therefore, the allocation process concerns fragments of the fact table and bitmap join indexes. Note that the number of generated fragments may be largely greater than the number of disks. To ensure a high parallelism degree and efficient load balancing, a *round robin allocation* of fact fragments over disks is used. Bitmap fragments belonging to the same fact fragment are placed onto *consecutive disks* to enable intra-query parallelism. In [19], an allocation tool called *Warlock* is proposed. The main particularity of this tool is that it allows data warehouse designer to control the number of generated fragments of the fact table to avoid its explosion [2].

Furtado [9] discussed partitioning strategies for node-partitioned data warehouse. He recommended partitioning the fact table based on the largest dimension tables. Each largest dimension table is partitioned using the *Hash mode* based on its primary key. The fact table is also partitioned using the *Hash mode* using foreign keys referencing the largest dimension tables. The generated fragments are allocated in a *round robin* and *random* ways. Smallest dimension tables are replicated over all nodes. This fragmentation does not take into account star join query requirements[2] and it does not care about the number of generated fragments as in [3,19].

In [12], a data allocation problem on a Database Cluster (DBC) is discussed. A DBC is a cluster of interconnected machines, each running an independent DBMS, that are linked together by a middleware layer so that they can collaboratively manage one database and process queries against this database. A key issue in this setting is how to place the data on the different machines. Two obvious approaches are to fully replicate the database on all machines or to partition the data on the machines. The authors present an approach that

---

[1] Dimension tables are virtually partitioned.

[2] A star join query is characterized by selections defined on dimension tables and joins between the fact and dimension tables.

combines partitioning and replication for OLAP style workloads. The fact table is partitioned and replicated across nodes using chained declustering; and the dimension tables are fully replicated. This enables the middleware layer to do load balancing among the replicas to improve query response time. They recommend using chained declustering for replicating fact table partitions without detailing how designer choose the number of replicas to use. As in [9], this work does not control the number of generated fragments of the fact table.

To summarize, the most existing works are mainly concentrated on partitioning and allocation phases done iteratively.

3   Steps of Iterative Design Approach

In this section, we briefly describe the two main steps of parallel data warehouse design which are *fragmentation* and *allocation*.

The data horizontal partitioning is the core of parallel design. It consists in fragmenting virtually/really dimension tables using selection predicates of queries and using their fragmentation schemes to partition the fact table. The fact table partitioning is known by *referential partitioning*, which was recently supported by Oracle11G [8]. Such a decomposition of the fact table may generate a large number of fragments [18,2,3]. There is some consensus on giving designers the possibility to control this number [18,3]. Consequently, horizontal partitioning problem (HPP) may be defined as follows:

Given a data warehouse with a set of d dimension tables $D = \{D_1, D_2, ..., D_d\}$ and a fact table F, a workload Q of queries $Q = \{Q_1, Q_2, ..., Q_l\}$ and a maintenance constraint W fixed by data warehouse designer that represents the maximal number of fact fragments that he/she wants. The HPP consists in splitting the fact table into NFF fragments based on fragmentation schemas of dimension tables, such that: **(a)** the cost of evaluating all queries is minimized and **(b)** $NFF \leq W$.

This problem is known as a NP-complete problem [3]. Consequently, several algorithms have been proposed to solve it [2,3,18,9]. Let $FF = \{F_1, ..., F_K\}$ be the set of generated fact fragments.

The data allocation problem consists in finding an efficient distribution of FF over M nodes of the parallel machine such as performance of queries *executed over nodes* is satisfied. This problem is known as a NP-complete [10]. As we said before (Section 1), the iterative approach does not consider the interdependency between fragmentation and allocation problems. To overcome this limitation, we propose a joint approach in the next section.

4   Our Joint Design Approach

Before presenting in details our design approach, we formalize it as an optimization problem.

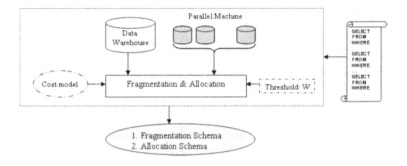

Fig. 1. Steps of our Joint Approach

4.1 Formalisation of the Joint Approach

Given:

- A shared nothing parallel machine with M nodes $NO = \{N_1, N_2, ..., N_M\}$.
- A transfer matrix indicating the communication cost between nodes. It contains nodes as rows and as columns.
- data warehouse schema composed of d dimension tables $\{D_1, ..., D_d\}$ and one fact table F. As in [9,12], we suppose that all dimension tables are replicated over the nodes of shared nothing parallel machine and are in their main memory.
- A set of queries $Q = \{Q_1, Q_2, ..., Q_l\}$ to be executed over the parallel machine. Each query Q_k $(1 \leq k \leq l)$ has an access frequency f_k.
- A maintenance constraint W representing the number of fragments that data warehouse designer considers relevant for his/her allocation process. This number shall be greater than the number of nodes (see recommendations of Stöhr et al. [18]).

Our problem of designing a parallel data warehouse consists in fragmenting the fact table into NFF of fragments and allocating them simultaneously so that the cost of executing all queries over all nodes is minimized. Figure 1 illustrates the steps of our approach. Contrary to the iterative approach that uses two cost models, our approach uses only one that monitors whether the generated fragmentation schema is useful for the allocation process.

In the next section, we propose an algorithm dealing with joint design problem.

4.2 Fragmentation and Allocation Procedures

Our design algorithm is composed of two procedures: one for fragmenting the data warehouse and other for allocating the generated fragments over nodes.

Partitioning Procedure. For the fragmentation, we adapt our genetic algorithm proposed in a centralized environment [2]. It generates a random population that contains several chromosomes [3]. For each chromosome CR_i, our algorithm checks if it satisfies the maintenance constraint ($NFF_i \leq W$), where NFF_i represents the number of fragments of CR_i. If it is the case, this chromosome is kept in the population, otherwise, merges operations are applied to reduce its number of fragments [3]. Once initiation population created, our genetic algorithm performs operators such as *crossover* and *mutation* in order to improve the quality of this population. The application of these operators is monitored by an evaluation function, which allocates the generated fragments of each valid chromosome over the nodes of the parallel machine. Once this allocation done, the cost of executing queries over nodes is calculated. At the end of this algorithm, the chromosome that offers the minimum cost represents the fragmentation schema. In the next section, we describe how the fragment allocation is done.

Fragment Allocation Procedure. To allocate each chromosome, we propose a new allocation procedure based on *affinity between fragments*. This idea is borrowed from Navathe et al.'s work [16] who developed an algorithm for vertically decomposing relational tables. The attributes with *high affinity* are grouped together to form a vertical fragment. The steps of our allocation procedure are:

1. *Construction of Fragment Usage Matrix (FUM)*: it indicates the usage of fragments according to the set of queries. FUM contains queries as rows and fragments as columns. The value $FUMij$ ($1 \leq i \leq l, 1 \leq j \leq NFF$) is equal 1 if the query Q_i uses the fact fragment F_j, otherwise, it is equal to 0. An additional column is added to represent the access frequency of each query.

2. *Construction of Fragment Affinity Matrix (FAM)*: it contains fragments as rows and as columns[4] It indicates the affinity between two fragments F_i and F_j. Each value $fam_{i,j}$ of FAM represents the sum of frequencies of queries accessing simultaneously the fragments F_i and F_j.

3. *Fragments Clustering*: in order to generate groups of fragments, we adapt Navathe et al.'s algorithm [16]. The FAM matrix is transformed into a complete graph called the *Affinity Graph* (*AG*). An edge in *AG* is labelled with a weight representing the affinity between its vertices. It starts by choosing randomly a fragment of *AG* and then, forming a linearly connected spanning tree, it generates all meaningful cycles simultaneously. Contrary to the original Navathe et al.'s algorithm [16], where attributes are grouped based on their high affinities, our approach groups fragments based on their low affinities to increase inter parallelism between nodes. At the end of this step, a set of cycles $C = \{C_1, ..., C_H\}$ is generated, where each one represents a sub set of fragments.

---

[3] Each chromosome represents a potential fragmentation schema.
[4] FAM is a symmetric matrix.

4. *Allocating Cycles over Nodes*: Cycles are placed in *round robin fashion* over nodes. Since each fragment belongs to one and only one cycle, it will be allocated to only one node (non redundant allocation). Our allocation procedure differs from [18,9], since our allocation unit is a cycle and not a single fragment.

The allocation schema of each chromosome may be represented by a matrix called, Fragment Allocation Matrix ($FAAM$). Rows and columns of this matrix represent respectively fragments and nodes. Each value of this matrix is defined as follows: $faam_{ij} = 1$ if the fragment F_i is allocated at the node N_j, 0 otherwise. Once allocation process done, we calculate the cost of executing the set of queries Q over M nodes in terms of number of inputs outputs (IOs)[5]. It is given by the following equation:

$$\sum_{k=1}^{l} Max_{1 \leq j \leq M} \left(\sum_{i=1}^{N} FUM_{ki} \times FAAM_{ij} \times |Fi| \right) \tag{1}$$

with l, M, N et $|Fi|$ represent, respectively the number of queries, nodes, fragments and the size of the fragments F_i in terms of number of pages ($|F_i| = \left\lceil \frac{||F_i|| \times Length}{PS} \right\rceil$, where $||F_i||$, $Length$ and PS represent respectively the cardinality of the fact fragment F_i, length of each fact tuple and page size of a node (we suppose that all nodes have the same page size). The final allocation schema is *the schema that minimizes the above cost*.

5   Query Processing and Load Balancing

The main advantage of our allocation algorithm is its simplicity and low complexity $O(NFF^2)$, where NFF represents the number of fact fragments[6]. Its main drawbacks is that it may suffer from the load balancing problem, since it takes into account only *query access frequencies* to generate cycles of fact fragments. In this section, we propose an algorithm that enhances the performance of our allocation algorithm and ensures the load balancing between nodes. Load balancing is the major resource allocation problem in order to effectively utilize all available resources. It can be applied for different workload granularities depending on the level of parallelism [17]. At the highest level, we have inter-query parallelism with a concurrent execution of independent and queries (multi-user mode). Intra-query parallelism requires additional forms of load balancing for assigning sub queries to nodes. In this work we concentrate on intra query parallelism.

To execute a given query, we should first identify its valid fragments and their localisations. To ensure a load balancing between nodes, a reallocation of

---

[5] The IOs cost for a given query executed over all nodes corresponds to the cost of the loaded node.

[6] Navathe et al.'s algorithm [16] has this complexity.

fragments is required when executing each query. The reallocation problem is formalized as follows: Given (i) a Fragment Allocation matrix $FAAM$, (ii) a set of M nodes, and (iii) a communication matrix. The redistribution problem consists in migrating fragments from high loaded nodes (HLN) to less loaded nodes (LLN). This migration shall minimize the communication and query processing costs. To obtain the optimal solution, we need to consider all reallocation combinations that are expensive and infeasible.

To resolve this problem, a measurement of the load of each node is required. To do so, we present the following definition:

Definition 1. *The loading level of a node N_i (denoted by $LL(N_i)$) corresponds to the number of sub queries executed on that node.*

Based on this definition, we can define a node to be highly loaded/lowly loaded if its loading level is greater/less than a threshold. This threshold may be fixed by the data warehouse designer. In this work, it is computed as follows:

$$Threshold = \left\lceil \frac{Number\ of\ all\ sub\ queries}{M} \right\rceil$$

Each node N_i of (HLN) is associated with a weight $(WHLN(N_i))$ representing the number of queries making it highly loaded. $WHLN(N_i)$ is computed as follows: $WHLN(N_i) = LL(N_i) - threshold$.

Similarly, each node of LLN is assigned to weight $WLLN(N_i)$ representing the number of queries that can receive. $WLLN(N_i)$ is computed as follows: $WHLN(N_i) = Threshold - LL(N_i)$. HLN and LLN are sorted based on their weights.

Based on this definition, our load balancing algorithm first identifies highly loaded nodes and lowly loaded nodes and then migrates nodes from HLN to LLN till all nodes become balanced.

6   Experimental Results

Intensive experimental studies are conducted to evaluate our parallel data warehouse design methodology that includes data partitioning, data allocation, query processing and load balancing.

Dataset: We use the dataset from the APB1 benchmark [5]. The star schema of this benchmark has one fact table *Actvars* (24 786 000 tuples, tuple length is 74 bytes) and four dimension tables: *Prodlevel* (9 000 tuples), *Custlevel* (900 tuples), *Timelevel* (24 tuples) and *Chanlevel* (9 tuples).

Workload: We have considered a workload of 55 single block queries (i.e., no nested sub queries) with 40 selection predicates defined on 9 different attributes (Class_Level, Group_Level, Family_Level, Line_Level, Division_Level, Year_Level, Month_Level, Retailer_Level, All_Level). The domains of these attributes are split into: 4, 2, 5, 2, 4, 2, 12, 4, and 5 sub domains, respectively to perform the

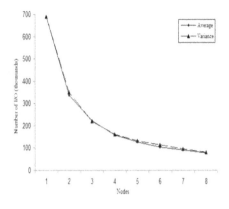

Fig. 2. Average vs. Variance

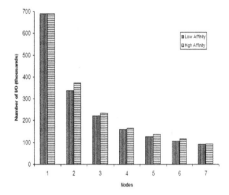

Fig. 3. Quality of Allocation using Low and High Affinities

genetic algorithm [2]. We do not consider update queries. Note that each selection predicate has a selectivity factor computed using SQL queries executed on the data set of APB1 benchmark.

We considered a shared nothing machine with 12 nodes. The page size is 65536 bytes. The crossover and mutation rates used by our experiments are 80% and 20%. These values are obtained by running our genetic algorithm several times.

Our algorithms are implemented using Java performed under Intel Pentium IV 2.8 GHZ with a memory of 1 Go.

The first experiment that we conducted studies the impact of the criteria used for generating cycles of fragments on performance of queries. Recall that our allocation unit is a cycle of fragments. Each cycle is generated as follows: we start with a random fragment of graph affinity, and we try to expand it by considering other fragments. A fragment can join a current cycle, if it has a reasonable affinity compare the existing ones. In this experiment, we evaluate two acceptance criteria *Average* and *Variance*: *a fragment can join the current cycle if its affinity is less than the average of affinities (or the variance) of the existing ones.* We set the fragmentation threshold to 200, and we vary nodes from 1 to 8. For each value, the cost of executing the 55 queries is computed. Figure 2 describes the obtained results. The two criteria give practically the same results, with a slight advantage to the average. Accordingly, we use this criterion in the following experiments.

To make sure that our grouping criterion that consists in forming cycles of fragments based on their low affinities is interesting than that used by [16], we conduct an experiment as follows: fragmentation threshold is set to 200 and the number of nodes is varied from 1 to 7. For each value, the number of inputs outputs required to execute the 55 queries is estimated in both variants of the allocation procedure (low affinity and high affinity). Figure 3 summarizes the results of this comparison. The low affinity criterion is slightly better than that with high affinity. This is because it increases the degree

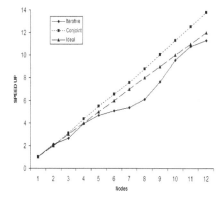

#Node	Iterative Approach (# IO)	Our Approach (# IO)	Reduction %
1	690 000	690 000	0,00%
2	325 000	337 000	-3,69%
3	262 000	222 000	15,27%
4	175 000	159 000	9,14%
5	148 000	126 000	14,86%
6	136 000	105 000	22,79%
7	128 000	91 000	28,91%
8	113 000	78 330	30,68%
9	90 000	68 660	23,71%
10	72 000	61 000	15,28%
11	64 000	55 000	14,06%
12	61 000	50 000	18,03%

Fig. 4. Speed up of Iterative and Conjoint Approaches

Fig. 5. Reduction offered by our Approach

of parallelism. Accordingly, we opt for this criterion for the remaining experiments.

Figure 4 compares the performance of our conjoint and iterative approaches. Iterative approach is done as follows: we fragment data warehouse schema using the genetic algorithm proposed in [2], and then we allocate the generated fragments over the nodes using our allocation procedure. This comparison is done based on the speed up factor. For a fragmentation threshold of 200, we vary the number of nodes from 1 to 12 and for each value; we calculate the speed up for each approach. We note that our approach has a *linear speed up*, whereas the iterative one has a *sub-linear speed up*. This result confirms that our approach is more suitable for designing a parallel data warehouse. Note that the speed up of our approach is not ideal, since the load balancing module is discarded. We conduct the same experiment, but instead of measuring the speed up, we calculate the IO cost required for each approach. Figure 5 shows the obtained results. Our approach outperforms largely the iterative one, where it offers 14%-31% savings.

Figure 6 studies the effect of the fragmentation threshold on the performance of our approach. To do this, we set the number of nodes to 10 and we vary the fragmentation threshold W from 50 to 300. We note that increasing the threshold improves significantly the performance of queries. Increasing this threshold allows more attributes to participate in the partitioning process.

Figure 7 studies the behaviour of our approach with taking into account the load balancing module. For a fragmentation threshold of 200, we varied the number of nodes from 1 to 12 and for each value we calculate the speed up to run the 55 queries before and after adding load balancing. We note that the speed up is linear for both cases. The speed up offered by our approach with load balancing is nearly perfect.

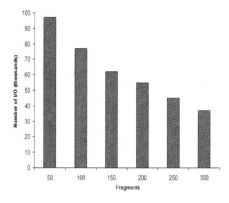
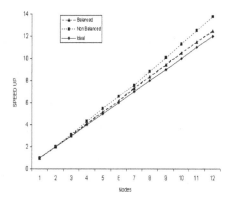

Fig. 6. Effect of Number of Fragments on Query Performance

Fig. 7. Speed up of Balanced and non Balanced Approaches

7   Conclusion and Perspective

The size of data warehouses does not cease increasing in recent years. To optimize complex OLAP queries, traditional optimization techniques are not sufficient. Consequently, parallel processing has become a necessity. The design of a parallel data warehouse passes through five stages: the choice of hardware architecture, data warehouse fragmentation, fragments allocation, query processing and load balancing among the nodes. When exploring the majority of works on designing parallel data warehouses, we figure out that fragmentation and allocation are usually treated in an isolation way. In this paper, we showed the interest of addressing the problems of selecting partitioning and allocation schemes simultaneously. We proposed an algorithm composed of two procedures, one for fragmenting the data warehouse (using a genetic technique) and another for allocating the fragments generated (using an affinity technique). The main particularity of fragmentation procedure is that it evaluates the quality of the generated partitioning schema based on its allocation effectiveness. Since our allocation procedure is based on the affinity between fragments, there is a risk that it does not balance the load between nodes. Consequently, we propose an algorithm for that. It consists in migrating fragments from highly nodes to lowly nodes in order to ensure intra query parallelism. Our proposed methodology is validated using data set of APB1 benchmark by considering a shared nothing parallel machine. The results are encouraging and show the interest of our approach.

In future work, it would be interesting to use other algorithms to partition the data warehouse schema, since, our genetic algorithm is costly in terms of computation time, and to conduct more experiments by considering more nodes.

References

1. Apers, P.M.G.: Data allocation in distributed database systems. ACM Transactions on database systems 13(3), 263–304 (1988)
2. Bellatreche, L., Boukhalfa, K., Abdalla, H.I.: Saga: A combination of genetic and simulated annealing algorithms for physical data warehouse design. In: Bell, D.A., Hong, J. (eds.) BNCOD 2006. LNCS, vol. 4042, pp. 212–219. Springer, Heidelberg (2006)
3. Bellatreche, L., Boukhalfa, K., Richard, P.: Data partitioning in data warehouses: Hardness study, heuristics and oracle validation. In: Song, I.-Y., Eder, J., Nguyen, T.M. (eds.) DaWaK 2008. LNCS, vol. 5182, pp. 87–96. Springer, Heidelberg (2008)
4. Chakravarthy, S., Muthuraj, J., Varadarajan, R., Navathe, S.B.: An objective function for vertically partitioning relations in distributed databases and its analysis. Distributed and Parallel Databases Journal 2(2), 183–207 (1994)
5. OLAP Council. Apb-1 olap benchmark, release ii (1998),
   http://www.olapcouncil.org/research/bmarkly.htm
6. DeWitt, D.J., Gray, J.: Parallel database systems: The future of high performance database systems. Communnications of the ACM 35(6), 85–98 (1992)
7. DeWitt, D.J.D., Madden, S., Stonebraker, M.: How to build a high-performance data warehouse, http://db.lcs.mit.edu/madden/high_perf.pdf
8. Eadon, G., Chong, E.I., Shankar, S., Raghavan, A., Srinivasan, J., Das, S.: Supporting table partitioning by reference in oracle. In: SIGMOD 2008 (2008)
9. Furtado, P.: Experimental evidence on partitioning in parallel data warehouses. In: DOLAP, pp. 23–30 (2004)
10. Karlapalem, K., Pun, N.M.: Query driven data allocation algorithms for distributed database systems. In: Tjoa, A.M. (ed.) DEXA 1997. LNCS, vol. 1308, pp. 347–356. Springer, Heidelberg (1997)
11. Bouganim, L., Florescu, D., Valduriez, P.: Dynamic load balancing in hierarchical parallel database systems. In: Proceedings of the International Conference on Very Large Databases, pp. 436–447 (1996)
12. Lima, A.B., Furtado, C., Valduriez, P., Mattoso, M.: Improving parallel olap query processing in database clusters with data replication. To appear in Distributed and Parallel Database Journal (2009)
13. Mahapatra, T., Mishra, S.: Oracle Parallel Processing. O'Reilly, Sebastopol (2000)
14. Mehta, M., DeWitt, D.J.: Data placement in shared-nothing parallel database systems. VLDB Journal 6(1), 53–72 (1997)
15. Menon, S.: Allocating fragments in distributed databases. IEEE Transactions on Parallel and Distributed Systems 16(7), 577–585 (2005)
16. Navathe, S.B., Ra, M.: Vertical partitioning for database design: a graphical algorithm. In: ACM SIGMOD, pp. 440–450 (1989)
17. Rahm, E., Marek, R.: Analysis of dynamic load balancing strategies for parallel shared nothing database systems. In: Proceedings of the International Conference on Very Large Databases, pp. 182–193 (1993)
18. Stöhr, T., Märtens, H., Rahm, E.: Multi-dimensional database allocation for parallel data warehouses. In: Proceedings of the International Conference on Very Large Databases, pp. 273–284 (2000)
19. Stöhr, T., Rahm, E.: Warlock: A data allocation tool for parallel warehouses. In: Proceedings of the International Conference on Very Large Databases, pp. 721–722 (2001)

Fast Loads and Fast Queries

Goetz Graefe

Hewlett-Packard Laboratories
Goetz.Graefe@HP.com

Abstract. For efficient query processing, a relational table should be indexed in multiple ways; for efficient database loading, indexes should be omitted. Moerkotte's "small materialized aggregates" can be used to alleviate this tension, notably in the form of Netezza's "zone maps." Their most significant advantageous characteristics are that (i) load bandwidth is maximized by avoiding the cost of index maintenance, (ii) there is no need for complex index tuning, and (iii) scans for typical queries are very fast. Their most significant limiting characteristics are that (iv) they are effective only for query predicates on columns correlated with the load sequence, (v) individual outlier values can sharply reduce their effectiveness, and (vi) they fail to improve search performance within a zone.

In this research, we introduce zone filters and zone indexes that address these limitations without reducing the advantages. The new data structures can be created as side effects of the load process, with all required analyses accomplished while a moderate amount of new data still remains in the buffer pool. Traditional sorting and indexing are not required. Nonetheless, query performance matches that of zxone maps where those apply, exceeds it for predicates for which zone maps are ineffective, and can be comparable to query processing in a database with traditional indexing, as demonstrated in our simulations.

1 Introduction

In relational data warehousing, there is a tension between load bandwidth and query performance, between effort spent on maintenance of an optimized data organization and effort spent on large scans and large, memory-intensive join operations. For example, appending new records to a heap structure can achieve near-hardware bandwidth. If, however, the same data must be integrated into multiple indexes organized on orthogonal attributes, several random pages are read and written for each record insertion. Some optimizations are possible, notably sorting new data and merging them into the old indexes, thus reducing random I/O but still moving entire pages to update individual records [GKK 01]. Nonetheless, load bandwidth will be a fraction of the hardware bandwidth.

The difference in query performance is just as clear. A fully indexed database permits efficient index-to-index navigation for many queries, with all memory available for the buffer pool. With index-only retrieval, non-clustered indexes may serve as vertical partitions similar to columnar storage. With a few optimizations, index

T.B. Pedersen, M.K. Mohania, and A M. Tjoa (Eds.): DaWaK 2009, LNCS 5691, pp. 111–124, 2009.

searches can be very fast. For example, IBM's Red Brick product supports memory-mapped dimension tables. If, on the other hand, the load process leaves behind only a heap structure with no indexes, query processing must rely on large scans, large hash joins, etc. Shared scans may alleviate the pain to some degree, but general sharing of more complex intermediate query results has remained a research idea without practical significance.

In the proposed design using zone filters and zone indexes, loading writes each item to disk only once and never revisits items with multi-pass algorithms such as external merge sort or B-tree reorganization. Nonetheless, predicates on all columns are supported. Compared to traditional B-trees, the data structure left after loading and used by query evaluation is less refined. The advantage of the technique is that there is no need for reorganization after the load is complete or for index tuning, yet query processing can be almost as efficient as in a fully indexed database, as demonstrated in our experiments.

2 Prior Work

2.1 Small Materialized Aggregates

Moerkotte [M 98] seems to have been the first to suggest parsimonious use of materialized view for query answering and for efficiency in query execution plans. The latter is based on a correlation of load sequence with time-related data attributes. Obvious examples include attributes of type "date" or "time;" less obvious examples with nonetheless strong correlation include sequentially assigned identifiers such as an order number, invoice number, etc. For example, if most or even all orders are processed within a few days of receipt, references to order numbers in shipments are highly correlated with the time when shipments occurred and were recorded in the database.

For those cases, Moerkotte proposed small materialized aggregates (SMAs) defined by fairly simple SQL queries. For example, for each day or month of shipments, the minimal and maximal order number might be recorded in a materialized view. Based on this materialized view, a query execution plan searching for a certain set of order numbers can quickly and safely skip over most days or months. In other words, the materialized view indicates the limits of actual order numbers within each day or month, and it can thus guarantee the absence of order numbers outside that range.

Figure 1 reproduces an illustration by Moerkotte [M 98]. It shows three buckets (equivalent to database segments or zones). Each bucket contains many records, with three records per bucket shown. Several small materialized aggregates are kept for each bucket, in this case three different aggregates, namely minimum and maximum values as well as record counts. The number of values for each of these aggregates is equal to the number of buckets. A query about a date range within the year 1998 can skip over all buckets shown based on the maximum date values in the appropriate file of small materialized aggregates.

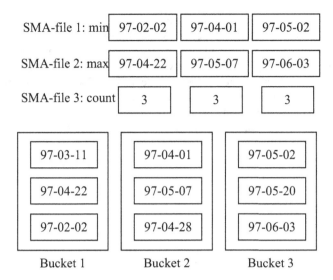

Fig. 1. Moerkotte's illustration

Omission of maintenance during deletion results in possible inaccuracy, e.g., deletion of the last instance of the lowest or highest value. Such inaccurate aggregates can still be used to guide query execution plans. Specifically, they indicate the lowest or highest value that ever existed, not the lowest or highest value that currently exists. Thus, they can guarantee that values outside the range currently do not exist, even if the current values may not span the entire range.

2.2 Zone Maps

Netezza employs a closely related technique that has proven itself in practice, both during proof-of-concept benchmarks and in production deployments. In order to simplify the software and to speed up query processing, only one record type is permitted within each database segment or "zone" (as far as we know). To avoid the need for manual tuning, zone maps cover all columns in a table. For each column and each physical database segment (e.g., 3 MB of disk space), minimal and maximal value are gathered during loading and retained in the database catalogs. During insertions into a database segment, the values are updated accurately. During deletion, maintenance of these values is omitted.

Query processing in data warehousing often involves query predicates on a time attribute. Often, there is a high correlation between the load sequence and data attributes capturing a time dimension, e.g., the date of sales transactions. In those cases, the information in the zone maps enables table scans to skip over much of the data.

Columns without correlation and predicates on those columns do not benefit from zone maps. For example, if there is a query predicate on product identifier, it is unlikely that the range of product identifiers within a database segment is substantially smaller than the entire domain of product identifiers. Thus, there will rarely be an opportunity to combine zone map information about multiple columns.

The date ranges of neighboring zone maps may overlap. As there usually is a strong correlation between load sequence and dates, these overlaps are limited in practice. The ranges of product identifiers overlap between zone maps; in fact, there does not seem to be any correlation between product identifiers and date or load sequence. Therefore, a range query on the date column likely benefits from the zone maps, whereas a range query on product identifiers typically does not.

After zone maps reduce the number of zones that must be inspected in detail, Netezza relies on hardware support (FPGAs) to scan zones and all their records quickly. For generality and portability, we propose a data organization within zones (or equivalent data segments) that enables efficient loading as well as efficient query processing.

2.3 Partitioned B-trees

Partitioned B-trees are normal B-trees with a hidden artificial leading key field to create flexibility during loading and query processing [G 03]. One of the design goals was to combine loading at hardware speed and query processing as in a fully indexed database.

Loading includes the logic for run generation during external merge sort and extends each B-tree being loaded with sorted partitions, which separated by an artificial leading key field containing partition identifiers or run numbers. Incremental index reorganization may exploit idle times between loading and query processing. The required logic is the same as merging in a traditional external merge sort.

Figure 2 illustrates a partitioned B-tree with 3 partitions. B-tree reorganization into a single partition uses the merge logic of external merge sort. Index search prior to full index optimization needs to probe all existing partitions just like a query with no restriction on a leading B-tree column [LJB 95]. If a partition is small, it might be more efficient to scan all its records than to probe it using standard B-tree search. B-tree optimizations for prefix and suffix truncation make it likely that partition boundaries coincide with leaf boundaries.

If multiple indexes exist for a table being loaded, one can either sort each memory contents once in each desired sort order and append the sorted records as a new segment to all appropriate indexes, or one can append the new data to only one index, say the clustered index of the table. In the latter case, the load process can achieve near-hardware bandwidth. Later, the incremental optimization steps not only move data from the appended segments to the main segment but also propagate the insertions to the table's other indexes.

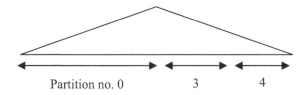

Fig. 2. Partitioned B-tree

2.4 In-Page Organization

When an entire database segment or zone is read and processed as a single unit, its internal organization is even more important than the internal organization of traditional database pages of much more moderate size. Nonetheless, several research efforts have focused on evaluation and improvement of the internal organization of database pages, both for unsorted heap files and for B-tree indexes. An advanced internal organization has been employed in at least one database product [BB 02].

For the former, the most promising direction seems to have been a columnar organization or vertical partitioning, with "mini-pages" dedicated to individual fields of the records stored on the page [ADH 01]. For the latter, the initial paper on B-trees suggested binary search [BM 70] whereas recent proposals include caching keys frequently used in binary search [GL 01, L 01], organizing those keys as a B-tree of cache lines [RR 00], or optimizing a page for interpolation search [G 06].

3 Zone Filters

Starting with this section, we introduce our proposals for zone filters and zone indexes. These techniques improve and integrate prior techniques and thus combine and exceed their benefits. Zone filters generalize small materialized aggregates and zone maps; and zone indexes speed up search within a database segment. Our design for zone filters differs from Moerkotte's small materialized aggregates by exploiting aggregates beyond those expressible in SQL. It differs from Netezza's zone maps in two ways: multiple low and high values and bit vector filters.

3.1 Multiple Extreme Values

For each zone and each column, our design retains the m lowest and the n highest values. If m = n = 1, this aspect of zone filters equals zone maps. By generalizing beyond a single value at each end of the range, zone filters can be effective in cases for which zone maps fail, without substantial additional effort during loading.

If the Null value for a domain "sorts low," it may be the "lowest" value for a column in many zones. By retaining at least m = 2 low values, the lowest valid value is always included in the zone filter even in the presence of Null values. Thus, queries with range predicates can always test whether a zone indeed contains values in the desired range of values, and Null values do not create a problem or inefficiency.

In addition, retaining m+n extreme values per zone and column permits efficient query processing even in the presence of outliers. For example, if the column in question describes the business value of sales transactions, a single sales transaction of high value might greatly expand the range as described by the very lowest and the very highest value. Many sets of real-world measurements include outliers, not only values but also sizes, weights, distances, intervals, etc. Even a few low and high values, ideally one more value than actual outliers, can ensure that query predicates can be handled effectively, i.e., only those zones are inspected in detail that probably contribute to the query result.

Data records

| 97-02-02; 4711; … |
| 97-03-11; 6400; … |
| 97-04-01; 0528; … |
| 97-03-31; 0911; … |
| 97-03-29; 2002; … |
| … |

2× low values
2× high values

| 97-02-02; 0528; … |
| 97-03-11; 0911; … |
| 97-03-31; 4711; … |
| 97-04-01; 6400; … |

| 97-04-22; 1200; … |
| 97-04-28; 5817; … |
| 97-05-02; 3333; … |
| 97-05-01; 1740; … |
| 97-04-24; 1492; … |
| … |

| 97-04-22; 1200; … |
| 97-04-24; 1492; … |
| 97-04-28; 3333; … |
| 97-05-02; 5817; … |

| 97-04-28; 6192; … |
| 97-05-20; 6400; … |
| 97-06-03; 0635; … |
| 97-05-01; 1795; … |
| 97-05-05; 1848; … |
| … |

| 97-04-28; 0635; … |
| 97-05-01; 1795; … |
| 97-05-20; 6192; … |
| 97-06-03; 6400; … |

Fig. 3. Zone filter with m = 2 and n = 2

Figure 3 shows a zone filter with m = n = 2 or four synopsis records per database segment or zone. Notice that there is only a single data record for month 97-02, which may therefore be considered an outlier. For a query with predicate "= 97-02-22" or "between 97-02-16 and 97-02-28," zone maps cannot exclude that zone, whereas the zone filters in Figure 3 do based on their multiple low and high values.

If there are fewer than m+n distinct values, some of the m+n values in the zone filter might be Null. The lowest value actually found in the domain is always retained as the lowest value in the zone filter.

In such cases, or if the search key is smaller than the largest among the m lowest values or larger than the smallest among the n highest values, the set of extreme values in the zone filter supports not only range ("<", "between") predicates but also equality ("=", "in") predicates, even for query constants that are within the range between low and high values. For example, if the m = 3 lowest values retained in a zone filter are (4, 7, 12), a search for value 9 can safely skip the zone. Thus, in the case of domains with few distinct values, the generalization from Moerkotte's SQL aggregates and Netezza's zone maps to m+n extreme values not only handles Null values and outliers but also offers new functionality and performance improvements for an additional set of queries.

Some readers might fear the cost of maintaining m+n values, in particular for nontrivial values of m and n. In that case, the load process should employ two priority queues for the lowest and highest values seen so far. The values are the roots of these priority queues are the mth lowest and the nth highest value seen so far. The priority queues are initialized with the first m+n distinct values in the load stream. Each subsequent value in the load stream is compared with these root values, and if necessary a pass through one priority queue is required with $\log_2(m)$ or $\log_2(n)$ comparisons. The two required comparisons are comparable to the 2 comparisons required while building a zone map.

3.2 Bit Vector Filters

If the values in a database column have no correlation with the load sequence of the table, the range between minimal and maximal actual value in each zone will approach the entire domain of the column. In those cases, even m+n extreme values will provide hardly any reduction in the number of zones that need to be scanned in detail.

Each zone might contain only a few distinct actual values, or at least substantially less than the entire domain. For those cases, we propose to include bit vector filtering in the zone filter. For each zone and for each column, a bit vector filter provides a synopsis of the actual values, and scans with equality predicates can exclude zones where the constant literal in the query predicate maps to an "off" bit in the bit vector filter.

For example, consider a sales table in an organization with seasonal products, or any other domain in which items are introduced and discontinued over time. Depending on the assignment of product identifiers for new products, zone maps may conceivably be effective for their highest values. Old products are usually discontinued in a more random numeric sequence, so zone maps would not help with the lowest values. Bit vector filters, on the other hand, are independent of the sort order of product identifiers, their introduction, and their discontinuance, whereas zone maps depend on the correlation between column values and the load sequence.

Figure 4 illustrates a zone filter with a bitmap per column; the m+n records with low and high values are omitted from the diagram. For this illustration, the last decimal digit of the date or the product identifier is used as bit position. For example, value 0528 in the second column maps to bit position 8 in the second bitmap within the zone filter.

It might be useful to reserve one bit position in the bitmap for Null values. By doing so, queries specifically searching for missing values (Null values) can be processed very efficiently, with little impact on all other queries. If the bit vector filter indicates the presence of a Null value, there is no need to include it among the lowest m values in the zone filter. Thus, an additional actual value can be included in the zone filter for a slight increase in effectiveness.

As there is no need to capture occurrence of a particular value twice in a zone filter, and as the zone filter is the combination of extreme values and bit vector filter, the m+n extreme values are not represented in the bit vector filter. Thus, if no more than m+n distinct values occur in a zone (in one column), the bit vector filter is entirely clear.

Data records   Bitmap per column

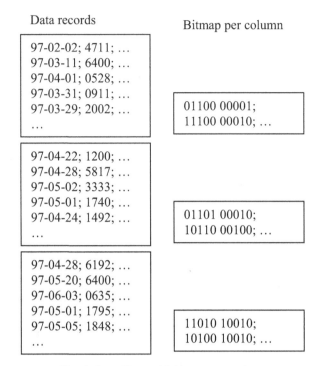

Fig. 4. Zone filters with bitmaps per column

Bit vector filters typically do not support range predicates very well. If a query range is small, it may be possible to enumerate the values in the range. For example, the predicate "between 4 and 6" can be rewritten to "in (4, 5, 6)" for an integer domain. In order to solve the problem more generally, some bit positions in the bit vector filter may be dedicated to ranges. For example, the column value 154 might be mapped to a bit position for the specific value as well as another bit position dedicated to the range 100-199. A query with the predicate "between 167 and 233" might be able to skip over many zones even if some data values such as 154 might create false positives.

Bit vector filters are really a special form of aggregate. Thus, as records are added to a zone during load processing, each record adds to the bit vector filter. They are different than traditional aggregates in that record deletions are not reflected in a bit vector filter. In that sense, a bit vector filter does not reflect the current actual values in a zone but the set of values that have existed in the zone and column since the bit vector filter was created. If deletions are frequent, it might be useful to recomputed bit vector filters every now and then in the affected database segments or zones.

4   Zone Indexes

While zone filters let scans skip over many parts of a table or database, zone indexes are a technique that enables efficient search within a zone.

The essence of zone indexes is to embed in each zone or database segment information comparable to a traditional index, limited in scope to the records within that zone or segment. Embedding zone indexes within the zone or database segment aids both load and query performance. During loading, a zone index can be created without accesses to external storage while the segment is assembled in the buffer pool. During query processing, any zone or segment not eliminated by zone filters can be searched efficiently using the indexes within the zone. Since the zone filters are embedded, they are loaded into the buffer pool in memory together with the detail data in the segment.

Our proposed default for zone indexes is to index every column. If the data records are sorted according to one column only, this is analogous to clustered and non-clustered indexes in traditional databases. The main difference to traditional indexes is that that all these indexes may share data values very much in the spirit of T-trees optimized for in-memory indexing [LC 86]. Thus, every value needs to be stored only once. Moreover, the indexes may be optimized for CPU caches [GL 01, L 01]. The reason these techniques are applicable is that entire database segments are moved between memory and disk as a unit during database loading and during query processing.

Like zone filters and unlike traditional indexes, zone indexes can be constructed quite inexpensively as part of the load process, with little processing effort and memory. Creation of a traditional index requires sorting the future index records; for a large index, an external merge sort requires external storage for intermediate runs and multiple passes over the data. Maintenance of such an index during a large load requires either many random insertions or sorting the change set followed by merging the change into the existing index. Partitioned B-trees do not reduce this effort; they merely create the ability to perform the work later and incrementally.

Encoding and compression may differ from one database segment or zone to another, for example, due to a different set of distinct values in a column. Even the set of indexes may differ among zones. The important points here are that indexes are applied within database segments used as contiguous disk storage and as in-memory data structures, they are created during high-bandwidth database loading with moderate processing and memory requirements, and storage formats may differ among database segments or zones.

5 Performance

In order to assess the value of zone filters and zone indexes, we implemented an approximation using a commercial database product and the "line item" table of the well-known TPC-H database. The following three sub-sections demonstrate the value of zone maps with low and high values, of zone filters with bit vector filters, and of zone indexes.

Instead of building multiple non-clustered indexes, we created only a clustered index on two of the three date columns in each table. Given that this particular database has no outlier values, we kept the number of low and high values to one each in the experiments ($m = n = 1$). Instead of bit vector filters, we used small tables.

The table contains about 60 M rows (~6 GB). It is stored in a clustered B-tree index organized on "commit date," with 2,466 distinct values and about 2½ MB of data per distinct commit date. The "ship date" is a secondary sort column in the clustered index. For the "receipt date," there is only an incidental order due to correlation; this is not exploited during query processing.

The hardware is a dual-core Intel T7200 CPU running at 2 GHz, 2 GB of RAM, a SATA drive with all system files, and a PATA drive with the test database (and its recovery log, but nothing else). Both drives use NTFS file systems recently defragmented. The database server is limited to 128 MB of workspace in RAM. While the system ran other applications concurrently, the CPU utilization was very low throughout all experiments and the times reported below reflect I/O times very accurately. Each statement started with a warm procedure cache (queries compiled ahead of time) and a cold buffer pool.

5.1 Effect of Zone Maps

To emulate a zone filter with lowest and highest values, the database also contains a materialized view with lowest and highest receipt date for each ship date. These columns have high correlation, but the ranges of receipt dates overlap for a number of neighboring commit dates. In other words, this is precisely the constellation for which Netezza's zone maps are designed, with the ordering on commit dates emulating a load sequence.

Table 1. Effect of zone maps

Experiment	SQL Text
Query using a clustered index 3.75 sec	Select count (*) From lineitem Where l_commitdate between '06/16/1994' and '06/30/1994'
Query without a useful index 682 sec	Select count (*) From lineitem Where l_receiptdate between '06/16/1994' and '06/30/1994'
Zone map creation 691 sec	Select l_commitdate as zone min (l_receiptdate) as low max (l_receiptdate) as high Into map_receiptdate From lineitem Group by l_commitdate
Zone map usage 54.8 sec	Select count (*) From map_receiptdate, lineitem Where low <= '6/30/1994' and high >= '6/16/1994' and l_commitdate = zone and l_receiptdate between '6/16/1994' and '6/30/1994' option (loop join)

Table 1 shows relevant SQL text and the observed performance. All queries in our experiments are variations on the first query shown. If this query can exploit a clustered index, it merely scans a contiguous key range and counts 374,382 rows at about 100,000 rows per second or about 10 MB per second.

The first variation of this query forces a complete scan of the table. Scanning 6 GB in 682 seconds indicates also about 9 MB per second. The difference in performance between a disk-order scan and an index-order scan is probably not significant as the database was recently defragmented.

Creation of a zone map employs a very simple plan here, effectively a complete scan of the clustered index with very moderate effort for aggregation calculations plus insertion into a new table serving as zone map in our emulation of this technology. The small performance difference between the prior query and the creation of a zone map indicates that creation of a zone as side effect of loading will hold up in practice.

The final query in Table 1 shows the effect of zone maps. The "option" syntax ensures the desired query execution plan. Using the auxiliary table as zone maps, the plan completes 10 times faster than the equivalent query without zone maps. This mirrors the fact that about 9% of all zones (220 of 2,466) contain receipt dates that satisfy the query predicate. The overhead of scanning the auxiliary table containing the zone maps is negligible.

5.2   Effect of Bit Vector Filtering

In order to emulate bit vector filtering in a zone filter, the database contains a materialized view that captures the distinct part numbers for each commit date. The performance difference between query execution plans that do or do not use this materialized view indicates the value of bit vector filtering in zone filters.

Table 2. Effect of zone filters

Experiment	SQL Text	
Zone filter creation 1,656 sec	Select	distinct l_commitdate as zone, checksum (l_partkey) % 64000 as value
	Into	filter_partkey
	From	lineitem
Zone filter usage 300 sec	Select	count (*)
	From	(select distinct zone from filter_partkey where value = checksum (1705409) % 64000) as filter, lineitem
	Where	l_commitdate = zone and l_partkey = 1705409
	option (loop join)	

Table 2 shows SQL statements for creation and usage of emulated zone filters. 64,000 bits might seem relatively large compared to a zone map with only minimum and maximum value, but 64,000 bits or 8 KB are a fairly dense synopsis for a database segment or zone of 2½ MB and a domain of 2,000,000 values.

In the emulation, however, due to the lack of an appropriate bitmap data type, individual records are created such that the materialized view contains 50,000,000 records or about 2,000 records per distinct value in the commit date column. The size of the table is about 1 GB. With the experimental system, a scan of 1 GB requires about 100 seconds. Nonetheless, even this poor zone filter eliminates a large number of database segments from the scan of the line item table, sufficient for a faster query execution time than a table scan without indexes and zone filters.

With proper bitmaps, the size of all zone filters would be about 2,466 × 8 KB = 20 MB. If not kept in memory, they could be scanned in 2 seconds. The observed query time would not be 300 seconds but about 200 seconds. Without bit vector filtering, all database segments or zones need to be scanned, and the query execution time equals that of the query without a useful index in Table 1. In other words, the benefit of zone filters with bit vector filters can be substantial, even for columns and predicates for which traditional zone maps completely fail.

5.3 Effect of Zone Indexes

In order to assess the value of zone indexes, another variation of the base query of Table 1 restricts the ship date, as shown in Table 3. This query is very similar to the last query of Table 1. Both queries exploit zone maps to skip over irrelevant database segments or zones. The difference is that a restriction on receipt date in Table 1 must scan each relevant database segment in its entirety, whereas a restriction on ship date can exploit the index where ship date is the minor key. A scan of the auxiliary table produces zones indicated as search keys for the first column of the index; for each zone, the index is searched as the remaining query predicate matches the second column of the index. This second index key is equivalent to an index on ship date within each zone or database segment.

The performance difference demonstrates the value of an index within each database segment, with 55.8 seconds versus 7.95 seconds. In fact, the elapsed time of the

Table 3. Effect of zone indexes

Experiment	SQL Text

Zone index usage 7.95 sec	Select count (*)
	From map_shipdate, lineitem
	Where low <= '6/30/1994'
	            and high >= '6/16/1994'
	            and l_commitdate = zone
	            and l_shipdate between
	            '6/16/1994' and '6/30/1994'
	option (loop join)

query in Table 3 is remarkable not only for the performance difference relative to the last query of Table 1 but also for the performance similarity relative to the first query of Table 1. Their difference is merely a factor of 2 in spite of the fact that creation and maintenance of a traditional clustered index requires effort and indeed reduces load bandwidth to a fraction of the hardware bandwidth. Zone filters and zone indexes, on the other hand, can easily be created as side effects as the unsorted load stream passes through memory.

6 Summary and Conclusions

In summary, we identified three advantageous characteristics and three limiting ones in Moerkotte's "small materialized aggregates" and Netezza's "zone maps." Our generalization of those prior designs overcomes their limitations yet retains their advantages.

Together, these techniques enable efficient query processing immediately after a high-bandwidth load. Load processing, with moderate memory and processing needs, creates these structures as a side effect. Intermediate database reorganization is not required. Query evaluation relies only on data structures left behind by the load operation yet it can skip many database segments in most cases and can search the remaining ones efficiently.

In addition to loading traditional data warehouses, the technique might prove useful in capturing and indexing very large data streams. The effort for sorting database segments in memory is comparable to creation of in-memory indexes, yet zone filters and zone indexes are equally suitable for disk storage and thus for large data streams.

Figure 5 illustrates the tradeoffs among alternative techniques. Multiple traditional indexes permit high query performance at the expense of poor load performance. Heaps without indexes permit loading at hardware write bandwidth but lead to poor query performance, in particular for highly selective queries. Partitioned B-trees combine high load bandwidth and high query performance but only at the expense of intermediate operations that reorganize and optimize the B-trees. Zone maps permit the load bandwidth of heaps and enable good query performance but only in cases of correlation between load sequence and columns in query predicates. Finally, a table with

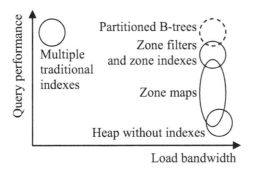

Fig. 5. Alternative techniques at a glance

zone filters and zone indexes can be loaded at hardware write bandwidth, requires no reorganization or optimization after loading, and instantly enables query performance comparable to multiple traditional indexes.

Acknowledgements

Harumi Kuno's many insightful suggestions on contents, text, and diagrams have improved this paper and are greatly appreciated.

References

[ADH 01] Ailamaki, A., DeWitt, D.J., Hill, M.D., Skounakis, M.: Weaving relations for cache performance. In: VLDB 2001, pp. 169–180 (2001)

[BB 02]   Bumbulis, P., Bowman, I.: A compact B-tree. In: SIGMOD 2002, pp. 533–541 (2002)

[BM 70]   Bayer, R., McCreight, E.M.: Organization and maintenance of large ordered indexes. In: SIGFIDET Workshop 1970, pp. 107–141 (1970)

[G 03]   Graefe, G.: Sorting and indexing with partitioned B-trees. In: CIDR 2003(2003)

[G 06]   Graefe, G.: B-tree indexes, interpolation search, and skew. In: DaMoN 2006, p. 5 (2006)

[GKK 01] Gärtner, A., Kemper, A., Kossmann, D., Zeller, B.: Efficient bulk deletes in relational databases. In: ICDE 2001, pp. 183–192 (2001)

[GL 01]   Graefe, G., Larson, P.-Å.: B-Tree indexes and CPU caches. In: ICDE 2001, pp. 349–358 (2001)

[L 01]   Lomet, D.B.: The evolution of effective B-tree page organization and techniques: A personal account. SIGMOD Record 30(3), 64–69 (2001)

[LC 86]   Lehman, T.J., Carey, M.J.: A study of index structures for main memory database management systems. In: VLDB 1986, pp. 294–303 (1986)

[LJB 95] Leslie, H., Jain, R., Birdsall, D., Yaghmai, H.: Efficient search of multi-dimensional B-trees. In: VLDB 1995, pp. 710–719 (1995)

[M 98]   Moerkotte, G.: Small materialized aggregates: A light weight index structure for data warehousing. In: VLDB 1998, pp. 476–487 (1998)

[RR 00]   Rao, J., Ross, K.A.: Making B+-trees cache conscious in main memory. In: SIGMOD 2000, pp. 475–486 (2000)

TidFP: Mining Frequent Patterns in Different Databases with Transaction ID

C.I. Ezeife* and Dan Zhang

School of Computer Science, University of Windsor,
Windsor, Ontario,
Canada N9B 3P4
cezeife@uwindsor.ca, zhang3d@uwindsor.ca
http://www.cs.uwindsor.ca/~cezeife

Abstract. Since transaction identifiers (ids) are unique and would not usually be frequent, mining frequent patterns with transaction ids, showing records they occurred in, provides an efficient way to mine frequent patterns in many types of databases including multiple tabled and distributed databases. Existing work have not focused on mining frequent patterns with the transaction ids they occurred in. Many applications require finding strong associations between transaction id (e.g., certain drug) and the itemsets (e.g., certain adverse effects) to help deduce some pertinent lacking information (like how many people use this product in total) and information (like how many people have the adverse effects).

This paper proposes a set of algorithms TidFPs, for mining frequent patterns with their transaction ids in a single transaction database, in a multiple tabled database, and in a distributed database. The proposed technique scans the database records only once even with level-wise Apriori-based mining techniques, stores frequent 1-items with their transaction id bitmap, outperforms traditional approaches and is extendible to other tree-based mining techniques as well as sequential mining.

Keywords: Data mining, Transaction id, Frequent Patterns, Distributed Mining, Multiple Table Mining.

1 Introduction

Mining frequent itemsets from a database table has been solved largely by algorithms that are Apriori based (e.g., the Apriori algorithm [1]) and those that are pattern-tree growth techniques (e.g., FP-tree [6]). Algorithms for mining frequent patterns from sequential databases also exist and include GSP [11], PrefixSpan [10], SPADE [12], SPAM [2], WAP [9] and PLWAP [4]. The focus of all these existing techniques does not include generating frequent patterns, showing the records where they occurred or with their transaction ids as may

---

* This research was supported by the Natural Science and Engineering Research Council (NSERC) of Canada under an Operating grant (OGP-0194134) and a University of Windsor grant.

T.B. Pedersen, M.K. Mohania, and A M. Tjoa (Eds.): DaWaK 2009, LNCS 5691, pp. 125–137, 2009.

Table 1. Example Drug/Side Effects Database Records

Tid (Drug)	Items (Side Effects)
D_1	1 3 4
D_2	2 3 5
D_3	1 2 3 5
D_4	2 5

be needed by some applications. Existing algorithms are also designed for single table mining and not for mining multiple related tables in a not necessarily normalized database. Assume a pharmacovigilance database table which contains reports about the adverse events of certain drugs from medical professionals as well as patients as depicted in Table 1 where the set of items (adverse side effects) I = {1, 2, 3, 4, 5} and the set of transaction ids (Drugs) Tids = {D_1, D_2, D_3, D_4}.

Mining all drugs that have similar frequent side effects at minimum support of 50% would require generating frequent itemsets (or frequent side effects) with the transaction id (or Drug id) in the format [< $itemset$ > Tid-list] that allows mining more informative large itemsets as L = { [< 1 > D_1D_3], [< 2 > $D_2D_3D_4$], < 3 > $D_1D_2D_3$], [< 5 > $D_2D_3D_4$], [< 1, 2 > $D_2D_3D_4$],[< 1, 3 > D_1D_3], [< 2, 3 > D_2D_3], [< 2, 5 > $D_2D_3D_5$], [< 3, 5 > D_2D_3], [< 2, 3, 5 > D_2D_3]}.

1.1 Contributions and Outline

This paper proposes a series of algorithms for mining frequent patterns with their transaction ids on different types of databases, including (i) from a single table, (ii) from a multiple database set of related tables, (iii) from a horizontally distributed database tables, and (iv) from a vertically distributed database tables. The objectives of the proposed techniques are:

1. Enabling more informative mining: For many applications, just producing the frequent patterns without linking them to the specific transactions they occurred in, may not be adequate. Also, enabling mining of multiple related tables either in a single or distributed database environment, provides answers to more complex queries.

2. Improving Mining Efficiency: This system aims at improving the mining efficiency by cutting down from several to one, the number of times the original database is scanned for purposes of support counting.

Section 2 presents related work, Section 3 presents the proposed systems: TidFP, TidFP-multi, TidFP-hordist and TidFP-vertdist miners for respectively mining single table, multiple tables, horizontally distributed tables, and vertically distributed tables. Section 4 describes the experimental results, while section 5 presents conclusions and future work.

2   Related Work

Association rule can be used to find correlation among items in a given transaction. Association rule mining was proposed in [7], where the formal definition of the problem is presented as: Let $I = \{i_1, \ldots, i_n\}$ be a set of literals, called items. Let database, D be a set of transaction records, where each transaction T is a set of items such that $T \subseteq I$. Associated with each transaction is a unique identifier, called its transaction id (TID). We say that a transaction T contains X, a set of some items in I, if $X \subseteq I$. An association rule is an implication of the form $X \rightarrow Y$, where $X \subseteq I$, $Y \subseteq I$, and $X \cap Y = \emptyset$. The rule $X \rightarrow Y$ holds in the transaction set D with confidence c if $c\%$ of transactions in D that contain X also contain Y. The rule $X \rightarrow Y$ has support s in the transaction set D if $s\%$ of transactions in D contain $X \cup Y$. An example database is shown in Table 1. Here, there are four transactions with TID D_1, D_2, D_3, and D_4. The rule $\{side\ effect\ 1\} \rightarrow \{side\ effect\ 2\}$ is an association rule because with a given minimum support of 50% or 2 out of 4 transactions, the 2-itemset $(1,2)$ which, this rule is generated from, has a support of 3/4 or 75%. The confidence for this rule is 1/2=50%.

Several important association rule mining algorithms including the Apriori [7], [1], and Fp-growth [6] exist. The basic idea behind the Apriori algorithm [7], [1], is to level-wise, use shorter frequent k-itemsets (L_k) to deduce longer frequent (k+1)-itemsets (L_{k+1}) starting from candidate 1-itemsets consisting of single items in the set I defined above, until either no more frequent itemsets or candidate itemsets can be found. Thus, the Apriori algorithm finds frequent k-itemsets L_k from the set of frequent (k-1)-itemsets L_{k-1} using the following two main steps involving joining the $L_{(}k-1)$ with $L_{(}k-1)$ Apriori-gen way to generate candidate k-itemsets C_k, and secondly, pruning the C_k of itemsets not meeting the Apriori property or not having all their subsets frequent in previous large itemsets. To obtain the next frequent L_k from candidate C_k, the database has to be scanned for support counts of all itemsets in C_k. A modified version of the Apriori algorithm called AprioriTid [1] avoids re-scanning the database to enumerate frequent patterns. AprioriTid maintains a candidate set C'_k. Every entry of C'_k has two parts. One is transaction ID and the other is a list of k-itemsets. Instead of scanning the database to count the support for every candidate itemset, the algorithm iterates C'_k. Simultaneously, C_{k+1} is generated to enumerate (k+1)-itemsets. Although transaction Ids for every frequent itemsets can be obtained with this algorithm, experiments show that AprioriTid approach slows down performance once it processes large datasets. Since level-wise candidate generation as well as numerous scans of the database had been seen as a limitation of this approach, optimization techniques in the literature and alternative tree-based solution proposal with Frequent pattern tree growth FP-growth [5], [6] had also been used. The FP-growth approach scans the database once to build the frequent header list, then, represents the database transaction records in descending order of support of the F_1 list so that these frequent transactions are used to construct the FP-tree. The FP-tree are now mined for frequent patterns recursively through conditional pattern base of the conditional FP-tree and suffix growing of the frequent patterns. None of the frequent itemset mining algorithms considers mining frequent patterns with their transaction ids.

Some existing sequential pattern mining algorithms with techniques using transactions IDs to generate frequent sequential patterns and count supports include SPADE [12] and SPAM [2]. SPADE uses a vertical id-list database format that associates each sequence to a list of objects in which it occurs along with the time-stamps and all frequent sequences are enumerated through temporal joins (or intersections) on id-lists. SPADE only needs to access the original database 3 times for support counting. Algorithm SPAM [2] has similar ideas as SPADE. However, instead of vertical representation of id-list, SPAM uses vertical bitmap representation of the entire database that fits in main memory. These sequential mining techniques are not focussed on generating Fps with their Tids and incur such limitations as inefficient memory utilizations and not suitable to scale to very large databases.

3   The Proposed TidFP Algorithms

Section 3.1 presents the main algorithm TidFp, being proposed for mining frequent patterns with the transaction ids where they occurred. Section 3.2 presents the version of the algorithm for mining multiple tables called TidFp-multi, section 3.3 presents the version of the algorithm for mining horizontally distributed database tables called TidFp-hordist, while section 3.4 provides the TidFp-vertdist for mining vertically distributed database tables.

3.1   TidFp for Mining Fps with Transaction Ids on a Single Table

Since an important goal of the TidFp algorithm is linking all frequent patterns to the database records or transactions where they came from, the TidFp algorithm represents each frequent k-itemset as an m-attribute tuple of the form $< F_{k_1}$, $T_{1k_1}, T_{2k_1}, \ldots, T_{mk_1} >$, where F_{k_1} is the first frequent k-itemset, and T_{mk_1} is the mth transaction id of the first frequent k-itemset. For example, given Table 1 and minimum support of 50%, the list of frequent 1-itemsets is $F_1 = \{< 1, D_1,$ $D_3 >, < 2, D_2, D_3, D_4 >, < 3, D_1, D_2, D_3 >, < 5, D_2, D_3, D_4 >\}$. This implies as well that the candidate 1-itemsets listed by this technique is in the same form as: $C_1 = \{< I_1, T_{11}, T_{21}, \ldots, T_{m1} >\}$, where I_1 is the the first candidate 1-itemset, and T_{m1} is the mth transaction id of the first candidate 1-itemset. For our example drug database, the candidate 1-itemset is given as $C_1 = \{< 1, D_1,$ $D_3 >, < 2, D_2, D_3, D_4 >, < 3, D_1, D_2, D_3 >, < 4, D_1 >, < 5, D_2, D_3, D_4 >\}$. Thus, with this TidFp technique, the database is scanned only once to obtain the candidate 1-itemsets with a list of their Tids. The Tids of each candidate itemset is implemented either as a bitmap stored for each itemset or as only one stored bitmap that itemsets point to. Then, the count of each candidate itemset's Tids is equivalent to the support of the itemset. The itemsets having support less than the minimum support are excluded from the frequent 1-itemset

list, leading to the itemset $< 4, D_1 >$ being deleted from the C_1 list to get F_1. In order to get the higher order candidate and frequent (k+1)-itemsets F_{k+1}, given a frequent k-itemset F_k, TidFp algorithm applies a modification of the Apriori-gen join function called the map-gen join function, which works on two components of the itemsets consisting of the itemset part and the transaction id part and obtaining higher order frequent (k+1)-itemsets does not require re-scanning the database for their supports as is needed with the Apriori-gen join. With the TidFp, the candidate (k+1)-itemsets C_k is obtained from the frequent k-itemsets for $k \geq 1$, by joining frequent k-itemsets F_k with itself mapgen way such that $C_{k+1} = F_k \bowtie F_k$. To join mapgen way, for each pair of itemsets M and $P \in F_k$ where each F_k itemset has the two parts "< itemset, transaction id list >", the following three conditions have to be satisfied: M joins with P to get itemset $M \cup P$ if the following conditions are satisfied.
(a) itemset M comes before itemset P in F_k,
(b) the first k-1 items in M and P (excluding just the last item) are the same,
(c) the transaction id list of the new itemset $M \cup P$ represented as $Tid_{M \cup P}$ is obtained as the intersection of the Tid lists of the two joined k-itemsets M and P and thus, $Tid_{M \cup P} = Tid_M \cap Tid_P$.
The formal algorithm TidFp is presented as Algorithm 1.

Algorithm 1. *(TidFp:Computing Frequent Patterns with Tids)*

Algorithm TidFp()
Input: A list of 1-items, Transaction Table of 1-items,
                  minimum support $s\%$.
Output: A list of frequent patterns Fps.
Other variables: candidate sets C_k, Frequent k-itemsets F_k, k = 1 initially.
begin
    1. Scan the DB once to compute
    $C_k = \{< item_{k1}, Tidlist_{item_{k1}} >, \ldots, < item_{km}, Tidlist_{item_{km}} >\}$.
    2. Compute frequent k-itemset F_k from candidate k-itemsets
    C_k as $F_k = \{$list of k-itemset with Tidlist count $\geq$ minsupport $\}$.
    3. While $(F_k \neq \emptyset)$ do
    begin
    3.1. k = k+1
    3.2. Compute the next candidate set C_{k+1} as F_k map-gen join F_k.
    i.e. Each itemset $M \in F_k$ joins with another itemset
    $P \in F_k$ if the following conditions are satisfied.
        (a) itemset M comes before itemset P in F_k
        (b) the first k-1 items in M and P (excluding just the last
        item) are the same.
        (c) Tid list of the two joined k-itemsets M and P is:
        $Tid_{M \cup P} = Tid_M \cap Tid_P$.
    3.3. For each itemset in C_k do
        3.3.1. Calculate all subsets and prune if not previously large.
    3.4. If $C_k = \emptyset$ then break and go to step 4
    end
    4. Compute all Frequent patterns as $FP = F_1 \cup \ldots F_k$
end

Table 2. Example Patient/Drugs Database Records

Patient	Drugs
P_1	D_1 D_2
P_2	D_1 D_2 D_3
P_3	D_3 D_4
P_4	D_1 D_2 D_4

3.2   TidFp-multi: Mining FPs on Multiple Tables with Transaction Ids

One advantage of mining frequent patterns with transaction ids is that it facilitates discovering more meaningful knowledge from not just a single database table but a database with a multiple of related tables. It can also be extended to distributed databases having partitioned tables distributed across a number of sites. For example, consider a pharmacovigilance database, which has reports about adverse events of certain drugs from medical professionals and patients and there is need to answer queries from two related database tables Drug/Side Effects (as in Table 1) and Patient/Drug shown as Table 2. We might be interested in answering with frequent pattern mining, questions like the following, which will not be easily answered with simple SQL or stored procedure queries.
1. How many people have various patterns and frequent patterns of adverse effects given minimum 50% total occurrence?
2. How many people use frequent combinations of products having minimum total occurrence of 50%?
3. Which drugs have dangerous combinations of adverse effects?
Answering query 1 above requires finding the TidFp of Table Drug/Side Effect (Table 1) to get $TidFp_{Drug}$, also finding TidFp of Table Patient/Drug (Table 2) to get $TidFp_{Patient}$ and getting the count of $TidFp_{Drug} \cap TidFp_{Patient}$. Query 2 is answered by mining TidFp on table Patient, while query 3 is answered by mining TidFp on Table Drug.

Thus, mining more complex knowledge from a multiple of related tables in a database, would normally entail applying the TidFp algorithm on the individual database tables and answering the queries by either integrating the mined FPs from different tables using set operations of intersection, union, minus as is suitable to answer the queries. Having the Tids with FPs makes this integration easy and possible.

3.3   TidFp-hordist: Mining Horizontally Distributed Tables with TidFp

Most existing algorithms on mining distributed databases including [3], [8] focus on privacy-preserving distributed mining of association rules, whereby the data at different sites are secure data that are not shared with other sites. Many applications belong to the same organization (e.g., an automobile company), located at different sites and collaborative mining of distributed data at different sites would provide both local and global knowledge for marketing promotions among others. There are two main techniques for partitioning global data or table T, belonging to an organization into f fragments of the table based on some fragmentation criteria, to be distributed at perhaps f locations. First method is horizontal fragmentation where each horizontal fragment, F_i, has a number n_{F_i} of records of the global table T such that $\sum_{i=1}^{f} n_{F_i} = |T|$ meaning that the sum of the number of records in all f fragments will give back the cardinality of the global table T. On the other hand, each vertical fragment of T has only some attributes of T but has the same cardinality as T giving that the sum of the arity (number of attributes) in each of the f vertical fragments F_i, is the same as the arity of the global table T, that is, $\sum_{i=1}^{f} arity_{F_i} = arity_T$. Thus, although the global data T, is distributed either horizontally or vertically, the goal of distributed frequent pattern mining given a minimum support s threshold is to find all global frequent patterns GFPs and local frequent patterns LFPs that meet the minimum support threshold.

An Existing Algorthm on Distributed Mining
The FDM algorithm [3] for distributed FP mining first generates global candidate k-itemset $CG_{i(k)}$ by apriori-gen joining of global large (k-1)-itemsets at site i, $GL_{i(k-1)}$ with itself. The global (k-1)-itemsets at each site i, $GL_{i(k-1)}$, are obtained by intersecting the global large (k-1)-itemsets with the local large (k-1)-itemsets $LL_{i(k-1)}$. Next, the local database is scanned to prune the itemsets in the candidate k-itemset $CG_{i(k)}$ whose local support is less than the minsupport $s\%$, while the rest are put in the local large k-itemsets at site i, $LL_{i(k-1)}$. Each site then broadcasts its local large k-itemsets $LL_{i(k)}$ to all sites, the union of all the $LL_{i(k)}$ will give the $LL_{i(k)}$ from where each site computes the support of items in $LL_{i(k)}$, which are broadcasted to all sites so that they can combine them to compute the global frequent itemsets G_k.

The Proposed TidFp-Hordist Algorithm
The difference between the proposed TidFp-Hordist algorithm approach for mining horizontally distributed table and other approaches like those of the FDM [3] summarized above, is that the TidFp-Hordist takes advantage of the Tid's of each Fp when forming global large itemsets and thus, requires only one initial broadcasting of the first local frequent 1-itemsets from all sites to each site and global frequent 1-itemset GFP_1 is computed as the union of all local $LFP_{i(1)}$ itemsets while the next global candidate (k+1)-itemset, C_{k+1} is computed as global GFP_k map-gen join GFP_k. Thereafter, subsequent global GFP_k and candidate C_k are computed without any further broadcast and support counting from local databases. The formal TidFp-Hordist algorithm is presented as Algorithm 2.

Algorithm 2. *(TidFp-Hordist:FPs with Tids in Horizontally Distributed DBs)*

Algorithm TidFp-Hordist()
Input: A list of 1-items, a number i of sites,
         a set of i horizontally fragmented Transaction
         Tables of 1-items DB_i, minimum support $s\%$.
Output: A list of global frequent patterns GFps.
Other variables: global candidate sets GC_k, global
         Frequent k-itemsets GF_k, local Frequent k-itemsets LF_k
         $k = 1$ initially.
   1. Scan each local DB_i once to compute
   $LC_{i(k)} = \{< item_{k1}, Tidlist_{item_{k1}} >, \ldots, < item_{km}, Tidlist_{item_{km}} >\}$.
   2. Compute each local i frequent k-itemset $LF_{i(k)}$
   from local candidate k-itemsets $LC_{i(k)}$ as
   $LF_{i(k)} = \{$list of k-itemset with Tidlist count $\geq$ minsupport$\}$.
   3. Let each site i broadcast its local frequent $LF_{i(k)}$.
   4. Compute global GF_k as itemsets in the union of all local $LF_{i(k)}$
   with support $\geq s\%$ of global $|DB|$.
   5. At each site i, while global $(GF_{i(k)} \neq \emptyset)$ do
   begin
   5.1. k = k+1
   5.2. Compute the next global candidate set GC_k from GF_{k-1}
   as GF_{k-1} map-gen join GF_{k-1}.
   5.3. For each itemset in global GC_k do
         5.3.1. Calculate all subsets and prune if not previously large.
   5.4. If $GC_k = \emptyset$ then break and go to step 6
   5.5. Compute GF_k as itemsets in GC_k with support count $\geq$ minsupport.
   end
   6. Compute all global Frequent patterns as
   $GFP = GF_1 \cup \ldots GF_k$
end

Application of the TidFp-Hordist Algorithm

EXAMPLE 2: Given the two horizontally fragmented tables shown as Tables 3 and 4, which are equivalent to the example database of Table 1, and a minimum support threshold of 50%, use the TidFp-Hordist algorithm to obtain the global frequent patterns GFP, across the two distributed tables at two sites.

SOLUTION 2: Applying the algorithm TidFp-Hordist to the two horizontally distributed database tables above at minsupport of 50% to mine global frequent patterns GFPs would entail executing the steps of the algorithm as follows: Step 1,

Table 3. Horizontally Distributed Drug Table 1

Tid (Drug)	Items (Side Effects)
D_1	1 3 4
D_2	2 3 5

Table 4. Horizontally Distributed Drug Table 2

Tid (Drug)	Items (Side Effects)
D_3	1 2 3 5
D_4	2 5

we compute the local for site 1 candidate $C_{1(1)} = \{ < 1, T_1 >, < 2, T_2 >, < 3, T_1, T_2 >, < 4, T_1 >, < 5, T_2 > \}$. Then, we compute the local for site 2 candidate $C_{2(1)} = \{ < 1, T_3 >, < 2, T_3, T_4 >, < 3, T_3 >, < 5, T_3, T_4 > \}$. Step 2, we compute the local frequent $LF_{i(k)}$ as: $LF_{1(1)} = \{ < 1, T_1 >, < 2, T_2 >, < 3, T_1, T_2 >, < 4, T_1 >, < 5, T_2 > \}$. $LF_{2(1)} = \{ < 1, T_3 >, < 2, T_3, T_4 >, < 3, T_3 >, < 5, T_3, T_4 > \}$. Step 3 entails each site having all local $LF_{i(k)}$ through broadcast. Step 4, we now get global $GF_1 = $ itemsets in $\cup_{i=1}^{2} LF_{i(k)}$ with support of 50% of global DB cardinality with count of at least 2. GF_1 is from $LF_{1(1)} \cup LF_{2(1)}$ $= \{ < 1, T_1, T_3 >, < 2, T_2, T_3, T_4 >, < 3, T_1, T_2, T_3 >, < 5, T_2, T_3, T_4 > \}$. Step 5 computes the next global candidate set at each site GC_2 as: GF_1 map-gen $GF_1 = \{ < 1, 2, T_3 >, < 1, 3, T_1, T_3 >, < 1, 5, T_3 >, < 2, 3, T_2, T_3 >, < 2, 5, T_2, T_3, T_4 >, < 3, 5, T_2, T_3 > \}$. The next global frequent GF_2 is computed from GC_2 as $\{ < 1, 3, T_1, T_3 >, < 2, 3, T_2, T_3 >, < 2, 5, T_2, T_3, T_4 >, < 3, 5, T_2, T_3 > \}$. Back to beginning of step 5, next global candidate $GC_3 = GF_2$ map-gen $GF_2 = \{ < 2, 3, 5, T_2, T_3 > \}$ and the frequent $GF_3 = \{ < 2, 3, 5, T_2, T_3 > \}$. The global GFPs $= \cup_{i=1}^{3} GFPi$, which are the same as the table mined undistributed.

3.4 TidFp-Vertdist: Mining Vertically Distributed Tables with TidFp

The version of the TidFp algorithm for mining vertically distributed database is very much like the one for mining the horizontally distributed database discussed in detail above. The difference is in how the tables are fragmented, which only affects how the supports of the local and global tables are computed. In the horizontally fragmented tables, the local support counts are different and based on the local cardinality, which is less or equal to the cardinality of the global database. On the other hand, in the vertically distributed tables, the support counts of the vertical fragments and the global database are the same. As defined in the previous section, a vertical fragment of a table has only some of the attributes of the original table but has all records of the table. For example, the example Drug Table T (Tid, Attr1, Attr2, Attr3, Attr4), can be fragmented vertically into two tables with schemas T_{v1} (Tid, Attr1, Attr2) and T_{v2} (Tid, Attr3, Attr4). Our application of the algorithm on the vertically fragmented tables produced the same correct results as well.

4 Experiments and Performance Analysis

To test the proposed TidFp algorithm, we conducted experiments to (1) determine performance gain in terms of CPU execution time gain of the TidFp in comparison with the Apriori algorithm, which also determines Tids.

In this case, we first ran the Apriori algorithm and then scanned the database for each frequent pattern to collect the TIDs they appeared in.

(2) determine memory usages of the proposed TidFp in comparison with the Apriori algorithm.

Comparing TidFp and Apriori Execution Times and Memory Usages
The two algorithms Apriori with Tid and our proposed TidFp algorithm were implemented in C++ with the same data structure in UNIX environment, where the programs are compiled with "g++ filename" and executed with "a.out". Then, the CPU execution times for the two algorithms were tested for transactional databases of different sizes of 250K (or 250 thousand) records, 500K, 750K, 1M (or 1 million), and 2M records generated with the IBM quest synthetic data generator publicly available at at http://www.almaden.ibm.com/cs/quest/ and used by other pattern mining research. The characteristics of the generated data are described as follows: $|D|$: Number of records in the database, $|C|$: Average length of the records, $|S|$: Average length of maximal potentially frequent itemset, $|N|$: number of items (attributes).

With the average length of records (C) in our data as 10, and number of attributes (N) as 10 with S as 5, a full description of one set of experimental data with number of records as 250 thousand is C10.S5.N20.D250K. All experiments are performed on on a more powerful multiuser UNIX SUN microsystem with a total of 16384 MB memory and 8 x 1200 MHz processor speed, which generally produces faster execution times than when run on microcomputer environment. The minimum support for the experiments range between 10% and 50%. The result of the experiments are summarized in five tables below to show:

(1) Execution time efficiencies of the algorithms at medium minimum support threshold of 40% and for different database sizes as shown in Table 5.
(2) scalability of the algorithms with a large database of 2 million records at different minimum supports with data C10.S5.N20.D2M as shown in Table 6.
(3) feasibility and scalability at a medium sized database of 500K records at different minimum supports with data C10.S5.N20.D500K as shown in Table 7.
(4) feasibility and scalability of the algorithms at a small sized database of 250K records at different minimum supports C10.S5.N20.D250K as shown in Table 8.
(5) Memory usages of the algorithms at medium minimum support threshold of 40% for different database sizes as shown in Table 9. The RES memory usage value is collected for the program process of "a.out" on UNIX with the "top -u" command.

It can be seen from the experiments that the proposed TidFp outperforms the Apriori algorithm to the tune of 25 times better in execution time and is more scalable than the Apriori algorithm and in particular, at low minimum support thresholds and large data sizes. From experiment 5 on Table 9, it can be seen though that the proposed TidFp algorithm requires more running memory than

Table 5. Execution times for different datasets at MinSupport of 40%

Algorithms	Runtime (in secs) for different Data sizes)				
	250K	500K	750K	1M	2M
TidFp	22	47	64	94	187
Apriori	542	1071	1623	2239	4495

Table 6. Execution times for dataset at different MinSupports (large data 2M)

Algorithms	Runtime (in secs) at different supports(%)				
	10%	20%	30%	40%	50%
TidFp	23236	1472	330	187	147
Apriori	crashed	39434	11571	4495	2141

Table 7. Execution times for dataset at different MinSupports (medium data 500K)

Algorithms	Runtime (in secs) at different supports(%)				
	10%	20%	30%	40%	50%
TidFp	6231	391	90	47	38
Apriori	crashed	8702	2729	1071	509

Table 8. Execution times for dataset at different MinSupports (small data 250K)

Algorithms	Runtime (in secs) at different supports(%)				
	10%	20%	30%	40%	50%
TidFp	2853	176	37	22	17
Apriori	crashed	4329	1348	542	267

Table 9. Memory Usages for Different Data Sizes at Minsupport of 40%

Algorithms	Memory Usages for Different Data Sizes				
	250K	500K	750K	1M	2M
TidFp	10MB	14MB	18MB	22MB	42MB
Apriori	2872KB	3664KB	4408KB	5424KB	10M

the Apriori algorithm but this is a reasonable tradeoff. This good performance of the TidFp algorithm is because the TidFp only needs to scan the database once, while the Apriori algorithm re-scans the database for every support counting. When the TidFp algorithm intersects the transaction IDs for 2 items sets, it uses bitmap representation. For example, the transaction ID bitmaps from the two F_1 itemsets $[< 1 > D_1D_3]$ and $[< 1 > D_2D_3D_4]$ obtained when the TidFp executes the operation $[< 1 > D_1D_3]$ mapgen-Join $[< 1 > D_1D_3D_4]$ are shown in Table 10. The Tid list of the resulting 2-itemset $< 1, 2 >$ obtained by

Table 10. Bitmaps for Itemsets $< 1 >$, $< 2 >$ and $< 1, 2 >$ Tids

Itemset	Transaction id Bitmap			
	D_1	D_2	D_3	D_4
$< 1 >$	1	0	1	0
$< 2 >$	0	1	1	1
$< 1, 2 >$	0	0	1	0

mapgen-joining the two 1-itemsets $< 1 >$ and $< 2 >$ is shown as the third row of Table 10. The Tid list of the resulting (k+1)-itemset is obtained from intersecting the Tid lists of the two joining k-itemsets, which is accomplished with bitmap AND operation.

5   Conclusions and Future Work

This paper proposes an intuitive approach for mining frequent patterns with transaction ids, which is useful for addressing the needs of several applications including mining multiple related tables in a database for more informative knowledge discovery. The paper also introduced versions of this algorithm for mining horizontally and vertically distributed databases and multiple related tables in a database. It has also been shown through experiments that TidFp execution time is up to 25 times better than the Apriori algorithm. It has also been shown that both number of database scans needed for support counting and communication costs are drastically reduced when this approach is employed. This approach is extendible to other types of mining like mining sequences and on pattern growth techniques for mining frequent patterns. Future work should also determine or analyze the gain made through saving broadcast delays and resources as well as execution time when mining distributed tables.

References

1. Agrawal, R., Srikant, R.: Fast Algorithms for Mining Association Rules in Large Databases. In: Proceedings of the 20th International Conference on very Large Databases Santiago, Chile, pp. 487–499 (1994)
2. Ayres, J., Flannick, J., Gehrke, J., Yiu, T.: Sequential Pattern Mining using A Bitmap Representation. In: Proceedings of the ACM SIKDD conference, Edmonton, Alberta, Canada, pp. 429–435 (2002)
3. Cheung, D.W.-L., Ng, V., Fu, A.W.-C., Fu, Y.: Efficient Mining of Association Rules in Distributed Databases. Transactions on Knowledge and Data Engineering 8(6), 911–922 (1996)
4. Ezeife, C.I., Lu, Y.: Mining Web Log sequential Patterns with Position Coded Pre-Order Linked WAP-tree. The International Journal of Data Mining and Knowledge Discovery (DMKD) 10, 5–38 (2005)
5. Han, J., Kamber, M.: Data Mining: Concepts and Techniques. Morgan Kaufmann Publishers, New York (2001)

6. Han, J., Pei, J., Yin, Y., Mao, R.: Mining Frequent Patterns without Candidate Generation: A Frequent-Pattern Tree approach. International Journal of Data Mining and Knowledge Discovery 8(1), 53–87 (2004)
7. Imielinski, T., Swami, A., Agarwal, R.: Mining association rules between sets of items in large databases. In: Proceeding of the ACM SIGMOD conference on management of data, Washington D.C., May 1993, pp. 207–216 (1993)
8. Kantarcioglu, M., Clifton, C.: Privacy-preserving Distributed Mining of Association Rules on Horizontally Partitioned Data. In: The proceedings of the ACM SIGMOD Workshop on Research Issues on Data Mining and Knowledge Discovery, DMKD 2002, pp. 24–31 (2002)
9. Pei, J., Han, J., Mortazavi-asi, B., Zhu, H.: Mining Access Patterns Efficiently from web logs. In: Proceedings, Pacific-Asia conference on Knowledge Discovery and data Mining, Kyoto, Japan, pp. 396–407 (2000)
10. Pei, J., Han, J., Mortazavi-Asl, B., Pinto, H., Chen, Q., Dayal, U., Hsu, M.C.: PrefixSpan: Mining Sequential Patterns Efficiently by Prefix-Projected Pattern Growth. In: Proceedings of the 2001 International Conference on Data Engineering (ICDE 2001), Heidelberg, Germany, pp. 215–224 (2001)
11. Srikanth, R., Aggrawal, R.: Mining Sequential Patterns: generalizations and performance improvements, Research Report, IBM Almaden Research Center 650 Harry Road, San Jose, CA 95120, 1–15 (1996)
12. Zaki, M.J.: SPADE: An Efficient Algorithm for Mining Frequent Sequences. Machine learning 42, 32–60 (2001)

Non-Derivable Item Set and Non-Derivable Literal Set Representations of Patterns Admitting Negation

Marzena Kryszkiewicz

Institute of Computer Science, Warsaw University of Technology
Nowowiejska 15/19, 00-665 Warsaw, Poland

Abstract. The discovery of frequent patterns has attracted a lot of attention of the data mining community. While an extensive research has been carried out for discovering positive patterns, little has been offered for discovering patterns with negation. The main hindrance to the progress of such research is huge amount of frequent patterns with negation, which exceeds the number of frequent positive patterns by orders of magnitude. In this paper, we examine properties of derivable and non-derivable patterns, including those with negated items. In particular, we establish important relationships among patterns admitting negation that have the same canonical variant. By analogy to frequent non-derivable itemsets, which constitute a concise lossless representation NDR of frequent positive patterns, we introduce frequent non-derivable literal sets lossless representation NDRL of frequent positive patterns admitting negation. Then we use the derived properties of literal sets to offer a concise representation NDIR of frequent patterns admitting negation that is built only from positive non-derivable itemsets. The relationships between the three representations are identified. The transformation of the new representations into not less concise lossless closure representations is discussed.

1 Introduction

Discovering of frequent patterns in large databases is an important data mining problem. The problem was introduced in [1] for a sales transaction database. Frequent patterns were defined there as sets of items that are purchased together frequently. Frequent patterns are commonly used for building association rules. For example, an association rule may state that 80% of customers who buy fish also buy white wine. This rule is derivable from the fact that fish occurs in 5% of sales transactions and set {fish, white wine} occurs in 4% of transactions. Patterns and association rules can be generalized by admitting negation. A sample association rule with negation could state that 75% of customers who buy coke also buy chips and neither beer nor milk. The knowledge of this kind is important not only for sales managers, but also in medical areas, where both the occurrence and lack of symptoms of illnesses is of importance. Similarly, the knowledge of co-occurrence and lack of co-occurrence of terms with a given term is useful in discovering synonyms and homonyms [27].

Admitting negation in patterns usually results in an abundance of mined patterns, which makes analysis of the discovered knowledge infeasible. It is thus preferable to discover and store a possibly small fraction of patterns, from which one can derive all

T.B. Pedersen, M.K. Mohania, and A M. Tjoa (Eds.): DaWaK 2009, LNCS 5691, pp. 138–150, 2009.
© Springer-Verlag Berlin Heidelberg 2009

other significant patterns when required. Lossless representations of frequent positive patterns were discussed e.g., in [5-8,11-14,18-21,23-25]. Among them, the generalized disjunction-free set representations [11-14,20-21] and the frequent non-derivable itemsets NDR [6-7], as well as their closures [18,24], are most concise ones. It is worth noting that these representations are even by two orders of magnitude more concise than the approximate ones, which were offered in [4,26]. However, to the best of our knowledge, only two concise lossless representations of frequent patterns admitting negation were offered in the literature [15-16]: the one based on generalized disjunction-free patterns admitting negation, and the one based on generalized disjunction-free positive patterns. The latter representation is by orders of magnitude more concise than the set of all frequent patterns with negation. As follows from the experiments [15], the direct discovery of all frequent patterns with negation from data sets is often impossible, but the discovery of the latter representation is usually feasible. Indeed, frequent patterns with negation that were not computable from data sets, were usually derivable from the representation.

Pattern representations are often used for efficient building of different types of rules, e.g. concise representations of association rules [9,12,25], dependence rules [28], decision rules [2,17], episode rules [10], and for clustering documents [22]. In particular, dependence rules are built only from such non-derivable patterns the frequencies of which are significantly different from their expected frequencies, which are calculated from the supports of their sub-patterns. In the case of derivable patterns, their real and expected frequencies are identical, and thus uninteresting.

In this paper, we examine properties of derivable and non-derivable patterns, including those with negated items. In particular, we establish important relationships among patterns admitting negation that have the same canonical variation. By analogy to frequent non-derivable itemsets, which constitute a concise lossless representation of frequent positive patterns, we introduce frequent non-derivable literal sets lossless representation NDRL of frequent positive patterns admitting negation. Then we use the derived properties of literal sets to offer a concise representation NDIR of frequent patterns admitting negation that is built only from positive non-derivable itemsets. The relationships between the three representations are identified. The transformation of the new representations into not less concise closure representations is discussed.

The layout of the paper is as follows: Section 2 recalls the notions of frequent itemsets, downward closed sets, generalized disjunctive rules. There, we show how these rules can be used to derive and/or estimate the supports of itemsets. Finally, we recall the non-derivable itemset representation. In Section 3, we generalize a notion of an itemset to a literal set, which admits negation, introduce a notion of a variation of a pattern and canonical pattern, and recall the relationship between certain generalized disjunctive rules and supports of literal set variations. Our main contribution in this paper is presented in Sections 4 and 5. In Section 4, we examine properties of derivable and non-derivable literal sets, which we use later to offer an efficient representation of frequent patterns admitting negation. In Section 5, we propose and examine properties of two variants of non-derivable patterns: 1) built from non-derivable literal sets, 2) built from non-derivable itemsets. We also note that instead of the new representations one may use the closures of their elements without any loss in the pattern derivation power. Section 6 concludes our work. Appendix contains proofs not included in the main part of the paper.

2 Basic Notions

2.1 Itemsets, Frequent Itemsets, Downward Closed Sets

Let $I = \{i_1, i_2, ..., i_m\}$, $I \neq \varnothing$, be a set of distinct *items*. In the case of a transactional database, a notion of an item corresponds to a sold product, while in the case of a relational database an item will be a pair (*attribute, value*). Any set of items is called an *itemset*. Let $\mathcal{D}$ be a set of transactions (or tuples, respectively), where each transaction (tuple) T is a subset of I. Without any loss of generality, we will restrict further considerations to transactional databases. *Support* of itemset X is denoted by $sup(X)$ and is defined as the number (or percentage) of transactions in $\mathcal{D}$ that contain X. Itemset X is called *frequent* if its support is greater than some user-defined threshold *minSup*, where $minSup \in [0, |\mathcal{D}|]$.

A set $X \subseteq 2^I$ is defined as *downward closed*, if for each itemset in X, all its subsets are also in X; that is, if $\forall X \in \mathcal{X}$, $Y \subset X \Rightarrow Y \in \mathcal{X}$. Please note that all supersets of an itemset which does not belong to a downward closed set X do not belong to X either.

Property 2.1.1 [1].
a) Let $X, Y \subseteq I$. If $X \subset Y$, then $sup(X) \geq sup(Y)$.
b) The set of all frequent itemsets is downward closed.

2.2 Generalized Disjunctive Rules

In this section, we recall the notion and properties of generalized disjunctive rules [12, 20], which are useful for reasoning about the supports of itemsets.

Let $Z \subseteq I$. $X \rightarrow a_1 \vee ... \vee a_n$ is defined a *generalized disjunctive rule based on* Z (and Z is the *base of* $X \rightarrow a_1 \vee ... \vee a_n$), if $X \subset Z$ and $\{a_1, ..., a_n\} = Z \backslash X$.

In the sequel, $\vee A$, where $A = \{a_1, ... ,a_n\}$, will denote $a_1 \vee ... \vee a_n$. One can easily note that $\{Z \backslash A \rightarrow \vee A | \varnothing \neq A \subseteq Z\}$ is the set of all distinct generalized disjunctive rules based on Z. Hence, there are $2^{|Z|-1}$ distinct generalized disjunctive rules based on Z.

Support of $X \rightarrow \vee A$ is denoted by $sup(X \rightarrow \vee A)$ and is defined as the number (or percentage) of transactions in $\mathcal{D}$ in which X occurs together with at least one item from A. Please note that e.g. $sup(X \rightarrow a) = sup(X \cup \{a\})$, and $sup(X \rightarrow a \vee b) = sup(X \cup \{a\}) + sup(X \cup \{b\}) - sup(X \cup \{ab\})$.

Table 1. Sample database $\mathcal{D}$

Id	Transaction
T_1	$\{abce\}$
T_2	$\{abcef\}$
T_3	$\{abceh\}$
T_4	$\{abe\}$
T_5	$\{aceh\}$
T_6	$\{bce\}$
T_7	$\{h\}$

Table 2. Generalized disjunctive rules based on $\{abc\}$

$r: X \rightarrow a_1 \vee ... \vee a_n$	$sup(X)$	$sup(r)$	$err(r)$	certain?
$\{ab\} \rightarrow c$	4	3	1	No
$\{ac\} \rightarrow b$	4	3	1	No
$\{bc\} \rightarrow a$	4	3	1	No
$\{a\} \rightarrow b \vee c$	5	5	0	Yes
$\{b\} \rightarrow a \vee c$	5	5	0	Yes
$\{c\} \rightarrow a \vee b$	5	5	0	Yes
$\varnothing \rightarrow a \vee b \vee c$	7	6	1	No

Theorem 2.2.1 [12, 20]. Let $X{\rightarrow}VY$ be a generalized disjunctive rule. Then:

$$sup(X{\rightarrow}VY) = \Sigma_{\emptyset{\neq}Z{\subseteq}Y} (-1)^{|Z|-1} \times sup(X{\cup}Z).$$

It follows from Theorem 2.2.1 that the support of $X{\rightarrow}VY$ depends on the supports of only non-empty subsets of $X{\cup}Y$.

Error of $X{\rightarrow}VA$ is denoted by $err(X{\rightarrow}VA)$ and is defined as the number (or percentage) of transactions containing X that do not contain any item from A; that is,

$$err(X{\rightarrow}VA) = sup(X) - sup(X{\rightarrow}VA).$$

$X{\rightarrow}VA$ is defined a *certain rule* if $err(X{\rightarrow}VA) = 0$. Clearly, if $X \rightarrow VA$ is a certain rule, then each transaction containing X contains also at least one item from A. Table 2 shows all generalized disjunctive rules based on sample itemset $\{abc\}$ that were found in database $\mathcal{D}$ from Table 1, and indicates which of the rules are certain.

Property 2.2.1. If $X{\rightarrow}VA$ is a certain rule, then $\forall X'{\supset}X$, $X'{\rightarrow}VA$ is certain and $\forall A'{\supset}A$, $X{\rightarrow}VA'$ is certain.

Corollary 2.2.1 [12, 20]. Let $X{\rightarrow}VY$ be a generalized disjunctive rule. The error of $X{\rightarrow}VY$ is derivable from the supports of subsets of $X{\cup}Y$:

$$err(X{\rightarrow}VY) = \Sigma_{Z{\subseteq}Y} (-1)^{|Z|} \times sup(X{\cup}Z).$$

In the sequel, we will denote the sub-expression $\Sigma_{Z{\subset}Y} (-1)^{|Z|} \times sup(X{\cup}Z)$, which is derivable only from proper subsets of $X{\cup}Y$, by $b(X{\rightarrow}VY)$.

Corollary 2.2.2. Let $X{\rightarrow}VY$ be a generalized disjunctive rule. Then:

$$err(X{\rightarrow}VY) = (-1)^{|Y|} \times sup(X{\cup}Y) + b(X{\rightarrow}VY).$$

Corollary 2.2.3. Let $X{\rightarrow}VY$ be a generalized disjunctive rule. Then:

$$err(X{\rightarrow}VY) = 0 \text{ iff } sup(X{\cup}Y) = (-1)^{|Y|+1} \times b(X{\rightarrow}VY).$$

Thus, if $X{\rightarrow}VY$ is a certain generalized disjunctive rule, then $sup(X{\cup}Y)$ is determinable from the supports of only proper subsets of $X{\cup}Y$.

Please note that Corollary 2.2.3 allows us to calculate the support of an itemset, provided we know at least one certain generalized disjunctive rule based on this itemset. In the sequel, we discuss the case of estimating the support of an itemset when we do not know if any generalized disjunctive rules based on it are certain.

2.3   Using Generalized Association Rules to Estimate Supports of Itemsets

Combining Corollary 2.2.2 with the observation that the error of any rule $X{\rightarrow}VY$ is greater than or equal to 0 allows us to formulate Corollary 2.3.1, showing the way of estimating the support of the base $X{\cup}Y$ of the rule by means of $b(X{\rightarrow}VY)$.

Corollary 2.3.1. Let $X,Y{\subset}I$ and $X{\rightarrow}VY$ be a generalized disjunctive rule. Then:

$$sup(X{\cup}Y) \geq (-1)^{|Y|+1} \times b(X{\rightarrow}VY) = -b(X{\rightarrow}VY), \text{ when } |Y| \text{ is even, and}$$

$$sup(X{\cup}Y) \leq (-1)^{|Y|+1} \times b(X{\rightarrow}VY) = b(X{\rightarrow}VY), \text{ when } |Y| \text{ is odd.}$$

Since an itemset B is the base of $2^{|B|-1}$ distinct generalized disjunctive rules based on it, $2^{|B|-1}$ inequalities bounding the support of B can be formed. Please recall that the bounds are derivable only from the proper subsets of B. One may also add trivial inequality bounding the support of itemset B; namely, $sup(B) \geq 0$. Hence, we obtain the following set of $2^{|B|}$ inequalities bounding $sup(B)$:

- $\forall Y \subseteq B$ ($|Y|$ is odd $\Rightarrow sup(B) \leq b(B \backslash Y \rightarrow \lor Y)$),
- $\forall Y \subseteq B$ ($Y \neq \emptyset$ and $|Y|$ is even $\Rightarrow sup(B) \geq -b(B \backslash Y \rightarrow \lor Y)$),
- $sup(B) \geq 0$.

Using these inequalities, we may define the following bounds:

Upper bound on $sup(B)$, $B \neq \emptyset$, denoted as $u(B)$, is defined as:

$$u(B) = min(\{b(B \backslash Y \rightarrow \lor Y)|\ Y \subseteq B \text{ and } |Y| \text{ is odd}\}).$$

Lower bound on $sup(B)$, $B \neq \emptyset$, denoted as $l(B)$, is defined as:

$$l(B) = max(\{-b(B \backslash Y \rightarrow \lor Y)|\ \emptyset \neq Y \subseteq B \text{ and } |Y| \text{ is even}\} \cup \{0\}).$$

Property 2.3.1. Let $B \subseteq I$. Then:

a) $l(B) \leq sup(B) \leq u(B)$.
b) If $l(B) = u(B)$, then $l(B) = sup(B) = u(B)$.

Theorem 2.3.1 [Appendix]. Let $\emptyset \neq Y \subseteq B \subseteq I$ and $err(B \backslash Y \rightarrow \lor Y) = 0$ Then:

a) If $|Y|$ is odd, then $sup(B) = b(B \backslash Y \rightarrow \lor Y) = u(B)$.
b) If $|Y|$ is even, then $sup(B) = -b(B \backslash Y \rightarrow \lor Y) = l(B)$.

Theorem 2.3.2 [Appendix]. Let $\emptyset \neq B \subseteq I$. Then:

a) $(\exists Y \subseteq B, err(B \backslash Y \rightarrow \lor Y) = 0 \wedge |Y|$ is odd$)$ iff $sup(B) = u(B)$.
b) Let $sup(B) \neq 0$. $(\exists Y \subseteq B, err(B \backslash Y \rightarrow \lor Y) = 0 \wedge Y \neq \emptyset \wedge |Y|$ is even$)$ iff $sup(B) = l(B)$.
c) $(sup(B) = 0 \vee (\exists Y \subseteq B, err(B \backslash Y \rightarrow \lor Y) = 0 \wedge Y \neq \emptyset \wedge |Y|$ is even$))$ iff $sup(B) = l(B)$.

Corollary 2.3.2 [Appendix]. Let $\emptyset \neq B \subseteq I$. Then:

a) If there is a certain generalized disjunctive rule based on B, then $\forall X \supset B$, $l(X) = sup(X) = u(X)$.
b) If $sup(B) = 0$, then $\forall X \supset B$, $sup(X) = 0$.
c) If $sup(B) = 0$ or there is a certain generalized disjunctive rule based on B, then $\forall X \supset B$, $l(X) = sup(X) = u(X)$.

Corollary 2.3.3. Let $\emptyset \neq B \subseteq I$. If $sup(B) = l(B)$ or $sup(B) = u(B)$, then $\forall X \supset B$, $l(X) = sup(X) = u(X)$.

Proof: By Corollary 2.3.2 and Theorem 2.3.2. (For a different proof see [6-7].)

2.4 Representing Frequent Itemsets with Non-derivable Itemsets

The equivalent formulae for the bounds $l(B)$ and $u(B)$ for $sup(B)$, which we introduced in the Subsection 2.3, were derived in a different way earlier in [6-7]. The bounds were used there to define *non-derivable* and *derivable* itemsets as follows.

An itemset X is defined as *non-derivable* if $l(X) \neq u(X)$; otherwise, it is defined as *derivable*. Beneath we provide an important property of derivable and non-derivable itemsets, which follows immediately from Property 2.3.1b and Corollary 2.3.3.

Property 2.4.1. Let $X \subseteq I$.

a) If X is derivable, then $\forall Y \supset X$, Y is derivable.
b) If X is non-derivable, then $\forall Y \subset X$, Y is non-derivable.

Non-derivable itemsets were used in [6-7] to define the NDR representation of frequent itemsets. NDR was defined there as the set of all frequent non-derivable itemsets stored altogether with their supports:

$$\text{NDR} = \{(X, sup(X))|\ X \subseteq I,\ u(X) \neq l(X)\ \text{and}\ sup(X) > minSup\}.$$

Please, note that for each itemset in NDR, all its proper subsets are in NDR and for each maximal itemset in NDR, all its proper supersets are either derivable or infrequent, or both. The determination whether an itemset X from outside NDR is frequent or not may be carried out as follows: first $l(X)$ and $u(X)$ are calculated (potentially in a recursive way). If $u(X) \neq l(X)$, then X is non-derivable infrequent. Otherwise, it is frequent and its support can be determined as $sup(X) = u(X)$ (or alternatively, $sup(X) = l(X)$). Hence, NDR guarantees the derivation of the supports of all frequent itemsets and potentially the supports of some infrequent itemsets.

3 Patterns Admitting Negation

3.1 Sets Admitting Negated Items

In this section, we introduce the notions related to patterns with negated items and their properties based on [15]. Let $L = I \cup \{-a\ |a \in I\}$. Each element in L will be called a *literal*. Elements in $L \setminus I$ will be called *negative literals*. By analogy, items in I will be also called *positive literals*. Each pair of literals a and $-a$ in L is called *contradictory*. For the sake of convenience, we will apply the following notation: if l stands for a literal, then $-l$ will stand for its contradictory literal.

A *literal set* (or briefly *liset*) is defined as a set consisting of non-contradictory literals in L. A *liset* is called *positive* if all literals contained in it are positive. A *liset* is called *negative* if all literals contained in it are negative. *Lisets* X and Y are called *contradictory* if $|X| = |Y|$ and for each literal in X there is a contradictory literal in Y. A liset contradictory to X will be denoted by $-X$.

Support of liset X is denoted by $sup(X)$ and defined as the number (or percentage) of transactions in $\mathcal{D}$ that contain all positive literals in X and do not contain any negative literal from X.

Instead of an original database $\mathcal{D}$, it is sometimes convenient to consider an extended database $\mathcal{D}'$ in which each transaction T in $\mathcal{D}$ is extended with all negative literals contradictory to the items that do not occur in T.

Table 3 is such an extended version of the database from Table 1. Clearly, all transactions in the extended database will be of the same size equal to $|I|$.

Table 3. Extended version $\mathcal{D}'$ of database $\mathcal{D}$ from Table 1

Id	Transaction
T_1	$\{(\ a)(\ b)(\ c)(\ e)(-f)(-h)\}$
T_2	$\{(\ a)(\ b)(\ c)(\ e)(\ f)(-h)\}$
T_3	$\{(\ a)(\ b)(\ c)(\ e)(-f)(\ h)\}$
T_4	$\{(\ a)(\ b)(-c)(\ e)(-f)(-h)\}$
T_5	$\{(\ a)(-b)(\ c)(\ e)(-f)(\ h)\}$
T_6	$\{(-a)(\ b)(\ c)(\ e)(-f)(-h)\}$
T_7	$\{(-a)(-b)(-c)(-e)(-f)(\ h)\}$

Using the extended database, the support of liset X can be calculated as the number (or percentage) of transactions containing all literals (both positive and negative) in X. Though we do not recommend this method for evaluating lisets, this interpretation allows us to obtain immediately concepts, properties and theorems related to patterns with negation by analogy to those presented in Sections 1-2; simply words *item* and *itemset* should be replaced by *literal* and *liset*, respectively.

A *canonical variation of a liset X* (denoted by $cv(X)$) is defined as an itemset obtained from X by replacing all negative literals in X by contradictory literals; that is,

$$cv(X) = P \cup (-N),$$

where P is the set of all positive literals in X and N is the set of all negative literals in X. Clearly, if X is a positive liset, then $cv(X) = X$.

All lisets having the same canonical variation as liset X are denoted by $\mathcal{V}(X)$; that is,

$$\mathcal{V}(X) = \{ Y \subseteq L | cv(Y) = cv(X) \}.$$

Each liset in $\mathcal{V}(X)$ is called a *variation of X*.

$\mathcal{V}(X)$ contains only one positive liset, which is $cv(X)$, and only one negative liset, namely $-cv(X)$. Clearly, the number of all variations of X equals $2^{|X|}$ and the sum of the supports of all variations of X equals $|\mathcal{D}|$.

Property 3.1.1. Let X be a liset.
a) $\mathcal{V}(Z) = \mathcal{V}(X)$ for any $Z \in \mathcal{V}(X)$.
b) $|\mathcal{V}(X)| = 2^{|X|}$.
c) $\Sigma_{Z \in \mathcal{V}(X)} sup(Z) = |\mathcal{D}|$.

Property 3.1.2. A liset may have at most $\lceil |\mathcal{D}| / minSup \rceil - 1$ frequent variations.

3.2 Errors of Generalized Disjunctive Rules and Supports of Liset Variations

There is an interesting relationship between the errors of rules for a liset and the supports of its variations, which we formulate beneath.

Property 3.2.1. Let A, X and Z be lisets such that A, $X \subset Z$ and $Z \setminus X = A$ and $A \neq \emptyset$. The error of rule $X \rightarrow \lor A$, which is based on Z, equals the number of transactions in the extended database $\mathcal{D}'$ in which X occurs and no literal from the set A occurs; that is,

$$err(X \rightarrow \lor A) = sup(X \cup (-A)).$$

Example 3.2.1. By Property 3.2.1, $err(\{ab\} \rightarrow c \lor d) = sup(\{ab(-c)(-d)\})$, $err(\{ab\} \rightarrow (-c) \lor (-d)) = sup(\{abcd\})$, $err(\{(-a)b\} \rightarrow (-c) \lor d) = sup(\{(-a)bc(-d)\})$. $\square$

By Property 3.2.1, the antecedent (or alternatively, consequent) of a generalized disjunctive rule r based on a liset Z uniquely determines the variation V of Z, $V \neq Z$, the support of which equals the error of r.

Corollary 3.2.1 [29]. Let $A \cup X \subseteq I$. $sup(X \cup (-A)) = \Sigma_{Z \subseteq A} (-1)^{|Z|} \times sup(X \cup A)$

Proof: By Property 3.2.1 and Corollary 2.2.1. $\square$

Corollary 3.2.2 [29]. Let X be an itemset. The supports of all variations in $V(X)$ that differ from X are computable from the supports of only X and its proper subsets.

Proof: Follows from Corollary 3.2.1.   □

4 Properties of Derivable and Non-derivable Lisets

By definition, a liset X is derivable if $l(X) = u(X)$. In this section, we formulate and prove another necessary and sufficient condition of being a derivable liset. The new condition will be expressed in terms of the supports of variations of X. In addition, we investigate the relationship between (non-)derivability of a liset and (non-)derivability of its variations. Eventually, we derive a bound on the length of a non-derivable liset.

Lemma 4.1 [Appendix]. Let B be a liset. If two variations of B differ from each other on an odd number of literals, then one of the variations differs from B on an odd number of literals and the other variation differs from B on an even number of literals.

Lemma 4.2 [Appendix]. Let $B{\subseteq}L$. If one of the variations of B differs from B on an odd number of literals and another variation of B differs from B on an even number of literals, then the two variations differ from each other on an odd number of literals.

Theorem 4.1. Let $B{\subseteq}L$. There are at least two variations of B that have supports equal to 0 and differ from each other on an odd number of literals iff B is derivable.

Proof: ($\Rightarrow$) Let Y, Z be variations of B that have supports equal to 0 and differ from each other on an odd number of literals. Hence, by Lemma 4.1, either $|B{\setminus}Y|$ is odd and $|B{\setminus}Z|$ is even, or $|B{\setminus}Y|$ is even and $|B{\setminus}Z|$ is odd. Without any loss of generality, we will further assume that $|B{\setminus}Y|$ is odd and $|B{\setminus}Z|$ is even.

We note that $B{=}(B{\cap}Y){\cup}(B{\setminus}Y)$ and $Y{=}(B{\cap}Y){\cup}(-(B{\setminus}Y))$. Since $sup((B{\cap}Y){\cup}(-(B{\setminus}Y))) = sup(Y) = 0$ and $|B{\setminus}Y|$ is odd (*), and by this $B{\setminus}Y$ contains at least one literal from B, Property 3.2.1 allows us to derive: $B{\cap}Y{\rightarrow}V(B{\setminus}Y)$ is a generalized disjunctive rules based on B and $err(B{\cap}Y{\rightarrow}V(B{\setminus}Y)) = 0$ (**). Hence, by (*), (**), and Theorem 2.3.2a, $sup(B) = u(B)$.

Now, we will consider the variation Z of B. In this case $|B{\setminus}Z|$ is even. We will consider two cases: 1) $|B{\setminus}Z| > 0$, 2) $|B{\setminus}Z| = 0$.

Case 1. By Property 3.2.1 and Theorem 2.3.2b, $sup(B) = l(B)$ (the proof is analogous as in the case of variation Y).

Case 2. $B = Z$. Hence, $sup(B) = sup(Z) = 0$. Thus, by Theorem 2.3.2c, $sup(B) = l(B)$.

Thus, we have proved that $l(B) = sup(B) = u(B)$, so B is derivable.

($\Leftarrow$) Since B is derivable, $l(B) = sup(B) = u(B)$. Since $sup(B) = u(B)$, by Theorem 2.3.2a, there is a proper subset of B, say Y, such that $|Y|$ is odd and $err(B{\setminus}Y{\rightarrow}VY) = 0$. Hence, $|Y|$ is odd and, by Property 3.2.1, $sup((B{\setminus}Y){\cup}(-Y)) = 0$ and $(B{\setminus}Y){\cup}(-Y) \in V(B)$ (*). Since $l(B) = sup(B)$, by Theorem 2.3.2c, 1) $sup(B) = 0$ or 2) there is a proper subset of B, say Y', such that $Y'{\neq}\varnothing$, $|Y'|$ is even, and $err(B{\setminus}Y'{\rightarrow}VY') = 0$. In the former case, B is its own variation that has support equal to 0 (**). In the latter case, $|Y'|$ is even and, by Property 3.2.1, $sup((B{\setminus}Y'){\cup}(-Y')) = 0$ and $(B{\setminus}Y'){\cup}(-Y') \in V(B)$ (***).

Hence, by (*), (**), and (***), and Lemma 4.2, there are variations of B that have supports equal to 0 and differ from each other on an odd number of literals.   □

Theorem 4.2. Let B be a liset.
a) B is derivable iff all variations of B are derivable.
b) B is non-derivable iff all variations of B are non-derivable.

Proof: Ad a) ($\Leftarrow$) Trivial.

($\Rightarrow$) Let B be derivable and $X \in \mathcal{V}(B)$. Then, by Property 3.1.1a and Theorem 4.1, there are lisets in $\mathcal{V}(B) = \mathcal{V}(X)$ that have supports equal to 0 and differ from each other on an odd number of literals. Hence, by Theorem 4.1, X is derivable.

Ad b) ($\Leftarrow$) Trivial.

($\Rightarrow$) Let B be non-derivable and $X \in \mathcal{V}(B)$. Then by Property 3.1.1a and Theorem 4.1, there are not zero support lisets in $\mathcal{V}(X) = \mathcal{V}(B)$ that differ from each other on an odd number of literals. Hence, by Theorem 4.1, X is non-derivable.   □

Now, we will derive the bound on the length of non-derivable lisets. Let Z be a non-derivable liset. Then, by Theorem 4.1, at most one of its variations has support equal to 0. Thus, at least $2^{|Z|}$-1 variations of Z have supports greater than 0. Hence, $2^{|Z|}$-1 cannot exceed the number of transactions in $\mathcal{D}$. Thus, $2^{|Z|}$-1 $\leq |\mathcal{D}|$, so $|Z| \leq \lfloor \log_2(|\mathcal{D}|+1) \rfloor$.

Corollary 4.1. A non-derivable liset contains at most $\lfloor \log_2(|\mathcal{D}|+1) \rfloor$ literals.

5 Representing Frequent Positive and Negative Patterns

First we propose a *non-derivable liset representation of frequent patterns admitting negation* (NDLR) as an analog of NDR. We define NDLR as the family of all frequent non-derivable lisets stored altogether with their supports:

$$\text{NDLR} = \{(X, sup(X)) \mid X \subseteq L \wedge u(X) \neq l(X) \wedge sup(X) > minSup)\}.$$

Clearly, it is a lossless representation of all frequent lisets, and can be used in the same way as NDR for deriving frequent lisets and theirs supports.

Now, we offer another representation of frequent lisets called *non-derivable itemset representation of frequent patterns admitting negation* (NDIR). NDIR is defined as non-derivable itemsets stored altogether with their supports each of which has at least one frequent variation:

$$\text{NDIR} = \{(X, sup(X)) \mid X \subseteq I \wedge u(X) \neq l(X) \wedge (\exists Z \in \mathcal{V}(X)\ sup(Z) > minSup)\}.$$

Property 5.1. $\text{NDIR} = \{(cv(X), sup(cv(X))) \mid X \subseteq L \wedge u(X) \neq l(X) \wedge sup(X) > minSup\}$.

Proof: $\{(cv(X), sup(cv(X))) \mid X \subseteq L \wedge u(X) \neq l(X) \wedge sup(X) > minSup\} =$
/* by Theorem 4.2b, $u(X) \neq l(X)$ iff $u(cv(X)) \neq l(cv(X))$ */ $=$
$\{(cv(X), sup(cv(X))) \mid X \subseteq L \wedge u(cv(X)) \neq l(cv(X)) \wedge sup(X) > minSup)\} =$
$\{(Y, sup(Y)) \mid X \subseteq L \wedge Y \subseteq I \wedge Y = cv(X) \wedge u(Y) \neq l(Y) \wedge sup(X) > minSup\} =$
$\{(Y, sup(Y)) \mid Y \subseteq I \wedge u(Y) \neq l(Y) \wedge (\exists X \in \mathcal{V}(Y), sup(X) > minSup)\} = \text{NDIR}.$ □

By Property 5.1, for any liset in NDLR, NDIR contains its canonical variation (and the support of this variation).

Property 5.2.
a) If $X \in$ NDIR, then $\forall Y \subset X$, $Y \in$ NDIR.
b) If $X \notin$ NDIR, then $\forall Y \supset X$, $Y \notin$ NDIR.

Proof: Follows from properties of frequent lisets and derivable lisets.   ☐

Theorem 5.1. NDIR is a lossless representation of all frequent lisets.

Proof: The supports of all variations of each itemset in NDIR are determinable (according to Corollary 3.2.1) from the supports of only the itemset and its proper subsets, which also belong to NDIR (by Property 5.2a) (*). By Property 5.1, for any liset in NDLR, NDIR contains its canonical variation, and the support of this variation. Hence, and by (*), the supports of all lisets in NDLR are determinable from the supports of itemsets in NDIR. Thus, NDIR is a lossless representation of NDLR, and by this, NDIR is a lossless representation of all frequent lisets.   ☐

Property 5.3.
a) NDR $\subseteq$ NDIR $\subseteq$ NDLR.
b) If $X \in$ NDIR, then at most $\lceil |\mathcal{D}| / minSup \rceil - 1$ variations of X belong to NDLR.
c) The number of literals in elements of NDR, NDIR, and NDLR does not exceed $\lfloor \log_2(|\mathcal{D}|+1) \rfloor$.
d) If $minSup = 0$, then NDIR = NDR.
e) If $minSup = 0$ and $X \in$ NDIR, then $2^{|X|}$ variations belongs to NDLR.

Proof: Ad b) By Property 3.1.2.
    Ad c) By Corollary 4.1.   ☐

Clearly, NDR, NDRL, and NDIR are downward closed sets. In addition, in the case of each of these representations, the support of an element not belonging to the representation can be determined from the supports of only its proper subsets. It was shown in [18] that all downward closed representations that satisfy the above condition can be transformed to even more concise representations, which preserve the original derivation power, by replacing all their elements with their *closures*, where a *closure* of a set X is defined as the greatest superset of X that occurs in the same transactions as X. Since several distinct sets may have the same closure, the transformed representations cannot have more elements than the original representations.

Corollary 5.1. The sets of the closures of the elements of NDR, NDIR, and NDLR are not more numerous representations than NDR, NDIR, and NDLR, respectively and preserve the derivation power of the original representations.
    Lack of space disallows us to present experimental results related to the conciseness of the new representations in detail. So, we conclude them only briefly. The conciseness of NDIR and NDLR is of the same order as respective representations based on generalized disjunction-free sets and lisets, respectively (please, see [15] for the experimental results). While the discovery of all frequent patterns with negation is often impossible even for minimal support threshold close to

100%, all the four types of representations are derivable even for threshold values close to 0%.

6 Summary and Conclusions

In this paper, we derived a number of properties of derivable and non-derivable lisets. In particular, we have proved that a liset is derivable if at least two of its variations have supports equal to 0 and differ from each other on an odd number of literals. We have also proved that whenever a liset is non-derivable, then all its variations are also non-derivable, and whenever a liset is derivable, then all its variations are also derivable. We have shown that, for a given data set $\mathcal{D}$, the number of literals in a non-derivable literal set does not exceed $\lfloor \log_2(|\mathcal{D}|+1) \rfloor$.

We have introduced two lossless representations of frequent patterns admitting negation: 1) NDLR that consists of all frequent non-derivable lisets, and 2) NDIR that consists of all non-derivable itemsets having at least one frequent variation. We found that for any value of $minSup$: NDR$\subseteq$NDIR$\subseteq$NDLR. Also, we found that for $minSup = 0$, NDIR = NDR and for each itemset X in NDIR, there are $2^{|X|}$ lisets in NDLR.

Based on [18], we concluded that the elements of NDIR and NDLR can be replaced with their closures and the resultant representations will be still lossless representations of all frequent patterns admitting negation, not less concise than NDIR and NDLR, respectively.

References

1. Agrawal, R., Imielinski, T., Swami, A.: Mining Associations Rules between Sets of Items in Large Databases. In: ACM SIGMOD, Washington, USA, pp. 207–216 (1993)
2. Baralis, E., Chiusano, S., Garza, P.: On Support Thresholds in Associative Classification. In: SAC 2004, Taipei, Taiwan, pp. 553–558. ACM, New York (2004)
3. Bastide, Y., Taouil, R., Pasquier, N., Stumme, G., Lakhal, L.: Mining Frequent Patterns with Counting Inference. ACM SIGKDD Explorations 2(2), 66–75 (2000)
4. Boulicaut, J.-F., Bykowski, A., Rigotti, C.: Approximation of frequency queries by means of free-sets. In: Zighed, D.A., Komorowski, J., Żytkow, J.M. (eds.) PKDD 2000. LNCS, vol. 1910, pp. 75–85. Springer, Heidelberg (2000)
5. Bykowski, A., Rigotti, C.: A Condensed Representation to Find Frequent Patterns. In: PODS 2001. ACM SIGACT-SIGMOD-SIGART, USA, pp. 267–273 (2001)
6. Calders, T., Goethals, B.: Mining All Non-Derivable Frequent Itemsets. In: Elomaa, T., Mannila, H., Toivonen, H. (eds.) PKDD 2002. LNCS, vol. 2431, pp. 74–85. Springer, Heidelberg (2002)
7. Calders, T., Goethals, B.: Non-Derivable Itemset Mining. In: Data Mining and Knowledge Discovery, vol. 14, pp. 171–206. Kluwer Academic Publishers, Dordrecht (2007)
8. Casali, A., Cicchetti, R., Lakhal, L.: Essential Patterns: A Perfect Cover of Frequent Patterns. In: Tjoa, A.M., Trujillo, J. (eds.) DaWaK 2005. LNCS, vol. 3589, pp. 428–437. Springer, Heidelberg (2005)
9. Gasmi, G., Ben Yahia, S., Mephu Nguifo, E., Bouker, S.: Extraction of Association Rules Based on Literalsets. In: Song, I.-Y., Eder, J., Nguyen, T.M. (eds.) DaWaK 2007. LNCS, vol. 4654, pp. 293–302. Springer, Heidelberg (2007)

10. Harms, S.K., Deogun, J., Saquer, J., Tadesse, T.: Discovering Representative Episodal Association Rules from Event Sequences Using Frequent Closed Episode Sets and Event Constraints. In: ICDM, San Jose, USA, pp. 603–606. IEEE Computer Society Press, Los Alamitos (2001)
11. Kryszkiewicz, M.: Concise Representation of Frequent Patterns Based on Disjunction–Free Generators. In: ICDM 2001, San Jose, California, USA, pp. 305–312 (2001)
12. Kryszkiewicz, M.: Concise Representations of Frequent Patterns and Association Rules. Publishing House of Warsaw University of Technology, Warsaw (2002)
13. Kryszkiewicz, M.: Reducing Infrequent Borders of Downward Complete Representations of Frequent Patterns. In: The First Symposium on Databases, Data Warehousing and Knowledge Discovery, Baden-Baden, Germany, pp. 29–42 (2003)
14. Kryszkiewicz, M.: Reducing Borders of k-Disjunction Free Representations of Frequent Patterns. In: SAC 2004, Nikosia, Cyprus, pp. 559–563. ACM, New York (2004)
15. Kryszkiewicz, M.: Generalized Disjunction-Free Representation of Frequent Patterns with Negation. J. JETAI, 63–82 (2005)
16. Kryszkiewicz, M.: Reasoning about Frequent Patterns with Negation. In: Kryszkiewicz, M. (ed.) Encyclopedia of Data Warehousing and Mining, pp. 941–946. Information Sc. Publishing, Idea Group (2005)
17. Kryszkiewicz, M.: Using Generators for Discovering Certain and Generalized Decision Rules. In: Hybrid Intelligent Systems (HIS), pp. 181–186 (2005)
18. Kryszkiewicz, M.: Closures of Downward Closed Representations of Frequent Patterns. In: Corchado, E., Wu, X., Oja, E., Herrero, Á., Baruque, B. (eds.) HAiS 2009. LNCS (LNAI), vol. 5572, pp. 104–112. Springer, Heidelberg (2009)
19. Kryszkiewicz, M., Cichoń, K.: Support Oriented Discovery of Generalized Disjunction-Free Representation of Frequent Patterns with Negation. In: Ho, T.-B., Cheung, D., Liu, H. (eds.) PAKDD 2005. LNCS, vol. 3518, pp. 672–682. Springer, Heidelberg (2005)
20. Kryszkiewicz, M., Gajek, M.: Concise Representation of Frequent Patterns Based on Generalized Disjunction-Free Generators. In: Chen, M.-S., Yu, P.S., Liu, B. (eds.) PAKDD 2002. LNCS, vol. 2336, pp. 159–171. Springer, Heidelberg (2002)
21. Kryszkiewicz, M., Rybiński, H., Gajek, M.: Dataless Transitions between Concise Representations of Frequent Patterns. J. Int. Inf. Systems 22(1), 41–70 (2004)
22. Kryszkiewicz, M., Skonieczny, Ł.: Hierarchical Document Clustering Using Frequent Closed Sets. In: Advances in Soft Computing, pp. 489–498. Springer, Heidelberg (2006)
23. Mannila, H., Toivonen, H.: Multiple Uses of Frequent Sets and Condensed Representations. In: KDD 1996, Portland, USA, pp. 189–194 (1996)
24. Muhonen, J., Toivonen, H.: Closed Non-derivable Itemsets. In: Fürnkranz, J., Scheffer, T., Spiliopoulou, M. (eds.) PKDD 2006. LNCS, vol. 4213, pp. 601–608. Springer, Heidelberg (2006)
25. Pasquier, N., Bastide, Y., Taouil, R., Lakhal, L.: Efficient Mining of Association Rules Using Closed Itemset Lattices. J. Inf. Systems 24(1), 25–46 (1999)
26. Pei, J., Dong, G., Zou, W., Han, J.: On Computing Condensed Frequent Pattern Bases. In: ICDM, Maebashi, Japan, pp. 378–385. IEEE Computer Society Press, Los Alamitos (2002)
27. Rybiński, H., Kryszkiewicz, M., Protaziuk, G., Jakubowski, A., Delteil, A.: Discovering synonyms based on frequent termsets. In: Kryszkiewicz, M., Peters, J.F., Rybiński, H., Skowron, A. (eds.) RSEISP 2007. LNCS (LNAI), vol. 4585, pp. 516–525. Springer, Heidelberg (2007)
28. Savinov, A.: Mining Dependence Rules by Finding Largest Itemset Support Quota. In: SAC, Taipei, Taiwan, pp. 525–529. ACM, New York (2004)
29. Toivonen, H.: Discovery of Frequent Patterns in Large Data Collections. Ph.D. Thesis, Report A-1996-5, University of Helsinki (1996)

Appendix - Proofs

Proof of Theorem 2.3.1. Ad a) Let $|Y|$ be odd. Hence, by Corollary 2.2.3, $sup(B) = b(B\backslash Y \rightarrow \lor Y)$, and by Corollary 2.3.1, the following holds for each rule $B'\backslash Y' \rightarrow \lor Y'$, where $\varnothing \neq Y' \subseteq B$ and $|Y'|$ is odd: $sup(B) \geq b(B'\backslash Y' \rightarrow \lor Y')$. Thus, $sup(B) = b(B\backslash Y \rightarrow \lor Y) = min(\{b(B'\backslash Y' \rightarrow \lor Y')| \varnothing \neq Y' \subseteq B$ and $|Y'|$ is odd$\}) = u(B)$.

Ad b) Proof is analogous to the proof of Theorem 2.3.1a.   □

Proof of Theorem 2.3.2. Ad a) ($\Rightarrow$) Follows immediately from Theorem 2.3.1a.

($\Leftarrow$) By definition of $u(B)$, there is a rule, say $B\backslash Y \rightarrow \lor Y$, where $Y \subseteq B$ and $|Y|$ is odd, such that $u(B) = b(B\backslash Y \rightarrow \lor Y)$ (*). Since $sup(B) = u(B)$, we conclude further $sup(B) = b(B\backslash Y \rightarrow \lor Y)$. Hence, by Corollary 2.2.3, $err(B\backslash Y \rightarrow \lor Y) = 0$ (**). The theorem follows from (*) and (**).

Ad b) Proof is analogous to the proof of Theorem 2.3.2a.

Ad c) Follows from Theorem 2.3.2b and the definition of $l(B)$.   □

Proof of Corollary 2.3.2. Ad a) Let $X \supset B$. Since there is a certain generalized disjunctive rule based on B, then by Property 2.2.1, there is a certain generalized disjunctive rule based on X with an even number of items in the consequent and there is a certain generalized disjunctive rule based on X with an odd number of items in the consequent. Hence, by Theorem 2.3.2a-b, $l(X) = sup(X) = u(X)$.

Ad b) By Property 2.1.1a.

Ad c) By Corollary 2.3.2a-b.   □

Proof of Lemma 4.1. Let $Y, Z \in \mathcal{V}(B)$ and Y differs from Z on an odd number, say k, of literals. Let $C = Y \cap Z$ and $D = Y \backslash Z$. Then, $(-D) = Z \backslash Y$ and $k = |D| = |(-D)|$, $Y = C \cup D$, $Z = C \cup (-D)$. Hence, $|Y \backslash B| = |C \backslash B| + |D \backslash B|$ and $|Z \backslash B| = |C \backslash B| + |(-D) \backslash B|$. Let $m = |C \backslash B|$, $k_1 = |D \backslash B|$, and $k_2 = |(-D) \backslash B|$. Hence, $|Y \backslash B| = m + k_1$ and $|Z \backslash B| = m + k_2$. We note that $k_1 + k_2 = |D \backslash B| + |(-D) \backslash B| = k$, which is odd. Thus, if k_1 is odd, then k_2 is even, and if k_1 is even, then k_2 is odd. Hence, if $|Y \backslash B| = m + k_1$ is odd, then $|Z \backslash B| = m + k_2$ is even, and if $|Y \backslash B| = m + k_1$ is even, then $|Z \backslash B| = m + k_2$ is odd.   □

Proof of Lemma 4.2. Let $Y, Z \in \mathcal{V}(B)$ and Y differs from B on an odd number of literals (i.e. $|B \backslash Y|$ is odd) (*) and Z differs from B on an even number of literals (i.e. $|B \backslash Z|$ is even) (**). Since $Y, Z \in \mathcal{V}(B)$, then $Y, B \in \mathcal{V}(Z)$ (***). By (*), (***), and Lemma 4.1, either $|Y \backslash Z|$ is odd and $|B \backslash Z|$ is even, or $|Y \backslash Z|$ is even and $|B \backslash Z|$ is odd. Taking into account (**), we conclude $|Y \backslash Z|$ is odd. Hence, the variations Y and Z of liset B differ from each other on an odd number of literals.   □

Which Is Better for Frequent Pattern Mining: Approximate Counting or Sampling?

Willie Ng and Manoranjan Dash

Centre for Advanced Information Systems,
Nanyang Technological University, Singapore 639798
{ps7514253f,AsmDash}@ntu.edu.sg

Abstract. We investigate the problem of finding frequent patterns in a continuous stream of transactions. In the literature two prominent approaches are often used: (a) perform approximate counting (e.g., lossy counting algorithm (LCA) of Manku and Motwani, VLDB 2002) by using a lower support threshold than the one given by the user, or (b) maintain a running sample (e.g., reservoir sampling (Algo-Z) of Vitter, TOMS 1985) and generate frequent itemsets from the sample on demand. Both approaches have their advantages and disadvantages. For instance, LCA is known to output all frequent itemsets (recall = 1) but it also outputs many false frequent itemsets (low precision). Sampling is fast, but it outputs a large number of false itemsets as frequent itemsets, particularly when sample size is not large. Although both approaches are known to be practically useful, to the best of our knowledge there has been no comparison between the two approaches. In addition, we propose a novel sampling algorithm (*DSS*). *DSS* selects transactions to be included in the sample based on histogram of single itemsets. An empirical comparison study between the 3 algorithms is performed using synthetic and benchmark datasets. Results show that *DSS* is consistently more accurate than LCA and Algo-Z, whereas LCA performs consistently better than Algo-Z. Furthermore, *DSS*, although requires more time than Algo-Z, is faster than LCA.

1 Introduction

In this paper, we focus on frequent pattern mining (FPM) over streaming data. FPM is extremely popular particularly amongst researchers of data mining. On static data, many algorithms on FPM have been proposed. This research has led to further efforts in various directions [1]. But for streaming data the advancement in research has not been so spectacular. Although for several years many researchers have been proposing algorithms on FPM over streaming data (first prominent paper appeared in VLDB 2002 [2] [1]), even the recent papers on the topic [5] show that FPM in streaming data is not trivial. Manku and Motwani nicely described this problem in their seminal work - Lossy Counting Algorithm (LCA) [2]. Their work led to many other similar papers that use approximate counting [6,7,8].

At the heart of approximate counting is the fact that for stream data one cannot keep exact frequency count for all possible itemsets. Note that here we are concerned with

---

[1] Prior to this work, Misra and Gries [3] proposed the deterministic algorithm for ϵ-approximate frequency counts. The same algorithm has been rediscovered recently by Karp et al [4].

T.B. Pedersen, M.K. Mohania, and A M. Tjoa (Eds.): DaWaK 2009, LNCS 5691, pp. 151–162, 2009.

datasets that have 1000s of items or more, not just toy datasets with less than 10 items. The power set of the set of items cannot be maintained in today's memory. To solve this memory problem, LCA maintains only those itemsets which are frequent in at least a small portion of the stream, but if the itemset is found to be infrequent it is discontinued. As LCA does not maintain any information about the stream that has passed by, it adds an error frequency term to make the total frequency of a "potentially" frequent itemset higher than its actual frequency. Thus, LCA produces 100% recall but suffers from poor precision. These facts are described in [9].

Sampling is another approach that is used in [10,11,12,13] to produce frequent item-sets. The idea is very simple: maintain a sample over the stream and when asked, run the frequent pattern mining algorithm, such as the Apriori algorithm, to output the frequent itemsets. Research in sampling focuses on how to maintain a sample over the streaming data. On one hand approximate counting such as LCA does not keep any information about the stream that has passed by but keeps exact information starting from some point in the stream. On the other hand sampling keeps information about the whole stream (one can use a decaying factor to decrease or increase the influence of the past) but only partially. So, unlike approximate counting, sampling will have both false positives and false negatives. A researcher would love to know how these two compare against each other. This is the focus of this paper.

We have an additional contribution in this paper. Simple random sampling (SRS), or its counterpart reservoir sampling in streaming data, is known to suffer from a few limitations: First, an SRS sample may not adequately represent the entire data set due to random fluctuations in the sampling process. This difficulty is particularly apparent at small sample ratios which is the case for very large databases with limited memory. Second, SRS is blind towards noisy data objects, i.e., it treats both *bona fide* and noisy data objects similarly. The proportion of noise in the SRS sample and the original data set are almost equal. So, in the presence of noise performance of SRS degrades.

In this paper a new sampling algorithm called *DSS* (Distance based Sampling for Streaming data) is proposed for streaming data. The main contributions of this proposed method are: *DSS* elegantly addresses both issues (very large size of the data or in other words, very small sampling ratio, and noise) simultaneously. Experiments in FPM are conducted, and in all three of them the results show convincingly that *DSS* is far superior than *SRS* at the expense of a slight amount of processing time. Later in the conclusion the trade-off analysis between accuracy and time shows that it is worthwhile to invest in *DSS* than other approaches including *SRS* particularly when the domain is very large and noisy. Experiments are done mainly using synthetic datasets from IBM QUEST project and benchmark dataset, Kosarak.

2 Preliminaries

The problem of mining frequent patterns online can be formally stated as follows: Let D denote a transactional data stream, which is a sequence of continuously arriving transactions, e.g., $t_1, t_2, ..., t_N$. We denote N as the number of transactions processed so far. Let $\mathcal{I} = \{i_1, i_2, \ldots, i_n\}$ be a set of distinct literals, called items. Each transaction has a unique identifier (*tid*) and contains a set of items. An itemset X is a set of items such that $X \in (2^{|\mathcal{I}|} - \{\emptyset\})$ where $2^{|\mathcal{I}|}$ is the power set of $\mathcal{I}$. Next, the frequency of an itemset X, denoted by $freq(X)$, is the number of transactions in D that contain X. The support of an itemset X, denoted by $\sigma(X)$, is the ratio of the frequency of X to the number of transactions processed so far, i.e., $\sigma(X) = freq(X)/N$. Given a pre-defined support threshold σ_{min}, an itemset X is considered a frequent itemset (FI) if its frequency, $freq(X)$, is more than or equal to $\sigma_{min} \times N$.

2.1 Approximate Counting

For online mining, one typically cannot obtain the exact frequencies of all itemsets, but has to make an estimation. In general, approximate solutions in most cases may already be satisfactory to the need of users. Indeed, when faced with an infinite data set to analyze, many existing works explicitly trade off accuracy for speed where the quality of the final approximate counts are governed by an error parameter, ϵ [2,6,8]. Unfortunately, we note that defining a proper value of ϵ is non-trivial [7]. Usually, we end up facing a dilemma. That is, by setting a small error bound, we achieve good accuracy but suffer in terms of efficiency. On the contrary, a bigger error bound improves the efficiency but seriously degrades the mining accuracy [9]. Note that ϵ is actually a minimum support threshold used to control the quality of the approximation of the mining result ($\epsilon \leq \sigma_{min}$). We refer the reader to the super exponential growth mentioned in [14]. The paper reported that even a slight decrease in the support threshold might create a deep impact on the performance of most mining algorithms. In this paper, LCA is chosen for discussion due to its popularity. In fact, several recent algorithms also adopted error bound approach of LCA [6]. We refer the reader to [2] for more detailed discussion on the technical aspects of LCA.

2.2 Reservoir Sampling

The issue of how to maintain a sample of a specified size over data that arrives online has been studied in the past. The standard solution is to use reservoir sampling proposed by J. S. Vitter [15]. The technique of reservoir sampling is, in one sequential pass, to select a random sample of n transactions from a dataset of N transactions where N is unknown and $n \ll N$. In [15], algorithm Z (Algo-Z) is introduced. If the time for scanning the dataset is ignored, Algo-Z has expected CPU time $O(n(1+log(\frac{N}{n})))$, which is optimum up to a constant factor. However, there are limitations of reservoir sampling. They are: (a) a reservoir sample may not adequately represent the entire dataset due to random fluctuations in the sampling process. This difficulty is particularly apparent at small sample ratios which is the case for data streams with limited memory; (b) reservoir sampling is blind towards noisy data objects, i.e., it treats both *bona fide* and noisy data objects similarly; and (c) Vitter's reservoir sampling cannot handle deletions.

3 Distance Based Sampling

We propose a distance based sampling that is designed to work "count" dataset, that is, dataset in which there is a base set of "items" and each data element (or transaction) is a vector of items. Usually, for distance based sampling, the strategy is to produce a sample whose "distance" from the complete database is minimal. The main challenge is to find an appropriate distance function that can accurately capture the difference between the sample (S) and all the transactions seen so far (D) in the stream.

For distance based sampling, the basic intuition is that if the distance between the relative frequency of the items in S and the corresponding relative frequency in D is small, then S is a good representative of D. Ideally, one needs to compare the frequency histograms of all possible itemsets: 1-itemsets, 2-itemsets, 3-itemsets, and so on. But experiments suggest that often it is sufficient to study the 1-itemset histograms [16,17,18].

We define a few distance functions which are all based on frequencies of each item. Here, the relative frequency of item $\mathcal{A}$ in S and D is given by $Sup(\mathcal{A}; S) = \frac{freq(\mathcal{A};S)}{|S|}$ and $Sup(\mathcal{A}; D) = \frac{freq(\mathcal{A};D)}{|D|}$, respectively. $freq(\mathcal{A}; U)$ is the frequency of $\mathcal{A}$ in a set of transactions U. In this paper, we use $Dist_2$ (also known as Euclidean distance) in our discussion–

$$Dist_2(S, D) = \sum_{\mathcal{A} \in \mathcal{I}} (Sup(\mathcal{A}; S) - Sup(\mathcal{A}; D))^2. \tag{1}$$

3.1 Distance Based Sampling for Streaming Data

Like any other reservoir sampling method designed for data stream, the initial step of *DSS* is to insert the first n transactions into a "reservoir". The rest of the transactions are processed sequentially; transactions can be selected for the reservoir only as they are processed. Because n is fixed, whenever there is an insertion of transaction, there is sure to be a deletion. In *DSS*, a local histogram $(Hist_{\mathcal{L}})$ and a global histogram $(Hist_{\mathcal{G}})$ are employed to keep track of the frequency of items generated by S and D respectively. Ideally, for a sample to be a good representation of the entire data, the discrepancy between $Hist_{\mathcal{L}}$ and $Hist_{\mathcal{G}}$ should be small. In other words, both $Hist_{\mathcal{L}}$ and $Hist_{\mathcal{G}}$ should look similar. Any insertion or deletion of transaction on the sample will affect the shape of $Hist_{\mathcal{L}}$.

To maintain the sample, *DSS* prevents an incoming transaction from entering S if its existence in S increases the discrepancy. In addition, *DSS* helps to improve the quality of the sample by deleting transaction whose elimination from S maximally reduces (or minimally increases) the discrepancy. Therefore, there is a ranking mechanism in *DSS* to rank the transactions in S so that the "weakest" transaction can be replaced by the incoming transaction if the incoming transaction is better. The transactions in the initial sample are ranked by "leave-it-out" principle, i.e., distance is calculated by leaving out the transaction. Higher ranks are assigned to the transaction removal of which leads to higher distance. Mathematically, the distance is calculated as follows for ranking:

$$Dist_t = Dist_2 (S - \{t\}, D) \tag{2}$$

where $t \in S$. For the initial ranking, both S and D contain the first n transactions. Table 1 shows the conceptual idea of ranking. *Tid* 10 is ranked highest because its removal produces the maximum distance of 25.0. Similarly, *Tid* 3 is ranked lowest because its removal produces the minimum distance of 2.0. Let *LRT* denotes the lowest ranked transaction.

Table 1. Conceptual Ranking of Transactions

Rank	Distance	Tid
1st	25.00	10
2nd	15.00	20
$\vdots$	$\vdots$	$\vdots$
n^{th}	2.00	3

When a new transaction t_{new} arrives, $Hist_{\mathcal{G}}$ is immediately updated. A decision is made whether to keep it in the sample by comparing the distances computed when $Hist_{\mathcal{L}}$ is 'with' and 'without' t_{new}. First, we use Eq. 3 to calculate the distance between $Hist_{\mathcal{G}}$ and $Hist_{\mathcal{L}}$ when t_{new} is absent. Next, LRT is temporarily removed from the sample. We use Eq. 4 to calculate the distance between $Hist_{\mathcal{G}}$ and $Hist_{\mathcal{L}}$ when t_{new} is present. Note that because of the incoming t_{new}, the D in Eq. 3 and 4 is updated ($D = D + t_{new}$). If $Dist_{without\_t_{new}} > Dist_{with\_t_{new}}$, then t_{new} is selected to replace LRT in the current sample. LRT is permanently removed. t_{new} will be ranked in the sample using $Dist_{without\_t}$ value. On the other hand, if $Dist_{without\_t_{new}} \leq Dist_{with\_t_{new}}$, then t_{new} is rejected and LRT is retained in the reservoir.

$$Dist_{without\_t_{new}} = Dist_2(S, D) \tag{3}$$

$$Dist_{with\_t_{new}} = Dist_2((S - LRT + t_{new}), D) \tag{4}$$

Ideally, all transactions in the current sample should be re-ranked after processing each incoming transaction because $Hist_{\mathcal{G}}$ is modified. Re-ranking is done by recalculating the distances using 'leave-it-out' principle for all transactions in the current sample and those which are rejected already. But this can be computationally expensive because of the nature of data stream. We need a trade-off between accuracy and speed. Thus, re-ranking is done after selecting $\mathcal{R}$ new transactions in the sample. We do not consider the rejected transactions while counting $\mathcal{R}$. For our later experiments, we set $\mathcal{R}$ equal to 10. *DSS* tries to ensure that the relative frequency of every item in the sample is as close as possible to that of the original data stream. In the implementation, *DSS* stores the sample in an array of pointers to structures which holds the transaction and its distance value. Initial ranking and re-ranking of the sample according to the distances involve two steps. The first step is to calculate the distances of the transactions in the sample and the second step is to sort the sample by increasing distances using the standard quick sort technique. When $Dist_2$ is well implemented, the distance based sampling can have computational cost at most $O(|t_{max}|)$, where $|t_{max}|$ is the maximal length of the transaction vector –see next section. We summarize *DSS* as follows:

1. Insert the first n transactions into S.
2. Initialize $Hist_{\mathcal{G}}$ and $Hist_{\mathcal{L}}$ to keep track of the number of transactions containing each item $\mathcal{A}$ in S and D.
3. Rank the initial S by 'leave-it-out' method using Eq. 2.
4. Read the next incoming t_{new}.
5. Include t_{new} into $Hist_{\mathcal{G}}$.
6. Compare the distances of S 'without' and 'with' t_{new} using Eq. 3 and 4.
   a. If $Dist_{without\_t_{new}} > Dist_{with\_t_{new}}$, then replace LRT with t_{new} and update $Hist_{\mathcal{L}}$.
   b. Else, reject t_{new}.

7. If $\mathcal{R}$ new transactions are already selected, re-rank the transactions in the current sample. Eg. $\mathcal{R} = 10$.
8. Repeat steps 4 to 7 for every subsequent incoming transaction t_{new}.

3.2 Complexity Analysis

When finding LRT, we need to compute $Dist_2$ for each $t \in S$. This is the distance between $Hist_G$ and $Hist_C$ when a transaction t is removed from S. Here, S_t denotes $S - \{t\}$. We let $\mathcal{F}_{AS}$, $\mathcal{F}_{AS_t}$ and $\mathcal{F}_{AD}$ represent the absolute frequency of item $\mathcal{A}$ in S, S_t and D respectively. Note that,

$$\mathcal{F}_{AS_t} = \begin{cases} \mathcal{F}_{AS} - 1 & \text{if } \mathcal{A} \in t \\ \mathcal{F}_{AS} & \text{else } \mathcal{A} \notin t. \end{cases}$$

To search for the weakest transaction in S, the set of n distances that are generated from Eq 2 has to be compared with one another such that

$$t^* = \underset{1 \leq t \leq n}{\operatorname{argmin}} \ Dist_2(S - t, D) = \underset{1 \leq t \leq n}{\operatorname{argmin}} \sum_{\mathcal{A} \in \mathcal{I}} \left(\frac{\mathcal{F}_{AS_t}}{|S_t|} - \frac{\mathcal{F}_{AD}}{|D|} \right)^2. \tag{5}$$

Unfortunately, the determination of t^* can be computationally costly. The worst case time complexity is $O(n.|\mathcal{I}|)$, where $|\mathcal{I}| \gg 1$. However, we note that even though removing a transaction from S will affect relative frequencies of all items, most absolute frequencies will remain unchanged. Only those items contained within the transaction t are affected. In addition, we can make use of the fact that, in general,

$$\underset{x \in U}{\operatorname{argmin}} \ f(x) = \underset{x \in U}{\operatorname{argmin}} \ cf(x) + d. \tag{6}$$

For any constant c and real number d that we introduced, the final outcome will still remain the same. With this understanding, we can rewrite Eq. 5 as

$$t^* = \underset{1 \leq t \leq n}{\operatorname{argmin}} \sum_{\mathcal{A} \in \mathcal{I}} \left(\left(\mathcal{F}_{AS_t} - \frac{\mathcal{F}_{AD}}{|D|}|S_t| \right)^2 - \left(\mathcal{F}_{AS} - \frac{\mathcal{F}_{AD}}{|D|}|S_t| \right)^2 \right)$$

$$t^* = \underset{1 \leq t \leq n}{\operatorname{argmin}} \sum_{\mathcal{A} \in t} \left(1 - 2\mathcal{F}_{AS} + 2\frac{\mathcal{F}_{AD}}{|D|}|S_t| \right)$$

From the above representation of t^*, it is possible to reduce the worst-case cost of ranking from $O(n.|\mathcal{I}|)$ to $O(n.|T_{max}|)$. Similarly, for comparing between LRT and t_{new}, we can apply the same strategy. The cost will then be $O(|T_{max}|)$.

3.3 Handling Noise in DSS

Noise here means a random error or variance in a measured variable. The data object that holds such noise is a noisy data object. Such noisy transaction have very little or no similarity with other transactions. This "aloof" nature of noise is used to detect and remove them. These noisy transactions can also be called outliers. Such outlier detection and removal has been studied abundantly, for example in [19]. These methods typically use some kind of similarity measurement and a similarity threshold to determine whether an object is noise, that is if a data object has lesser similarity than the threshold, it is considered noise.

In this paper the focus is to maintain a small sample of transactions over a streaming data. *DSS* not only maintains a good representative sample, it also removes noise by not selecting them in the sample. *DSS* banks on the *aloofness* of the transactions to determine whether an incoming transaction is corrupted by noise. Removal of a noisy transaction will induce smaller distance than removal of a *bona fide* transaction. As *DSS* selects transactions into the reservoir sample by the distance it induces, it naturally rejects the noisy transactions. Later experimental results show that *DSS* has a better ability to handle noise than LCA and Algo-Z.

4   Performance Study

This section describes the experimental comparison between *DSS* and LCA in the context of FPM. In addition, we also compared *DSS* with simple random sampling (SRS) using Algo-Z. All experiments were performed on a 1.7GHz CPU Dell PC with 1GB of main memory and running on the Windows XP platform. We used the code from the IBM QUEST project [21] to generate the datasets. We set $\mathcal{I} = 10k$ unique items. The two datasets that we generated are $T10I3D2000K$ and $T15I7D2000K$. Note that $T15I7D2000K$ is a much denser dataset than $T10I3D2000K$ and therefore requires more time to process. In addition, to verify the performance of all the 3 algorithms on real world dataset, we used a coded log of a clickstream data (denoted by Kosarak) from a Hungarian on-line news portal [22]. This database contains 990002 transactions with average size 8.1. We fixed the minimum support threshold $\sigma_{min} = 0.1\%$ and the batch size $B_{size} = 200K$. For comparison with LCA, we set $\epsilon = 0.1\sigma_{min}$ which is a popular choice for LCA.

4.1   Performance Measure

To measure the quality of the mining results, we use two metrics: the recall and the precision. Given a set AFI_{true} of all true frequent itemsets and a set AFI_{appro} of all approximate frequent itemsets obtained in the output by some mining algorithms, the recall is $\frac{|AFI_{true} \cap AFI_{appro}|}{|AFI_{true}|}$ and the precision is $\frac{|AFI_{true} \cap AFI_{appro}|}{|AFI_{appro}|}$. If the recall equals 1, the results returned by the algorithm contains all true results. This means no false negative. If the precision equals 1, all the results returned by the algorithm are some or all of the true results, i.e., no false positive is generated. To evaluate the quality of the sample, we can apply the symmetric difference $SD = \frac{|(AFI_{true} - AFI_{appro}) \cup (AFI_{appro} - AFI_{true})|}{|AFI_{true}| + |AFI_{appro}|}$. Alternately, we define the overall accuracy with $Acc = 1 - SD$.

4.2   Time and Accuracy Measurements

Figure 1 depicts the number of frequent itemsets uncovered during the operation of the 3 algorithms for the 3 different datasets. Note that for *DSS* and Algo-Z, the size of the reservoir is 5k transactions. In the figure, ORG indicates the true size of frequent itemsets generated when the Apriori algorithm operates on the entire dataset. Interestingly, we can observe that the general trends of the sampled datasets resemble the true result.

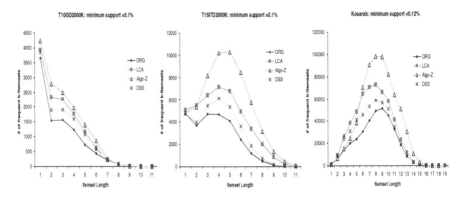

Fig. 1. Itemset size vs Number of Frequent Itemsets

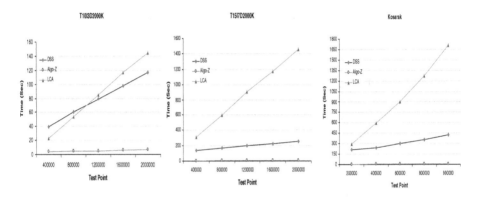

Fig. 2. Execution time on $T10I3D2000K$, $T15I7D2000K$ and Kosarak

This is also similar with the output from LCA. However, Algo-Z tends to drift very far away from ORG. LCA is caught in between the two sampling algorithms. *DSS* gives the most appealing results in term of fidelity. The graphs from *DSS* are the closest to ORG. Figure 2 shows the execution time on the 3 datasets. The results of the algorithms are computed as an average of 20 runs. Each run corresponds to a different shuffle of the input dataset. Note that for *DSS* and Algo-Z, the execution time consists of the time spent for obtaining the sample as well as using the Apriori algorithm to generate the frequent patterns (AFI_{appro}) from the sample. As we can see, the cumulative execution time of *DSS*, Algo-Z and LCA grows linearly with the number of transactions processed in the streams. In particular, Algo-Z is the fastest because it only needs to maintain a sample by randomly selecting transaction to be deleted or inserted in the reservoir. Unlike Algo-Z, *DSS* processes every incoming transaction by computing its distance in the sample. As a result, its processing speed is slower than Algo-Z. However, the slowest algorithm is LCA. In all 3 datasets, LCA spent the most amount of time. Figure 3 displays the *Acc* of the 3 algorithms against the number of transactions processed. As expected, Algo-Z achieved the worst performance. As the number of transactions to be processed increases, we see that its accuracy drops significantly. Since the reservoir is a

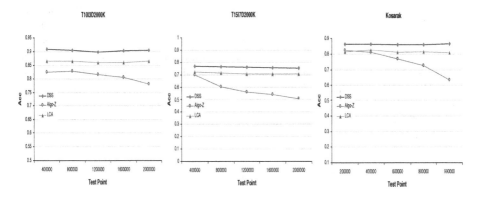

Fig. 3. Accuracy on $T10I3D2000K$, $T15I7D2000K$ and Kosarak

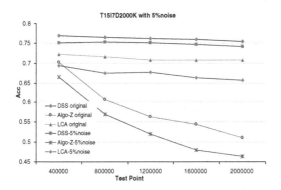

Fig. 4. Accuracy on Noisy Data Set

fixed size, the sampling ratio decrease at every test point. *DSS* achieved good accuracy even for small sampling ratio. Its accuracy remains stable for all the test points. Although LCA guarantees to produce 100% recall, its performance was heavily affected by its poor precision. This can be clearly seen in the figure.

4.3 Handling Noise

Removing transactions that are corrupted with noise is an important goal of data cleaning as noise hinders most types of data analysis. This section shows how *DSS* is not only able to produce high quality sample from normal dataset, but it is also able to cope with dataset having corrupted transactions and thus preventing these transactions from being inserted into the reservoir. To demonstrate the robustness of *DSS* against noise, we let the 3 algorithms to operate on a noisy dataset. For this experiment, we added 5% of noise to $T15I7D2000K$. Noise was added using the $rand$ function. With a probability of 5%, we corrupt an item in a transaction by replacing it with any item in $\mathcal{I}$. Similar to the previous experiments, we made use of the true frequent patterns uncovered from the original dataset to compare with the approximate frequent patterns uncovered from the corrupted dataset. Figure 4 illustrates the performances of the 3 algorithms against noise. For reference, the results for the algorithms operating on the noise free dataset are also included in the plot. From the graph, *DSS* suffers the least in term of accuracy when noise was added to the data. Its overall accuracy is maintained at about 75% and its maximun drop in performance is at most 2% when the test point is at 400k. However, for LCA and Algo-Z, the gap between the original result without noise and the result with noise is wider when compared with the one by *DSS*. The greatest drop in accuracy is by 5% for LCA when the test point is at 2000k and by 6% for Algo-Z when the test point is at 1600k.

4.4   Comparison with Theoretical Bounds

In [12], it was shown how to apply Chernoff bounds for FPM. Denote by $\mathbf{X}$ as the number of transactions in the sample containing the itemset I. Random variable $\mathbf{X}$ has a binomial distribution of n trials with the probability of success σ_{min}. For any positive constant, $0 \leq \varepsilon \leq 1$, the Chernoff bounds state that

$$P(\mathbf{X} \leq (1 - \varepsilon)n\sigma_{min}) \leq e^{-\varepsilon^2 n\sigma_{min}/2} \tag{7}$$

$$P(\mathbf{X} \geq (1 + \varepsilon)n\sigma_{min}) \leq e^{-\varepsilon^2 n\sigma_{min}/3} \tag{8}$$

Chernoff bounds provide information on how close is the actual occurrence of an itemset in the sample, as compared to the expected count in the sample. *Accuracy* is given as $1 - \varepsilon$. The bounds also tell us the probability that a sample of size n will have a given accuracy. We call this aspect *confidence* of the sample (defined as 1 minus the expression on the right hand side of the equations). The first equation gives the lower bound – the probability that the itemset occurs less often than expected and the second one gives the upper bound – the probability that the itemset occurs more often than expected, for a desired accuracy.

The following plots in Figure 5 show the results of comparing theoretical Chernoff bound with experimentally observed results. We show that for the databases we have considered the Chernoff bound is very conservative. We can obtain the theoretical confidence value by evaluating the right hand side of the equations. For example, for the upper bound the confidence $C = 1 - e^{-\varepsilon^2 n\sigma_{min}/3}$. We can obtain experimental confidence values as follows. We take s samples of size n, and for each item we compute the confidence by evaluating the left hand side of the two equations as follows. Let i denote the sample number, $1 \leq i \leq s$. Let $l_I(i) = 1$ if $(n\sigma_{min} - \mathbf{X}) \geq n\sigma_{min}\varepsilon$ in sample i, otherwise 0. Let $h_I(i) = 1$ if $(\mathbf{X} - n\sigma_{min}) \geq n\sigma_{min}\varepsilon$ in sample i, otherwise 0. The confidence can then be calculated as $1 - \sum_{i=1}^{m} h_I(i)/s$, for the upper bound. We take $s = 100$ samples for both Algo-Z and *DSS* for each of the three datasets. We cover our discussion on all single itemsets. Using the theoretical and experimental approaches we determine the probabilities (1-confidence) and plot them in the following figures. Figure 5 compares the distribution of experimental confidence of simple random sampling and *DSS* to the one obtained by Chernoff upper bounds. The graphs show the results using $T15I7D2000K, n = 2000$ with $\varepsilon = 0.01\%$. From the figure, Chernoff bounds, with a mean probability of 99.95%, suggests that this sample size is 'likely' unable to achieve

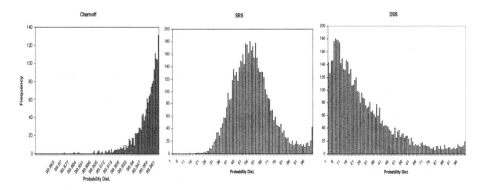

Fig. 5. Probability Distribution

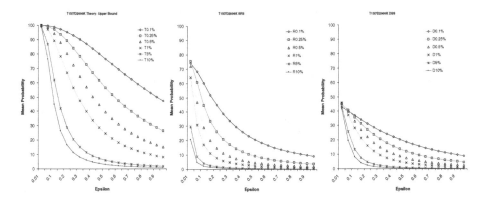

Fig. 6. $T15I7D2000K$: Epsilon vs. mean Probability

the given accuracy. Obviously, this is very pessimistic and over conservative. In actual case, Algo-Z and *DSS* gave a mean probability of 75% and 43% respectively.

Figures 6, provides a broader picture of the large discrepancy between Chernoff bounds and experimental results. Using $T15I7D2000K$, we plot the mean of the probability distribution for different Epsilon (ε). Different values of sample size are used (from 0.1% to 10%). The higher the probability, the more conservative the approach is. So we can see that *DSS* samples are the most reliable, followed by Algo-Z sample, and the theoretical bounds are the most conservative.

5   Conclusions

In this paper we compared approximate counting (lossy counting LCA) with random sampling (reservoir sampling Algo-Z) over data streams. Our results show that LCA is more accurate than Algo-Z. But LCA requires more time than Z. We also proposed a novel histogram based sampling algorithm *DSS*. *DSS* outperformed the other two significantly in accuracy. It required more time than Algo-Z but less time than LCA.

References

1. Goethals, B.: Survey on frequent pattern mining (manuscript) (2003)
2. Manku, G., Motwani, R.: Approximate frequency counts over data streams. In: VLDB, pp. 346–357 (2002)
3. Misra, J., Gries, D.: Finding repeated elements. Scientific Computing Programming 2(2), 143–152 (1982)
4. Karp, R.M., Shenker, S., Papadimitriou, C.H.: A simple algorithm for finding frequent elements in streams and bags. ACM Transactions on Database Systems 28(1), 51–55 (2003)
5. Calders, T., Dexters, N., Goethals, B.: Mining frequent itemsets in a stream. In: Perner, P. (ed.) ICDM 2007. LNCS, vol. 4597, pp. 83–92. Springer, Heidelberg (2007)
6. Cheng, J., Ke, Y., Ng, W.: A survey on algorithms for mining frequent itemsets over data streams. An International Journal of Knowledge and Information Systems (2007)
7. Cheng, J., Ke, Y., Ng, W.: Maintaining frequent itemsets over high-speed data streams. In: Proceedings of the 10th Pacific-Asia Conference on Knowledge Discovery and Data Mining, pp. 462–467 (2006)
8. Giannella, C., Han, J., Pei, J., Yan, X., Yu, P.: Mining frequent patterns in data streams at multiple time granularities. In: Kargupta, H., Joshi, A., Sivakumar, K., Yesha, Y. (eds.) Next Generation Data Mining, pp. 191–212. AAAI/MIT (2003)
9. Ng, W., Dash, M.: Efficient approximate mining of frequent patterns over transactional data streams. In: Song, I.-Y., Eder, J., Nguyen, T.M. (eds.) DaWaK 2008. LNCS, vol. 5182, pp. 241–250. Springer, Heidelberg (2008)
10. Toivonen, H.: Sampling large databases for association rules. In: VLDB 1996: Proceedings of the 22th International Conference on Very Large Data Bases, pp. 134–145 (1996)
11. Mannila, H., Toivonen, H., Verkamo, A.I.: Efficient algorithms for discovering association rules. In: Fayyad, U.M., Uthurusamy, R. (eds.) AAAI Workshop on Knowledge Discovery in Databases (KDD 1994), pp. 181–192 (1994)
12. Zaki, M., Parthasarathy, S., Li, W., Ogihara, M.: Evaluation of sampling for data mining of association rules. In: Seventh International Workshop on Research Issues in Data Engineering, RIDE 1997 (1996)
13. Yu, X., Chong, Z., Lu, H., Zhou, A.: False positive or false negative: Mining frequent itemsets from high speed transactional data streams. In: Int. Conf. on VLDB (2004)
14. Kohavi, Z.Z.R.: Real world performance of association rule algorithms. In: ACM SIGKDD (2001)
15. Vitter, J.: Random sampling with a reservoir. ACM Transactions on Mathematical Software 11, 37–57 (1985)
16. Chen, B., Haas, P.J., Scheuermann, P.: A new two-phase sampling based algorithm for discovering association rules. In: KDD, pp. 462–468 (2002)
17. Bronnimann, H., Chen, B., Dash, M., Haas, P., Scheuermann, P.: Efficient data reduction with ease. In: Proceedings of ACM SIGKDD International Conference in Knowledge Discovery and Data Mining, pp. 59–68 (2003)
18. Chuang, K.-T., Chen, M.-S., Yang, W.-C.: Progressive sampling for association rules based on sampling error estimation. In: Ho, T.-B., Cheung, D., Liu, H. (eds.) PAKDD 2005. LNCS, vol. 3518, pp. 505–515. Springer, Heidelberg (2005)
19. Kubica, J., Moore, A.: Probabilistic noise identification and data cleaning. In: Proceedings of International Conference on Data Mining, ICDM (2003)
20. Zhu, X., Wu, X., Khoshgoftaar, T.M., Shi, Y.: Empirical study of the noise impact on cost-sensitive learning. In: Proceedings of International Conference on Joint COnference on Artificial Intelligence (IJCAI) (2007)
21. Agrawal, R., Srikant, R.: Fast algorithms for mining association rules. In: Proc.of the 20th VLDB conf. (1994)
22. Bodon, F.: A fast apriori implementation. In: Proceedings of the IEEE ICDM Workshop on Frequent Itemset Mining Implementations, FIMI 2003 (2003)

A Fast Feature-Based Method to Detect Unusual Patterns in Multidimensional Datasets

Minh Quoc Nguyen, Edward Omiecinski, and Leo Mark

College of Computing,
Georgia Institute of Technology,
Atlanta, GA 30332, USA
{quocminh,edwardo,leomark}@cc.gatech.edu

Abstract. We introduce a feature-based method to detect unusual patterns. The property of normality allows us to devise a framework to quickly prune the normal observations. Observations that can not be combined into any significant pattern are considered unusual. Rules that are learned from the dataset are used to construct the patterns for which we compute a score function to measure the interestingness of the unusual patterns. Experiments using the KDD Cup 99 dataset show that our approach can discover most of the attack patterns. Those attacks are in the top set of unusual patterns and have a higher score than the patterns of normal connections. The experiments also show that the algorithm can run very fast.

1 Introduction

Outlier detection is an interesting topic in data mining. Historically, outliers were considered noise that would adversely affect the quality of the data analysis process and would need to be removed from the dataset. However, the deviation of outliers from other data may indicate interesting or fraudulent activities that require our attention instead of simply discarding them [8]. In practice, an outlier detection method can be used as an unsupervised technique for fraudulent activity detection, network instrusion detection and system monitoring. Outlier detection methods can be divided into two categories: statistical-based [8] and distance-based [5]. The statistical based approach can discover outliers by computing the probability of the observation from the underlying distributions of the dataset. However, those distributions are usually unknown in many data mining applications. In constrast, the distance-based approach [5] can detect the outliers without knowing the underlying distributions. Breunig et al [2] introduced a local outlier factor (LOF) that can detect density-based outliers which can not be detected by the former method. The main advantage of these methods is that it can detect local outliers without knowing the distribution of the dataset. However, their methods only focus on detecting individual outliers.

In this paper, we introduce a method to detect unusual patterns very fast. From our perspective, the outliers may simply be noise due to data sparsity

T.B. Pedersen, M.K. Mohania, and A M. Tjoa (Eds.): DaWaK 2009, LNCS 5691, pp. 163–176, 2009.

in high dimensions [1,17]. Thus, it may be more interesting to detect patterns of unusual observations instead of individual outliers. In the following section, we will show that the unsual patterns in our definition possess such unique characteristics that they can not be detected by using clustering algorithm or applying clustering algorithm on the detected outliers. We will call this feature-based unsual patterns.

2   Motivation

Table 1a consists of four items with three attributes A_1, A_2 and A_3. In table 1b, each column contains the statistics for the items in the column. The rows in the table show the k-distance [2] with k = 3, the average distance of all KNNs (3 nearest neighbors) and the LOF factors of the items. We assume that the set of four items is a subset of a larger dataset with the ranges of the attributes from 0 to 1000. Thus, the normalization is unneccessary. According to table 1b, X_1 and X_2 are highly ranked outliers due to their LOF score and X_4 is the lowest ranked outlier in the group. X_1 and X_2 are more unusual than X_4. A closer look at table 1a shows that this is not necessarily true. According to the table, the values for attributes A_2 and A_3 are actually almost similar for all the items. The values vary uniformly from 150 to 350. The items are similar in term of A_2 and A_3. For atttribute A_1, there are two distinct groups: the group of X_1, X_2 and X_3 with the mean of 9 and a group of X_4. The first attribute of X_4 deviates significantly from the other items in the table. It is five times greater than the average of group of X_1, X_2 and X_3. Despite this abnormality, X_4 is considered less of an outlier than X_1 and X_2 according to the definition of LOF. Even though the range of A_1 is smaller than those in A_2 and A_3, the deviation is significant. In this example, X_4 is unusual whereas X_1, X_2 and X_3 are normal. We consider an alternative approach to discover the unusual item X_4. The approach is to compute the outliers on all subspaces. However, the number of subspace is exponential. In addition, it can not detect this type of feature-based anomalies. The example shows the scenarios where unusual records are undetectable by the traditional approach. In the next section, we formally define the problem and introduce a method to detect the feature-based unsual patterns.

Table 1.

ID	A_1	A_2	A_3
X_1	8	250	300
X_2	9	250	250
X_3	10	350	150
X_4	50	300	200

(a) Example

	X_1	X_2	X_3	X_4
k-distance	180.29	141.42	180.29	119.43
Mean	116.58	91.06	134.32	94.14
LOF Score	1.55	1.55	1.34	1.27

(b) Metrics

3 Formal Definitions

We extend the definition of the relational algebra to incorporate the concept of an interval tuple and the operations on an interval tuple. Recall that a relation R consists of n attributes $A_1, \ldots, A_n$, a tuple t of R is an ordered list of values corresponding to the attributes. The notation $t.A_i$ refers to attribute A_i of tuple t. Where applicable, we drop the attribute name and use the subscript, i.e. t_i, to refer to the attribute in order to simplify the definitions. In addition, we use the terms feature and attribute interchangably.

An interval is a set of real numbers bounded by two end points, which can be represented by $[a, b]$. An interval-tuple I is an order list of intervals $\{I_i\}$. The interval selection operation $\sigma_I(R)$ selects a subset S of the tuples from R such that each t in the subset S satisfies the following condition: $t_i \in I_i, \forall I_i \neq NULL$. We say that S is covered by I, I is a cover interval-tuple of S and I_i is a cover interval of S. If I has only one interval, say I_i, $\sigma_{I_i}(R)$ can be used. In this case, we say $\sigma_{I_i}(R)$ is an interval selection operation on interval I_i for R.

In example 1a, a relation R consists of X_1, X_2, X_3 and X_4. The operation $\sigma_I(R)$ and $\sigma_{I_1}(R)$ on $I \equiv <[7, 10], [250, 250], [250, 350]>$ return $\{X_1, X_2\}$ and $\{X_1, X_2, X_3\}$ respectively. $S \equiv \{X_1, X_2\}$ is covered by I.

We also define the function $\omega(S)$ that returns the smallest intervals that cover S. It is an inverse function of the function σ. In the example above, $\omega(S)$ returns $J \equiv <[8, 9], [250, 250], [250, 300]>$.

Definition 1. α *function between two interval-tuples on attribute i is defined by:*

$$\alpha(I_i, J_i) = \frac{\max(\inf I_i, \inf J_i) - \min(\sup I_i, \sup J_i)}{\min(\sup I_i, \sup J_i)}, \tag{1}$$

$$if\ I_i \cap J_i \equiv \varnothing$$

$$and\ \alpha(I_i, J_i) = 0\ if\ I_i \cap J_i \neq \varnothing \tag{2}$$

The function α measures the dissimarity between two intervals on an attribute i based on the ratio of the difference between two intervals instead of the distance between two intervals. *When the ratio on all the attributes for two interval-tuples I and J is small, they are considered close.* We then can measure the closeness between any two sets as follows:

Definition 2. *Two sets S and S' are close under α_c, denoted by close$(S, S'/\alpha_c)$ $\equiv$ true if and only if $\alpha(\omega_i(S), \omega_i(S')) < \alpha_c, \forall i \in 1 \ldots n$.*

After generating rules using the feature-based samples (which will be discussed later), we will try to combine the related rules into common patterns. Two rules will be combined into a pattern (definition 3) if they are close to each other as follows:

Definition 3. *Given $P = \{S_i\}$, if $\forall S_i \in P$, there is at least one $S_j \in P$ such that S_i and S_j are close under α_c, we say $\{S_i\}$ forms a pattern P under α_c. We define $supp(P) = \sum |S_i|$ as the support of P.*

Definition 3 implies the chain property that can result in very high support patterns. We can define the normal and unsual patterns as follows:

Definition 4. *We say a pattern P is normal under N_u if supp(P) > N_u, where N_u is a user defined value.*

Definition 5. *We say a pattern P is unusual under N_u if supp(P) $\leq N_u$.*

We exploit the chain property to combine observations into groups of related observations. If an observation does not belong to any large group, it is unsual. We also observe that the normal observations should account for the majority of the dataset.

In our paper, we need to produce rules to learn normal patterns. A rule is a set of strongly correlated observations. We will use an interval splitting function f^S to generate rules. First, the split function on a set R divides each column into at least k intervals. Each interval is split into smaller intervals if the ratio of the difference between two consecutive ordered values from R on the same column in the interval is greater than α_s. The combinations of the intervals from different attributes produces the interval-tuples. We use $f^S(R/k, \alpha_s)$ to denote the set of interval-tuples returned from the split function on R.

The sampling is then performed based on an attribute as follows:

Given an interval I_i on attribute i, we select a sample R from dataset D such that $R \equiv \sigma_{I_i}(D)$. We then perform the split function f^S on R to obtain the set of interval-tuples $V \equiv f^S(R/k, \alpha_s)$. For each interval tuple $I \in V$, we create a set S of tuples from R where $S \equiv \sigma_I(R)$. If S does not satisfy the condition $|f_i^S(S/\alpha_s)| = 1, \forall i$, which means S can be divided into smaller sets, we split interval tuple I into smaller interval tuples by performing the split function $f^S(S/k = 1, \alpha_s)$ on S (the value of one for k means that we only split an interval when there is a change of the values in the interval). We say $\{S\}$ are the rules generated from sample R. The rules are used to create patterns. In the algorithm, *we compare the unusual patterns against other normal patterns and try to merge the unusual patterns.* Therefore, the normal observations that are mistakenly flagged as unusual will be grouped into the normal patterns.

Definition 6. *The o-score function between two interval tuples is defined by*

$$o\text{-}score(I,J) \equiv \sqrt{\sum \alpha^2(I_i, J_i)} \tag{3}$$

Definition 6 defines the score between two interval tuples. The score of a pattern against another pattern o-score(P,P') is the smallest score between the two interval tuples of two patterns. As we see, the score of an unusual pattern is used to measure the deviation of an unusual observation from the normal observation. Thus, the $u - score$ of an unusual pattern P is the smallest score between it and the normal patterns.

```
 1: procedure RULEBASE(D)
 2:     for all i ∈ n do
 3:         U ← makeSamples(D, i)
 4:         for all R ∈ U do
 5:             V ← f^S(R/k, α_s)
 6:             for all I ∈ L do
 7:                 S ← makeRules(R, I)
 8:                 put S into RP
 9:             end for
10:             makePatterns(RP, NP, UP)
11:             for all p ∈ UP do
12:                 compute o-score(p, NP)
13:             end for
14:             P ← matchNorm(NP, P)
15:             UP ← matchUnusual(UP, P)
16:             P ← combine(P, UP)
17:         end for
18:     end for
19: end procedure
```

4   Framework

The outline for the algorithm is shown in the procedure RULEBASE. The algorithm consists of n rounds where n is the number of attributes. In the ith round, the makeSamples function creates a set of samples from dataset D on attribute i. In this implementation, the samples are created as follows. The data set is sorted on attribute i and divided into chunks of the size N_c. Each chunk R is a sample from which to learn rules on attribute i.

The split function $f^S(R/k, α_s)$ outputs a set V of interval-tuples. The set of rules $\{S\}$ are created for each interval-tuple $I ∈ V$. The makePatterns function merges all the rules and outputs a set NP of normal patterns and a set UP of unusual pattern candidates according to definitions 3, 4 and 5. The unusual score for all unusual pattern candidates is also computed against set NP in lines 11 to line 13.

The algorithm from line 3 to line 13 creates the normal patterns and unusual pattern candidates for each sample R. As discussed in the introduction, we need to rule out the normal patterns. Variable P at line 14 contains the set of unusual pattern candidates. For each candidate $p ∈ P$, p will be removed from P if there exists a normal pattern $p_n ∈ UP$ such that p and p_n are close under $α_c$ (line 14). This is done since those candidates are shown to be normal in another pattern. For the first round, no action is performed on UP from line 15 to 16 and all the unusual pattern candidates are put into P. P contains the set of candidates for the first attribute. The items which do not belong to any pattern in this set of candiates are normal.

For the next rounds, the unusual pattern candidates will be removed from UP if they are not in P (line 15) because they were flagged as normal. The function at line 16 combines the unusual pattern candidates into larger patterns according to the close function under parameter α_c. The candidates which are normal after the combination are removed from P. The new unusual score of each new pattern is the lowest score from the candidates for which the new pattern is created. When the computation is finished for all rounds, all the unsual pattern candidates in P that do not match any normal pattern or do not have the support sufficient enough to be normal are unsual patterns.

The close and score functions are computed from the α function which requires the dividends to be non zero. For each round, we replace the zeros with the average of the next c items when the dataset is sorted on attribute i. The zeros can be replaced by 0.5 if the dataset contains only 0 and 1 for attribute i.

4.1   Parameter Setting

The algorithm consists of five parameters: α_c, α_s, k, N_c, N_u, which can be determined in a straightforward manner. N_u is a user-defined parameter indicating the size of a pattern to be considered unusual. N_c is the sample size. k is choosen based on the number of possible patterns in a sample. The number of patterns increases in the sample when k increases. However, if those patters are similar, they will be combined together eventually. The value of k does not impact the output significantly. Typically, k can be around 4 and N_c can be computed from $k \times N_u$. Besides k, α_s is used to split an interval if there is an abnormal change in the interval. The parameter α_c defines the cutoff point for the unusual observations. Heuristically, we can choose α_c and α_s between 0.3 and 0.6.

4.2   Running-Time Complexity

We denote $|NP|$, $|UP|$ and $|P|$ as the total number of items in NP, UP and P respectively. From line 5 to 9, it takes $O(N_c)$ time to make the rules. The makePattern function needs to combine all the patterns and has a worst case of $O(N_c^2)$. The running time from line 11 to 13 is $O(|UP| \times |NP|)$. We can use hashtables to store patterns for NP, UP and P in order to execute the matching functions (line 14, 15) in linear time with respect to the number of items. The execution time for the two lines are $O(min(|NP|, |P|))$ and $O(min(|UP|, |P|))$ respectively. Since $|NP| + |UP| = N_c$, the execution time for them is less than $O(N_c)$. The running time for line 16 is $O(|UP| \times |P|)$. The total running time from line 3 to 16 is $O(N_c + N_c^2 + |UP| \times |NP| + N_c + |UP| \times |P|)$. Since N_c and $|UP| * |NP|$ are less than N_c^2, the formula can be reduced to $O(max(N_c^2, |UP| \times |P|))$. In general N_c^2 is small and does not grow with the dataset and $|UP| \times |P|$ is small on the average. The running time can be considered constant, $O(c)$.

There are n rounds. It takes $O(NlogN)$ time to sort the data according to the attribute of the current round. Each round has $\frac{N}{N_c}$ chunks and the execution time for each chunk is $O(c)$. Therefore, the total execution time is $O(c \times n \times \frac{N}{N_c} + n \times NlogN)$. Since $(\frac{c}{N_c \times logN} + 1)$ is small and inversely proportional to N, we replace it by ς , the formula can be written as $O(\varsigma \times n \times N \times logN)$. The formula shows that the algorithm can be executed very fast.

Table 2. List of Attacks

Type	#attacks	Type	#attacks	Type	#attacks
pod	210	teardrop	179	loadmodule	9
ipsweep	209	back	168	ftp write	8
warezclient	203	guess passwd	53	multihop	7
portsweep	200	bufferoverflow	30	phf	4
nmap	200	land	21	perl	3
smurf	200	warezmaster	20	spy	2
satan	197	imap	12		
neptune	196	rootkit	10		

5   Experiments

We ran experiments using the KDD CUP 99 Network Connections Data Set from the UCI repository [16]. The data was compiled from a wide variety of intrusions simulated in a military network environment prepared by MIT Lincoln Labs. Each record has 42 features containing the network connection information. Among them, 34 features are continuous and 8 features are symbolic. The last feature labels the type of connection. About 80% of the data set are normal connections and 20% are attack connections. Because the rate of attack connections is very high for regular network activities, we follow the data generation approach in [4] to create a new dataset with a very low number of attacks randomly drawn from the KDD Cup Dataset to test whether the small patterns would be discarded by the sampling method in a large dataset. The new dataset contains 95,174 normalized connections in total and the total number of attacks account for only 2.2% of the dataset. To make the experimental results less biased by the possible special characteristics of some types of attack, the data set contains 22 types of attack with the number of records for each attack type varying from 2 to 210. The attack with the largest size accounts for only 0.2% of the dataset. The details of the number of connections for each attack are shown in table 2.

5.1   Competing Methods

First, we run an oultier detection algorithm on the dataset to obtain the list of outliers. In this experiment, we use the well-known outlier detection method LOF [2] to compute outliers in the experiments. The local outlier factor of a point p is the ratio between the kdist(p) and the average of the kdists of all the KNNs of p where kdist(p) is the distance from p to the kth nearest neighbor of p. The records with high LOF scores are considered as outliers. In LOF, min_pts^{th} is the number of nearest neighbors to compute kdist.

Figure 1a shows the LOF precision/recall curve for different values of min_pts. Generally, $min\_pts = 100$ performs better than the other two values of min_pts. As we see, LOF obtains the highest precision rate (56%) when the recall rate = 2.6%. When the recall rate reaches 20%, the precision rate drops dramatically to 10%.

Since the precision is low for a high recall rate, we apply different clustering algorithms on the outliers to group them into their corresponding attack types. First, we ran KMEAN on the top outliers which account for 10% of the dataset. There are 40 clusters. We then filtered the output by removing clusters with the size greater than 250. There are 30 such clusters. Ten of them have at least 50% attacks. Table 3a shows the corresponding rank of the first five clusters that contain the attack connections. As we can see, 13 out of 18 top clusters do not contain any attack.

In the next experiment, we used a shared nearest neighbor (SNN) clustering algorithm to cluster the top outliers. SNN returns one cluster of size 3814 and 292 clusters of size less than 143. There are 19 clusters with the size from 50 to 143. Table 3b shows the top three clusters that contains the attack connections. The first attack cluster ranks 9th and its size is 78. In those two experiments, the attack connections are not shown as strong unusual patterns.

The clustering of the top outliers could not detect all the top ten groups of attacks. Because of this, we want to cluster the entire dataset with the hope that those groups can be discovered. It is difficult to set the parameter k for KMEAN in this large dataset. With large k, KMEAN will return many clusters whereas small values of k may group the normal connections with attack connections. Therefore, we used SNN for this experiment since it can discover small clusters with different densities in large datasets [6] without requiring the number of clusters as input. The algorithm returns 131 clusters with the size from 50 to 250. Among them, there are six clusters that contain attack connections (see table 4). The first cluster that contains the attack connections is ranked 35th. As we can see, the attack and normal connections are divided into smaller clusters. The attack patterns are not shown clearly in this experiment. The result shows that even though SNN can discover small clusters, they will return many of them.

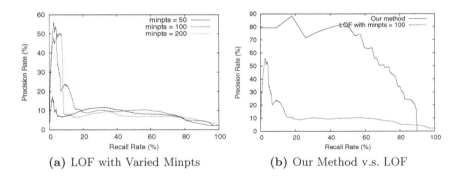

(a) LOF with Varied Minpts   (b) Our Method v.s. LOF

Fig. 1. Precision-Recall (PR) Curve

5.2   Our Method

In the next experiment, we ran our method on the dataset to discover the unusual patterns. We set the parameters according to the parameter setting from the algorithm section ($N_u = 250$, $\alpha_c = 0.5$, $k = 4$, $\alpha_s = 0.5$, and $N_c = 1200$). The algorithm returns 20 unsual patterns with the size of at least 50. Table 5a shows the top 10 unusual patterns by size. The size of those patterns varies from 67 to 243. The first two patterns are of the attack type. Seventy percent of those patterns contain 100% attacks. The other 9 smaller patterns containing attacks are shown in table 5b. According to table 5a, we see that satan, neptune and portsweep follow strong patterns. In the table, we see that warezclient attacks also follow a pattern but its score is low (6.5) relative to those attack types, which means that its pattern is slightly different from normal patterns.

As mentioned above, the data set contains 10 attack patterns with a size of at least 100. In our method, eight of ten are identified in the top unusual patterns (by size).

Table 7 shows the recall rate for each attack type found in table 5a. With the low false alarm rate, we still get a high recall rate.

We then take all the unusual patterns with the size of at least 50 and order them by score. Table 6 shows the ranking, score and attack type of the patterns. According to the table, our method correctly identifies some of the attacks in the first six unusual patterns. All of these patterns have the detection rate of 100%. Among them, the attack type of Satan has the highest score which is 203.5. The score of the first normal connection pattern in the table is only 29.6 and its size is only 54.

In ranking either by size or by score (after the patterns with very low sizes are pruned), we can see that the attack types of satan, portsweep, neptune, and nmap are discovered by our approach. The results imply that these types of attack follow some patterns which are strongly different from normal connections.

Table 3. Top outliers (10%)

ID	Rank	Size	Connections	Rate(%)
1	7	159	neptune	95.0
			portsweep	1.3
2	12	114	satan	0.9
			teardrop	89.5
			ipsweep	0.9
3	13	110	nmap	100.0
4	16	107	smurf	73.8
5	18	101	satan	1.0
			ipsweep	4.0
			portsweep	87.1

(a) Clustered by KMEAN

ID	Rank	Size	Connections	Rate(%)
1	9	78	neptune	100.0
2	15	58	smurf	77.6
3	16	56	smurf	100.0

(b) Clustered by SNN

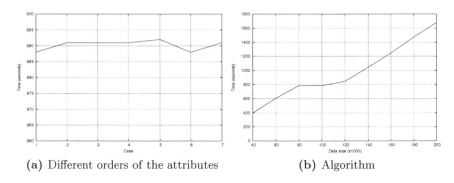

(a) Different orders of the attributes   (b) Algorithm

Fig. 2. Running Time

Figure 1b shows the PR curve of our method versus LOF. The figure shows that our method yields a precision rate of 80% at very low recall rates. The method outperforms the LOF approach. Since we quickly prune the patterns with an unusual score below the cutoff threshold, the attacks with a very low unusual score are removed from the output. That is why the figure only shows the recall rate up to 90%. However, this does not affect the result much since the precision rate is usually low at this recall rate.

According to the experiments, the following interesting observations are found. *Most unusual patterns with normal connections do not form highly supported patterns.* Even though they may have high scores, they appeared as very low support patterns, whereas the attack connections form patterns with high support. *The normal connections that can form patterns with high support tend to have a very low score.* Few discovered unusual patterns have mixed results. Another important observation is that the number of unusual patterns with high support is very small.

In conclusion, we have performed a variety of experiments with different combinations of outlier detection and density-based clustering algorithms. The precision is low when the recall rate increases to 20%. For the clustering algorithms, the normal connections were also grouped into small clusters. However, our

Table 4. SNN on the entire dataset

ID	Rank	Size	Connections	Rate(%)
1	35	112	nmap	100.0
2	57	89	warezclient	73.0
			rootkit	1.1
3	74	79	neptune	100.0
4	90	65	ipsweep	100.0
5	112	59	smurf	78.0
6	118	58	smurf	100.0

Table 5. Top Unusual Patterns Ordered by Size

Rank	Size	Score	Types	Rate(%)
1	243	24.1	smurf	79.4
			normal	20.6
2	192	43.6	nmap	100
3	170	12.9	normal	100
4	169	203.5	satan	100
5	150	104.1	neptune	100
6	129	26.1	ipsweep	100
7	114	14.2	back	100
8	84	6.5	warezclient	100
9	72	15.6	normal	100
10	67	107.7	portsweep	100

(a) Top 10

Rank	Size	Score	Types	Rate(%)
13	66	11.8	teardrop	100
14	64	54.	pod	100
20	50	48.8	pod	100
22	48	83.1	guesspwd	100
23	44	156.6	neptune	100
33	31	20.0	teardrop	100
35	29	30.0	back	100

(b) Others

algorithm grouped the attack and normal connections almost correctly according to their connection type even though there were 22 types of attacks of small size. They are shown clearly as unusual patterns in terms of size and score.

6 Performance

We implement a memory-based version of the algorithm in Java. For each round, the dataset can be sorted in $O(NlogN)$ time. We use a hashtable data structure to store the list of unusual candidate items so that the matching functions, matchNorm and matchUnusual, can be performed in constant time. At first, we ran the program with different initial attributes. The number of unusual candidates vary from 30K to 60K. During the first round, we don't combine the unusual patterns, therefore, the performance is not affected. In the next few

Table 6. Unusual Patterns Ordered by Score

Rank	Type	Score	Size	Rank	Type	Score	Size
1	satan	203.5	169	11	smurf	24.1	243
2	portsweep	107.7	67	12	normal	15.6	72
3	neptune	104.1	150	13	normal	14.9	64
4	pod	54.6	64	14	back	14.2	114
5	pod	48.8	50	15	normal	12.87	170
6	nmap	43.6	192	16	teardrop	11.8	66
7	normal	29.6	54	17	normal	10.5	67
8	normal	26.2	63	18	normal	9.7	56
9	ipsweep	26.1	127	19	warezclient	6.6	84
10	normal	25.9	59	20	normal	4.2	67

Table 7. The Recall Rate of the Attack Types in Top 10 Patterns

smurf	nmap	satan	neptune	back	ipsweep	warezclient	portsweep
96.5	96	85.8	76.5	67.9	61.2	41.4	33.5

rounds, the matching functions reduce the number of unusual items dramatically before the combining step.

Figure 2a shows the performance of the algorithm with different random orders of the attributes. According to the figure, we see that the order of the attributes does not affect the running time significantly. Also, the results are almost the same for different orders of the attributes. The attack connections are consistently in the top unusual patterns.

In the next experiment, we ran the program on the KDD dataset with the size varying from 40K to 200K. Figure 2b shows that the execution time of the program is linear with the growth of the data size. By replacing the memory-based hashtable data structure and merge sort algorithm with the disk-based versions, the algorithm can be used for any large dataset.

7   Related Work

Outlier detection has been extensively studied in the field of statistics [8]. The method relies on using the underlying distribution of the dataset to detect the outliers. The limitations of the statistical-based approach is that the underlying distribution is usually unknown and that approach does not perform well in high dimensions. The distance-based approach can discover outliers without knowing the underlying distribution [5,2]. However, we have shown that there are cases where the unusual observations can not be detected by the distance-based approach.

A method [4] was introduced to discover anomalous records in categorical datasets by performing the conditional probability tests on combinations of the attribute values. The test requires that the values must be discretized into the set of values. Hence, it is more suitable for a categorical dataset. Spiros et al [18] introduces LOCI method to detect outliers by using multi-granualarity deviation factor (MDEF). The authors then propose an approximate version to speed up the method. The method is mainly based on the modification of an approximate nearest neighbor search algorithm (quad-tree) in order to avoid the cost of computing the MDEF scores for all the points in data set. Thus, the method will depend on the performance of the index tree in order to speed up the method. Minh et al [15] introduced the adaptive dual-neighbor method in order to detect small groups of outliers. The advantage of the method is that it reduces the number of noise clusters. However, its time complexity is $O(n^2)$ and it can not detect the feature-based unusual patterns. Recently, Kriegel et al [12] have proposed the angle-based method that computes outlier scores based on the angles of the points with respect to their local regions. The method aims to provide more accurate rankings of the outliers in high dimensions. The major limitation of the algorithm is its time complexity. The naive implementation of the algorithm runs in $O(n^3)$, whereas the approximate $O(n^2)$ version only produces top 1 outliers in high dimensions. In addition, the method can not detect outliers if they are surrounded by other points.

Density-based clustering methods cluster the dataset based on the local density of the nearby items. The nearby items with the same density are clustered together [6,9]. Density-based clustering can discover the clusters with different sizes and shapes [6]. The main difference between this method and our method is that this method focuses on clustering the dataset, whereas our method focuses on discovering unusual patterns based on individual features. We use the similarity in the values of items in an individual feature to generate patterns. The items that belong to large patterns are quickly removed from the learning process. As a result, our algorithm can run fast.

8   Conclusion

We have introduced a fast method that can discover unusual observations by generating rules from the feature-based samples. A rule is a set of strongly related observations. The rules are then combined into patterns by using the α function. When a pattern gains high support, it is considered normal. Otherwise, we will try to merge the pattern with other patterns. If they still can not gain enough support. It is considered unusual. The unusual patterns are ranked based on their supports. In the experiments, we have compared our algorithm with other possible alternatives, namely outlier detection and clustering, to discover unusual patterns. According to the results, our approach yields the highest detection rate with the most unusual items grouped into the correct corresponding patterns. We have also introduced the score function to measure the degree of deviation of the unusual patterns from the normal patterns. The running time complexity of the method is $O(\varsigma \times n \times N \times logN)$. The experiments confirm that our method can run very fast in large datasets.

References

1. Beyer, K.S., Goldstein, J., Ramakrishnan, R., Shaft, U.: When is nearest neighbor meaningful? In: Beeri, C., Bruneman, P. (eds.) ICDT 1999. LNCS, vol. 1540, pp. 217–235. Springer, Heidelberg (1999)
2. Breunig, M.M., Kriegel, H.-P., Ng, R.T., Sander, J.: LOF: identifying density-based local outliers. SIGMOD Rec. 29(2), 93–104 (2000)
3. Chawla, N.V., Lazarevic, A., Hall, L.O., Bowyer, K.W.: SMOTEBoost: Improving prediction of the minority class in boosting. In: Lavrač, N., Gamberger, D., Todorovski, L., Blockeel, H. (eds.) PKDD 2003. LNCS, vol. 2838, pp. 107–119. Springer, Heidelberg (2004)
4. Das, K., Schneider, J.: Detecting anomalous records in categorical datasets. In: KDD 2007: Proceedings of the 13th ACM SIGKDD international conference on Knowledge discovery and data mining, pp. 220–229. ACM Press, New York (2007)

5. Knorr, E.M., Ng, R.T.: Algorithms for mining distance-based outliers in large datasets. In: VLDB 1998: Proceedings of the 24rd International Conference on Very Large Data Bases, pp. 392–403. Morgan Kaufmann Publishers Inc., San Francisco (1998)

6. Ertöz, L., Steinbach, M., Kumar, V.: Finding clusters of different sizes, shapes, and densities in noisy, high dimensional data. In: Proceedings of the third SIAM international conference on data mining, pp. 47–58. Society for Industrial and Applied, Philadelphia (2003)

7. Fan, H., Zaïane, O.R., Foss, A., Wu, J.: A nonparametric outlier detection for effectively discovering top-N outliers from engineering data. In: Ng, W.-K., Kitsuregawa, M., Li, J., Chang, K. (eds.) PAKDD 2006. LNCS, vol. 3918, pp. 557–566. Springer, Heidelberg (2006)

8. Hawkins, D.: Identification of outliers. Chapman and Hall, London (1980)

9. Jarvis, R.A., Patrick, E.A.: Clustering using a similarity measure based on shared near neighbors. IEEE Transactions on Computers C-22(11), 1025–1034 (1973)

10. Ke, Y., Cheng, J., Ng, W.: Mining quantitative correlated patterns using an information-theoretic approach. In: KDD 2006: Proceedings of the 12th ACM SIGKDD international conference on Knowledge discovery and data mining, pp. 227–236. ACM Press, New York (2006)

11. Korn, F., Pagel, B.-U., Faloutsos, C.: On the 'dimensionality curse' and the 'self-similarity blessing'. IEEE Transactions on Knowledge and Data Engineering 13(1), 96–111 (2001)

12. Kriegel, H.-P., Hubert, M.S., Zimek, A.: Angle-based outlier detection in high-dimensional data. In: KDD 2008: Proceeding of the 14th ACM SIGKDD international conference on Knowledge discovery and data mining, pp. 444–452. ACM, New York (2008)

13. Lazarevic, A., Kumar, V.: Feature bagging for outlier detection. In: KDD 2005: Proceeding of the eleventh ACM SIGKDD international conference on Knowledge discovery in data mining, pp. 157–166. ACM Press, New York (2005)

14. Mannila, H., Pavlov, D., Smyth, P.: Prediction with local patterns using cross-entropy. In: KDD 1999: Proceedings of the fifth ACM SIGKDD international conference on Knowledge discovery and data mining, pp. 357–361. ACM, New York (1999)

15. Nguyen, M.Q., Mark, L., Omiecinski, E.: Unusual Pattern Detection in High Dimensions. In: The Pacific-Asia Conference on Knowledge Discovery and Data Mining (2008)

16. Newman, C.B.D., Merz, C.: UCI repository of machine learning databases (1998)

17. Shaft, U., Ramakrishnan, R.: Theory of nearest neighbors indexability. ACM Trans. Database Syst. 31(3), 814–838 (2006)

18. Papadimitriou, S., Kitagawa, H., Gibbons, P.B., Faloutsos, C.: LOCI: Fast outlier detection using the local correlation integral. In: Proceedings of the 19th International Conference on Data Engineering: 2003, pp. 315–326. IEEE Computer Society Press, Los Alamitos (2003)

19. Steinwart, I., Hush, D., Scovel, C.: A classification framework for anomaly detection. J. Mach. Learn. Res. 6, 211–232 (2005)

Efficient Online Aggregates in Dense-Region-Based Data Cube Representations

Kais Haddadin[1] and Tobias Lauer[2]

[1] Jedox AG, Freiburg, Germany
kais.haddadin@jedox.com
[2] Institute of Computer Science, University of Freiburg, Germany
lauer@informatik.uni-freiburg.de

Abstract. In-memory OLAP systems require a space-efficient representation of sparse data cubes in order to accommodate large data sets. On the other hand, most efficient online aggregation techniques, such as prefix sums, are built on dense array-based representations. These are often not applicable to real-world data due to the size of the arrays which usually cannot be compressed well, as most sparsity is removed during pre-processing. A possible solution is to identify dense regions in a sparse cube and only represent those using arrays, while storing sparse data separately, e.g. in a spatial index structure. Previous dense-region-based approaches have concentrated mainly on the *effectiveness* of the dense-region detection (i.e. on the space-efficiency of the result). However, especially in higher-dimensional cubes, data is usually more cluttered, resulting in a potentially large number of small dense regions, which negatively affects query performance on such a structure. In this paper, our focus is not only on space-efficiency but also on time-efficiency, both for the initial dense-region extraction and for queries carried out in the resulting hybrid data structure. We describe two methods to trade available memory for increased aggregate query performance. In addition, optimizations in our approach significantly reduce the time to build the initial data structure compared to former systems. Also, we present a straightforward adaptation of our approach to support multi-core or multi-processor architectures, which can further enhance query performance. Experiments with different real-world data sets show how various parameter settings can be used to adjust the efficiency and effectiveness of our algorithms.

1 Introduction

Online analytic processing (OLAP) allows users to view aggregate data from a data warehouse displayed on demand, using a model that is usually referred to as the *data cube* [5], which includes operations such as slicing and dicing as well as roll-up and drill-down along hierarchies defined over dimensional attributes. Depending on the architecture of the OLAP system, the aggregate values are either pre-computed and stored or calculated from the base data on the fly, i.e. only when the respective value is requested. (Combinations of both approaches are also very common.) The latter strategy may result in longer times for retrieving and aggregating the necessary base values – especially if they have to be loaded from secondary storage media – but it is

T.B. Pedersen, M.K. Mohania, and A M. Tjoa (Eds.): DaWaK 2009, LNCS 5691, pp. 177–188, 2009.

usually faster regarding the changes of cell values, as expensive re-computation of stored aggregates is avoided. In addition, this approach is used by most in-memory OLAP databases which try to store all of the base data in RAM and calculate all aggregate values online (except for recently computed aggregates residing in a cache).

Efficient online aggregation can be achieved by transforming the base data according to some pre-processing strategy. Well-known examples are the prefix-sum approach [8] and its variants, which allow the computation of arbitrary range queries in constant time, usually at the expense of update costs. More sophisticated strategies provide a variety of tradeoffs between query and update times [4, 9, 12]. The iterative data cube (IDC) [12] allows the combination of several of those strategies by choosing a separate one for each dimension.

One common feature of these methods is that that they are based on an (multidimensional) array representation of the data. The advantage is a convenient and efficient access to each cell through its coordinates. However, a serious drawback of the above strategies is that the pre-processing step usually requires filling also the empty cells and hence effectively turns sparse cubes into dense ones. This prevents an efficient compression of the arrays, which thus require much more space than the original sparse data. In particular, this means that data which might otherwise easily fit into main memory cannot be accommodated any more in such a representation.

As a way out of this dilemma, dense-region-based representations have been proposed [2, 3, 11]. These approaches try to exploit what has been called the "dense-region-in-sparse-cube" property [2]: in many real-world datasets, the majority of filled cells are not distributed evenly within the universe of all possible cells but are clustered in certain regions. Each of these regions can be represented as a sub-cube using any of the above methods, while outliers (filled cells that do not belong to any dense region) can be stored separately. Previous approaches have concentrated mainly on effectiveness, i.e. the reduction of sparsity, but have neglected efficiency, both for the pre-processing step of detecting dense regions and for the queries carried out on the resulting data structure.

2 Preliminaries and Related Work

For the purposes of this work, we consider a *data cube* C as a d-dimensional hyper-rectangle of cells. Each dimension D_i is a discrete range of all n_i possible base values in that dimension; for simplicity, we assume $D_i = \{1, \ldots, n_i\}$. We call n_i the length or cardinality of dimension D_i.

It is obvious that a data cube C consists of $\prod_{i=1}^{d} n_i$ cells. We refer to this number the *capacity* of C. The capacity must be distinguished from the *size* of C, which is defined as the number of cells which actually hold a value (other than zero). The size thus corresponds to the number of base records stored in the fact table for the cube. The *density* of C is the ratio of filled cells to all cells, i.e. $density(C) = size(C) / capacity(C)$. The density of most cubes in real-world OLAP scenarios is very low (usually much less than 1%); this is what is referred to as cube *sparsity*.

Dense-region-based OLAP representations aim at identifying dense sub-cubes within a sparse cube. Each of them can be represented efficiently using the methods discussed above. They must then be maintained in some index structure allowing quick access to them during queries. An additional data structure is required for storing and accessing the outlier cells, which are not located inside any dense region.

Clearly the main goal of any dense-region-based approach is the reduction of sparsity, i.e. the effective identification of clusters of cells. The sparsity reduction can be measured by the *global density* of the resulting set S of sub-cubes, which is defined as

$$density(S) = \frac{\sum_{c \in S} size(c)}{\sum_{c \in S} capacity(c)}$$

Many clustering methods for identifying dense regions have been proposed, especially in the data mining literature, and we refer to [13] for an overview. Since in our scenario, however, the regions will be represented as hyper-rectangles, we aim at identifying rectangular regions and can thus use more straightforward methods.

One of first approaches to dense-region-based OLAP was given in [2]. It proposes extracting dense sub-cubes through an algorithm called *ScanChunk* and then inserting these sub-cubes in an R-tree [7] structure, while outliers are indexed in ROLAP table form. However, the *ScanChunk* algorithm is not efficient for large data cubes. The reason is that its time complexity depends on the capacity of the data cube. The capacity of cubes can be extremely large, especially in higher dimensions. Our goal is to find a procedure whose complexity is a function of the *size* of the data cube rather than its capacity. Also, the clustering result is measured only by the global density. This overlooks another important factor, the number of resulting sub-cubes, which is a decisive factor for query performance.

3   Data Structure and Query Processing

Our data structure for dense-region OLAP is a combination of two variants of the R-tree and the IDC [12]. The basic idea is to transform the extracted dense sub-cubes into IDCs and make them accessible through an R*-tree [1]. The outlier cells are stored in an aR-tree [10] with additional aggregate information in the inner nodes.

An aggregate query on our proposed date structure is split into two sub-queries; one on the set of sub-cubes, the other one on the outliers. A range query in the R*-tree returns a list L of all the sub-cubes that intersect with the query range. The query is then translated into the appropriate range query for each cube in L. An IDC representation of these cubes will answer each of these queries efficiently, in our implementation in time $O(1)$, independent of the size of the sub-cube or the query range. The combined aggregate of the results is the answer of the first sub-query.

The second sub-query is carried out on the outlier cells, which are stored in the leaf nodes of the aR-tree (see Figure 1). A range query in this augmented tree does not have to return all cells in the query range. Instead, if the MBR of a whole subtree is contained in the query range, the aggregate value stored in the inner node pointing to this subtree can be used, making the query more efficient. In the example in Figure 1, if the MBR of entry P4 is fully contained in the range of a query, the aggregate value 18 stored with P4 is used directly instead of descending further down the tree.

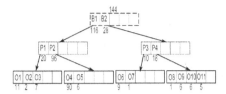

Fig. 1. Outliers are strored in an aR-tree [10]

A straightforward way of further enhancing query processing, especially on multi-core architectures, is the parallel computation of the two parts of a query in separate threads, since the two queries are completely independent of each other. Of course the workload in an arbitrary query will often not be divided evenly between the two sub-queries. However, queries usually arrive in bulks because users request complete "views" usually consisting of hundreds or even thousands of range queries. Hence, each thread can process its share of all queries before the final result is created. We have achieved significant speedups in several of our tests using this strategy.

4   Efficient Sub-cube Extraction

Our approach is based on the method proposed in [3, 11], which divides the procedure into four basic steps – splitting, shrinking, merging, and filtering – where the merge step aims at reducing the number of sub-cubes. However, we identified several short-comings in this step, which are avoided in our new merge procedure. In addition, we dispose of the shrinking step and improve the splitting algorithm, turning it more time-efficient, especially in higher-dimensional cubes. The tests described in [3, 11] were done using artificial data of rather low dimensionality ($d \leq 5$). In our own experiments, we were able to use real-world data cubes provided by an industrial partner, where dimensionalities ranged from 6 to 13. Table 1 and Figure 2 illustrate the overall effect of the extraction for four example cubes.

The table lists the global capacities (which corresponds the required space) before and after the extraction. The changes in (global) density through the sub-cube extraction can be seen in Figure 8. Obviously, the sparsity is reduced dramatically (except for DC_1, which was very dense already). In the following subsections, we outline the different steps of the extraction procedure.

Table 1. Capacities before/after extraction

	DC_1	DC_2	DC_3	DC_4
Original capacity	171,000	3.9×10^{14}	5.4×10^9	8.1×10^8
Sum of subcube capacities	164,160	361,033	25,078,968	5,257,635

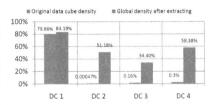

Fig. 2. Densities before/after sub-cube extraction

4.1 Recursive Splitting

We adopt the basic splitting method from [11], where for each dimension of the cube a histogram is computed which divides the dimension in dense and sparse intervals. The histogram h_i of D_i is an array of length n_i, where the value $h_i[j]$ is the number of filled cells in the j-th slice of the cube when sliced along D_i (see Figure 3). The histograms are then used to detect dense intervals in each dimension by comparing the values against a predefined threshold α. The dense intervals then will only include cells with value greater than or equal to α. In Figure 2, using $\alpha = 1$ will produce two dense intervals in each dimension of the initial cube.

The Cartesian product of all dense intervals over all dimensions creates a set of sub-cubes, which are candidates for dense regions. The extraction procedure in [11] instantiates all these sub-cubes and then invokes the same splitting procedure recursively for them. Apparently, many of these candidate cubes will be empty, and computing the histograms for them is neither necessary nor useful. Our initial tests found that even checking them for emptiness to filter them out takes a lot of time. Moreover, the method suggested in [11] for computing the histograms using a Boolean array is not very efficient either. First, the array has a size proportional to the capacity of cube C (rather than its size) and hence should be compressed in order to be handled. Second, in order to compute a histogram, all cells (filled or not) of the current cube are checked, which greatly decreases performance. (Recall that the histograms count *filled* cells, hence ideally only they should be looked at.)

Our approach proceeds the opposite way: a coordinate list of only the filled cells for the current cube is created and processed. For each one, we identify the dense interval to which it belongs in each dimension, using the histograms. We then check whether or not that sub-cube has already been created. If yes, we add the cell to the coordinate list of that sub-cube. If not, the cube must be created before. This way, only those cubes are initialized which actually contain any filled cells, which dramatically reduces execution time. After all existing sub-cubes have been created, the same procedure is called recursively for each of them and a variable μ storing the current recursion depth is adjusted (cf. Figure 3). The recursive splitting stops if a sub-cube is *continuous*, i.e. if all its dimensions consist of only one dense interval.

The main bottleneck in our procedure is the time to check whether a sub-cube already exists, since this involves searching a potentially large collection of created cubes. This is especially relevant in the initial call of the recursive procedure, when the histograms contain a large number of dense intervals. To reduce the number of possible sub-cubes, a technique known as *histogram flattening* can be used. Flattening reduces the differences between adjacent histogram cells by averaging the value of each cell with those of its neighbours. The reduction is controlled by a flattening factor f, which defines the number of neighbours considered on each side. Details about flattening can be found in [11], where it is also proposed. However, it is neither motivated there why flattening should be used, nor does the procedure improve the experimental results there; on the contrary, the sub-cube extraction always deteriorated the quality of the output (i.e. the global density) in the tests. In our own experiments we found that flattening can greatly reduce the time needed for splitting, especially during the first split of the initial cube. Therefore, in our work, a new factor called the flattening threshold μ is introduced to the flattening process. Flattening is applied only

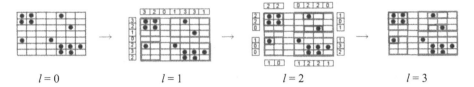

$l = 0$ $l = 1$ $l = 2$ $l = 3$

Fig. 3. Recursive splitting of a cube (l represents the recursion level)

to sub-cubes whose depth l in the recursion is less than or equal to μ. So, if $\mu = 0$, there is no flattening at all. For $\mu = 1$, we apply flattening only to the original cube. If $\mu = 2$, besides the original sub-cube flattening is applied to the sub-cubes resulting from the first split. Setting $\mu = \infty$ will apply flattening to all sub-cubes on all levels.

The effects of flattening vary for different values of μ. The speedup achieved for splitting is big when flattening is applied to the first level of splitting ($\mu = 1$), but does hardly increase further when splitting is called to further levels of the split procedure. The reason is that sub-cubes produced at a deeper level of splitting will not be split into many new ones. On the other hand, applying flattening only to the sub-cubes at the first levels of splitting will not reduce the final number of sub-cubes and outliers much. The number of sub-cubes and outliers resulting from the splitting is reduced as flattening is applied to sub-cubes down to deeper or all levels in the split procedure (μ is big or $\mu = \infty$). This enhances the query performance but requires more space as the resulting sub-cubes are less dense. Flattening when applied to only the first levels of the recursive split tree (i.e. small μ) does not significantly increase the overall capacity of the resulting sub-cubes. Detailed results on flattening are given in section 5.1.

4.2 Sub-cube Merging

In order to reduce the number of sub-cubes, nearby cubes can be merged into one if the resulting cube is not too sparse, i.e. if its density is over a certain given threshold. The result of merging a set S of sub-cubes is the smallest cube encompassing all cubes in S, i.e. the minimum bounding region (MBR) of S. The merge step is very important, as a lower number of sub-cubes improves query performance. Hence, the choice of the density threshold is one way of trading memory for query performance.

A merge step is also proposed in [3] and [11]; however, we identified some serious shortcomings there. First of all, only *pairs* of sub-cubes are considered for merging. Such a method may miss potential candidates if, for instance, a set of k sub-cubes together could be merged but no single pair of them would reach the density threshold after merging just those two. The second shortcoming is that no method is described for efficiently detecting "nearby" candidate pairs. An exhaustive comparison of all possible pairs of sub-cubes will result in a number of checks that is quadratic in the number of sub-cubes before merging. This number will of course be much higher if triples or even larger subsets of sub-cubes are also considered as candidates. Third, the method does not check for potential overlaps of the merged cubes with other existing sub-cubes (in fact, the merge condition in [11] would not even detect if another sub-cube is completely inside the merged area and thus increase its density). While overlapping sub-cubes will not result in incorrect answers if handled appropriately, it is surely undesirable, as reserving space for the same cells twice wastes memory.

We propose the following improved method for merging sub-cubes, which uses an important property of the R*-tree, in which the set of sub-cubes is maintained: objects in close spatial proximity are likely to be stored in the same subtree [1]. Hence, on the leaf level we can expect that nearby sub-cubes will be stored as siblings.

Algorithm. *merge(set of subcubes S ,χ)*

---

1. insert subcubes from S into R*-tree r_1
2. $r_2 =$ **new** R*-tree
3. **repeat**
4.   **for** each leaf node l in r_1 do
5.   **if** *density(l.MBR) $\geq \chi$* **and** r_1.intersect(l.MBR) = l.entries
           and r_2.intersect(l.MBR) = $\emptyset$ **then**
6.   subcube $c =$ merge(*l.entries*)
7.   insert c in r_2
8.   **else**
9.   insert *l.entries* in r_2
10.   **end if**
11.   **end for**
12. $r_1 = r_2$
13.   empty r_2
14. **until** no more merging during *for*-loop
15. **return** set of subcubes in r_1

First, the sub-cubes resulting from the previous steps are inserted into an R*-tree r_1. An additional R*-tree r_2 is created for the newly merged sub-cubes and the sub-cubes that will not be merged. Each leaf in r_1 is checked for merging. For each leaf MBR (i.e. the minimum bounding hyper-rectangle containing all its entries), we compute the density (sum of sizes of the contained sub-cubes divided by the MBR capacity). If it is greater than or equal to the merge threshold χ, we also check whether the leaf MBR overlaps any of the newly merged sub-cubes (stored in r_2) or if it overlaps any entries of other leaves in its own R*-tree r_1. If no merging is done (the leaf MBR was not dense enough or it overlaps with other sub-cubes), the leaf entries are inserted in r_2 without merging. Otherwise, the leaf MBR forms a new sub-cube and is inserted in r_2. When all the leaves in r_1 have been checked (lines 4-11), we replace r_1 with r_2 and empty r_2. If any cubes were merged in the *for*-loop, the procedure is repeated. Otherwise, the set of sub-cubes in r_1 is reported as the result of the merge. (To restrict the time taken for this step, the procedure could alternatively terminate after a fixed number of iterations given as an upper limit.)

This method for merging overcomes the shortcomings of the approach in [11]. Instead of checking all pairs of sub-cubes, our method only checks candidate sub-cubes that are spatially close, based on the properties of the R*-tree. This reduces the number of candidate sets to be checked in each iteration from $\lceil n^2/2 \rceil$ to $\lceil n/m \rceil$, where n is the number of sub-cubes before merging and m is the minimum number of entries in each node of the R*-tree. In addition, there is no limitation to pairs of cubes in our approach. If a leaf is dense enough, it will be merged regardless of how many entries

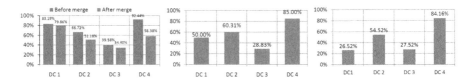

Fig. 4. Effects of merging: change in density (left), reduction rate of sub-cubes (centre), and speedup rate of query performance (right)

it includes. Finally, no overlapping occurs in our merge procedure. This check is done quite efficiently through two R*-tree range queries (one each in r_1 and r_2). Still, the performance of our procedure can be affected negatively if the size of the R*-tree gets very big and the insertion and search becomes slower. This can happen if hundreds of thousands of sub-cubes result from splitting. However, this problem can be handled by reducing the input set through the use of histogram flattening during the split step.

The effects of merging can be seen in Figure 4. We chose $\chi = 0.5$ (i.e. newly merged sub-cubes must be at least 50% dense). As can be seen, merging causes the global density of the sub-cubes to decrease. This was expected since merging nearby sub-cubes will usually add sparse regions. However, this reduction of density is not nearly as drastic as the gain in density due to the splitting step.

Apparently, merging can significantly reduce the number of sub-cubes (which is the main goal of this step). As can be seen in Figure 4 (centre), a reduction by 28% to 85% was achieved in our tests. (This number can be increased by lowering the merge threshold, if memory is available.) The effects of this reduction on query performance are significant, as shown in Figure 4 (right). The speedup for queries is almost proportional to the reduction rate. Hence, significant speedup can be obtained through a reduction of sub-cubes, achieved in a trade-off with extra space. One might expect DC_1 to give a better speedup ratio than DC_3, because it had a bigger sub-cubes reduction rate. The reason it did not occur is that in DC_1 the number of sub-cubes was just reduced from two to one through merging, which hardly changed the R*-tree size.

4.3   Filtering

Since splitting is recursive, it will often produce very small sub-cubes, many of which may still be left after merging. If the size of such a cube is below a certain threshold δ, it makes more sense to consider the filled cells as outliers than to maintain a large number of such small sub-cubes. The value of δ is crucial for the balance between the number of sub-cubes and the number of outliers. A good choice of this threshold is mostly based on the strategy chosen for the representation of sub-cubes but can also depend on the distribution of the data. Figure 5 shows experimental results for three different cubes (the merge threshold used was $\chi = 0.5$ in all cases).

Obviously, with increasing δ the number of sub-cubes (left column) decreases and the number of outliers (centre column) goes up. The right column shows the time taken for 10,000 aggregate queries in the resulting structure (the R-trees were *not* queried in parallel in this test). For the first two cubes, a good balance between sub-cubes and outliers is reached at $\delta = 32$ and $\delta = 16$, respectively, while for the third one

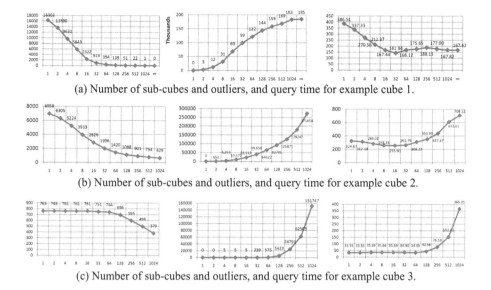

(a) Number of sub-cubes and outliers, and query time for example cube 1.

(b) Number of sub-cubes and outliers, and query time for example cube 2.

(c) Number of sub-cubes and outliers, and query time for example cube 3.

Fig. 5. Effects of filtering with different values of δ on the number of sub-cubes (left), number of outliers (centre), and processing time (in sec.) for 10,000 queries (right)

(bottom row), the number of (small) sub-cubes is so low that filtering with $\delta \leq 64$ has hardly any effect.

5   Further Experimental Results

5.1   Histogram Flattening

To test all the effects of flattening method mentioned in section 4.1, we performed a set of experiments on an extremely sparse cube (density $\approx 5.2 \times 10^{-16}$) with 13 dimensions, which produced more than 385,000 sub-cubes during the splitting step. In this experiment, we applied flattening using different values of the flattening level threshold μ. The horizontal axis in all figures represents different values for μ. Again, $\mu = 0$ means no flattening, $\mu = 1$ means that flattening is applied only to the original cube, and so on. In order to reduce the large number of sub-cubes, we also aimed to increase the number of merges by chose a very low merge threshold $\chi = 0.01$.

Figure 6 illustrates the change in the number of sub-cubes after each step and the final number of outliers after merging. As shown in this figure, the number of sub-cubes and outliers decreases as flattening is applied down to deeper levels in the split procedure. From Figures 6a and 6b, we can see that the reduction rate in the number of sub-cubes between μ values 4, 5, 6 and ∞ is higher than between values 0, 1, 2 and 3. The reason why the number of outliers also goes down with deeper flattening is that many outliers are "swallowed", as the sub-cubes grow larger. This can be seen in Figure 7, which shows how the capacities increase as the flattening is applied further to deeper levels of splitting. However, the capacity is still far away from that of the original data cube ($\approx 6.1 \times 10^{21}$).

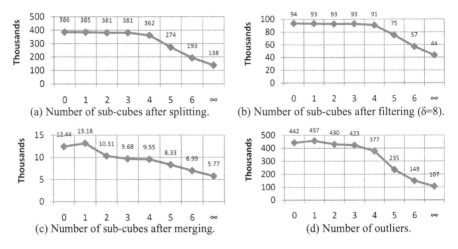

(a) Number of sub-cubes after splitting.

(b) Number of sub-cubes after filtering (δ=8).

(c) Number of sub-cubes after merging.

(d) Number of outliers.

Fig. 6. Effects of flattening with different values for μ

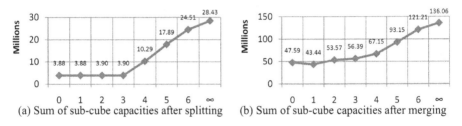

(a) Sum of sub-cube capacities after splitting

(b) Sum of sub-cube capacities after merging

Fig. 7. Effects of flattening on capacities

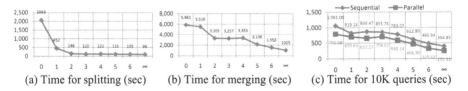

(a) Time for splitting (sec)

(b) Time for merging (sec)

(c) Time for 10K queries (sec)

Fig. 8. Effects of flattening on performance

As mentioned before, when flattening is only applied to a few levels (up to 3), it does hardly affect the overall capacity. However, as is illustrated in Figure 8, even when only flattening at the first two levels, it greatly reduces the time required for the split (a) and merge (b) procedures. Also, the time for querying in the final data structure decreases due to the lower number of sub-cubes (c).

To summarize, flattening has effects on both efficiency and effectiveness of our extraction procedure. If it is applied with a small value for μ, we can expect a faster splitting with no significant change in the output of the procedure. Hence, we advice to use flattening at least with *flattening level* $\mu = 1$. If memory limits are less important, flattening gives us the ability to use some extra space to reduce the number of sub-cubes after splitting; this speeds up the extraction procedure (especially the merge

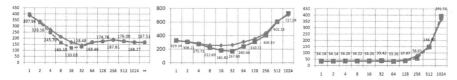

Fig. 9. Comparison between sequential (blue) and parallel (red) query computation times (in sec) for different values of δ in the same data cubes as in Figure 5

step) and most importantly enhances the performance of queries. Hence, the flattening level threshold μ is another way to adjust the trade-off between query time and space.

5.2 Parallel Query Computation

Several experiments were performed to test the effects of carrying out sets of queries in parallel. First, a set of 10,000 "bad case" queries was created for each cube and executed sequentially. Then a parallel version was carried out using two separate threads, where one thread computed the outputs from the sub-cube R*-tree and the other thread the outputs from the outlier aR-tree. Four cubes were tested. The results are shown in Figures 9 and 10. Just like in Figure 5, the horizontal axis represents the different values of filtering threshold *minimum size* δ.

As can be seen in Figure 9, if either (almost) only sub-cubes or only outliers exist (i.e. very small or very big δ), the parallel version will not provide any significant speedup, as one sub-query always dominates the search time by far. However, even if the parallel version does not necessarily enhance the performance at the optimal point of the sequential computation, the parallel version is never slower at that point. Thus, if we have a good configuration in the sequential version of the program with both outliers and sub-cubes, the parallel version can be expected to give a better or at least the same performance (but not worse).

For the sparsest cube in our test, the parallel computation of queries showed the biggest effects, as can be seen in Figure 10. For example, when $\delta = 8$ and $\chi = 0.01$, the time required by the parallel version is about 30% less than for the sequential version. Hence, our approach seems to be effective especially in cases with high sparsity of the original cube and a low merge threshold.

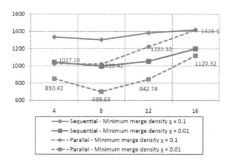

Fig. 10. Sequential vs. parallel queries in a very sparse cube

6 Conclusions

We have presented a dense-region-based data structure for in-memory OLAP, which can be constructed a lot more efficiently than earlier approaches. The resulting global density can be influenced by two parameters, the flattening level and the minimum merge threshold, which allows for an adjustable trade-off between query performance and memory usage. In addition, a filtering threshold can be used to fine-tune the ratio of sub-cubes to outliers for faster query processing. Tests with several real-life data cubes have confirmed the usefulness of our approach, especially for very sparse cubes, and have shown good choices of the parameters for different cubes.

The most important future work involves a possible automation of the selection of parameter values on the basis of heuristics derived from our tests, which will greatly improve the usability of the approach when integrated in commercial OLAP systems.

References

[1] Beckmann, N., Kriegel, H.-P., Schneider, R., Seeger, B.: The R*-tree: An efficient and robust access method for points and rectangles. In: Proceedings of ACM SIGMOD, pp. 322–331 (1990)

[2] Cheung, D.W., Zhou, B., Kao, B., Kan, H., Lee, S.D.: Towards the building of dense-region-based OLAP system. Data and Knowledge Engineering 36(1), 1–27 (2001)

[3] Chun, S., Chung, C.-W., Lee, S.-L.: Space-efficient cubes for OLAP range-sum queries. Decision Support Systems 37(1), 83–102 (2004)

[4] Geffner, S., Agrawal, D., El Abbadi, A., Smith, T.: Relative prefix sums: an efficient approach for querying dynamic OLAP data cubes. In: Proceedings of International Conference on Data Engineering, Sydney, Australia, pp. 328–335 (1999)

[5] Gray, J., Chaudhuri, S., Bosworth, A., Layman, A., Reichart, A.D., Venkatrao, M., Pellow, F., Pirahesh, H.: Data cube: A relational aggregation operator generalizing group-by, cross-tab, and sub-totals. In: Data Mining and Knowledge Discovery, pp. 29–53 (1997)

[6] Gupta, H., Harinarayan, V., Rajaraman, A., Ullman, J.: Index selection for OLAP. In: Proceedings of the 13th International Conference on Data Engineering, pp. 208–219 (1997)

[7] Guttman, A.: R-trees: A dynamic index structure for spatial searching. In: Proceedings of ACM SIGMOD, pp. 47–57 (1984)

[8] Ho, C.-T., Agrawal, R., Megido, N., Srikant, R.: Range queries in OLAP data cubes. In: Proceedings of ACM SIGMOD, pp. 73–88 (1997)

[9] Lauer, T., Mai, D., Hagedorn, P.: Efficient range-sum queries along dimensional hierarchies in data cubes. In: Proceedings of DBKDA, Cancún, Mexico (2009)

[10] Mamoulis, N., Bakiras, S., Kalnis, P.: Evaluation of top-k OLAP queries using aggregate R-trees. In: Bauzer Medeiros, C., Egenhofer, M.J., Bertino, E. (eds.) SSTD 2005. LNCS, vol. 3633, pp. 236–253. Springer, Heidelberg (2005)

[11] Lee, S.-L.: An effective algorithm to extract dense sub-cubes from a large sparse cube. In: Tjoa, A.M., Trujillo, J. (eds.) DaWaK 2006. LNCS, vol. 4081, pp. 155–164. Springer, Heidelberg (2006)

[12] Riedewald, M., Agrawal, D., El Abbadi, A.: Flexible data cubes for online aggregation. In: Proc. of the International Conference on Database Theory, pp. 159–173 (2001)

[13] Witten, I.H.: Data Mining: Practical Machine Learning Tools and Techniques. Addison-Wesley, Reading (2000)

[14] Zhao, Y., Deshpande, P., Naughton, J.: An array-based algorithm for simultaneous multidimensional aggregates. In: Proc. ACM SIGMOD, Arizona, USA, pp. 159–170 (1997)

BitCube: A Bottom-Up Cubing Engineering

Alfredo Ferro, Rosalba Giugno, Piera Laura Puglisi, and Alfredo Pulvirenti

Department of Mathematics and Computer Science, University of Catania
{ferro,giugno,lpuglisi,apulvirenti}@dmi.unict.it

Abstract. Enhancing on line analytical processing through efficient cube computation plays a key role in Data Warehouse management. Hashing, grouping and mining techniques are commonly used to improve cube pre-computation. BitCube, a fast cubing method which uses bitmaps as inverted indexes for grouping, is presented. It horizontally partitions data according to the values of one dimension and for each resulting fragment it performs grouping following bottom-up criteria. BitCube allows also partial materialization based on iceberg conditions to treat large datasets for which a full cube pre-computation is too expensive. Space requirement of bitmaps is optimized by applying an adaption of the WAH compression technique. Experimental analysis, on both synthetic and real datasets, shows that BitCube outperforms previous algorithms for full cube computation and results comparable on iceberg cubing.

Keywords: Data Mining, Cubing, Data Warehouse, Bitmap.

1 Introduction

Since the introduction of Data Warehouse and OLAP systems, research has focused in the design of efficient algorithms for cube pre-computation. Given a base table $\mathcal{R}$, the data cube can be defined as a generalization of the standard GROUB-BY operator in which the aggregation of every combinations of attributes, appearing in the *group-by* clause, are computed. A *n-dimension data cube* is composed of cells of the following form $(v_1, v_2, \cdots, v_n, c)$ where c is a list of measures. A cell can have a value v_j or a * symbol to indicate that all values of that dimension have been grouped. A cell is called *m-dimensional*, if and only if there are exactly m ($m \leq n$) values among $(v_1, v_2, \cdots, v_n)$ which are not *. It is called a *base* cell if $m = n$ and *aggregate* cell if $m = 0$.

Different kinds of cubing are possible. The full cube computation allows the pre-computation of all parts (cuboids) of a data cube [6]; the iceberg cube introduces conditions to materialize only a subset of a cube satisfying them [3]; the closed cube compresses a cube by representing only closed cells and predicting the remaining from them [14]; and shell-fragment cube selects in advance the dimensions of interest [7]. In [13], full and iceberg cubing computations are recognized as fundamental step for other categories. However, iceberg cubes can be efficiently computed only for measures having the anti-monotonic property [2]. For iceberg conditions involving non antimonotonic complex measures, such as

T.B. Pedersen, M.K. Mohania, and A M. Tjoa (Eds.): DaWaK 2009, LNCS 5691, pp. 189–203, 2009.

AVERAGE, pruning strategies cannot be applied. Therefore, full cube materialization is needed. Nevertheless, in the cases in which full materialization is inefficient, complex measures can be transformed into weaker, but antimonotonic measures (i.e. average can be changed into top-k average measure [13]).

Cubing algorithms perform aggregations following two major approaches: *top-down* and *bottom-up*. In a top-down model, group-bys with a small number of dimensions are obtained from the ones with a higher number of dimensions through caching intermediate computations [1,15,10,4]. Generally, these algorithms are not suitable for iceberg cubing, since before pruning, the whole cube must be computed. In the bottom-up approach, nodes associated with a small number of dimensions are ancestors of nodes with a higher number of dimensions. This model (introduced by BUC [3]) uses the a-priori based strategy [2] to compute only cubes satisfying the iceberg condition. Other works include hybrid algorithms [13], cube compressions [5], and cubes for very high dimensional data [7]. More recently, in the case of Relation-OLAP, the management of hierarchies has been taken into account [8], see [9] for a survey also.

In this paper, *BitCube*, a bottom-up algorithm for full and iceberg cube computation over sparse relations is introduced. It partitions the base table into horizontal fragments with respect to leftmost ordered dimension and unlike BUC, it does not recursively sort the remaining dimensions into partitions. BitCube uses bitmaps as lookup tables, that is inverted indexes [7], to identify values shared by the same records in a partition, allowing fast aggregation computations. Bitmaps are compressed by using an adaptation of the Word-Aligned Hybrid (WAH) [12] algorithm. Such a compression leads to a significative improvement of the performance of BitCube in terms of both space requirements and running time when skewed data is used. BitCube has been compared with full and iceberg cubing algorithms (BUC [3], MM-Cubing [11], StarCubing [13] and CURE [8]) on real and synthetic datasets. Experiments show that BitCube outperforms all other compared systems on full cubing and results comparable on iceberg cubing.

The paper is organized as follows. Section 2 briefly reviews previous methods. Section 3 introduces BitCube algorithm. Experiments and comparisons are reported in Section 4. Section 5 concludes the paper and draws future research directions.

2   Related Work

In what follows, the main algorithms for full and iceberg cubing are summarized. Algorithms are described according to their computation strategy: top-down, bottom-up and mixed. Finally, approaches to treat very large amounts of data and to deal with hierarchies are reviewed.

Top-down Approaches. Following a top-down spanning tree, group-bys with a small number of dimensions are obtained from the ones with a higher number of dimensions through caching intermediate computations. Top-down methods include PipeSort, PipeHash, Overlap [1], MultiWay Array Cube [15],

PartitionCube [10], and Grouping Set Query [4]. MultiWay is usually taken as a representative example. The main idea is the following: since group-bys need at the time only a portion of the data, the base table is compressed and partitioned into subcubes (chunks) small enough to fit into main memory. Morever, different group-bys are computed in one pass by properly defining chunks computation order. Such a strategy results inefficient when the dimensionality is high and data are sparse. Generally, top-down cube approaches are not suitable for iceberg cubing, since, in order to compute all cuboids in parent nodes, it requires to know if children cuboids exceed minimum support.

Bottom-up Approaches. In the bottom-up spanning tree, nodes associated with a small number of dimensions are ancestors of nodes with a higher number of dimensions. This model (introduced by BUC [3]) allows to prune unnecessary computations by introducing a-priori based strategy [2] (iceberg condition). If a node does not meet the minimum support, its descendants do the same. BUC reads the base table and partitions it based on the ordered values of a dimension. For each partition, all its cuboids are calculated and recursively the algorithm continues on the remaining dimensions. Full cubing is straightforward computed by setting the iceberg condition to 1. The method proposed here, BitCube, follows also a bottom-up approach. It partitions the base table into fragments with respect to leftmost ordered dimension. However, unlike BUC, it does not recursively sort the remaining dimensions into partitions.

Mixed Approaches. Star-Cubing [13] combines top-down and bottom-up approaches. Inspired by MultiWay [15], it simultaneously aggregates multiple dimensions. Then, it partitions parent nodes in a bottom-up fashion and prunes descendants that do not satisfy the threshold. It organizes input tuples in hyper-tree structure (star-tree). To build the star-tree, the star-table is needed. The star-table is constructed by traversing the associated subtree and counting the frequencies of each attribute.

Approaches for Very Large Amount of Data. Range Cube [5] allows the construction of compressed cubes without any loss of precision. Range Cube exploits correlations among attributes values and uses them to compress the base table. It uses a hyper-tree structure (range-trie). Unlike star-tree, where one node represents one dimension, in a range-trie a node represents shared attributes values, i.e., a distinct group of tuples to aggregate. In [7], a new computational model for high dimensional datasets is proposed. The method is based on the pre-computation of "shell fragments" (i.e. vertical partitions of the dataset) together with their materializations. Efficiency is obtained by using "inverted indexes" (i.e. for each value in each dimension, a list of record-ids associated with it is stored). Finally, intersection among fragments is performed online by computing full group-bys on desiderated dimensions. BitCube uses inverted indexes also and following suggestion in [7], it optimizes them by using bitmaps. However, unlike previous methods, BitCube uses bitmaps also as lookup tables to identify values shared by the same records. Therefore, this allows to speed-up the aggregation process by reducing the number of possible candidates to group.

Hierarchical Relational OLAP. Most of the existing Relation OLAP cubing systems focus on flat datasets and do not include hierarchies in the dimensions. Recently in [8], authors proposed CURE, a method to construct data cubes on large datasets with arbitrary hierarchies. Hierarchies make cube computation harder since the search space becomes bigger. Authors designed a framework based on the fact that cubes contain a large amount of redundant data. Their algorithm is able to efficiently compute cubes on flat and hierarchical data.

3   BitCube

BitCube is a bottom-up cubing algorithm. In what follows full and iceberg cube computation are presented.

3.1   Full Cubing

Let $\mathcal{R}$ be a base table on dimensions $d_1, d_2, \ldots, d_n$. Let C_i be the cardinality of d_i for $i = 1, \ldots, n$. BitCube partitions the base table into fragments with respect to the leftmost ordered dimension.

For each dimension d_j the values present in a partition and the number of their occurrences are stored in a *dynamic vector* called D_{d_j}. For example, in Figure 1, a partition of a dataset having 4 dimensions (A,B,C,D) is reported. Each dimension in the dataset has cardinality 9. The dataset was ordered with respect of dimension A. The dynamic vector for B contains the values of B

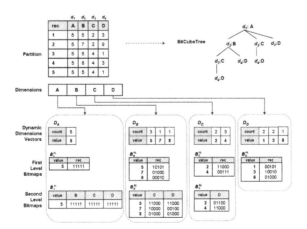

Fig. 1. BitCube Data Structures. A partition of a dataset having 4 dimensions (A,B,C,D).

present in the partition, $\{5, 7, 8\}$, together with the number of records containing them, i.e. count of 5 is 3. BitCube uses bitmaps (i) as inverted indexes to locate positions of dimension values into records (first level bitmaps) and (ii) as lookup tables to identify values shared by the same records (second level bitmaps).

Definition 1 (First Level Bitmap). *Let P be a partition of k records of a base table $\mathcal{R}$. For each dimension d_i and for each value v in P, a first level bitmap $\mathcal{B}_{d_i}^{FL}[v]$ is a bitmap of k bits where $\mathcal{B}_{d_i}^{FL}[v][j] = 1$ if the dimension d_i has value v in the j-th record of P.*

In Figure 1, the first level bitmap of B, $\mathcal{B}_B^{FL}$, is defined for each value that B assumes in the partition. $\mathcal{B}_B^{FL}[5] = 10101$. Since the first, the third, and the fifth bits are set, value 5 appears in records 1, 3, and 5.

Definition 2 (Second Level Bitmap). *Let P be a partition of k records of a base table $\mathcal{R}$. Let D_{d_i} be the dynamic vector of d_i with respect to P. For each value v of d_i and for each dimension d_t of cardinality C_t, $t > i$, a second level bitmap $\mathcal{B}_{d_i}^{SL,d_t}[v]$ is a bitmap of h bits, $h = min(C_t, k)$, where $\mathcal{B}_{d_i}^{SL,d_t}[v][j] = 1$ if both $D_{d_t}[j]$ and v appear in at least one record of P.*

For example, in Figure 1, $\mathcal{B}_B^{SL,C}[5] = 11000$. Since the bit in position 2 is set, value 5 of B is present together with the second value of the dynamic vector of C, that is 4 ($D_C[2] = 4$), in at least one record. Notice that, in the partition, the values 5 of B and 4 of C are present together in two records. This can not be deduced by looking at the second level bitmap only. The intersection of $\mathcal{B}_B^{FL}[5] = 10101$ and $\mathcal{B}_C^{FL}[4] = 00111$ will yield the records sharing values 5 and 4.

BitCube uses a tree, called *BitCubeTree*, to guide the aggregation process. It is defined as follows.

Definition 3 (BitCubeTree). *Given $\mathcal{R}(d_s, \ldots, d_n)$ a BitCubeTree is a bottom-up spanning tree T, with the following properties:*

- *the root of T contains d_1;*
- *a node in the tree storing the dimension d_i has n-i children representing $d_{i+1}, \ldots, d_n$;*
- *a node of level i allows the aggregation of $(i+1)$ dimensions stored in the path from the node up to the root (i.e., level 1 means group-by on pairs, level 2 group-by on triples and so on).*

Figure 1 reports a bitCube computation on a partition of a dataset having four dimensions.

A *Header-table* dynamically maintains the dimension values of the BitCube-Tree path under analysis. Moreover, for each entry, it stores a bitmap resulting from the intersection of the first level bitmaps of the values in the path from that entry up to the root. In Figure 4, the header table related to the aggregation of ABCD on values $(5, 5, 4, 1)$ is given; the quadruple is present two times in the partition (see Figure 1).

Figure 2 (top left), reports the details of BitCube algorithm. BitCube iterates for $k = n, \ldots, 1$ over the relation $\mathcal{R}(d_k, \ldots, d_n)$. At each time a BitCubeTree with root d_k is built (Line:3), records of $\mathcal{R}$ with respect to d_k are sorted, and partitioned into horizontal fragments $P_1, P_2, \ldots P_{C_k}$ (Line:4). For each value of d_i its aggregation value is stored (Line:7). Next, first and second level bitmaps are constructed (Lines:8-12) and the aggregation process runs on the current partition (Line:13). Notice that, the second level bitmaps can not be constructed for the dimension d_n (Lines:9-12) since there are no further dimensions with which it can be combined. Moreover, when the partition is made for the last dimension d_n, no further aggregations need to be computed and the computation stops (Line:8). In Figure 2 (top right), details of the Aggregate procedure are reported. The algorithm recursively computes all the group-bys inside the partition through an in-order visit of the BitCubeTree (see Figure 3 for all aggregations related to partition in Figure 1). At each step the aggregation of the dimensions in the path from the root to the current dimension is computed. The aggregation of k dimensions is obtained by using knowledge from the grouping of the previous $k - 1$ dimensions in the following way. Let d_k be the dimension of the child of the root (Line:2). Candidate values to be aggregated are obtained by computing the intersection of second level bitmaps B^{SL,d_k} of d_k

```
BitCube(R, d1)
// R: Base table
// d1: Dimension used to start computation
1    for k = n to 1 do
2        T ← BitCubeTree(dk);
3        r ← root[T]; // the dimension of r is dk
         // each part. has the same value on dk
4    Partition(R, dk);
5    for each partition Pi do
6        if |Pi| ≥ minSup then //iceberg cond
7            WriteOutputRec;
8        if k < n then
9            for j = k to n-1 do
10               BFL_dj ← Build first level bitmap;
11               BSL_dj ← Build second level bitmap;
12           BFL_dn ← Build first level bitmap;
13           Store in a Header-Table BFL_dk and the
                 value of the dimension dk;
14       Aggregate(r,|Pi|);
     END BitCube.
```

```
Aggregate(r, size)
1    for each child c of r do
2        dk ← dimension of c;
         // Comp. the candidates vals to aggregate
3        Compute I^{SL}_{dk} as the AND of B^{SL,dk} of
             dk ancestors current values;
     //Verify candidates with first lev. bitmaps
4    for each candidate value v of dk in I^{SL}_{dk} do
5        Compute I^{FL}_{dk} as the AND of B^{FL}_{dk}[v] and
             I^{FL} of the father of dk;
6        Store I^{FL}_{dk} in the entry dk of the header table;
7        count ← # of bits 1 in I^{FL}_{dk};
8        if count ≥ minSup then //iceberg cond
9            WriteOutputRec;
10       if c is not a leaf then
11           if count = 1 then
12               DirectAggregate(I^{FL}_{dk});
13           else
14               Aggregate(c, size);
     END Aggregate.
```

```
AggregateFirstLevel(c, size)
     // dk is the dimension of c
1    for each value v of D_{dk} such that
             count of v ≥ minSup do //iceberg
2        WriteOutputRec;
3        if node c is not a leaf then //iceberg
4            if count of v is 1 then
5                DirectAggregate(dk, v);
6                continue;
7            else
8                Aggregate(c, size);
     END AggregateFirstLevel.
```

Fig. 2. BitCube Algorithm and the Aggregate procedure called by BitCube. On the bottom, the procedure to aggregate pairs. This can be added in the pseudo-code of Aggregate and executed if the level of the child node is 1.

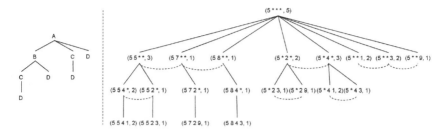

Fig. 3. BitCubeTree and all aggregations related to partition in Figure 1. Notice that, the tree is visited from right to left.

ancestors current values (Line:3). This allows to fast identify dimension values that may be shared by the same records in the current partition. For example, in Figure 4, the intersection of $B_B^{SL,D}[5] = 11000$ and $B_C^{SL,D}[4] = 11000$ yields that values of D in position 1 and 2 in the dynamic vector, that are 1 and 3 respectively, are the candidates to be aggregated with the triple (5,5,4). In order to discard false positives, for each candidate value v of d_k, its first level bitmap $B_{d_k}^{FL}[v]$ is intersected with the bitmap I^{FL} of its father dimension (Lines:4-5). The resulting bitmap, $I_{d_k}^{FL}$, is then stored in the entry header table related to current value v of d_k (Line:6). In Figure 4, by intersecting $B_D^{FL}[3] = 10010$ and $I_C^{FL} = 00101$ the bitmap 00000 is obtained. This means that no record in the partition has the tuple (5,5,4,3). Therefore value 3 is a false positive. If the number of bits in $I_{d_k}^{FL}$ equal to 1 satisfies the iceberg condition, the aggregation is written (Lines:7-8). The algorithm proceeds recursively until the current node is not a leaf (Line:14). However, if this number is 1, the aggregation of v with the ancestors values $(v_s, \ldots, v_{k-1}, v)$ is present in one record only (in Figure 3, these cases are represented by paths in which all nodes have a single child in the tree). This allows to obtain all higher dimensional cuboids directly from candidates as follows. The tuple $(v_s, \ldots, v_{k-1}, v)$ is grouped with all values of the remaining dimensions d_t $(t > k)$ having position j in their dynamic vector for all j such that $B_{d_k}^{SL,d_t}[v][j] = 1$ (Line:12). Notice that, since bitmaps for the root dimension are trivial, by slightly modifying the AGGREGATE algorithm, their computation can be avoided.

The aggregation procedure discussed above has been optimized when the algorithm analyzes nodes of level 1 of the BitCubeTree. Here group-bys on pairs are computed. Since the first dimension (root) in each partition assumes only one value, aggregations with children dimensions can be simply computed using the count of occurrences of their values stored in the dynamic vector. If the count of a value v satisfies the iceberg condition the aggregation is written. For example, the measure of the aggregation pair (5,5,*,*), see Figure 3, can be directly obtained by the count of value 5 for B (see Figure 1). Figure 2 (bottom) reports the pseudo-code for pair aggregation. This can be added in pseudo-code of Figure 2 and executed if the level of the child node is 1.

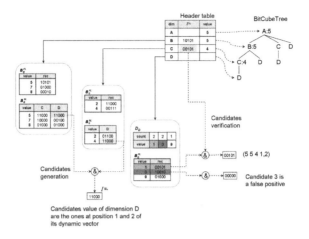

Fig. 4. BitCube aggregation major steps

Notice that, the above algorithm and the related examples uses the COUNT operator as aggregation measure. The proposed method can be extended to support other aggregation functions in the following way. During the AND operations, the offset of set bits in the first level bitmaps will be used to retrieve the measure values of the records in the partition. These values will be used to compute the needed aggregation functions (SUM, MAX or MIN of measure values).

3.2   Iceberg Cubing

In the iceberg cube construction, BitCube allows to prune unnecessary computations by introducing a-priori based strategy. It discards partitions with a number of records less than the minimum support. For each value of dimension d_i meeting the constraint (Figure 2 top left, Lines:5-6), its aggregation value is stored (Line:7).

During the iceberg cube computation, the visit of the BitCubeTree is performed from right to left. Such a visit allows to compute pair aggregations which will be used to speed-up higher dimensional aggregations located at the left side of the tree. For example, in the aggregation tree of Figure 3, dimensions A and D are first aggregated. Then, this knowledge is used to optimize the aggregation of A, C and D in the following way. First, a third level bitmap (iceberg bitmap) is introduced. It is associated to each dynamic vector of each dimension d_i.

Definition 4 (Iceberg Bitmap). *Let P be a partition of a base table $\mathcal{R}$ with respect to a value w. Let D_{d_i} be the dynamic vector of d_i with respect to P. Let t be the size of D_{d_i} in the partition. An iceberg bitmap $B_{d_i}^{iceberg}$ is a bitmap of t bits where $B_{d_i}^{iceberg}[v] = 1$ if the group-by of the pair (w,v), where v is a value of d_i, satisfies the iceberg condition.*

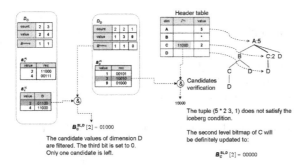

Fig. 5. BitCube Iceberg aggregation with iceberg condition count(*)>1. The aggregating process of dimensions ACD on values (5, *, 2, 3) is reported.

An example of iceberg bitmap follows. Let consider the partition of Figure 1 in which 5 is the value of the dimension A, and let the iceberg condition be count(*)>1. In Figure 5, $B_D^{iceberg}[1] = 1$ because the pair group-by (5,1) satisfies the iceberg condition. The same applies for the pair (5,3), whereas the bit $B_D^{iceberg}[9] = 0$ because the count of group-by (5,9) is 1. During the execution of AGGREGATEFIRSTLEVEL procedure (Figure 2 bottom), the iceberg bitmap of dimension d_k is computed. Moreover, iceberg bitmaps of dimensions d_i, for $k < i \leq n$ have been already computed. Before starting the tuple aggregation (line 8 of Figure 2), the entries of $B_{d_k}^{SL}$ in the current value of d_k, corresponding to values of dimensions d_i, for $k < i \leq n$, which do not satisfy the iceberg condition, are set to 0. This is done by intersecting $B_{d_k}^{SL}$ in the current value of d_k with the iceberg bitmaps of dimensions d_i, for $k < i \leq n$. Such an intersection yields the set of candidates of each dimension d_i, $k < i \leq n$, that must be verified. The verification will be done through the aggregate procedure (Figure 2, Lines 4-9). In the running example, Figure 5, the group-by between dimension A and C in the values (5,2), satisfies the iceberg condition (recall that the condition is count(*) >1). The candidate elements of dimension D which can be aggregated with A and C (in the values (5,2)) are the ones obtained by intersecting $B_C^{SL,D}[2] = 01100$ with $B_D^{iceberg} = 110$. The filtered bitmap will be $B_C^{SL,D}[2] = 01000$. The bit set in $B_C^{SL,D}[2]$ corresponds to value 3 of dimension D. Candidates are verified through the procedure of Figure 2 (lines 4-9). That is, $B_C^{FL}[2] = 01100$ is intersected with $B_D^{FL}[3] = 10010$. The tuple (5 * 2 3,1) does not satisfy the iceberg condition. Therefore, $B_C^{SL,D}[2] = 00000$. This implies that value 2 of dimension C cannot be further aggregated.

3.3   Management of Large Bitmaps

A non trivial shortcoming of the proposed method is memory requirement. Large datasets generate big bitmaps which influence the behavior of the proposed system in terms of used space. For example, in the case of skewed data, long runs of zeros, which can be compressed, are stored inside bitmaps. A variation of the Word-Aligned Hybrid (WAH) [12] bitmap compression algorithm has been applied. WAH is an efficient compression method based on the run-length encoding. A bitmap is represented using a counter and a mixture of 0s and 1s. The basic idea of run-length code is to represent list of consecutive identical bits by its length together with its value. The leftmost bit of a word (a word has 32 bits) is used to distinguish a literal word or a fill word. A literal word contains a list of 31 bits values. The second bit of a fill word indicates the fill bit (0/1) and the 30 lower bits store the length. Our variation of WAH maintains the words literal only. For each run of literals (one or more word literals) a pair of integer values are stored (compressed literal range). The first represents the position of the word literal in the non compressed bitmap, the second stores the number of consecutive words literal present in the run increased by the length of the previous run of literals. Such an algorithm is able to produce smaller bitmaps and has been equipped with the AND boolean operator. In the next section, the compressed version of BitCube will be referred as CBitCube.

4   Results

To evaluate the efficiency of the proposed method, experimental analysis on both synthetic and real-world datasets has been performed. BitCube and CBitCube have been compared with full and iceberg cubing algorithms (BUC [3], MM-Cubing [11] and StarCubing [13]) using the C++ implementations of the Illimine package[1]. BitCube has been compared with the implementation of CURE [8] provided by the authors. Comparisons of BitCube with MultiWay [15] are not provided since its implementation is no longer available. However, StarCubing, in [13], has been shown to be more efficient than MultiWay in full cube computation. Experiments were performed on a server HP Proliant DL380 with 4GB RAM, equipped with Linux Debian Operating System. BitCube was implemented in C++ language[2]. In the experiments, S represents the data skewness (zipf) of the data (i.e. as bigger is S as more skewed are the data, $S = 0$ represents uniform data). Synthetic datasets have been obtained using the generator provided in the Illimine package. Following [13], two kinds of distribution have been considered, uniform and skewed with Zipf. Input data have been generated as flat files of integer ranging from 0 to 10000. Algorithms running times include output writing. Data reading is not considered because it is negligible. All the algorithms were executed using the COUNT operator as aggregation measure.

Concerning real dataset the Weather Dataset[3] and the Retail Dataset have been used. The Weather dataset consists of 9 dimensions and 1,015,367 tuples. It represents weather conditions in September 1985. The following dimensions with the respective cardinalities have been selected: station-id (7037), longitude (352), solar-altitude (179), latitude (152), present-weather (101), day (30),

---

[1] http://illimine.cs.uiuc.edu/download/index.php

[2] All the experiments together with the source code will be available at: http://ferrolab.dmi.unict.it/bitcube/

[3] http://cdiac.ornl.gov/ftp/ndp026b/

weather-change-code (10), hour (8), and brightness(2). Many of the attributes were highly skewed, and some of them were also significantly correlated (e.g. latitude, longitude). The Retail Dataset[4] was obtained from a small enterprise operating in the field of information systems for retail. It is a sparse relation consisting of 6 dimensions and 300,000 tuples. It represents detail of transactions of December 2004. The following dimensions with the respective cardinalities have been selected: customer-id (10439), transaction-id (55323), year-month-day-hour (58628), quantity (15), item-id (8456), and item-description (8428). Two of these attributes were significantly correlated (e.g. item-id, item-description).

In what follows, comparisons with CURE are not reported because CURE is released on Windows system whereas our system is optimized for Linux. However, we tried, on uniform data and for flat cube computation, the behavior of the two methods on the two different platforms. Experiments showed that they are comparable.

Full cube computation. In the first set of experiments (Figure 6, Figure 7) BitCube is compared with all other algorithms with respect to the full cube computation on uniform synthetic datasets. Cardinalities and dimensions have been analyzed in order to measure sparseness and the dimensionality effects. Figure 6 depicts algorithms behavior varying the dimensions from 3 to 10 with cardinalities from 10 to 10000. Experiments showed that BitCube performances increased with respect to the sparsity of the relations and the number of dimensions. For dense relations, it outperformed the other methods with a number of dimensions greater than 6. For example, with 10 dimensions (cardinalities 10 and 100) it showed a speed up, compared to StarCubing, ranging from 9% to 13%. Notice that, when a line of a competing algorithm extends off the top of the graph means that the algorithm was stopped when its running time exceeded such a time. In sparse relations (from cardinality 1000), results clearly showed that BitCube outperformed the compared algorithms yielding a speed up ranging from 30% to 90%.

Figure 7 measures the algorithms behavior with respect to the effect of sparseness (by increasing the cardinality) using respectively 5 and 10 dimensions. BitCube clearly outperformed the compared algorithms and it was not sensitive to the effect of sparseness. Moreover, it considerably improved as the cardinality of each dimension increased. This was due to the fact that since the sizes of partitions decrease the bitmaps operations got faster. From these experiments, BitCube clearly outperforms the compared algorithms. Note that BitCube and CBitCube have almost the same behavior, however, for uniform data, the compression does not lead to any space saving.

Iceberg cube computation. For iceberg cubing, the anti-monotonicity of COUNT operator is exploited by all the compared algorithms. The algorithms behavior were analyzed on uniform dataset of 1M tuples, varying the cardinalities, dimensions and min_sup. In Figure 8 the effect of sparseness together with min_sup is

---

[4] The dataset is available from the authors un request.

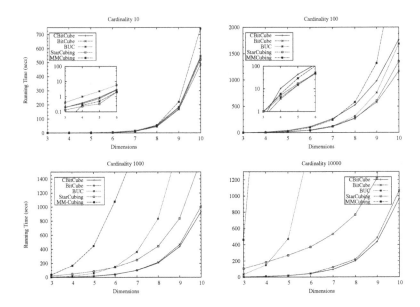

Fig. 6. Comparisons of full cubing on uniform datasets with 1M tuples

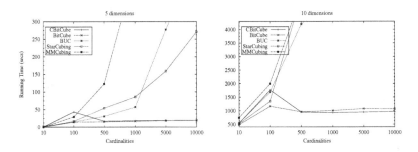

Fig. 7. Comparisons of full cubing on uniform datasets with 1M tuples on dimensions 5 and 10 varying cardinalities

reported. Algorithms were executed varying the dimensions from 5 to 13 with cardinalities from 10 to 100, and min_sup 100 and 1000. In higher dimensions BitCube exhibited the best behavior, showing a running time 19% faster than StarCubing with min_sup 100, and 55% faster than MM-Cubing with min_sup 1000. Also for the iceberg cube case, BitCube and CBitCube show almost the same behavior in the running time but with no any gain in the memory usage.

Data Skew. In this section, experiments show how data skewness affected the performances of cubing algorithms. Zipf has been used to control the skew of the data, varying from 1 to 2. In Figure 9 (a) and (b) experiments on full cubing are

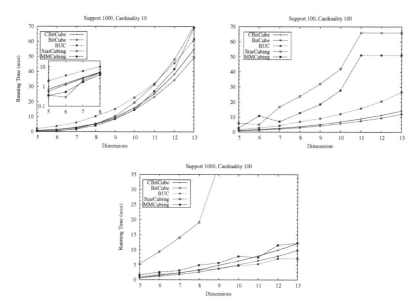

Fig. 8. Comparisons on Iceberg cubing on uniform datasets having 1M tuples

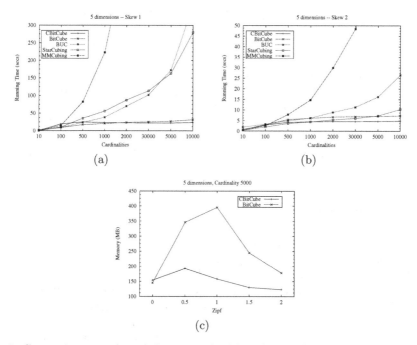

Fig. 9. Comparisons on skewed dataset with 1M tuples, on dimension 5, varying the cardinality from 10 to 10000, with Zipf 1 (a) and 2 (b). (c) Memory used by the two versions of the BitCube algorithm.

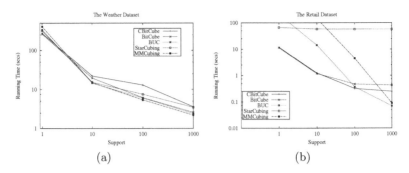

Fig. 10. Comparisons using real datasets: the Weather Dataset (a) and the Retail Dataset (b)

reported. These were done on dimension 5 with Zipf 1 and 2, varying the cardinality from 10 to 10000. Both BitCube and CBitCube perform better than the compared systems scaling very well with the cardinality of the data. Furthermore using skewed data CBitCube has a better behavior than BitCube on both running time and memory usage. In Figure 9 (c), the memory used by the two versions of BitCube is reported. CBitCube shows a compression ratio with respect BitCube up to 60%. BitCube requires more memory than competitors however, by making use of WAH compression, the space overhead is not significant.

Real Data. Concerning real data, Figure 10 reports algorithm comparisons on the two described datasets. BitCube outperformed all other algorithms for full cube computation. Concerning iceberg cubing BitCube resulted comparable with the other algorithms (the same behavior has been observed on synthetic data). Experiments (Figure 10 (a)) show that BitCube is negatively affected by skewness together with low cardinalities. For such kind of data MM-Cubing outperformed the other systems. However, for sparse relations (Figure 10 (b)) with medium cardinality and moderate skewness, BitCube resulted more efficient.

5   Conclusions

BitCube, a bottom-up algorithm for full and iceberg cube computation over sparse relation was presented. It partitions the base table into horizontal fragments with respect to leftmost ordered dimension. It uses bitmaps as inverted indexes to locate positions of dimensions values into records. Unlike previous methods, it does not recursively sort the remaining dimensions into partitions. Moreover, bitmaps are also used as lookup tables to identify values shared by the same records in a partition and are compressed using and adaptation of the WAH algorithm. Experiments showed that BitCube outperforms all the compared systems on full cubing and results comparable on iceberg cubing. Furthermore, the compression of the bitmaps on skewed data resulted very effective. Finally the extension of BitCube for approximate and compressed cubing computations will be exploited. Algorithms for cube maintenance will be investigated.

References

1. Agarwal, S., Agrawal, R., Deshpande, P., Gupta, A., Naughton, J.F., Ramakrishnan, R., Sarawagi, S.: On the computation of multidimensional aggregates. In: Proceedings of 22th International Conference on Very Large Data Bases (VLDB 1996), pp. 506–521 (1996)
2. Agrawal, R., Srikant, R.: Fast algorithms for mining association rules. In: Proc. of the 20th VLDB Conf., pp. 487–499 (1994)
3. Beyer, K., Ramakrishnan, R.: Bottom-up computation of sparse and iceberg cubes. In: Proceedings of the 1999 ACM SIGMOD international conference on Management of data, pp. 359–370 (1999)
4. Chen, Z., Narasayya, V.: Efficient computation of multiple group by queries. In: Proceedings of the 2005 ACM SIGMOD international conference on Management of data, pp. 263–274 (2005)
5. Feng, Y., Agrawal, D., Abbadi, A.E., Metwally, A.: Range cube: Efficient cube computation by exploiting data correlation. In: Proceedings of the 20th International Conference on Data Engineering (ICDE 2004), pp. 658–669 (2004)
6. Gray, J., Chaudhuri, S., Bosworth, A., Layman, A., Reichart, D., Venkatrao, M., Pellow, F., Pirahesh, H.: Data cube: A relational aggregation operator generalizing group-by. Data Mining and Knowledge Discovery 1, 29–54 (1997)
7. Li, X., Han, J., Gonzalez, H.: High-dimensional olap: A minimal cubing approach. In: Proc. Int'l Conf. Very Large Data Bases (VLDB 2004), pp. 528–539 (2004)
8. Morfonios, K., Ioannidis, Y.: Cure for cubes: cubing using a rolap engine. In: VLDB 2006: Proceedings of the 32nd international conference on Very large data bases, pp. 379–390 (2006)
9. Morfonios, K., Konakas, S., Ioannidis, Y., Kotsis, N.: Rolap implementations of the data cube. ACM Comput. Surv. 39(4), 12 (2007)
10. Ross, K.A., Srivastava, D.: Fast computation of sparse datacubes. In: Proceedings of 23rd International Conference on Very Large Data Bases (VLDB 1997), pp. 116–125 (1997)
11. Shao, Z., Han, J., Xin, D.: Mm-cubing: Computing iceberg cubes by factorizing the lattice space. In: Proc. 2004 Int. Conf. on Scientific and Statistical Database Management (SSDBM 2004), pp. 213–222 (2004)
12. Wu, K., Otoo, E.J., Shoshani, A.: A performance comparison of bitmap indexes. In: CIKM 2001: Proceedings of the tenth ACM international conference on Information and knowledge management, pp. 559–561 (2001)
13. Xin, D., Han, J., Li, X., Shao, Z., Wah, B.W.: Computing iceberg cubes by top-down and bottom-up integration: The starcubing approach. IEEE Transaction on Knowoledge and Data Engineering 19(1), 111–126 (2007)
14. Xin, D., Shao, Z., Han, J., Liu, H.: C-cubing: Efficient computation of closed cubes by aggregation-based checking. In: Proc. Int'l Conf. Data Eng (ICDE 2006), vol. 4 (2006)
15. Zhao, Y., Deshpande, P.M., Naughton, J.F.: An array-based algorithm for simultaneous multidimensional aggregates. In: Proceedings of the 1997 ACM SIGMOD international conference on Management of data, pp. 159–170 (1997)

Exact and Approximate Sizes of Convex Datacubes

Sébastien Nedjar

Laboratoire d'Informatique Fondamentale de Marseille (LIF),
Aix-Marseille Université - CNRS
Case 901, 163 Avenue de Luminy, 13288 Marseille Cedex 9, France
lastname@lif.univ-mrs.fr

Abstract. In various approaches, data cubes are pre-computed in order to efficiently answer OLAP queries. The notion of data cube has been
explored in various ways: iceberg cubes, range cubes, differential cubes
or emerging cubes. Previously, we have introduced the concept of convex
cube which generalizes all the quoted variants of cubes. More precisely,
the convex cube captures all the tuples satisfying a monotone and/or antimonotone constraint combination. This paper is dedicated to a study
of the convex cube size. Actually, knowing the size of such a cube even
before computing it has various advantages. First of all, free space can
be saved for its storage and the data warehouse administration can be
improved. However the main interest of this size knowledge is to choose
at best the constraints to apply in order to get a workable result. For
an aided calibrating of constraints, we propose a sound characterization,
based on inclusion-exclusion principle, of the exact size of convex cube
as long as an upper bound which can be very quickly yielded. Moreover
we adapt the nearly optimal algorithm HYPERLOGLOG in order to provide a very good approximation of the exact size of convex cubes. Our
analytical results are confirmed by experiments: the approximated size
of convex cubes is really close to their exact size and can be computed
quasi immediately.

1 Introduction and Motivations

The data cube is a key concept for data warehouse management. Research work
has proposed different variations around this concept. For instance, iceberg cubes
are partial cubes inspired from frequent patterns. They capture only sufficiently
significant trends by enforcing minimal threshold constraints over measures [1].
Range cubes can be seen as extending the previous ones because measures are
constrained in order to belong to a given range [2]. Users are then provided with
trends fitting in a particular "window". New trends appearing (or established
trends disappearing) when a data warehouse is refreshed are exhibited by differential cubes [3]. The latter can be perceived as the result of a set difference
between two cubes. Depending on the order of the two operands, appearing or
disappearing trends are exhibited. Finally, emerging cube [4] captures trends

T.B. Pedersen, M.K. Mohania, and A M. Tjoa (Eds.): DaWaK 2009, LNCS 5691, pp. 204–215, 2009.
© Springer-Verlag Berlin Heidelberg 2009

which are not significant at a moment but which grow significant later. In a dual way, it can exhibit relevant trends which become irrelevant. In addition with the appearing or disappearing trends of the differential cube, the emergent cube provides the decision maker with trend reversals. Such a knowledge is strongly required in multidimensional analysis of data stream [5] and OLAP.

Frequently these different types of cubes, by starting with the original data cube itself [6], have not been grasped as concepts but rather as the result of queries or more efficient algorithms. This is why, we have proposed a unifying structure which offers a soundly founded framework for characterizing the various quoted cubes : The Convex Datacube [7]. The convex cube gathers all the tuples of a datacube satisfying a combination of monotone and/or antimonotone constraints. Moreover it can be represented in a condensed way through its borders (the two sets of tuples which surround the solution space).

Knowing the size of convex cubes, without computing them, can be of great interest. First of all, this size is a critical parameter for the data warehouse administrator who must assess that free space is available for result storage before running a costly computation. It is also relevant when choosing the cuboid set to be materialized. Another important reason behind the knowledge of the convex cube size is related to the choice of the constraints. Actually if the constraints are not suitable to collected data, the user can be provided with a huge amount of tuples difficult or impossible to manage. In contrast, if too strong constraints are applied, only very few exceptional trends can be isolated. Thus in any case, knowing the size of the convex cube can really help the user for calibrating the constraints and the underlying thresholds without useless computations. Finally an alternative method for computing convex cubes using borders (and not by applying directly the constraints) would be workable and fruitful under the condition that borders reduce significantly the size of manipulated data. Once more, predicting the result size is very attractive in order to know what method can apply at best.

Our study of the convex cube size encompasses the following contributions:

(i)   we provide a sound characterization of the exact size of convex cubes. It is based on the concept of order ideal and makes use of the borders of the convex cube. We give a method for computing the size of an order ideal based on the inclusion-exclusion principle. However, computing such an exact size is particularly costly. This is why we continue our work as follows;

(ii) we give an upper bound for the convex cube size and its analytical estimation. This result can be yielded to the user very quickly. Thus he can have quasi immediately a rough idea of the size of the expected cube;

(iii) we also investigate an approach in order to approximate the exact size of convex cubes. Our probabilistic estimation adapts the nearly optimal algorithm HYPERLOGLOG [8]. The method has two main advantages. On one hand the estimated size of convex cubes is obtained very quickly and on the other hand it is really close to the exact size (observed error is typically less than 5%);

(iv) finally we perform various experiments with a twofold objective: comparing the exact size, the approximated size and the upper bound for the convex cubes and measuring the response time necessary to achieve these results. Experiments are convincing. They show that the approximated size is really suitable to the user expectations: it is really very close to the exact size of convex cubes. Moreover it can be obtained nearly immediately. Thus if a user is not satisfied by the returned size (too voluminous or in contrast too small), he can modify the expressed constraints by strengthening or relaxing them.

The article is organized as follows. Section 2 presents the background of our proposal by describing convex cubes and characterizing various variants of cubes on the basis of convex cube. Section 3 is dedicated to the study of the convex cube size. Experimental evaluations are reported in section 4. The conclusion resumes the strenghs of our approach.

2   Convex Datacubes

Let us consider a relation r with a set of attributes dimensions $\mathcal{D}$ (denoted by $D_1, D_2, \ldots, D_n$) and a set of measure (noted $\mathcal{M}$). The Convex Datacube characterization fits in the more general framework of the cube lattice of the relation r: $CL(r)$ [2]. The latter is a suitable search space which is to be considered when computing the data cube of r. It organizes the tuples, possible solutions of the problem, according to a generalization / specialization order, denoted by $\preceq_g$ [9]. These tuples are structured according to the attributes dimensions of r which can be provided with the value ALL [6]. Moreover, we append to these tuples a virtual tuple which only encompasses empty values in order to close the structure. Any tuple of the cube lattice generalizes the tuple of empty values. For handling the tuples of $CL(r)$, the operator $+$ is defined. This operator is a specification of GLB (Greatest Lower Bound) operator applied to the cube lattice framework [2]: provided with a couple of tuples, it yields the most specific tuple in $CL(r)$ which generalizes the two operands.

Example 1. Let us consider the relation DOCUMENT (*cf.* Table 1) giving the quantities of books sold by Type, City, Publisher and Language. In CL(DOCUMENT), let us consider the sales of Novels in Marseilles, whatever the publisher and language are, *i.e.* the tuple (Novel, Marseilles, ALL, ALL). This tuple is specialized by the two following tuples of the relation: (Novel, Marseilles, Collins, French) and (Novel, Marseilles, Hachette, English). Furthermore, (Novel, Marseilles, ALL, ALL) $\preceq_g$ (Novel, Marseilles, Collins, French) exemplifies the generalization order between tuples. Moreover we have (Novel, Marseilles, Hachette, English) + (Novel, Marseilles, Collins, French) = (Novel, Marseilles, ALL, ALL).

Table 1. Relation example DOCUMENT

Type	City	Publisher	Language	Quantity
Novel	Marseilles	Collins	French	100
Novel	Marseilles	Hachette	English	100
Textbook	Paris	Hachette	French	100
Essay	Paris	Hachette	French	600
Textbook	Marseilles	Hachette	English	100

Definition 1 (Measure function)
Let f be an aggregation function, r a database relation and t a tuple (or cell) of $CL(r)$. We denote by $f_{val}(t, r)$ the value of the aggregation function f associated to the tuple t in $CL(r)$.

Example 2. If we consider the Novel sales in Marseilles, for any Publisher and Language, *i.e.* the tuple (Novel, Marseilles, ALL, ALL) of CL(DOCUMENT) we have: $SUM_{val}((\text{Novel, Marseilles, ALL, ALL}), \text{DOCUMENT}) = 200$.

In the remainder of this section, we study the cube lattice structure faced with conjunctions of monotone and antimonotone constraints according to the generalization order. We show that this structure is a convex space which is called convex cube. We propose condensed representations (with borders) of the convex cube with a twofold objective: defining the solution space in a compact way and deciding whether a tuple t belongs or not to this space.

We take into account the monotone and antimonotone constraints the most used in database mining [10,11]. They are applied on:

- measures of interest like pattern frequency, confidence, correlation. In these cases, only the dimensions of $\mathcal{D}$ are necessary;
- aggregates computed from measures of $\mathcal{M}$ and using statistic additive functions (COUNT, SUM, MIN, MAX).

We recall the definitions of convex space, monotone and/or antimonotone constraints according to the generalization order $\preceq_g$.

Definition 2 (Convex Space). *Let $(\mathcal{P}, \leq)$ be a partial ordered set, $\mathcal{C} \subseteq \mathcal{P}$ is a convex space [12] if and only if $\forall x, y, z \in \mathcal{P}$ such that $x \leq y \leq z$ and $x, z \in \mathcal{C}$ then $y \in \mathcal{C}$. Thus $\mathcal{C}$ is bordered by two sets: (i) an "Upper set", noted U, defined by $U = \max_{\leq}(\mathcal{C})$, and (ii) a "Lower set", noted L and defined by $L = \min_{\leq}(\mathcal{C})$.*

Definition 3 (Monotone/antimonotone constraints).

1. *A constraint Const is monotone according to the generalization order if and only if: $\forall\, t, u \in CL(r) : [t \preceq_g u \text{ and } Const(t)] \Rightarrow Const(u)$.*
2. *A constraint Const is antimonotone according to the generalization order if and only if: $\forall\, t, u \in CL(r) : [t \preceq_g u \text{ and } Const(u)] \Rightarrow Const(t)$.*

Notations: We note *cmc* (*camc* respectively) a conjunction of monotone constraints (antimonotone respectively) and *chc* an hybrid conjunction of constraints. By resuming the symbols U and L according to the considered case, the introduced borders are indexed by the type of the constraint in question. For instance, U_{camc} symbolizes the set of the most specific tuples satisfying the conjunction of antimonotone constraints.

Example 3. - In the multidimensional space of our relation example DOCUMENT (*cf.* Table 1), we would like to know all the tuples for which the measure value is greater than or equal to 300. The constraint " SUM(*Quantity*) $\geq$ 300 " is antimonotone. If the amont of sales by Type, City and Publisher is greater than 300, then the quantity satisfies this constraint at a more aggregated granularity level *e.g.* by Type and Publisher (all the cities merged) or by City (all the types and publishers together). In a similar way, if we aim to know all the tuples for which the quantity is lower than 600, the underlying constraint " SUM(*Quantity*) $\leq$ 600 " is monotone.

Theorem 1. - *The cube lattice with monotone and/or antimonotone constraints is a convex space which is called convex cube, $CC(r)_{const} = \{t \in CL(r)$ such that $const(t)\}$. Its upper set U_{const} and lower set L_{const} are:*

$$1.\ if\ const = cmc,\ \begin{cases} L_{cmc} = \min_{\preceq_g}(CC(r)_{cmc}) \\ U_{cmc} = (\emptyset, \ldots, \emptyset) \end{cases}$$

$$2.\ if\ const = camc,\ \begin{cases} L_{camc} = (ALL, \ldots, ALL) \\ U_{camc} = \max_{\preceq_g}(CC(r)_{camc}) \end{cases}$$

$$3.\ if\ const = chc,\ \begin{cases} L_{chc} = \min_{\preceq_g}(CC(r)_{chc}) \\ U_{chc} = \max_{\preceq_g}(CC(r)_{chc}) \end{cases}$$

The upper set U_{const} represents the most specific tuples satisfying the constraint conjunction and the lower set L_{const} the most general tuples respecting such a conjunction. Thus U_{const} and L_{const} result in a condensed representation of the convex cube faced with a conjunction of monotone and/or antimonotone constraints.

The characterization of the convex cube as a convex space makes it possible to know whether a tuple satisfies or not the constraint conjunction by only knowing borders of the convex cube. Actually if a conjunction of antimonotone constraints holds for a tuple of *Space(r)* then any tuple generalizing it also respects the constraints. Dually if a tuple fulfils a monotone constraint conjunction, then all the tuples specializing it also satisfy the constraints.

In [7] we give different variants of data cubes and their characterization as convex cube. For each one, we give the SQL query which computes it, its characterization as a convex cube and the expression of its borders. Let us underline that the original data cube can be achieved from the iceberg cube definition by setting the minimal threshold to 1.

3   Sizes of Convex Datacubes

In this section we address the issue of characterizing and computing the size of convex cubes. The main interest of providing such a knowledge to the user is to help him during the constraint calibrating process.

3.1   Exact Size

In this section, we characterize the exact size of the convex cube by using the borders and the concept of order ideal [13].

Definition 4 (Order Ideal)
Let $T \subseteq CL(r)$ be a set of tuples. An order ideal generated by T is denoted by $\downarrow T$ and encompasses all the tuples generalizing at least a tuple of T. $\downarrow T$ is defined as follows:

$$\downarrow T = \{t \in CL(r) \mid \exists t' \in T \text{ such that } t \preceq_g t'\}$$

Definition 5 (Order Filter)
Let $T \subseteq CL(r)$ be a set of tuples. An order filter generated by T is denoted by $\uparrow T$ and encompasses all the tuples generalized by at least a tuple of T. $\uparrow T$ is defined as follows:

$$\uparrow T = \{t \in CL(r) \mid \exists t' \in T \text{ such that } t' \preceq_g t\}$$

Provided with the maximal tuples satisfying an anti-monotone constraint, the whole solution set is the order ideal generated by such maximal tuples. Conversely, the order filter generated by the minimal tuples verifying a monotone constraint gives the solution space.

By using this feature, we characterize the exact size of the Convex Cube through the following proposition.

Proposition 1. *Let $[L, U]$ be borders of the convex cube $CC(r)$. The size of the latter can be expressed in the following way according to the used couple of borders:*

$$|CC(r)| = |\downarrow U \cap \uparrow L|$$

Adapting the inclusion-exclusion principle to the cube lattice, we give a method for compute the size of the order ideal generated by a set of tuples $T = \{t_1, t_2, \ldots , t_n\}$. We define before two tool sequences (E) and (I) :

$$\begin{cases} E_0 = \emptyset \\ E_1 = \{t_1\} \\ \vdots \\ E_n = \{t_1, t_2, \ldots, t_n\} \end{cases} \qquad \begin{cases} I_1 = \emptyset \\ I_2 = \{t_2 + t_1\} \\ \vdots \\ I_n = \{t_n + t \mid t \in E_{n-1}\} \end{cases}$$

Proposition 2. *Let $T \subseteq CL(r)$ be a set of tuples. The following recursive function computes the cardinality of the order ideal generated by T :*

$$\begin{cases} |\downarrow T| \; = |\downarrow E_n| \\ |\downarrow E_n| = |\downarrow E_{n-1}| + |\downarrow t_n| - |\downarrow I_n| \\ |\downarrow t_n| \; = 2^{|Attr(t)|} \end{cases}$$

Unfortunately, computing the exact size of an order ideal is as costly as computing an iceberg cube. Thus the size expressions that we introduce should be approximated with a good accuracy. To meet this objective, we give two upper bounds and adapt an estimation method to our research context.

3.2 Upper Bound

A standard result for estimating the size of a data cube when data are supposed to be uniformly distributed is the following [14,15]: if P elements are chosen uniformly and at random from a set of N elements, the expected number of distinct elements is $N - N(1 - 1/N)^P$.

The formula can be adapted to quickly find the upper bound of the size of Convex Cube since in the worst case the convex cube is the data cube of r. Thus we have:

$$\begin{aligned} |CC(r)| \quad &\leq |DataCube(r)| \\ |DataCube(r)| &\leq \sum_{X \subseteq \mathcal{D}} \min(N_X - N_X(1 - \tfrac{1}{N_X})^{|r|}, |r|) \\ |CC(r)| \quad &\leq \sum_{X \subseteq \mathcal{D}} \min(N_X - N_X(1 - \tfrac{1}{N_X})^{|r|}, |r|) \end{aligned}$$

where $N_X = \prod_{D_i \in X} |r(D_i)|$ and $r(D_i)$ the projection of r on the dimension D_i and N_X is the size estimation of the cuboid according to a dimension set X, of course $N_X \leq |r|$

3.3 Probabilistic Estimation

Computing the exact size of convex cubes is hard and as costly as yielding the convex cubes. This is why we propose another upper bound which is more refined than the previous one, and which can be approximated in an efficient and accurate way.

Proposition 3. *Let $[L, U]$ be borders of the convex cube $CC(r)$. The size of the latter can be approximated in the following way according to the used couple of borders:*

$$|CC(r)| \leq |\downarrow U| - |\downarrow L| + |L|$$

In order to estimate a multi-set size, [8] proposes a cardinality estimator devised to perform suitable and concise enough observations on the hashed values

Algorithm 1. HYPERLOGLOG adapted in order to estimate the cardinality of an order ideal generated by a set of tuples T

Input: T ($T \subseteq CL(r)$)
Output: Cardinality estimation of $\downarrow T$
  define $m = 2^b$ with $b = 6$
  and $\alpha_m = 0.709$
  let M be a collection of m registers initialized to 0
  let $\rho(y)$ be the rank of the first 1-bit from the left in y
  let $h : CL(r) \rightarrow \{0, 1\}^{32}$ be a good hashing function
  for all $t \in T$ **do**
    for all t' such that $t' \preceq_g t$ **do**
      $x := h(t')$
      $j := \langle x_1, x_2, \ldots, x_b \rangle_2$ {the first b bits of x}
      $w := x_{b+1}, x_{b+2}, \ldots$
      $M[j] := max(M[j], \rho(w))$
    end for
  end for
  $E := \alpha_m m^2 . \left(\sum_{j=0}^{m-1} 2^{-M[j]} \right)^{-1}$
  $E^* := E$
  if $E \leq \frac{5}{2}m$ **then**
    let V be the number of registers equal to 0
    if $V \neq 0$ **then** $E^* := m.log(m/V)$
  end if
  if $E \geq \frac{1}{30} 2^{32}$ **then**
    $E^* := 2^{32} log(1 - E/2^{32})$
  end if
  return E^*

---

computed from the input multi-set. The method works fine because on one hand it is particularly efficient and requires a single pass over data and a very small amount of memory and on the other hand it estimates the cardinality of the multi-set with a very good accuracy (observed error is typically less than 5%).

We are provided with borders which condense a convex cube. By applying the algorithm HYPERLOGLOG (*cf.* algorithm 1) over all the tuples generalizing an element of the borders, we can estimate the cardinality of an order ideal generated by U or L and thus obtain the proposed upper bound (*cf.* proposition 3). The intuition behind the algorithm is the following. The used hashing function randomizes data in order that it becomes binary and nearly uniform and independent. Within the binary word obtained, the probability to find the first 1-bit at the position n is: 2^{-n}. By observing this value for all the tuples, a rough idea of the cardinality can be obtained. By simulating the effect of m experiments the cardinality is refined until achieving a good approximation (stochastic averaging). The algorithm returns an approximate cardinality with a very convenient accuracy. Once this cardinality is yielded, we use proposition 1 to get the estimated cardinality of the convex cube.

4   Experimental Evaluations

In order to validate our whole approach, we perform various experiments. For these experiments, we choose to compute a type of convex cube: the range cube which encompasses all the tuples provided with a measure value in the range $[\mathcal{M}_1; \mathcal{M}_2]$. The experimental evaluations are conducted on data issued from a large and various scope of domains similar to the data sets used in [16]. It is well known that synthetic data are weakly correlated [17,18] while many real and statistical databases are highly correlated. For synthetic data[1], we use the following notations to describe the relations: $\mathcal{D}$ the number of dimensions, $\mathcal{C}$ the cardinality of each dimension, $\mathcal{T}$ the number of tuples in the relation, $\mathcal{M}_1$ and $\mathcal{M}_2$ the thresholds corresponding to the range constraints and $\mathcal{S}$ the skewness or zipf of the data. When $\mathcal{S}$ is equal to 0, the data are uniform. When $\mathcal{S}$ increases, the data are more skewed. $\mathcal{S}$ is applied to all the dimensions in a particular database relation. For real data, we use the weather relations SEP83L.DAT and SEP85L.DAT used by [19], which have 1,002,752 tuples with 8 selected dimensions. The attributes (with their cardinalities) are as follows: year month day hour (238), latitude (5260), longitude (6187), station number (6515), present weather (100), change code (110), solar altitude (1535) and relative lunar luminance (155).

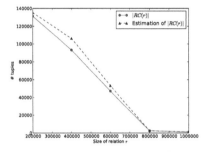

(a) Exact and approximate sizes of the Range Cube with $\mathcal{D} = 10$, $\mathcal{C} = 100$, $\mathcal{S} = 0$

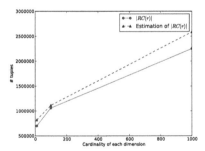

(b) Exact and approximate sizes of the Range Cube with $\mathcal{D} = 10$, $\mathcal{T} = 1000K$, $\mathcal{S} = 0$

First of all, we evaluate and compare the exact size, the upper bound and the approximated size of range cubes. We compute the cubes by adapting the algorithm Star-Cubing [16] in order to achieve their real size. We use the algorithm HYPERLOGLOG in order to yield their approximate size and we also compute the analytical upper bound. For each database, we study the influence, on the computed size or bound, of various parameters like the number of tuples in the original relations, the dimension cardinality, the data skewness and the user given thresholds. The obtained results are presented in figures 1a to 1b. In

---

[1] The synthetic data generator is given at: http://illimine.cs.uiuc.edu/

any case, the exact size and the approximate one are very close. These results reinforce our idea that a good and quick approximation works fine.

Then we compare the execution times required to achieve the previous results, and of course the approximation method is more efficient. But above and over that, our intention is to show that our strategy for calibrating the constraints is workable since the approximate size is achieved specially quickly (very less than 1 second).

We measure and compare the execution times necessary to on one hand achieve the range data cube and its size and on the other hand obtain its estimated size. The gain factor varies from 300 to 500 (*cf.* figures 1a to 1b).

5   Conclusion

In this paper, we take benefit of the unifying structure of the convex cube in order to study the size of a family of datacubes: Iceberg, range, differential, emergent cubes and the datacube itself. The convex cube is expected to have a workable size in order that the knowledge discovery is fruitful. However this size depends on the constraints specified by the user. It is thus especially relevant to predict the size of the convex cube. Supplied with this information, the user can adjust the constraints and choose relevant thresholds. The condition behind this prediction is to get a quasi-immediate result, then the thresholds can be calibrated at best. In order to meet this objective, we propose a very fast computation of an upper bound of the size in question. Then we soundly characterize the exact size of convex cubes. Such a characterization is based on the borders U and L. We also adapt the algorithm HYPERLOGLOG which implements a probabilistic counting method. Its input is the borders of the convex cube and its result an approximate size of such a cube with a good accuracy. Finally, to validate our whole method, we perform experiments. We compute and compare the exact and approximate sizes of convex cubes. Obtained results are convincing: computing the exact size is costly (even with algorithms proved to be efficient) while achieving the approximate size is specially efficient and gives results really close to the exact size.

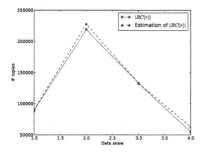

(a) Exact and approximate sizes of the Range Cube with $\mathcal{D} = 10$, $\mathcal{C} = 100$, $\mathcal{T} = 1000K$

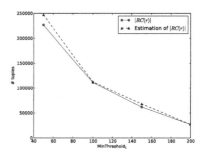

(b) Exact and approximate sizes of the Range Cube for weather relations

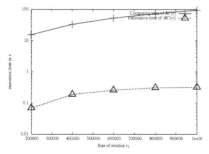

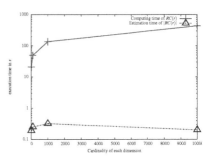

(a) Estimation and computation time of Range Cube with $\mathcal{D} = 10$, $\mathcal{C} = 100$, $\mathcal{S} = 0$

(b) Estimation and computation time of Range Cube with $\mathcal{D} = 10$, $\mathcal{T} = 1000K$, $\mathcal{S} = 0$

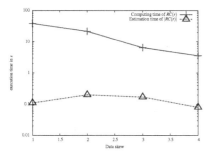

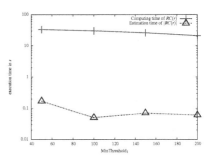

(a) Estimation and computation time of Range Cube with $\mathcal{D} = 10$, $\mathcal{C} = 100$, $\mathcal{T} = 1000K$

(b) Estimation and computation time of Range Cube for weather relations

References

1. Beyer, K.S., Ramakrishnan, R.: Bottom-up computation of sparse and iceberg cubes. In: Delis, A., Faloutsos, C., Ghandeharizadeh, S. (eds.) SIGMOD Conference, pp. 359–370. ACM Press, New York (1999)
2. Casali, A., Cicchetti, R., Lakhal, L.: Cube lattices: A framework for multidimensional data mining. In: Barbará, D., Kamath, C. (eds.) SDM. SIAM, Philadelphia (2003)
3. Casali, A.: Mining borders of the difference of two datacubes. In: Kambayashi, Y., Mohania, M., Wöß, W. (eds.) DaWaK 2004. LNCS, vol. 3181, pp. 391–400. Springer, Heidelberg (2004)
4. Nedjar, S., Casali, A., Cicchetti, R., Lakhal, L.: Emerging cubes for trends analysis in olap databases. In: Song, I.-Y., Eder, J., Nguyen, T.M. (eds.) DaWaK 2007. LNCS, vol. 4654, pp. 135–144. Springer, Heidelberg (2007)
5. Han, J., Chen, Y., Dong, G., Pei, J., Wah, B.W., Wang, J., Cai, Y.D.: Stream cube: An architecture for multi-dimensional analysis of data streams. Distributed and Parallel Databases 18(2), 173–197 (2005)
6. Gray, J., Chaudhuri, S., Bosworth, A., Layman, A., Reichart, D., Venkatrao, M., Pellow, F., Pirahesh, H.: Data cube: A relational aggregation operator generalizing group-by, cross-tab, and sub totals. Data Min. Knowl. Discov. 1(1), 29–53 (1997)

7. Casali, A., Nedjar, S., Cicchetti, R., Lakhal, L.: Convex cube: Towards a unified structure for multidimensional databases. In: Wagner, R., Revell, N., Pernul, G. (eds.) DEXA 2007. LNCS, vol. 4653, pp. 572–581. Springer, Heidelberg (2007)

8. Flajolet, P., Fusy, E., Gandouet, O., Meunier, F.: Hyperloglog: the analysis of a near-optimal cardinality estimation algorithm. In: Proceedings of the Conference on Analysis of Algorithms, AofA 2007, pp. 127–146 (2007)

9. Lakshmanan, L.V.S., Pei, J., Han, J.: Quotient cube: How to summarize the semantics of a data cube. In: VLDB, pp. 778–789. Morgan Kaufmann, San Francisco (2002)

10. Han, J., Kamber, M.: Data Mining: Concepts and Techniques. Morgan Kaufmann, San Francisco (2006)

11. Pei, J., Han, J., Lakshmanan, L.V.S.: Pushing convertible constraints in frequent itemset mining. Data Min. Knowl. Discov. 8(3), 227–252 (2004)

12. Vel, M.: Theory of Convex Structures, vol. (50). North-Holland, Amsterdam (1993)

13. Stumme, G., Taouil, R., Bastide, Y., Pasquier, N., Lakhal, L.: Computing iceberg concept lattices with titanic. Data Knowl. Eng. 42(2), 189–222 (2002)

14. Feller, W.: An Introduction to Probability Theory and Its Applications, vol. 2. John Wiley & Sons, Chichester (1971)

15. Shukla, A., Deshpande, P., Naughton, J.F., Ramasamy, K.: Storage estimation for multidimensional aggregates in the presence of hierarchies. In: Vijayaraman, T.M., Buchmann, A.P., Mohan, C., Sarda, N.L. (eds.) VLDB, pp. 522–531. Morgan Kaufmann, San Francisco (1996)

16. Xin, D., Han, J., Li, X., Shao, Z., Wah, B.W.: Computing iceberg cubes by top-down and bottom-up integration: The starcubing approach. IEEE Trans. Knowl. Data Eng. 19(1), 111–126 (2007)

17. Zaki, M.J., Hsiao, C.J.: Charm: An efficient algorithm for closed itemset mining. In: Grossman, R.L., Han, J., Kumar, V., Mannila, H., Motwani, R. (eds.) SDM. SIAM, Philadelphia (2002)

18. Pasquier, N., Bastide, Y., Taouil, R., Lakhal, L.: Efficient mining of association rules using closed itemset lattices. Information Systems 24(1), 25–46 (1999)

19. Xin, D., Shao, Z., Han, J., Liu, H.: C-cubing: Efficient computation of closed cubes by aggregation-based checking. In: Liu, L., Reuter, A., Whang, K.Y., Zhang, J. (eds.) ICDE, p. 4. IEEE Computer Society, Los Alamitos (2006)

Finding Clothing That Fit through Cluster Analysis and Objective Interestingness Measures

Isis Peña[1], Herna L. Viktor[1], and Eric Paquet[1,2]

[1] School of IT and Engineering, University of Ottawa, Ottawa, Canada
ipena011@uottawa.ca, hlviktor@site.uottawa.ca
[2] National Research Council of Canada, Ottawa, Canada
eric.paquet@nrc-cnrc.gc.ca

Abstract. Clothes should fit consumers well, be aesthetically pleasing and comfortable. However, repeated studies of customers' levels of satisfaction indicate that this is often not the case. For example, more robust males often find it difficult to find pants that are the correct length and fit their waists well. What, then, are the typical body profiles of the population? Would it be possible to identify the measurements that are of importance for different sizes and genders? Furthermore, assuming that we have access to an anthropometric database would there be a way to guide the data mining process to discover only those relevant body measurements that are of the most interest for apparel designers? This paper describes our results when addressing these questions through cluster analysis and interestingness measures-based feature selection. We explore a database containing anthropometric measurements as well as 3-D body scans, of a representative sample of the Dutch population.

Keywords: Utility-based data mining, anthropometry, interestingness measures, feature selection, cluster analysis, classification, relational and 3-D data.

1 Introduction

Apparel manufacturers develop sizing systems with the goal of satisfying consumers' needs for apparel that fits. Sizing is the process used to establish a size chart of key body measurements for a range of apparel sizes. To produce garments that fit the population, it follows that the sizes must correspond to real grouping. However, this is often not the case, as indicated by the results obtained by Shofield and LaBat [1]. Their study of forty size charts for women's clothing showed that the different sizes are defined using arbitrary constant intervals between sizes, all vertical measurements increase as the size increases and that the differences between the principal girths are constant for all sizes. Considering this situation, it is easily understandable that repeated studies of the degree of satisfaction with apparel show that consumers' needs are not being met. For example, a North American study found that about 50% of women and 62% of men cannot find satisfactorily fitting clothes [2]. According to Ashdown et al. [3] two main issues have limited the ability of the apparel companies to produce garments with quality fit. First, there has been a lack of up-to-date

T.B. Pedersen, M.K. Mohania, and A M. Tjoa (Eds.): DaWaK 2009, LNCS 5691, pp. 216–228, 2009.
© Springer-Verlag Berlin Heidelberg 2009

anthropometric data to describe the civilian population. Second, there is a lack of information about the principal aspects to consider when designing garments for a variety of body sizes and shapes.

Recent work has addressed, to some extent, these problems. Anthropometric surveys such as the *CAESAR^{TM} project* [4], *Size USA* [5] and *Size UK* [6] have been carried out on civilian populations. The *CAESAR^{TM} project* [4] includes several body measurements such as waist circumference, hip circumference, height, weight, etc. together with 3-D body scans of each participant. In *Size USA* [5] and *Size UK* [6] subjects were scanned in 3-D, and the body measurements were then extracted from the 3-D scans. Some recent studies attempt to find the most important aspects to be taken into account when designing garments. Viktor et al. [7] finds body size groupings in a sample of the North American male population. Veitch et al. [8] aim to produce a well fitting bodice for Australian women. After selecting twelve out of fifty-four measures and applying Principal Component Analysis (PCA), they define thirty-six categories: twelve sizes and three body shapes within each size. Hsu et al. [9] identify three body types and thirty-eight sizes for the female adult Taiwanese population by applying PCA on eleven anthropometric measures.

Although the abovementioned work attempt to address the problem of identifying the main aspects that should be consider for the design of garments, they focus only either on a specific body part, or on a gender. Moreover, they do not account for the economic factors of the data mining process. That is, they do not attempt to find the subset of body measurements that would be of the most interest when designing apparel for different sizes within each gender. This paper addresses this issue. Our goals are as follows. We aim to understand the typical consumers' body profile by identifying the natural body size groups and their distinctive characteristics. Also, we attempt to find the most important body measurements that define each size, and study how these measurements interrelate. Importantly, we aim to reduce the cost of the mining process, and the subsequent cost of apparel design, by reducing the number of body measurements to be used. To this end, we employ interestingness measures to identify the minimal sets of body measurements that are relevant for the different sizes, within each gender. In this way, we obtain reduced body measurements (both anthropometric and 3-D) of high utility, to be used to optimize apparel design.

This paper is organized as follows. Section 2 introduces utility-based data mining and, in particular, objective interestingness measures, which is used to guide our data mining process. This section explains the need for, and usefulness of, these measures when evaluating the results of the data mining process and when used for feature selection. This is followed, in Section 3, with an introduction to the CAESAR^{TM} anthropometric database. This section also explains our methodology and results when characterizing the population. In this section, we describe the cluster analysis of both the anthropometric measurements and 3-D shapes. In Section 4, we present the approach followed when reducing the number of body measurements, through the use of interestingness measures-based feature selection together with feature extraction. This section also discusses the results obtained for the various body sizes, within each gender. Section 5 concludes the paper.

2 Finding Interesting Patterns

Over the past years, research in data mining mainly focused on producing accurate models with little consideration of the economic aspects of the data mining process. For most real scenarios such as our case study, however, it becomes important to consider economic aspects of the mining process. This has lead to the emerging field of *utility-based data mining* which aims to consider all utility aspects in the data mining process, and maximize the utility of the entire process [10, 11]. In particular, a subset of utility-based approaches focuses on reducing the time and mining space cost by using *interestingness measures*. Interestingness measures are "measures" that narrow the search space and find the truly interesting patterns allowing pruning, evaluating, selecting and ranking patterns according to their potential interest to the user. Thus, using interestingness measures to identify interesting patterns allows the time and space cost of the mining process to be reduced [12]. Some criteria have been proposed to determine whether a pattern is interesting. These criteria are: *Generality, Reliability, Conciseness, Peculiarity, Diversity, Novelty, Surprisingness, Utility* and *Actionability* [12, 13]. The first five criteria are considered objective while the remaining four are considered subjective. Based on these criteria, interestingness measures are then divided into two categories: *objective* and *subjective*. For classification rules, the most important role of objective interestingness measures is to act as heuristics. However, interestingness measures such as *Information gain*, *Gain ratio* and *Chi-squared* are not only used as heuristics, but are also used for *feature selection* [14, 15]. In our case study, we consider the results of the data mining process to be interesting if it is able to correctly and accurately characterize the bodies of different sizes, within each gender. Furthermore, our aim is to find only those body measurements which are crucial when designing a garment for a size and therefore need special consideration to ensure that the clothes fit well.

3 Characterization of the Population

One of the most important challenges for the clothing industry is to produce garments with quality fit. Poor fitting garments may never be sold or customers may return them. In order to produce better fitting garments, accurate and up-to-date measurements need to be further analyzed in order to be able to better characterize the population [3]. To address the aforementioned issue, we aim to find the natural body size groupings using the anthropometric measurements and the 3-D shapes as contained in the CAESAR^{TM} anthropometric database. From these groups we identify size *Archetypes* and their most important characteristics.

The CAESAR^{TM} database [4] includes traditional anthropometric measurements of a large number of individuals from North America, Italy and the Netherlands. This number is forty-four (44) for the males, and forty-five (45) for the females, since under bust circumference is not appropriate for the male subjects. These measurements include height, weight, acromial height, waist circumference, thigh circumference and foot length, amongst others, which were recorded by domain experts. Additionally, the shape of each person was scanned in three dimensions using a full body scanner. That is, a laser scanning device measured and recorded detailed geometry of the subjects' body surface. The 3-D body scans were described using a global shape-based descriptor, which is an abstract and compact representation of the three-dimensional shape of the corresponding body. In essence, each scan is represented by a set of three histograms, which constitutes a 3-D shape index or descriptor for the human body [16]. This index characterizes the radial and angular distribution of the surface elements associated with a given body. The index is designed to be orientation invariant and robust against pose variation [16].

All our experiments are implemented in WEKA, a collection of machine learning algorithms for data mining tasks [17]. We use the Cleopatra visual data mining tool [16] to determine whether our results correspond to reality. We proceed as follows. Firstly, we use the anthropometric data to find the clusters therein, as described in Section 3.1. Next, we determine the Centroids of each cluster. For each Centroid, we use the Cleopatra system to retrieve the human body corresponding to it, together with the N nearest neighbours, where N is the number of bodies in this cluster. We verify the cluster membership by visually inspecting the results. This process is repeated for the 3-D body scans, as shown in Section 3.2. Next, we apply dimensionality reduction, as discussed in Section 4, to find reduced sets of anthropometric body measurements, and 3-D descriptors, respectively. In this study we consider the Dutch population, both males and females. The data was then first separated based on the gender of the subjects. The resulting sets consist of 567 males and 700 females.

3.1 Clustering of Anthropometric Measurements

In order to identify the natural body size groupings, we first apply cluster analysis techniques to the anthropometric data. Cluster analysis is an unsupervised learning data mining technique used to partition a set of physical or abstract objects into subsets or clusters based on data similarity [14, 17].

In the context of tailoring, the ideal scenario is to cover the greatest number of people with the fewest number of sizes [9]. Therefore, we aim to find the minimum number of clusters that fully characterize the population. Since three clusters is the minimum number that makes sense from a tailoring point of view, i.e. *small, medium*

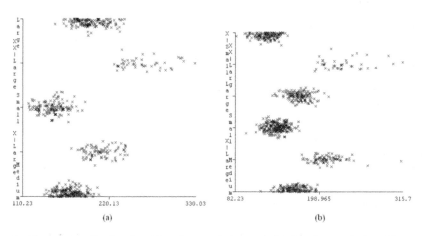

Fig. 1. Cluster visualization for (a) male population and (b) female population. The y axis represent the clusters, the x axis is the weight range.

and *large*, we start partitioning the data into three clusters. Then, by inspecting the-cluster distribution we decide whether is worth splitting the clusters, as described in [17]. This process is repeated until the clusters appear well-defined.

For our cluster analysis experimentation, a number of clustering algorithms was considered. These included partitioning, hierarchical, density-based, model-based and grid-based approaches. By inspection of the cluster distribution and through the analysis of the results using Cleopatra, we found that the male population is best characterized with five clusters and the female population with six clusters. We also observed that the best partition is achieved using a *density-based* algorithm with k-means components. The clusters obtained with this algorithm are shown in figure 1, where it may be observed that the clusters are compact and well defined. In order to visually validate the quality of the clusters produced by the density-based algorithm with k-means components, we use the cluster information to navigate through the 3-D body scans using the Cleopatra system and retrieve the body scans of the Archetypes or Centroids, as identified from the anthropometric data. Tables 1 and 2 indicate some of the characteristics of the Centroids of the male and female populations, respectively. Figures 2 and 3 show the 3-D body scans of the human subjects that correspond to these measurements, highlighting the difference in body types of the clusters. Thus, by inspecting the mean values in tables 1 and 2, the cluster distribution, and through the analysis of the results using Cleopatra, we observe that the anthropometric clusters discriminate between the different body sizes.

Table 1. Body measurements of male Centroids. The table shows the means (in cm), the standard deviation in parenthesis, and the number of subjects on each cluster.

	Small	Medium	Large	X-Large	XX-Large
Chest Circumference	93.1 (6.0)	96.5 (4.8)	107.8 (6.7)	104.9 (6.2)	121.2 (7.0)
Waist Circumference	82.5 (6.9)	86.5 (5.7)	96.3 (7.4)	97.4 (5.7)	112.5 (9.1)
Hip Circumference	95.3 (4.6)	98.5 (3.6)	103.1 (4.5)	108.1 (4.0)	116.3 (6.2)
NeckBase Circumference	45.4 (2.4)	47.4 (2.5)	50.3 (2.7)	51.0 (2.3)	55.2 (3.4)
Shoulder to Wrist	60.3 (3.3)	65.4 (2.6)	62.4 (3.0)	67.2 (3.2)	65.6 (3.6)
Stature	173.0 (6.6)	184.8 (4.6)	176.1 (5.6)	193.2 (6.2)	188.1 (8.3)
Shoulder Breadth	44.0 (1.6)	46.6 (1.8)	47.8 (2.0)	48.8 (2.0)	52.0 (3.4)
Weight (lbs)	149.9(14.8)	171.5(12.2)	198.9(17.5)	214.4(16.9)	264.1(29.2)
Num. of Subjects	126 (22%)	173 (31%)	139 (25%)	82 (14%)	47 (8%)

Table 2. Body measurements of female Centroids. The table shows the means (in cm), the standard deviation in parenthesis, and the number of subjects on each cluster.

	X-Small	Small	Medium	Large	X-Large	XX-Large
Bust Circumf.	90.7(5.9)	92.1 (5.0)	98.4 (6.2)	106.1(6.7)	117.5(8.3)	120.6(9.2)
Waist Circumf.	74.1 (6.4)	76.4 (5.9)	83.5 (7.1)	90.7 (8.0)	103.3(9.4)	107.9(11.0)
Hip Circumf.	97.2 (5.1)	101.2(5.4)	106.4(5.1)	110.1(5.8)	117.5(7.7)	121.6(10.4)
Shoulder to Wrist	55.1 (2.4)	58.5 (1.8)	62.0 (2.2)	56.8 (1.9)	58.6 (2.2)	63.0 (2.0)
Stature	160.0(4.7)	169.7(3.7)	176.9(4.8)	162.4(5.1)	167.1(4.3)	176.9(5.6)
Shoulder Breadth	40.3(1.7)	41.7(1.7)	44.0(2.1)	43.6(2.4)	47.0(2.9)	47.6(2.9)
Weight (lbs)	126.5(13.1)	141.1(12.4)	165.3(14.5)	171.1(13.3)	210.0(21.8)	234.3(28.9)
Num. of Subjects	130 (19%)	198 (28%)	125 (18%)	125 (18%)	83 (12%)	39 (6%)

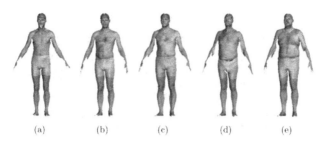

Fig. 2. Cluster Centroids for the male population. (a) Small, (b) Medium, (c) Large, (d) X-Large and (e) XX-Large.

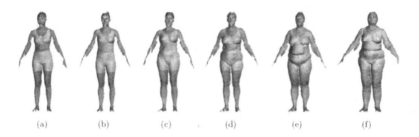

Fig. 3. Cluster Centroids for the female population. (a) X-Small, (b) Small, (c) Medium, (d) Large, (e) X-Large and (f) XX-Large.

By inspecting the body measurements of the male Centroids shown in table 1, it may be observed that the *XX-Large* subjects are generally shorter than the *X-Large* subjects. This is an important feature to consider when designing, for example, pants; the legs should have short lengths for the *XX-Large* size. We also observe that the *Large* and *X-Large* subjects have similar waist circumference, but the former have slightly wider chest. Subsequently, when designing jackets or shirts for the *Large* size subjects, these should be wider and shorter than the ones designed for the *X-Large* subjects.

For the female population is noticeable that the tallest individuals are among the *Medium* and *XX-Large* sizes. Also, the longest arm length is among the *Medium* and *XX-Large* populations, surpassing importantly the *Large* and *X-Large* size subjects. This may be interpreted as follows. The *Large* and *X-Large* subjects do not have the same constitution as the *Medium* size individuals. They are shorter and more robust than the *Medium* size subjects. This provides valuable information about garment design, for different sizes. For the *Medium* size, the clothes have to be designed mainly for tall subjects, while the design of garments for the *Large* and *X-Large* sizes, the girths are the primary aspect to take into account.

3.2 Clustering of 3-D Shapes

Following the same approach as with the anthropometric data, we subsequently performed cluster analysis on the 3-D data, which allows for the analysis of individuals

Table 3. Number of subjects per cluster

	Males	Females
X-Small	-	132 (20%)
Small	135 (25%)	63 (10%)
Medium	79 (15%)	103 (16%)
Large	166 (31%)	149 (22%)
X-Large	106 (20%)	147 (22%)
XX-Large	56 (10%)	69 (10%)

based on their 3-D shape. There are many motivations for using the 3-D shapes directly. For products for which the shape is of critical importance like protective equipments, considering primarily the 3-D scans may yield superior results. In this case, one is most interested in determining which 3-D *shape features* are potentially the most relevant. We proceeded as follows. The 3-D data was separated into two sets based on the gender of the subject. The resulting sets consist of 542 males and 663 females and we applied the density-based algorithm with k-means components to the data. The number of subjects per cluster is given in table 3. We verify the cluster membership through querying the 3-D scans using the Cleopatra system, in order to determine whether our results correspond to the reality. The 3-D body scans of the Centroids of the male and female populations are shown in figures 4 and 5, respectively. From the figures, it may be observed that the Centroids highlight the difference in body types of the different clusters. Thus, the results indicate that the 3-D clusters distinguish between the different body sizes.

By inspecting the number of subjects per cluster in tables 1 to 3, we notice the number of subjects is not the same when the clustering is performed using the anthropometric measurements and the 3-D data. In our attempt to find the reasons for this situation, we analyze the anthropometric measurements and 3-D scans information of a set of representative individuals. Our results indicate that the clustering of 3-D body scans, as may be expected, group together subjects with similar body shape. On the other hand, when the clustering is performed on the anthropometric data, the algorithm finds the clusters containing those subjects with similar height, weight, bust circumference, and so on. A possible "change" in size, by a subject, is thus due to the fact that even though two subjects may have similar body measurements; their *body shape* may be different. For example, two female subjects may have the same bust circumference. However, one

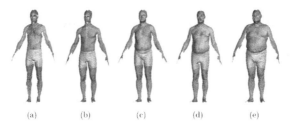

(a)   (b)   (c)   (d)   (e)

Fig. 4. Cluster Centroids from 3-D data for the male population. (a) Small, (b) Medium, (c) Large, (d) X-Large and (e) XX-Large.

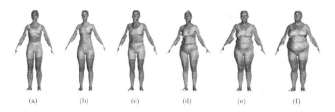

Fig. 5. Cluster Centroids from 3-D data for the female population. (a) X-Small, (b) Small, (c) Medium, (d) Large, (e) X-Large and (f) XX-Large.

of the subjects has a wide back and shoulders, while the other has a narrow back and-large bust, making their body shapes different. Thus, the clustering of the 3-D data and the clustering of anthropometric data provide a *different perspective* on the subject characterization and are therefore complementary to one another.

In the clothing industry, the design and manufacture of garments is based on body measurements. Therefore, the clustering obtained from the anthropometric measurements is most suitable for the direct application to tailoring and garments design. The clustering of 3-D shapes is, on the other hand, more useful when designing e.g. masks for protection against hazardous materials or custom made products such as artificial legs or helmets, where the goal is to produce a good fit for different shapes. Here, the anthropometric measurements are not enough to constraint the shape of an individual and consequently cannot replace a description based on the 3-D shape. Consequently, it is important, from the start, to determine if the particular application is based on measurements or shape.

4   Reducing the Number of Measurements

Recall from Section 2 that utility-based data mining accounts for the economic aspects that impact the mining process, and aims to maximize the utility of the process. In the previous section, we used in our analysis the total number of body measurements. We now aim to reduce the costs and time of the mining process by utilizing interestingness measures to reduce the number of body measurements and the size of the 3-D index. Our reasoning is as follows. The CAESAR^{TM} database contains a large number of attributes, making difficult for domain experts to interpret. Furthermore, even though it is the most comprehensive study of its kind to date, our aim is to develop a scalable solution for future studies which may involve a very large number of participants. Reducing the number of measurements increases the efficiency of the learning process, enhances comprehensibility of the learned results and improves the learning performance (predictive accuracy) [14]. Moreover, the reduced set of body measurements may help identify the measures that require special attention when designing garments, thus potentially reducing the cost and complexity of this process. To this end, we perform two kinds of dimension reduction techniques, namely *feature selection* and *feature extraction*. These techniques allow the reduction of the dataset size by removing irrelevant or redundant dimensions, or attributes, according to some metric [14].

Recall, from section 2, that interestingness measures are used in feature selection to remove the attributes with little or no predictive information. In our case study, this means that we use interestingness measures to identify the subset of the body measurements, or the 3-D descriptor, which is of most importance when describing an Archetype. For feature selection we thus apply *Information Gain, Gain Ratio, Chi Squared*, the *Consistency subset evaluator* and the *CFS subset evaluator*. These are measures that have been widely used in the context of feature selection and have been found to produce good results [18]. For feature extraction we use *Principal Component Analysis* (PCA), a well-known feature extraction method.

In order to verify our results, we consider the full set of anthropometric measurements (forty-four for the males, and forty-five for the females since under bust circumference was only recorded for the females) and the subsets produced by the feature selection and feature extraction. To this end, we constructed a number of classifiers, where the clusters we discovered during the characterization phase acted as class labels. For our experimentation, we consider three different classifiers, namely *RIPPER, C4.5* and *PART*. In the experiments we used ten-fold cross validation to test the accuracy of the classification models. The results of applying PCA and feature selection on the anthropometric data are shown in tables 4 and 5 for males and females, respectively. Shown are the predictive accuracy and, in parenthesis, the number of attributes in the subset.

From table 4 it may be seen that for the males, PCA and all feature selection methods produce good results. That is, the subsets contain a small number of attributes (except for the subsets obtained using the CFS evaluator) and improve the accuracy we obtained using the full (original) set of attributes. We observe that, in general; the highest accuracy is achieved using the subsets produced by Gain Ratio and PCA. Although PCA produces accurate results, its application in a tailoring scenario presents additional challenges, because PCA do not produce a subset of the original attributes. Instead, PCA produces a linear combination of the original set of attributes, preventing the direct application of PCA results in the tailoring process. We therefore select the subset containing six attributes produced by Gain Ratio, because this produces the best trade-off between accuracy and the number of attributes.

Table 4. Results of the attribute reduction for the male population

	Original	PCA	Info Gain	Gain Ratio	Chi Squared	Consistency Subset	CFS Subset
PART	80.3%	82.9%	82.7%	82.2%	81.1%	80.3%	81.3%
		(12)	(13)	(14)	(14)	(12)	(25)
RIPPER	80.6%	81.5%	81.3%	82.4%	81.3%	82.0%	80.3%
		(7)	(9)	(14)	(13)	(8)	(25)
C4.5	78.5%	82.0%	81.8%	82.2%	80.6%	80.4%	80.3%
		(7)	(9)	(6)	(13)	(12)	(31)

Table 5. Results of the attribute reduction for the female population

	Original	PCA	Info Gain	Gain Ratio	Chi Squared	Consistency Subset	CFS Subset
PART	81.1%	80.4%	83.9%	82.3%	84.7%	83.1%	81.9%
		(19)	(12)	(13)	(7)	(7)	(35)
RIPPER	77.9%	77.3%	83.4%	82.9%	82.7%	81.3%	81.9%
		(19)	(13)	(17)	(7)	(7)	(25)
C4.5	77.4%	77.4%	82.9%	83.3%	82.7%	81.7%	79.3%
		(19)	(13)	(11)	(7)	(7)	(25)

For the females (table 5), the best results are obtained using Information Gain, Gain Ratio and Chi Squared. These three interestingness measures produced subsets that highly improve the accuracy. However, the number of attributes in the subsets generated by Information Gain and Gain Ratio is larger than the number of attributes in the subset produced by Chi Squared. We therefore select the subset with seven attributes produced by Chi Squared. The reduced sets of body measurements for both males and females are presented in table 6. For the males, the reduced set of measurements indicates that the most significant measurements are the waist circumference, the chest girth at scye and the vertical trunk circumference. When tailoring shirts, sweaters or jackets, for the male population, these measurements should be considered carefully to produce garments that fit this population properly. For the females, the most important measurements are the bust circumference and, as in the case of the males, the chest girth at scye and the vertical trunk circumference. Therefore, when tailoring clothes for the Dutch females, the bust circumference requires special attention in order to design garments that fit the population better.

We also apply the abovementioned approach to the 3-D shapes. The goal here is to determine which part of the radial-angular distribution of the surface elements is the most relevant, both in terms of design and in terms of time to search. It follows that, in very large databases; the smaller the index, the faster the search. Tables 7 and 8 show the results for the males and females, respectively. The original number of indices used to describe the 3-D body scans is hundred and twenty for both males and females. Shown are the predictive accuracy and, in parenthesis, the number of attributes in the subset.

Table 6. Reduced set of anthropometric body measurements

Males	Females
1. Chest Girth at Scye	1. Arm Length (Spine-Wrist)
2. Hip Breadth Sitting	2. Bust Circumference
3. Stature	3. Chest Girth at Scye
4. Vertical Trunk Circumference	4. Stature
5. Waist Circumference	5. Thumb Tip Reach
6. Weight	6. Vertical Trunk Circumference
	7. Weight

Table 7. Results of the attribute reduction of 3-D shapes for the male population

	Original	PCA	Info Gain	Gain Ratio	Chi Squared	Consistency Subset	CFS Subset
PART	78.8%	81.6%	85.2%	81.0%	83.1%	82.3%	81.0%
		(19)	(58)	(56)	(51)	(24)	(52)
RIPPER	76.8%	80.6%	80.8%	79.3%	81.2%	78.8%	79.2%
		(15)	(54)	(54)	(51)	(8)	(52)
C4.5	79.3%	78.8%	80.3%	79.5%	80.4%	80.1%	80.4%
		(21)	(56)	(56)	(53)	(8)	(72)

Table 8. Results of the attribute reduction of 3-D shapes for the female population

	Original	PCA	Info Gain	Gain Ratio	Chi Squared	Consistency Subset	CFS Subset
PART	78.6%	75.5%	78.4%	79.2%	78.0%	78.0%	81.9%
		(32)	(57)	(54)	(54)	(44)	(59)
RIPPER	75.1%	76.0%	80.4%	80.1%	78.4%	79.8%	79.0%
		(9)	(55)	(55)	(55)	(8)	(88)
C4.5	78.7%	74.1%	77.5%	77.2%	77.1%	78.1%	79.8%
		(12)	(58)	(54)	(54)	(24)	(59)

Table 7 shows that for the male population, Information Gain and Chi Squared achieve the best accuracy. However, the number of attributes in the subsets produced by these measures is much higher than the number of attributes in the subsets produced by the Consistency subset evaluator. Moreover, there is a set produced by the Consistency subset evaluator that contains only eight indices, and the accuracy achieved by this set it is just slightly lower than the one obtained using the subsets generated by Information Gain and Chi Squared. The best trade-off between accuracy and the number of attributes is then achieved by the subset produced using the Consistency subset evaluator containing eight indices.

For the females (table 8), the highest accuracy is achieved by the subsets produced by the CFS subset evaluator. However, the numbers of attributes in these subsets are big. We also notice that PCA and the Consistency subset evaluator produced subsets containing a smaller number of attributes. Moreover, the accuracy, in the case of the subset produced by the Consistency subset evaluator containing eight indices, is comparable to the accuracy obtained using the subsets produced by the CFS subset evaluator. Therefore, the subset produced by Consistency subset containing eight indices, shows the best trade-off between the number of attributes and accuracy. Recall also that PCA provides linear combinations of the various elements of the descriptor, which might be difficult both to interpret and implement in the production line. Consequently, CFS would be preferred in most industrial applications. The subset of elements obtained with CFS provides us not only with a more compact index which higher descriptive power. In addition, it also gives us a set of distances and directions that are the most relevant in general in a 3-D shape; i.e. the distances and directions on which a design may be based. This information may subsequently be used in the scanning process, in order to generate an "intelligent" scan in which the most relevant regions of the body are scanned with higher resolution.

By inspecting the results of applying PCA and feature selection techniques on the anthropometric and 3-D data, it may be observed that in both anthropometric and 3-D data the accuracy is improved using a smaller set of attributes. When comparing the reduction achieved on the anthropometric data, the reduction of the 3-D data is higher. The reduced sets of attributes contain around eight attributes in both the anthropometric and 3-D data, but the original number of indices used to describe the 3-D body scans is hundred and twenty, while the original number of body measurement is forty-five for the females and forty-four for the males. Even though a smaller reduction ratio is achieved on the anthropometric data compared with the 3-D data, this is still significant, since approximately 80% of the body measurements are of less importance. Moreover, this reduction improves the efficiency of the classification results and reduces the complexity of the mining process by using only around 20% of the original attributes.

5 Conclusions

One of the greatest challenges for the apparel industry is to produce garments that fit the customers properly, are aesthetically pleasing and comfortable. In order to produce garments that fit us well, better characterizations of our populations are needed. Furthermore, the different sizes must correspond to real body shapes, in the sense that one or more archetypes should represent the individuals belonging to the same size accurately. Consequently, it is important to define clusters that may be characterized by one archetype, i.e. a truly representative of all other individuals that belong to the same cluster. Based on the assumption that the cluster has a convex or quasi convex symmetry, the archetype then corresponds to the closest individual to the Centroid of the cluster. We might also choose one of the individuals belonging to the sub-region with the highest density in terms of number of individuals. If the resulted clusters are not convex, more than one archetype might be necessary to fully characterize the cluster. In the context of tailoring, however, the optimal scenario is to cover the largest number of people with the fewest number of sizes. In this context then, it is preferred to have only one archetype, since each new size or sub-size involves more tailoring and increases the complexity in the manufacturing.

The method we utilize in this work satisfies the aforementioned requirements, since we were able to group the individuals into clusters with a well defined Centroid. Our verification, when using the Cleopatra system, indicates that the cluster membership corresponds to the reality, in the sense that the bodies correspond to our expectations of the cluster membership. Also, our results indicate that the number of body measurements may be significantly reduced by applying interestingness measure-based feature selection and feature extraction. Moreover, these new sets of reduced body measurements improve the predictive accuracy. These sets therefore contain the most important body measurements for defining the body sizes, and may be used in garment design to identify those body measurements that require special attention, when tailoring clothes for a specific population and gender.

What then, are the general conclusions and lessons to be learned from this data mining effort? Our experience shows that techniques, such as PCA, that combine attributes are difficult to interpret or, at least, to apply in a real world situation, such as when tailoring clothes. In our case study, we obtained better accuracy through dimensional reduction while improving the understandability of the model. It follows that using less attributes is more efficient from a "production" perspective, both from a monetary and procedural point of view. Furthermore, when considering both descriptive (relational) and multimedia data, the relationship is not necessarily intuitive. That is, it not clear if one should replaced the other and which one are more suitable for use; rather, they seem complimentary.

References

1. Schofield, N.A., LaBat, K.L.: Exploring the relationships of grading, sizing and anthropometric data. Clothing and Textiles Research Journal 23(1), 13–27 (2005)
2. DesMarteau, K.: CAD: Let the fit revolution begin. Bobbin 42, 42–56 (2000)
3. Ashdown, S., Loker, S., Rucker, M.: Improved Apparel Sizing: Fit and Anthropometric 3-D Scan Data. Annual Report NTC Project: S04-CR01-07. National Textile Center (2007)
4. Robinette, K.M., Blackwell, S., Daanen, H., Fleming, S., Boehmer, M., Brill, T., Hoeferlin, D., Burnsides, D.: Civilian American and European Surface Anthropometry Resource (CAESAR), Final Report, Volume I: Summary. AFRL-HE-WP-TR-2002-0169, United States Air Force Research Laboratory, Human Effectiveness Directorate, Crew System Interface Division, 2255 H Street, Wright-Patterson AFB OH 45433-7022 (2002)
5. Size USA. The US National Size Survey. Resource, http://www.sizeusa.com/
6. Size UK. UK National Sizing Survey. Resource, http://www.size.org/
7. Viktor, H.L., Paquet, E., Guo, H.: Measuring to fit: Virtual tailoring through cluster analysis and classification. In: Fürnkranz, J., Scheffer, T., Spiliopoulou, M. (eds.) PKDD 2006. LNCS, vol. 4213, pp. 395–406. Springer, Heidelberg (2006)
8. Veitch, D., Veitch, L., Henneberg, M.: Sizing for the Clothing Industry Using Principal Component Analysis - An Australian Example (2007)
9. Hsu, C.-H., Lin, H.-F., Wang, M.-J.: Developing female size charts for facilitating garment production by using data mining. Journal of Chinese Institute of Industrial Engineers 24(3), 245–251 (2007)
10. Zadrozny, B., Weiss, G., Saar-Tsechansky, M.: UBDM 2006: Utility-Based Data Mining 2006 workshop report. SIGKDD Explor. Newsl. 8(2), 98–101 (2006)
11. Weiss, G., Saar-Tsechansky, M., Zadrozny, B.: Report on UBDM-05: Workshop on Utility-Based Data Mining. SIGKDD Explor. Newsl. 7(2), 145–147 (2005)
12. Geng, L., Hamilton, H.J.: Interestingness measures for data mining: A survey. ACM Comput. Surv. 38(3), 9 (2006)
13. McGarry, K.: A survey of interestingness measures for knowledge discovery. Knowl. Eng. Rev. 20(1), 39–61 (2005)
14. Han, J., Kamber, M.: Data Mining: Concepts and Techniques, 2nd edn. Morgan Kaufmann, San Francisco (2006)
15. Kim, Y., Street, W.N., Menczer, F.: Feature selection in data mining. Data mining: opportunities and challenges, 80–105 (2003)
16. Paquet, E., Robinette, K.M., Rioux, M.: Management of three-dimensional and anthropometric databases: Alexandria and Cleopatra. Journal of Electronic Imaging 9, 421–431 (2000)
17. Witten, I.H., Frank, E.: Data Mining: Practical machine learning tools and techniques, 2nd edn. Morgan Kaufmann, San Francisco (2005)
18. Cunningham, P.: Dimension Reduction. Technical Report UCD-CSI-2007-7, University College Dublin, pp. 1–24 (2007)

Customer Churn Prediction for Broadband Internet Services

B.Q. Huang[1], M-T. Kechadi[1], and B. Buckley[2]

[1] School of Computer Science and Informatics, University College Dublin, Belfield,
Dublin 4, Ireland
[2] Eircom Limited, 1 Heuston South Quarter, Dublin 8, Ireland

Abstract. Although churn prediction has been an area of research in the
voice branch of telecommunications services, more focused studies on the
huge growth area of Broadband Internet services are limited. Therefore,
this paper presents a new set of features for broadband Internet customer
churn prediction, based on Henley segments, the broadband usage, dial
types, the spend of dial-up, line-information, bill and payment information, account information. Then the four prediction techniques (Logistic
Regressions, Decision Trees, Multilayer Perceptron Neural Networks and
Support Vector Machines) are applied in customer churn, based on the
new features. Finally, the evaluation of new features and a comparative
analysis of the predictors are made for broadband customer churn prediction. The experimental results show that the new features with these
four modelling techniques are efficient for customer churn prediction in
the broadband service field.

1 Introduction

Services companies of telecommunication service businesses in particular suffer
from a loss of valuable customers to competitors; this is known as customer
churn. In the last few years, there have been many changes in the telecommunications industry, such as, the liberalisation of the market opening up competition
in the market, new services and new technologies. The churn of customers causes
a huge loss of telecommunication service and it becomes a very serious problem.

Recently, data mining techniques have emerged to tackle the challenging problems of customer churn in telecommunication service field [4,16,15,3,11,7,17].
As one of the important measures to retain customers, churn prediction has
been a concern in the telecommunication industry and research [3]. Until now
the majority of churn prediction has been focused on voice services available
over mobile and fixed-line networks. Most of the literature introduces the usage of variables/features (which are customer demographics, contractual data,
customer service logs, call details, complain data, bill and payment, structure
of monthly service fees, as so on)[3,8,11,15,16,18], and the common modelling
techniques (which are Logistic Regressions (LR) Decision Trees (DT), Artificial
Neural Networks (ANN) and Random Forest (RF)) [7,8,11,16,19].

T.B. Pedersen, M.K. Mohania, and A M. Tjoa (Eds.): DaWaK 2009, LNCS 5691, pp. 229–243, 2009.

Broadband Internet services are potentially one of the greatest sources of revenue for providers and consequently feature highly in their marketing campaigns. However, the above techniques have not been applied to the specific area of churn prediction in broadband service field. Until now either very little churn prediction has been carried out on the broadband Internet services over fixed-line networks, or the literature of churn prediction in telecommunication does not provide the details of methodologies for churn prediction using broadband information [4,16,15,3,11,7,17]. Therefore, it is necessary to investigate the churn prediction in Broadband Internet service field.

This paper presents a new set of features with four modelling techniques for customer churn prediction in one telecommunication service field – broadband Internet. The new set of features are extracted from Henley segmentation, broadband usage, dial types, the spend of dial-up, line-information, bill and payment information, account information, call details and service log data. The modelling techniques used to predict churns are LR, DT, ANN and Support Vector Machines (SVM). Finally, based on the proposed features and the modelling techniques, experiments are carried out. The experimental results show that the presented features with the modelling techniques are efficient for broadband customer churn prediction.

The rest of this paper is organised as following: next section introduces the evaluation criterias of churn prediction systems. Section 3 describes our methodology which includes the techniques of feature extraction, normalisation and prediction. Experimental results with discussion are provided in Section 4, and the conclusion of this paper and future works are made in Section 5.

2   Evaluation Criterias

After a classifier or predictor is available, it will be used to predict the further behaviour of customers. As one of important step to ensure the model generalise well, the performance of the predictive churn model have to be evaluated. Table 1 shows a confusion matrix [10], where a_{11} is the number of the correct predicted churners, a_{12} is the number of the incorrect predicted churners, a_{21} is the number of the incorrect predicted nonchurners, and a_{22} is the number of the correct predicted nonchurners. From the confusion matrix, the most common evaluation criterias for a predictive model are introduced as follows:

- The overall accuracy (AC) is the proportion of the total number of predictions that were correct, calculated by $\frac{a_{11}+a_{22}}{a_{11}+a_{12}+a_{21}+a_{22}}$.

Table 1. Confusion Matrix

		predicted	
		CHUN	NONCHU
Actual	CHU	a_{11}	a_{12}
	NONCHU	a_{21}	a_{22}

- The accuracy of true nonchurn (TN) is the proportion of nonchurn cases that were correctly identified, written as $\frac{a_{22}}{a_{21}+a_{22}}$.
- The accuracy of true churn (TP) is defined as the proportion of churn cases that were classified correctly, calculated by $\frac{a_{11}}{a_{11}+a_{12}}$.
- The false churn rate (FP) is the proportion of nonchurn cases that were incorrectly classified as churn, written as $\frac{a_{21}}{a_{21}+a_{22}}$.
- The false nonchurn rate (FN) is the proportion of churn cases that were incorrectly classified as nonchurn, written as $\frac{a_{12}}{a_{12}+a_{11}}$.

There are other evaluation criterias and the details of them can be found in [10]. In this paper, we are more interested in the high accuracy of true churn and the low false churn rate.

3   Methodology

The proposed churn prediction system for broadband Internet consists of sampling data, preprocessing, and classification/prediction phases. Data sampling randomly selects a set of customers and their relative information, according the definition of churn. The preprocessing (also called data preparation) includes data cleaning, feature extraction and normalisation steps. The main task of data cleaning is to remove the irrelevant information which includes wrong spelling words caused by human errors, special mathematical symbols, missing values, strings "NULL", duplicated information, and so on. The task of feature extraction is to select features to address customers. The process of normalisation is to normalise the values of features into a range. The task of prediction phase is to predict the further behaviour of customers. The following subsections describe the features/variable extraction, normalisation and prediction/classification steps.

3.1   Feature/Variable Extraction

The feature extraction plays the most important role which can directly influence the performance of predictive models in the term of prediction rates. If a robust set of features is extracted in this phase, a significant improvement will be yielded. However, it is not easy to obtain such a set of features. Until now, most of the feature sets have been introduced for churn prediction in mobile telecoms industry [8,11,16,3,15] and fixed-line telecommunication [3,18]. However, in these existing feature sets, the broadband Internet information is not included. Thus, it is difficult to use the existing feature set for churn prediction in broadband Internet service field. Based on broadband Internet service information, the following features are selected for broadband Internet churn prediction in telecommunication:

- **Demographic profiles**: describe a demographic grouping or a market segment and the demographic information contains likely behaviour of customers. Usually, this information includes age, social class bands, gender, etc. The available demographic information for this research is gender and country. Therefore, these information may be useful for predicting the further behaviour of a customer.

- **Information of grants**: Some customers have obtained some special grants resulting in their bills being paid fully or partly by other parties. For example, a customer with a disability or over 80 are more unlikely to churn from that service.
- **Account information**: includes the account status, creation date, the bill frequency, the service packages, the account balance, payment types, dial-up types, dial-up cost, broadband opening date, download and upload capacity, total duration usage, average download and upload speeds, and general service usage information which includes the summarised call duration, the number of calls and standard prices, current outstanding charges and charges paid. Account information is very useful for predicting the customer behaviour for the next observation period.

Based on these new features, the average of call duration, the number of calls, the standard prices and the actual fees paid for 30 days (note the duration is a number of minutes) are also considered as new features. Let "$\overline{D_N}$", "$\overline{C_N}$", "$\overline{SP_N}$" and "$\overline{FP_N}$" be the average of call duration, the number of calls, the standard prices and the fees payed in 30 days of the most recent bill, respectively. They are obtained by equation 1:

$$\overline{C_N} = \frac{nCalls\_M}{nDays} * 30$$
$$\overline{D_N} = \frac{Duration\_M}{nDays} * 30$$
$$\overline{SP_N} = \frac{Fees\_M}{nDays} * 30 \tag{1}$$
$$\overline{FP_N} = \frac{Fee\_C}{nDays} * 30$$

where "$nCalls\_M$" is the number of calls in the most recent bill, "$Duration\_M$" is the duration of the most recent bill, "$Fees\_M$" is the fees of the most recent bill, "Fee_C" is the fees from customers, and "$nDays$" is the number of day of the bill, which can be obtained by equation 2.

$$nDays = endDate - startDate \tag{2}$$

where "$endDate$" and "$startDate$" are the dates of bill starting and ending. In addition, the ratio between the actual fees that should be pay and the call-duration of the current bill is extracted as a new feature, which is written as equation 3.

$$R\_AMNT\_DUR = \frac{Fees\_M}{Duration\_M} \tag{3}$$

- **Service orders**: describe the services ordered by the customer. The quantity of the ordered services, the rental charges are selected as new features.
- **Henley segments**: The algorithm of Henley segmentation [2] splits customers and potential customers into different groups or levels according to characteristics, needs, and commercial value. There are two types of Henley segments: the individual and discriminant segments. The individual segment includes ambitious Techno Enthusiast (ATE) and Comms Intense Families (CIF) Henley segments. The discriminant segments are the mutually exclusive segments (DS) and can represent the loyalty of customers. The Henley segments ("DS", "ATE" and "CIF") of the most recent 2 six-months are selected as new input features. Similarly, the missing information of the Henley segments are replaced by neutral data.
- **Broadband Internet and telephone line information**: this includes information about voice mail service (provided or not), the number of broadband lines, the number of telephone lines, and so on. The customers who have more telephone or broadband Internet lines might prefer the services more and they might be more willing to continues using the services. This information cant be useful for a prediction model. Therefore, the number of telephone and broadband lines, and the voice mail service indicator are selected as part of new features.
- **The historical information of payments and bills** : this concerns the billing information for each customer and service for a certain number of years. Each bill includes the total cost, prices, rental charges, call duration, charges paid so far, etc. Attributes monthly cost, rental charges, call duration and paid charges are extracted as new features. They are denoted by "mnCost", "mnRent_fees", "mnDur" and "paidfee", respectively. New other features are also created; the changed-cost, changed call-duration and rental changed-fees and are included in the set of new features. They are obtained by equation 4.

$$changed\_cost_{i,i-1} = \frac{|mnCost_i - mnCost_{i-1}|}{\sum_{j=2}^{T} |mnCost_j - mnCost_{j-1}|}$$

$$changed\_Duration_{i,i-1} = \frac{|mnDur_i - mnDur_{i-1}|}{\sum_{j=2}^{T} |mnDur_j - mnDur_{j-1}|} \quad (4)$$

$$changed\_rental\_Fees_{i,i-1} = \frac{|mnRent\_fees_i - mnRent\_fees_i|}{\sum_{j=2}^{T} |mnRent\_fees_j - mnRent\_fees_j|}$$

where $mnCost_i$, $mnDuration_i$ and $mnRent\_fees_i$ are the cost, call-duration and rental fees of the bill for the month i^{th}.

- **Broadband monthly usage information**: This information is used to record the details of the broadband monthly usage for each customer. Monthly information can show frequency of broadband use, total upload/download, connection duration. The following types of customers may often churn: i) those who have short online sessions, ii)Those who have small upload/download totals, iii) those who have greatly fluctuating monthly usage figures. Therefore features which capture this information must be included.

Therefore, some new features should be extracted from the usage information of broadband Internet for churn prediction in telecommunication service fields, especially in broadband Internet service field. Based on this information, the new extracted features are the sizes of the information downloaded and uploaded, the duration of using Internet every month, the changed sizes of the information downloaded and uploaded, and the changed the online duration of every consecutive two month, and the ratio between the total sizes of information downloaded/uploaded and the duration of online broadband Internet for a month.

Consider the sizes of the information downloaded and uploaded, the duration for month i are "DOW_i", "UP_i" and "ONT_i", respectively. If the change sizes of information downloaded and uploaded, and the duration of online on Internet are "$CH\_DOW_{i,i-1}$", "$CH\_UP_{i,i-1}$" and "$CH\_ONT_{i,i-1}$" respectively, they can be calculated by equations (5), (6) and (7), respectively.

$$CH\_DOW_{i,i-1} = \frac{|DOW_i - DOW_{i-1}|}{\sum_{j=2}^{M'} |DOW_j - DOW_{j-1}|} \tag{5}$$

$$CH\_UP_{i,i-1} = \frac{|UP_i - UP_{i-1}|}{\sum_{j=2}^{M'} |UP_j - UP_{j-1}|} \tag{6}$$

$$CH\_ONT_{i,i-1} = \frac{|ONT_i - ONT_{i-1}|}{\sum_{j=2}^{M'} |ONT_j - ONT_{j-1}|} \tag{7}$$

Consider the ratio between the total sizes of information downloaded/ uploaded and the duration of online broadband Internet for month i is "$R\_GB\_ONT_i$". The ratio can be calculated by equation (8).

$$R\_GB\_ONT_i = \frac{DOW_i + UP_i}{ONT_i} \tag{8}$$

– **Call details**: If the customers did not use the services, he might cease the services in the future. If the fees of services are suddenly increased or decreased, the customer might cease the services sooner. Call-details can reflect this information – how often customer have used the services with relative payment, and so on. The use of call details in churn prediction is reported in [16,18]. The call-detail contain call duration, price and types of call (e.g. International or local call) of every call. It is difficult to store all call details of every call every month for every customer. Most of the telecommunication companies keep the call details of a few months. The limited call details can be used for churn prediction in telecommunication.

Based on these month call details, the aggregated number of calls, duration and fees are extracted as new features. The basic idea for extracting features is to segment the call details into a number of defined periods, then to aggregate the duration, fees and the number of calls for each period for every customer. In literature [16,18], it is reported that the call-details of every 15 or 20 days are efficient. In this paper, the six-month call details are segmented into 15-day period, then number of calls, duration and fees of each 15-day period are aggregated for each customer. For a segment i of a customer's call details, let the aggregated number of calls, duration and fees be "CALL_N_i", "DUR$_i$" and "COST$_i$", respectively. The changed number of calls, changed-duration and changed-cost between two consecutive segment of call details can be obtained by Equation 9.

$$CH\_DUR_{i,i-1} = \frac{|DUR_i - DUR_{i-1}|}{\sum_{j=2}^{M'}|DUR_j - DUR_{j-1}|}$$

$$CH\_N_{i,i-1} = \frac{|CALL\_N_i - CALL\_N_{i-1}|}{\sum_{j=2}^{M'}|CALL\_N_j - CALL\_N_{j-1}|} \quad (9)$$

$$CH\_C_{i,i-1} = \frac{|COST_i - COST_{i-1}|}{\sum_{j=2}^{M'}|COST_j - COST_{j-1}|}$$

where M' is the number of call-detail segments; i and j are the indexes of call-detail segment, and $2 =< i, j <= M'$. In addition, the increment rates of the number of calls, duration and fees are calculated Equation 10.

$$R\_DUR_{i,i-1} = \frac{CH\_DUR_{i,i-1}}{CH\_DUR_{i,i-1} + DUR_{i-1}}$$

$$R\_N_{i,i-1} = \frac{CH\_N_{i,i-1}}{CH\_N_{i,i-1} + CALL\_N_{i-1}} \quad (10)$$

$$R\_C_{i,i-1} = \frac{CH\_C_{i,i-1}}{CH\_C_{i,i-1} + COST_{i-1}}$$

Thus, the new features includes the number of calls, duration, fees, the changed number of calls, changed-duration, changed-fees, the rates of the increased number of calls, the rates of the increased duration and the rates of the increased fees.

3.2   Normalisation

In the extracted features (See subsection 3.1), some predictors or classifiers (e.g. Artificial Neural Networks) have difficulties in accepting the string values of features, such as, genders, county names. The value of a feature was rewritten into binary strings.

In addition, the values of each of these features (e.g the number of lines, the sizes of information downloaded/uploaded, the duration of online on Internet, "R_GB4_ONT$_i$", "$\overline{C_N}$", "CALL_N_1", "DUR$_1$", "COST$_1$", "CALL_N_M", "mnDur$_1$" and "paidfee$_1$", etc.), lie in different dynamical ranges. The large values of features have larger influence over the cost functions than the small ones. However, it cannot reflect that the large values are more important in classifier design. To solve this problem, the values of these features can be normalised into a similar range by Equation 11.

$$\bar{x}_j = \frac{1}{N} \sum_{i=1}^{N} x_{ij}, \; j = 1, 2, \cdots, \iota$$

$$\sigma_j^2 = \frac{1}{N-1} \sum_{i=1}^{N} (x_{ij} - \bar{x}_j)$$

$$y = \frac{x_{ij} - \bar{x}_j}{r\sigma_j}$$

$$\tilde{x}_{ij} = \frac{1}{1 + e^{-y}} \tag{11}$$

Where x_j is the feature j^{th}, ι is the number of features, N is the number of instances or patterns and r is a Constant parameter which is defined by a user. In this study, r is set by one.

3.3   Prediction/Classification

Many techniques have been proposed for churn prediction in telecommunication. Three popular modelling techniques (Logistic Regression, Multilayer Perceptron neural networks and Decision Tree C 4.5) and one promising modelling technique (Support Vector Machines), are selected as predictors from the broadband churn prediction. These four modelling techniques are outlined as follows:

Logistic Regressions: Logistic regression [9] is a widely used statistical modelling technique for discriminative probabilistic classification. Logistic regression estimates the probability of a certain event taking places. The model can be written as:

$$prob(y = 1) = \frac{e^{\beta_0 + \sum_{k=1}^{K} \beta_k x_k}}{1 - e^{\beta_0 + \sum_{k=1}^{K} \beta_k x_k}} \tag{12}$$

where Y is a binary dependent variable which presents whether the event occurs (e.g. y=1 if event takes place, y=0 otherwise), $x_1, x_2, \cdots, x_K$ are the independent inputs. $\beta_0, \beta_1, \cdots, \beta_K$ are the regression coefficients that can be estimated by the maximum likelihood method, based on the provided training data. The detail of the logistic regression models can be found in [9].

Decision trees: A method known as "divide and conquer" is applied to construct a binary tree. Initially, the method starts to search an attribute with best information gain at root node and divide the tree into sub-tree. Summarily, the sub-tree is further separated recursively following the same rule. The partitioning stops if the leaf node is reached or there is no information gain. Once the tree is created, rules can be obtained by traversing each branch of the tree. The details of Decision Trees based on C4.5 algorithm are in literature [13,12].

Artificial neural networks: A MLP is a supervised feed-forward neural network and usually consists of input, hidden and output layers. Normally, the activation function of MLP is a sigmoid function. If an example of MLPs with one hidden layer, the network outputs can be obtained by transforming the activation functions of the hidden unit using a second layer of processing elements, written as follows:

$$Output_{net}(j) = f(\sum_{l=1}^{L} w_{jl} f(\sum_{i}^{D} w_{li} x_i))$$

$$j = 1, \cdots, J \qquad (13)$$

where D, L and J are total number of units in input, hidden and output layer respectively, and f is a activation function. The Back-Propagation (BP) or quick back-propagation learning algorithms would be used to train MLP. The more details with learning algorithm can be found on [14].

Support Vector Machines: An SVM classifier can be trained by finding a maximal margin hyper-plane in terms of a linear combination of subsets (support vectors) of the training set. If the input feature vectors are nonlinearly separable, SVM firstly maps the data into a high (possibly infinite) dimensional feature space by using the kernel trick [5], and then classifies the data by the maximal margin hyper-plane as following:

$$f(\boldsymbol{x}) = sgn\left(\sum_{i}^{M} y_i \alpha_i \phi(\boldsymbol{x_i}, \boldsymbol{x}) + \delta\right) \qquad (14)$$

where M is the number of samples in the training set, $\boldsymbol{x_i}$ is a support vector with $\alpha_i > 0$, ϕ is a kernel function, $\boldsymbol{x}$ is an unknown sample feature vector, and δ is a threshold.

The parameters $\{\alpha_i\}$ can be obtained by solving a convex quadratic programming problem subject to linear constraints [6]. Polynomial kernels and Gaussian radial basis functions (RBF) are usually applied in practice for kernel functions. δ can be obtained by taking into account the Karush-Kuhn-Tucker condition [6], and choosing any i for which $\alpha_i > 0$ (i.e. support vectors). However, it is safer in practice to take the average value of δ over all support vectors.

4   Experimental Results and Discussion

The 139000 customers were randomly selected from the real-world database provided by Eircom[1] in our experiments. In the training dataset, there are 6000 churners, 94000 nonchurners and total 100000 customers. In the testing dataset, there are 39000 customers which includes 2000 churners and 37000 nonchurners. Each customer is represented by the features which are described in Section 3.1. Based on the datasets, three sets of experiments were carried out in this papers, independently.

In the first set of experiments, a number of different feature subsets were used. The features that describe demographic profiles, information of grants, account

Algorithm 1. The procedure of an experiment for a feature subset

1. Select the subset of features
2. Base on the selected feature subset, load the data from the training set.
3. When the predictive model is available, load the data from testing set, based on the selected subset of features.
4. Evaluate the the outputs from the predictive model.

information, henley segments, broadband Internet and telephone line information, 6-month call details and details of broadband usage information depending on the number of months selected, were used. The broadband monthly usage information for a number of months is formed using the current months data in addition to all previous months data e.g. the 3-month data subset contains the data for month 3, 2, and 1 and the 7-month data subset contains the information for month 7, 6, 5, 4, 3, 2 and 1. In the first set of experiments, the number of months is between 1 and 11. Thus, 11 different subsets of features were used. For each subset of features, the general procedure of an experiment which is described by Algorithm 1 was carried out. Four prediction modelling techniques LR, DT, MLP and SVM were used for each subset of features.

For the second set of experiments, the features that describe the details of broadband usage information depending on the number of months selected (including the summary information of broadband usage on the bills), were used. Similarly, the same procedure of selecting feature subsets and the same number of months (from 1 to 11) used in the first set of experiments, were used for the second set of experiments. Therefore, the second set of experiments also used 11 different subsets of features. For each subset of features, the general procedure described by Algorithm 1 was applied to each experiment. In addition, in the second set of experiments, the same modelling techniques were used for each subset of features.

For the third set of experiments, the features without broadband usage information were used. For this subset of features, four prediction modelling techniques (LR, DT, MLP and SVM) were used. Based on this subset of features, the procedure (see Algorithm 1) was carried out for each of these modelling techniques.

In each subset of features, LR, DT, MLP and SVM were trained and tested. The training and testing datasets were not normalised for the DT, but were normalised for the LR, MLP or SVM. All the predictors were trained by 10 folds of cross-validations in each experiment.

In each set of experiments, each MLP with one hidden layer was trained. The number of input neurons of a MPL network is the same as the number of the dimensions of a feature vector. The number of output neurons of the network is the number of classes. Therefore, the number of output neurons is two in this application: one represents a nonchurner, the other represents a churner. If the numbers of input and output neurons are n and m, respectively, the number of hidden neurons of the MLP is $\frac{m+n}{2}$. The sigmoid function is selected as the activation function for all MLPs in the experiments. Each MLP was trained by 3 folds of cross-Validation and BP learning algorithm with learning rate

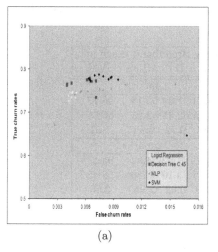 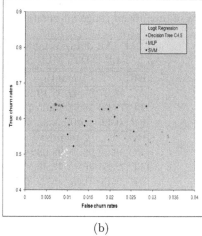

(a)   (b)

Fig. 1. ROC plot of false and true churn rates of the different number of months, where (a) and (b) presents the results of the first and second set of experiments, respectively

0.1, maximum cycle 1800 and tolerant error 0.05 were used to train the MLPs, based on the training dataset. The number of training cycles to yield the highest accuracy is about 600 for the MLPs.

Based on the extracted and normalised features, each SVM was trained to find the separating decision hyper-plane that maximises the margin of the classified training data. Two sets of values: the regularisation term $C \in \{2^8, 2^7, \cdots, 2^{-8}\}$ and $\sigma^2 \in \{2^{-8}, 2^{-7}, \cdots, 2^8\}$ of radial basis functions (RBF) were attempted to find the best parameters for the churn prediction. All together, 289 combinations of C and σ^2 with 3 folds of cross-validation were used for training each SVM. The optimal parameter sets (C, σ^2) yielding a maximum classification accuracy of standard SVMs were $(2^{-6}, 2^8)$ for each set of experiments. The optimal parameter sets (C, σ^2) yielding a maximum classification accuracy of SVMs were $(2^{-6}, 2^7)$ for the first and second set of experiments. For the third set of experiments, the optimal parameter sets (C, σ^2) yielding a maximum classification accuracy of SVMs were $(2^{-5}, 2^8)$.

Table 3 shows the prediction rates (AC, TP, FP) for the third set of experiments, and Table 2 shows the prediction rates (AC, TP, FP) for the first and second sets of experiments. The prediction rates on the left hand side of Table 2 summarise the results of the four techniques (LR, DT, MLP and SVM) performed on the first set of experiments. The results on the right hand side of the Table 2 were obtained from the second set of experiments (that used broadband usage information only). Based on Table 2, Figures 1(a) and 1(b) plot the receiver operating characteristics (ROC) graphs, which are TP against FP for the first and second sets of experiments, respectively. A point in the plots presents a pair of prediction rates (FP, TP) from a modelling technique based on a subset of features. Figures 1(a) and Figures 1(b) present the prediction rates for the first and second sets of experiments, respectively. The results of LR, DT, MLP

Table 2. Prediction rates based on different datasets, where BU presents number months of broadband usage information

	All data/variables				Only Broadband usage info.			
	LR	DT	MLP	SVM	LR	DT	MLP	SVM
BU	1 month				1 month			
AC	98.27	98.42	98.19	98.26	96.59	97.39	95.18	96.77
TP	73.90	76.55	75.05	77.90	49.15	62.45	53.95	55.65
FP	0.42	0.40	0.55	0.64	0.84	0.72	2.59	1.01
BU	2 months				2 months			
AC	98.22	98.38	98.03	98.24	96.55	97.52	94.82	95.33
TP	73.55	76.75	75.45	77.60	51.15	63.10	53.50	56.45
FP	0.45	0.45	0.75	0.65	0.99	0.62	2.95	2.56
BU	3 months				3 months			
AC	98.29	98.41	98.03	98.22	96.62	97.46	95.78	96.06
TP	74.70	76.25	75.45	77.35	50.00	63.80	54.20	63.20
FP	0.43	0.40	0.75	0.65	0.86	0.72	1.97	2.16
BU	4 months				4 months			
AC	98.27	98.26	97.33	98.19	96.60	97.48	95.44	95.42
TP	74.80	77.50	76.70	77.70	50.30	64.10	54.60	63.55
FP	0.46	0.62	1.55	0.70	0.90	0.71	2.35	2.85
BU	5 months				5 months			
AC	98.22	98.24	97.83	98.17	96.61	97.47	95.66	96.37
TP	74.60	77.65	76.55	78.00	51.05	64.10	55.30	62.70
FP	0.50	0.65	1.02	0.74	0.93	0.73	2.16	1.81
BU	6 months				6 months			
AC	98.27	98.27	97.92	98.12	96.59	97.44	94.71	96.23
TP	74.60	77.55	75.60	77.90	51.40	63.90	56.40	62.75
FP	0.45	0.61	0.88	0.79	0.97	0.75	3.22	1.96
BU	7 months				7 months			
AC	98.26	98.26	97.98	98.05	96.60	97.38	94.53	95.98
TP	74.55	77.90	75.30	78.25	51.05	63.65	55.40	60.60
FP	0.46	0.64	0.79	0.88	0.94	0.80	3.35	2.11
BU	8 months				8 months			
AC	98.27	98.16	98.12	98.05	96.58	97.32	94.49	96.52
TP	74.80	77.25	74.95	78.05	50.60	63.75	54.90	58.05
FP	0.46	0.71	0.63	0.87	0.94	0.86	3.36	1.41
BU	9 months				9 months			
AC	98.22	98.21	98.22	98.05	96.49	97.28	94.40	96.55
TP	74.60	77.20	74.20	77.70	49.70	63.45	53.50	59.35
FP	0.51	0.65	0.49	0.85	0.98	0.89	3.39	1.44
BU	10 months				10 months			
AC	98.17	98.20	98.19	97.96	96.44	97.03	94.89	96.40
TP	73.40	77.15	74.85	77.65	48.05	60.05	53.00	59.25
FP	0.49	0.66	0.55	0.95	0.94	0.97	2.84	1.59
BU	11 months				11 months			
AC	98.21	97.96	98.06	96.60	96.37	96.87	94.82	96.47
TP	72.75	73.45	67.15	64.70	47.05	58.25	50.10	52.35
FP	0.42	0.71	0.27	1.68	0.96	1.04	2.76	1.14

Table 3. Prediction rates based on the data without broadband usage information

	LR	DT C4.5	MLP	SVM
AC	97.941	97.723	96.828	97.744
TP	69.600	72.300	69.200	72.950
FP	0.527	0.903	1.678	0.916

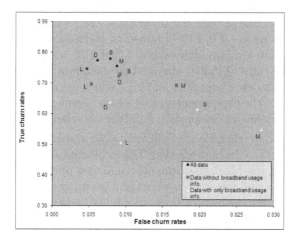

Fig. 2. ROC plot of false and true churn rates vs. the three different sets of data

and SVM models are shown in yellow, red, green and black, respectively, in these Figures. Table 2 Figures 1(a) and 1(b) show that:

1. The number of months of broadband usage information is between 3 to 9 to obtain better prediction rates
2. For the same subset of features, which type of modelling techniques would get higher prediction rates (TP) with lower prediction rates (FP) (e.g. the DT and SVM would get higher prediction rate (TP) than the LR and MLP; the DT and LR would get lower prediction rates (FP) than the SVM and MLP; the SVM would get slight high rates (TP) than than DT, etc.)
3. because the prediction rate (TP) is more significant when the FP is not very different, the SVM and DT outperform the LR and MLP.

As mentioned above, the number of months of broadband usage information is between 3 to 9 to obtain better prediction rates. In order to compare the efficiency of the feature subset that excludes broadband usage information with the feature subsets that include the broadband usage information, the average of prediction rates (FP, TP) for the months from 3 to 9 for each modelling technique were calculated. These pairs of the average prediction rates and the pairs of the prediction rates (FP,TP) from the third set of experiments are plotted in Figure 2: the yellow points are from the second set of experiment, the black points are from the first set of experiments, and the magenta points are from the third set of experiments. Figure 2 shows that:

1. For all the modeling techniques, the prediction rates for TP are reasonably high (above 73%) and the prediction rates for FP are very low (about max 1%)in the case of the first set of experiments.
2. The prediction rates (TP) are higher and the prediction rates (FP) are lower when all the information were used in the first set of experiments.
3. when the same data were used, the SVMs and DTs can obtain the highest prediction rates (TP) and lowest rates (FP); LR can get the lowest rates (TP), and MLP can provide the lower rates (TP) with highest rates (FP).
4. The prediction rates (TP, FP) obtained on the information without broadband information are also high (about TP of 71% and FP of 1.1%).
5. The prediction rates (TP, FP) obtained on the information with only broadband information are about 50% and 2%), which are quite good, considering the condition of the experiments, in which the number of customer that are churning is very low in comparison with the non-churners.

The computation overhead is very different when the different modelling techniques were used for the churn prediction. The most expensive computational cost was spent on using the MLP, the computational cost of using the SVM is more expensive, the lower and lowest ones were spent on using the DT and LR respectively. In addition, the outputs of these modelling techniques are different. DT can provide churn reasons without likelihood. LR and MLP can give the likelihood/probability for customer behaviour. The SVM can provide only binary output which presents churn or nonchurn. Therefore, which types of modelling techniques should be used depends on the objectives of an application. For examples, if interested in churn reasons, the DT should be used; if the probabilities of churns and nonchurns is required, the MLP or LR might be more suitable to use.

5   Conclusions

This paper presented a new set of features, based on Henley segmentation, the broadband usage, dial types, the spend of dial-up, line-information, bill and payment information, account information, call details and service log data. Four modelling techniques (LR, DT, MLP and SVM) were used for customer churn prediction in telecommunication service field, especially broadband Internet. Finally, based on this new set of features, the comparative experiments of the four modelling techniques were carried out. The experimental results showed that the high true churn of 77% with the low false churn rate of 2% can be achieved using the proposed features. Experimental results also showed which modelling technique is more suitable for broadband churn prediction depends on the objective of the churn prediction. For examples, DT and SVM should be used if interested in the true churn rate and false churn rate; the logistic regressions might be used if looking for the churn probability.

However, there are some limitations with our proposed techniques. In the future, other information (e.g. complain information, contract information, more fault reports, etc.) should be added into the new feature set in such a way to improve features. The dimensions of input features also should be reduced by using the principal components analysis methods and genetic algorithms. In addition, because the imbalance classification problem takes place in this application, the methods of imbalance classifications should be focused in the future.

Acknowledgements

This research was partly supported by Eircom of Ireland.

References

1. http://www.eircom.ie/cgi-bin/bvsm/bveircom/mainPage.jsp
2. http://www.henleymc.ac.uk/
3. Customer Churn Prediction Based on the Decision Tree in Personal Handyphone System Service (June 2007)
4. Au, W., Chan, C., Yao, X.: A novel evolutionary data mining algorithm with applications to churn prediction. IEEE Transactions on Evolutionary Computation 7, 532–545 (2003)
5. Boser, B., Guyon, I., Vapnik, V.: A training algorithm for optimal margin classifiers. In: Pro. the 5th Annual ACM Workshop on Computational Learning Theory, Pittsburgh,PA, July 1992, pp. 144–152. ACM Press, New York (1992)
6. Burges, C.J.C.: A tutorial on support vector machines for pattern recognition. Data Mining and Knowledge Discovery 2(2), 121–167 (1998)
7. Coussement, K., den Poe, D.V.: Churn prediction in subscription services: An application of support vector machines while comparing two parameter-selection techniques. Expert Systems with Applications 34, 313–327 (2008)
8. Hadden, J., Tiwari, A., Roy, R., Ruta, D.: Churn prediction: Does technology matter? International Journal of Intelligent Technology 1(2) (2006)
9. Hosmer, D., Lemeshow, S.: Wiley, New York (1989)
10. Japkowicz, N.: Why question machine learning evaluation methods? In: AAAI Workshop (2006)
11. John, H., Ashutosh, T., Rajkumar, R., Dymitr, R.: Computer assisted customer churn management: State-of-the-art and future trends (2007)
12. Quinlan, J.R.: C4.5: Programs for machine learning (1993)
13. Quinlan, J.R.: Improved use of continuous attributes in c4.5. Journal of Artificial Intelligence Research 4, 77–90 (1996)
14. Rumelhart, D., Hinton, G., Williams, R.: Learning internal representations by error propagation, vol. 1. MIT Press, MA (1986)
15. Wang, H.-Y., Hung, S.-Y., Yen, D.C.: Applying data mining to telecom churn management. Expert Systems with Applications 31, 515–524 (2006)
16. Wei, C., Chiu, I.: Turning telecommunications call details to churn prediction: a data mining approach. Expert Systems with Applications 23, 103–112 (2002)
17. Yan, L., Wolniewicz, R., Dodier, R.: Customer behavior prediction - it's all in the timing. Potentials, IEEE 23(4), 20–25 (2004)
18. Zhang, Y., Qi, J., Shu, H., Li, Y.: Case study on crm: Detecting likely churners with limited information of fixed-line subscriber. In: 2006 International Conference on Service Systems and Service Management, October 2006, vol. 2, pp. 1495–1500 (2006)
19. Zhao, Y., Li, B., Li, X., Liu, W., Ren, S.: Customer churn prediction using improved one-class support vector machine. In: Li, X., Wang, S., Dong, Z.Y. (eds.) ADMA 2005. LNCS, vol. 3584, pp. 300–306. Springer, Heidelberg (2005)

Mining High-Correlation Association Rules for Inferring Gene Regulation Networks*

Xuequn Shang**, Qian Zhao, and Zhanhuai Li

Institute of Computer Science and Engineering, Northwestern Polytechnical
University, P.O. Box 168 Shaanxi 710072, China
{shang,lizhh}@nwpu.edu.cn

Abstract. Construction gene regulation networks can provide insights into the understanding the molecular mechanisms underlying important biological processes. We present a novel association rule mining for building large-scale gene regulation networks from microarray data. Gene expression microarray data typically contains a very high gene dimension and a very low sample size, rendering a great challenge for existing association rule mining algorithms. In this paper, we develop a novel algorithm, $HCMiner$, to mine high-correlation association rules from microarray data. $HCMiner$ initially overlapping partitions the dimension of genes according to their correlations and introduces the support-free framework for mining association rules. Several experiments on Yeast dataset show that the proposed algorithm outperforms existing algorithms with respect to scalability and effectiveness.

1 Introduction

The recent development of high-throughput bio-techniques for functional genomics has generated a large amount of gene expression microarray data. Analyzing microarray data can provide novel insights in understanding basic mechanisms controlling cellular processes. Various data mining techniques have been employed to extract useful biological information from the huge and fast-growing gene expression data. One main objective of such data mining tasks is to uncover gene regulation networks from microarray data in order to deeper understanding the underlying complex genetic regulatory process, which has important implications in the pharmaceutical industry, complex disease treating, and many other biomedical fields.

One of the current widely used methods to derive gene regulatory from microarray data is association rule mining [5,9,13,21]. Association rules can capture biological correlation between genes, as well as reveal the direction of relationships. An association rule is an implication of the form $LHS \Rightarrow RHS$, where

* This research is partly supported by the National Natural Science Foundation of China (No. 60703105) and the Natural Science Foundation of Shaanxi Province (No. 2007F27). All opinions, findings, conclusions and recommendations in this paper are those of the authors and do not necessarily reflect the views of the funding agencies.
** Corresponding author.

T.B. Pedersen, M.K. Mohania, and A M. Tjoa (Eds.): DaWaK 2009, LNCS 5691, pp. 244–255, 2009.

LHS and *RHS* are sets of genes and relevant facts describing the cellular environment of the genes, the *RHS* set is likely to occur whenever the *LHS* set occurs. An example of an association rule mined from expression data might be *geneA* ⇒ *geneB*, meaning that, for the expression profile experimental data set, when gene A was measured as expressed, gene B is likely to be expressed.

While association rules can facilitate analysis of gene expression data sets, there are several limitations with the existing algorithms: (1) existing association rule mining techniques employ global analysis of microarrays which may not adequately capture co-regulation of genes. The rules generated may involve uncorrelated genes. (2) existing algorithms still suffer from the inherent dimensionality problem in microarray data. The traditional association rule mining algorithms work by enumerating the relationships among columns and often result in itemset explosion due to the high dimensionality of microarray data. Although the concept of row-enumeration have been introduced to efficiently prevent itemset explosion [7,18], a common problem of these algorithms is the limitation of support-confidence framework. For instance, many rules that a biologist would consider highly interesting are pruned if the support is set too low, as well as the rules with low support but correlated items are missed out if the support is set too high. (3) The rules generated by the existing association rule mining algorithms do not reveal the complex regulation relations from microarray data. According to biological interpretations, there are three types of regulation relations between genes: activation, inhibition, and dependency. However, the rules generated do not consider the inhibition relations.

Motivated by these concerns, we develop a novel algorithm, *HCMiner*, to mine highly correlated association rules from microarray data. Initially we propose a overlapping partitioning method to detect the groups within the dimension of genes according to their correlations. The items (genes) within a group are highly correlated and are considered as the candidates of co-regulated genes. Then, we propose a support-free framework for mining association rules within groups. The technical contributions of our work are summarized as follows:

– We investigate the complexity of the situation in high-dimensional gene expression data, and propose an overlapping partitioning method to allow one gene to be assigned to several groups according to their correlations.
– We develop a support-free framework for mining association rules and exploit new effective techniques to prune the search space.
– We consider both activation and inhibition relations between genes and deliver more valuable regulation information of gene network.

The remainder of this paper is organized as follows. In Section 2, we describe our investigate on the presence of groups within genes based on correlation relationships and present the method for overlapping partitioning gene expression data. The methods for mining highly correlated association rules are examine in Section 3. We report, in Section 4, evaluation results of applying the method to several datasets obtained from experimental observations of the gene expression data of yeast. Related work is illustrated in Section 5. We then summary the paper in Section 6 and discuss the possible future work.

2   Overlapping Partitions of Genes Expression Data

Association analysis has been extensively studied over the last decades, which is a task of find correlations between items in a dataset [2]. Recent results [1,20] have shown that correlated patterns may not always exist globally due to items do not correlate with each other uniformly. For instance, genes are not correlated in full dimensional space but correlated only in subset of dimensions. This phenomena can help get extra insights by finding association rules in small groups consisting of items with significant intra-group correlations but insignificant inter-group correlations. By partitioning the high-dimensional gene expression data we form the genes into groups. In addition, one gene may be involved in multiple pathways [12], thus identifying overlapping partitions is important in biological applications. In the following section, we describe an overlapping partitioning method to detect the groups within the dimension of genes according to their correlations.

2.1   Problem Formulation

Technically, gene expression data can be modeled using an undirected, unweighted gene coherence graph $G = (V, E)$ of a set of vertices V and a set of edges E, where each vertex represents a unique gene and each edge is a pair of genes if and only if their expression patterns are similar according to a user-specified similarity measure (e.g, Euclidean distance or Pearsons correlation coefficient). The degree of a vertex v_i, denoted as $Deg(v_i)$, is the number of edges incident to the vertex v_i.

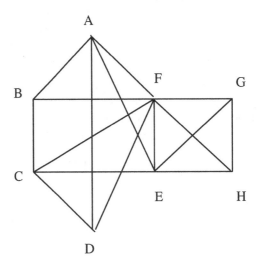

Fig. 1. An example graph G

Definition 1: Let $V = \{v_1, v_2, ..., v_n\}$ be a set of all vertices in Graph G. The correlation between two vertices v_i and $v_j \in V$, $Cor_{(v_i, v_j)}$, is defined as follows.

$$Cor_{(v_i, v_j)} = \frac{m}{n_1 \times n_2}$$

where m represents the number of common interaction neighbors shared by two vertices, i and j. n_1 and n_2 denotes the degree of the vertex v_i and v_j, respectively. Considering a pair of vertices $\{A, B\}$ in Figure 1, $Cor_{(A,B)} = \frac{1}{4 \times 3} = 0.08$.

Definition 2: A pair of vertices is highly-correlated if their correlation is not less than a predefined minimum threshold.

2.2   Partitioning Method

The procedure to partition the dimension of genes is based on the determination of highly-correlated genes. A naïve approach for this procedure is described as the following operations:

1. Calculate the correlation between all pairs of genes;
2. Keep the edges of G that the corresponding vertices is highly-correlated;
3. Find all the connected partitions of G. Each partition corresponds to a group of the genes.

There is an issue remaining in the naïve approach, that is, one gene only holds in one group. Such operation could lead to unbalanced partitions especially for gene coherence graph with the presence of highly connected hub nodes that are connected to a large number of other nodes [4,3]. To avoid the above bias, we propose an overlapping partitioning method, *Partitioning*, as outlined in Algorithm 1. *Partitioning* consists of two phases. In phase 1, the vertices are ordered in the degree descending order in order to avoid repeatedly checking pairs of vertices and speed up the process. For two vertices, v_i and v_j, *Partitioning* do the following examinations:

1. If v_i and v_j belong to the same group, there is no need to examine the pair;
2. If both v_i and v_j do not appear in an existing group, create a new group;
3. If only v_i belongs to an existing group, check whether v_j is high correlation with the other vertices in the groups that v_i belongs to; If yes, combine v_i and v_j to the same group. Otherwise, create a new group including v_i and v_j.

In phase 2, *Partitioning* do the merging process to combine small groups for the purpose of identifying more meaningful partitions and speeding up the process. For two groups, if the number of common genes shared by two groups satisfies a threshold, the two groups should be merged. As demonstrated in the example in Figure 1, *Partitioning* is able to discover two partitions: F,E,G,H and F,A,C,B,D if we set the correlation threshold is 0.09, which naïve approach cannot find.

Algorithm 1: *Partitioning*
begin
  sort the vertices in a given graph $G = (V, E)$ in degree
  descending order;
  for each vertex v_i in V do
    if v_i belongs to an existing partitions then
        for each vertex v_j connected to the vertex v_i
              and $Deg(v_j)$ ¡ $Deg(v_i)$ do
          if v_i and v_j belong to the same group then
            do nothing;
          else if v_j is high correlation with the other vertices
                in the groups that v_i belongs to then
              combine v_i and v_j to the same partition;
          else if
              create a new partition including v_i and v_j;
        end for
    else if create a new partition including v_i;
  end for
  call Merging
end

Fig. 2. Algorithm for *Partitioning*

3   *HCMiner*: Mining Highly Correlated Association Rules

For the set of genes in each partition we have to find association rules with confidence higher than the predefined minimum threshold. Most of existing association rule mining algorithms adopted support-confidence framework and consisted of two steps: discovery of frequent itemsets that satisfy with the predefined support threshold and then generation association rules that satisfy with the predefined confidence threshold from frequent itemsets. As we discussed above, there is a fundamental limitation of support-confidence framework especially to the case with microarrays. To overcome the aforementioned limitations of existing algorithms for mining association rule from microarray data, we propose support-free association rule mining algorithm, *HCMiner*, to extract gene activation and inhibition relations from microarray data.

Before the algorithm is presented, let us give some definitions. According to biological interpretations, there are three types of regulation relations between genes A and B: activation, inhibition, and dependency. They can be respectively defined as follows:

1. A activates B ($A \uparrow B$) if A is expressed ($A \uparrow$), then B is expressed ($B \uparrow$); or if A is depressed ($A \downarrow$), then B is depressed ($B \downarrow$);
2. A inhibits B ($A \downarrow B$) if A is expressed ($A \uparrow$), then B is depressed ($B \downarrow$); or if A is depressed ($A \downarrow$), then B is expressed ($B \uparrow$);
3. A and B is independent if A is expressed (or depressed), then B shows both expressing and depressing, or even remains unchanged.

Definition 3: Given two genes, A and B, the confidence of an activation rule in form of $A \uparrow B$ is defined as follows:

$$\text{conf}(A \uparrow B) = \frac{\|A \uparrow and B \uparrow\| + \|A \downarrow and B \downarrow\|}{\|A \uparrow or A \downarrow\|}$$

Definition 4: Given two genes, A and B, the confidence of an inhabitation rule in form of $A \downarrow B$ is defined as follows:

$$\text{conf}(A \downarrow B) = \frac{\|A \uparrow and B \downarrow\| + \|A \downarrow and B \uparrow\|}{\|A \uparrow or A \downarrow\|}$$

3.1   Algorithm Description

The *HCMiner* is an algorithm which can generate all activation and inhabitation rules with confidence above predefined confidence threshold. Figure 3 gives an overview of the *HCMiner*.

HCMiner exploits two confidence-based pruning methods that allows us to effectively prune the search space without the support requirement. Unlike support, confidence does not have the anti-monotone property. But confidence has weak anti-monotone property which is based on the following definitions.

Definition 5: Given an itemset L, a rule r is an $L - consisting$ $rule$ if $antecedent(r) \cup consequent(r) = L$ and $|antecedent(r)| = 1$. For example, the itemset $L = A, B, C, D$, the rule $A \Rightarrow BCD$ is an $L - consisting$ $rule$.

Definition 6: Given an itemset $L = i_1, i_2, ..., i_m$, the maximum confidence of the $L - consisting$ $rule$ is defined as $\max\{conf(i_1 \Rightarrow i_2, ..., i_m), conf(i_2 \Rightarrow i_1, i_3, ..., i_m), ..., conf(i_m \Rightarrow i_1, ..., i_{m-1})\}$.

Definition 7: Given an itemset I, let a rule $r = X \Rightarrow Y$, where $X \cup Y = I$. The set of *extensionrules* ER is the set of all rules generated from the rule r such that for each $er \in ER$, $antecedent(er) = antecedent(r)$ and $consequent(er) = consequent(r) \cup Z$, where Z is an itemset and $Z \subseteq I$. For example, the rule r is $A \Rightarrow B$, the rule $A \Rightarrow BC$ is an extension rule of r.

There are two pruning techniques which can enforce the algorithm *HCMiner*. To take advantage of the pruning power of minimum confidence threshold,

Algorithm 2: *HCMiner*
begin
  get size-1 items, sort all size-1 items in support ascending order;
  for size of items in (2,3,...,k-1) do
    generate candidate confidence rules
    prune based on the $L - consisting$ Pruning
    prune based on the *Extension* Pruning
    generate confidence rules
  end for
end

Fig. 3. Algorithm for *HCMiner*

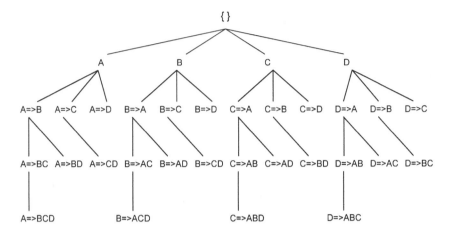

Fig. 4. Rule enumeration tree

HCMiner adopts a bottom-up search method and a rule enumeration tree, as shown in Figure 4.

1. *L-consisting* Pruning. We can prune rules generated from the same *L-consisting rule* by the anti-monotone property. If the maximum confidence of a *L-consisting rule* < *minconf* (the predefined minimum confidence threshold), then the set of the *L-consisting rules* are not confident and are pruned. For instance as shown in Figure 4, the itemset $L = A, B, C, D$. If the maximum confidence of *L-consisting rules* is conf($A \Rightarrow BCD$), which is less than the minimum threshold, then $B \Rightarrow ACD$, $C \Rightarrow ABD$, and $D \Rightarrow ABC$ is not confident and are pruned.
2. *Extension* Pruning. We can prune *extensionrules* by the anti-monotone property. If a rule is not confident, then the extension rules are not confident and pruned. For instance, the rule r is $A \Rightarrow B$. If $A \Rightarrow B$ is not confident, then $A \Rightarrow BC$, $A \Rightarrow BD$, and $A \Rightarrow BCD$ are not confident and are pruned.

4   Performance Evaluation

In this section, we present extensive experiments to evaluate the effectiveness of the proposed methods on gene expression data. We demonstrate the effect of the overlapping partition method and a computation performance comparison between the *HCMiner* algorithm, standard row-set enumeration algorithm, and standard column-set enumeration algorithm. All experiments were performed on a PC with a 2.53GHz Interl(R)2CPU and 2G RAM.

4.1   Datasets

In this study, we use the gene expression data CDC28 downloaded from http://cellcycle-www.stanford.edu. The CDC28 data set [10] generated by temporal microarray experiments records the mRNA transcript levels of the budding yeast S. cerevisiae during the cell cycle. In the data, the number of genes is 6178, and the number of transactions is 82. The reasons we use this dataset are as follows: (1) The accuracy of identification co-regulated genes from coexpressed microarray experiments is high in the case of the number of microarray experiments at between 50 and 100 [24]. (2) The time-series microarray data is widely used for inferring gene regulation schemes and metabolic pathways due to the capability of capture the dynamicity of biological networks. (3) The evaluation of the biological significance of the mining results is straightforward since the extensive study on the gene regulation schemes of S. cerevisiae. Missing values are filled using BPCAfill by Oba et al. [16] that performs much better than the existing methods for missing value estimation. The expression level of each gene is discretized by binning the log_2 of the expression level into the three levels, i.e. up-regulated, normal, or down-regulated, and represented by one single variable. For example, for gene A, an expression level can be represented as up-regulated (A=1), normal (A=0), or down-regulated (A=-1), respectively.

4.2   Performance Comparison

In this set of experiments, we demonstrate the performance of $HCMiner$ in comparison with row-set enumeration algorithm and column-set enumeration algorithm, by varying several parameters, such as support threshold and confidence threshold. Row-set enumeration methods have recently emerged to facilitate the mining of microarray data. The results of these are illustrated in Figure 5(a) and (b). Notice that, the execution time for $HCMiner$ consists of partitioning time and mining time. In Fig. 5(a), we can observe that the running time can be increased with the decrease of support thresholds. As expected, $HCMiner$ outperforms row-set enumeration and column-set enumeration for lower support thresholds. However, row-set enumeration has better performs than $HCMiner$ when the support threshold is set high. The major reason is that the cost of partitioning does not payoff for high support thresholds since the mining time are very small. In Fig. 5(b) for the support of 0, the running time of the column-set enumeration were so large that we had to abort the runs in the case of confidence threshold of 0. This is because that, for the column-set enumeration approach, a lower support threshold leads to a long computational time to extract a huge number of frequent patterns and the subsequent rule generation process is unable to proceed due to the limitation of memory. In addition, $HCMiner$ is significantly more efficient than row-set enumeration and column-set enumeration when support threshold is free.

We compare the association rules extracted by $HCMiner$ approach with those extracted by row-set enumeration approach to evaluate the capability of capture co-regulation of genes by overlapping partitioning method. The results show that row-set enumeration identifies rules containing uncorrelated genes. For example, the rule between Gene YBR088C and Gene YCL040W is generated by row-set enumeration approach. However, there is no such rules discovered by $HCMiner$. Looking into the original data of Gene YBR088C and Gene YCL040W in Fig. 6(a), we say that the changing tendency of those two

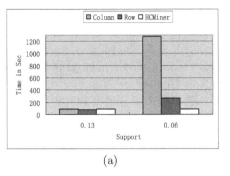
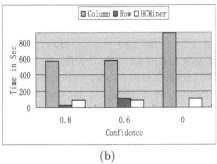

(a)   (b)

Fig. 5. (a) The run time vs. support threshold when the confidence threshold is 0.8, (b) the run time vs. confidence threshold when the support threshold is 0

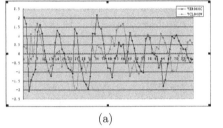
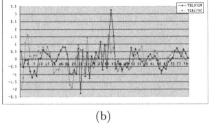

(a)   (b)

Fig. 6. (a) Gene YBR088C and Gene YCL040W, (b) Gene YBL032W and Gene YGR279C

genes are not co-regulated (the Pearson correlation coefficient between them is 0.0905056). So does the rule between Gene YBL032W and Gene YGR279C (the Pearson correlation coefficient between them is 0.361191), as shown in Fig. 6(b). This indicates that the association rule mining on gene expression data under row enumeration without overlapping partition may generate the rules on uncorrelated genes.

To evaluate the biological significance of discovered rules from gene microarray data, we use the Gene Ontology (GO) [8] to see if our rules are contained in GO gene relationships. If a rule contains an antecedent gene that shares a GO annotation with any genes in the consequent items, we say the rule describes biologically meaning relationships. We found that the higher percentage of rules contained a GO relationship as minconf is increased. These results strengthen our argument that support-free framework is effective for identifying relationships from microarray data sets.

5   Related Work

Association rule mining was originally designed to market analysis on transaction databases [2]. Recently some researches applying association rule mining to extract knowledge from biological data sets have already been completed. Oyama et al. [17] and Kotlyar et al. [14] combined multiple protein features such as sequence motifs and functional annotation to protein-protein interactions. Each transaction consists of a protein pair annotated with various features. When the form of dataset is given, various kind of rules can be expected. For example, let α and β denote the set of domains that appear in the LHS and RHS, respectively. Then, one of the found rules looks like: $\alpha \rightarrow \beta$. In fact, such rules enables us to make a predictions like "the protein having domain α will probably interact with the protein having domain β" or access the reliability of proposed interactions. A major reason that association rule mining has been used in these applications is due to the need to take a large number of variables into consideration. In another study [23], a new type of association mining, hyperclique pattern discovery, was applied to protein complex data to find functional modules. Gaurav et al. [19] use the concept of hyperclique to transform the original protein interaction networks by removing or adding edges in order to produce new graphs that are more suitable for protein function prediction. Recently, the association rule mining algorithms based on row-enumeration for mining interesting gene relationships from classified microarray data sets have been proposed. These include CARPENTER [18] and RERII [7]. Other data mining techniques, such as Bayesian network learning [15,11,13], neural network [22], bicluster models [6] have been employed to uncover the biological relationships of genes from microarray data. However, the poor scalability of most existing data mining analysis techniques has still limited the empirical value of gene microarray analysis.

6   Summary and Conclusion

Microarray data renders new challenges that make many traditional data mining techniques infeasible for mining the hidden gene relationships due to its high density. In this paper, we give the insight about the presence of groups determined by correlations between the genes and present an overlapping partition method to detect the groups within the dimension of genes. Additionally, we describe a support-free framework for mining association rules. Several experiments on Yeast dataset show that the proposed algorithm outperforms existing algorithms with respect to scalability and effectiveness.

However, the problem of analyzing gene microarray data sets is still challenging. There remain lots of further investigations. We will plan to investigate the proposed method on other more microarray data sets to evaluate its performance in terms of scalability and the biological significance of discovered rules. It is known that these data sets may contain high levels of errors in terms of both false positives and false negatives. Integration of different gene microarray data provides a way to improve data quality and recover missing data. We will plan to explore the proposed method to mine association rules from multiple microarray data sets for inferring gene regulation network.

References

1. Aggarwal, C., Procopiuc, C., Yu, P.: Finding localized associations in market basket data. IEEE Trans. Knowl. Data Eng. 14(1), 51–62 (2002)
2. Agrawal, R., Imielinski, T., Swami, A.N.: Mining association rules between sets of items in large databases. In: Proc. ACM SIGMOD International Conference on Management of Data, pp. 207–216 (1993)
3. Albert, R.: Scale-free networks in cell biology. Journal Cell Sci. 118, 4947–4957 (2005)
4. Barabasi, A., Oltvai, Z.: Network biology: understanding the cells functional organization. Nat. Rev. Genet. 5, 101–113 (2004)
5. Becquet, C., Blachon, S., Jeudy, J.B.B., Gandrillon, O.: Strong-association-rule mining for large-scale gene-expression data analysis: a case study on human sage data. Genome Biology 3(12) (2002)
6. Cheng, Y., Church, G.M.: Biclustering of expression data. In: Proceedings of the Eighth International Conference on Intelligent Systems for Molecular Biology, pp. 93–103 (2000)
7. Cong, G., Tan, K.L., Tung, A., Pan, F.: Mining frequent closed patterns in microarray data. In: Proc. Fourth IEEE Intl. Conf. Data Mining (ICDM), pp. 363–366 (2004)
8. The Gene Ontology Consortium. The gene ontology (go) database and informatics resource. Nucleic Acids Research 32, 258–261 (2004)
9. Creighton, C., Hanash, S.: Mining gene expression databases for association rules. Bioinformatics 19, 79–869 (2003)
10. Cho, R.J., et al.: A genome-wide transcriptional analysis of the mitotic cell cycle. Molecular Cell 2(1), 65–73 (1998)
11. Friedman, N., Linial, M., Nachman, I., Pe'er, D.: Using bayesian network to analyze expression data. Journal of Computational Biology 7, 601–620 (2000)
12. Gasch, A., Eisen, M.: Exploring the conditional coregulation of yeast gene expression through fuzzy k-means clustering. Genome Biol. 3 (2002)
13. Huang, Z., Li, J., Su, H., Watts, G., Chen, H.: Large-scale regulatory network analysis from microarray data: modified bayesian network learning and association rule mining. Decis. Support Syst. 43(4), 1207–1225 (2007)
14. Kotlyar, M., Jurisica, I.: Predicting protein-protein interactions by association mining. Information Systems Frontiers 8(1), 37–47 (2006)
15. Murphy, K., Mian, S.: Modeling gene expression data using dynamic bayesian networks. In: Technical Report, Computer Science Division, University of California, Berkeley (1999)
16. Oba, S., Sato, M., Takemasa, I., et al.: A bayesian missing value estimation method for gene expression profile data. Bioinformatics 19(16), 2088–2096 (2003)
17. Oyama, T., Kitano, K., Satou, K., Ito, T.: Extraction of knowledge on protein protein interaction by association rule discovery. Bioinformatics 18(5), 705–714 (2002)
18. Pan, F., Cong, G., Tung, K., Yang, J., Zaki, M.: Carpenter: Finding closed patterns in long biological datasets. In: Proc. ACM SIGKDD Intl. Conf. Knowledge Discovery and Data Mining (KDD), pp. 637–642 (2003)
19. Pandey, G., Steinbach, M., Gupta, R., Garg, T., Kumar, V.: Association analysis-based transformations for protein interaction networks: A function prediction case study. In: Proc. ACM SIGKDD International Conference on Knowledge Discovery and Data Mining (KDD), pp. 540–549 (2007)

20. Tsay, Y., Chang-Chien, Y.: An efficient cluster and decomposition algorithm for mining association rules. Inf. Sci. 160, 161–170 (2004)
21. Tuzhilin, A., Adomavicius, G.: Handling very large numbers of association rules in the analysis of microarray data. In: Proc. of the Eighth ACM SIGKDD International Conference on Knowledge Discovery and Data Mining, pp. 23–26 (2002)
22. Wahde, M., Hertz, J.: Modeling genetic regulatory dynamics in neural development. Journal of Computational Biology 8, 14863–14868 (2001)
23. Xiong, H., He, X., Ding, C., Zhang, Y., Kumar, V., Holbrook, S.R.: Identification of functional modules in protein complexes via hyperclique pattern discovery. In: Proc. Pacific Symposium on Biocomputing (PSB), pp. 221–232 (2005)
24. Yeung, K., Medvedovic, M., Bumgarner, R.: From co-expression to co-regulation: how many microarray experiments do we need? Genome Biol. 5(7) (2004)

Extend UDF Technology for Integrated Analytics

Qiming Chen, Meichun Hsu, and Rui Liu [*]

HP Labs, Palo Alto, California, USA and HP Labs, Beijing China
{qiming.chen,meichun.hsu,liurui}@hp.com

Abstract. Running analytics computation inside database engines through the use of UDFs (User Defined Functions) has been extensively investigated, but not yet become a scalable approach due to two major limitations. One limitation lies in that the existent UDFs are not relation-in, relation-out and schema-aware, unable to model complex applications, and cannot be composed with relational operators in a SQL query. Another limitation lies in the difficulty of programming UDFs for efficient interaction with query processing, since that requires hard-to-follow system knowledge beyond the analytics expertise. These limitations actually keep away most users from using UDFs for their analytics applications.

To solve these problems, we extend the UDF technology in both semantic and system dimensions. We first expand our investigation on Relation Valued Functions (RVFs) with the goal of having RVF executions tightly integrated with query processing, but allowing RVF developers to be liberated from DBMS internal details. We separate an RVF into two parts: ***RVF shell*** that contains the system utilities, and ***user-function*** that contains application logic only. We provided focused system support based on the notion of *invocation pattern*, and developed the mechanism for *generating an RVF-shell automatically* based on the schemas of its argument and return relations, the well understood invocation pattern, and the common data conversion protocol. A complete RVF is made by plugging the "user function" in the RVF-shell.

We have prototyped the proposed approach on the open-sourced database engine Postgres. Our experience reveals its advantages in making UDF tightly integrated with the query executor but relieving analytics users from dealing with system details – a fundamental data engineering requirement to make UDF technology practically usable for converging data intensive analytics and data management.

1 Introduction

Running data-intensive analytics computations outside database causes significant overhead in data access and transfer, which has been recognized as the major performance bottleneck in business intelligence applications, and has given rise to the need of pushing down data-intensive computations to the database engine. To reach

---

[*] Corresponding author.

T.B. Pedersen, M.K. Mohania, and A M. Tjoa (Eds.): DaWaK 2009, LNCS 5691, pp. 256–270, 2009.
© Springer-Verlag Berlin Heidelberg 2009

the analytics operations which are beyond the standard relational operations, we rely on User Defined Functions (UDFs) [1, 10,14].

1.1 The Challenge

However, the current UDF technology has several limitations. One limitation lies in the lack of formal support of relational input and output. Existing SQL systems offer scalar, aggregate and table functions, where a scalar or aggregate function cannot return a set; a table function does return a set but its input is limited to a single-tuple argument. These types of UDFs are not relation-in, relation-out and schema-aware, unable to model complex applications, and cannot be composed with relational operators in a SQL query. Further, they are typically executed in the tuple-wise pipeline of query processing, which may prohibit in-function batch and parallel processing. Although the notion of relational UDF has been studied by us [3] and others [1,10], it is not yet realized on any product due to the cumbersome in interacting with the query executor.

Next, there exists a conflict between UDF execution efficiency and easy-coding. A UDF is run in the query processing environment with a number of interactions with the query executor, for parsing parameter, converting data, switching memory contexts, etc. Coded using DB engine internal data structures and system calls, a UDF can be executed efficiently, but the analytics users have to deal with the hard-to-follow system details, which is often beyond their discipline; and this situation actually keeps them away from using the UDF technology.

1.2 Related Work

Integrating data-intensive analytics and data management in terms of UDFs has been extensively investigated [2-9], and the notion of relational UDF has been studied by [1,10] as well as by us [3] in different contexts. However, how to realize RVF efficiently has not yet been explored.

To ease the development and utilization of UDFs, some systems such as SQL Server, convert UDF's input data to strings from their system internal formats, which causes significant overhead in converting data and parsing parameters. With such a mechanism, as reported in [12,13], on SQL Server, no matter how simple a UDF is, it sharply underperforms compared with a system function or expression.

On the other hand, in some database systems such as Postgres, UDFs are coded in exactly the same way as system functions, which allows the UDFs to be executed efficiently, but requires the UDF developer to deal with tedious DBMS internal data structures and system calls, which, in fact, significantly contrasts to the easy coding of map() and reduce() functions on a MapReduce platform such as Hadoop where the system support are completely transparent to users [1,8].

1.3 The Proposed Solutions

We extend our previous work on Relation-Valued Functions (RVF) [3] by addressing the key implementation issues. We classify RVFs based on their "invocation patterns", namely, the mechanisms for dealing with input (e.g. passing in an input relation tuple by tuple, or as a whole) and return values (e.g. per-tuple return or set return). A well-defined invocation pattern with designated input-mode and return-mode underlies well-understood behavior and system interface, as well as focused system support.

Without any constraint, in a function body, the code for system utilities and for application logic may be interleaved in multiple ways. However, constrained by a specific invocation pattern, the steps of system interaction can be made deterministic, making it possible to single out the application logic from the system utilities, and to abstract high-level APIs for interacting RVF execution with query processing.

To convert relation objects from their DBMS internal data structures into simplified ones to be manipulated by applications, we introduced the Simple Relation Object Mapping (SROM) protocol. SROM also covers the User Defined Types (UDTs) composed from primitive types and collections. Based on SROM, the data structure declarations (in header files) of the involved relation objects, which are much simpler their DBMS internal formats, can be generated from the corresponding relation schemas.

In order to distinguish user's responsibility and system's responsibility, we separate an RVF into two parts: the RVF shell and the "*user-function*"; the *RVF-shell* contains the system utilities for running the RVF in the query processing environment, and the "*user-function*" contains application logic only without DBMS internal system calls. An RVF is made by plugging a user function in its shell. Since coding RVF shells requires the familiarity of system internal details which is in general beyond the expertise of analytics users, we developed the system utilities to generate an RVF shell from its function declaration, data mapping scheme and the designated invocation pattern.

Our solution stack is shown in Fig 1, where an RVF shell is generated, together with the above header files, The RVF developer is only responsible for providing the plugged in *user-function* which is free of DBMS internal data structures and system calls.

The proposed solutions have been prototyped on the open-sourced Postgres database engine, and we plan to transfer the implementation to a commercial and proprietary parallel database engine. We show how RVFs can be used intelligently to alleviate the shortcomings of SQL being cumbersome in expressing data flow logic, and in gaining high performance. Our experiment also reveals the system intelligence for separating RVF shell and user-function and for generating the shell automatically, which greatly scales the UDF applications for data-intensive analytics.

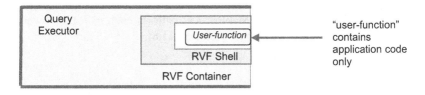

Fig. 1. SQL RVF on DB Engine

The rest of this paper is organized as follows: Section 2 outlines the need and characteristics of RVF; Section 3 describes RVF invocation patterns wrt input and return modes; Section 4 discusses the separation of *RVF-shell* and *user-function*; Section 5 introduces the RVF-shell generation approach; Section 6 illustrates experimental results; Section 7 concludes.

2   Extend Relation Operation with Relation Valued Function

In order to have data intensive applications executed inside database engine for fast data access and reduced data move, and composed with relational operators to form integrated dataflow pipelines, we extend our investigation on relation-input, relation-output and schema-aware Relation Valued Functions (RVFs) [3]. Although RVFs coded in non-SQL language such as C, have not yet supported by any database product at the SQL level, it is not our intention to reinvent the wheel. Here, for the sake of completeness, we provide a short description of this notion, motivate and justify our choice of using RVF to support data-intensive analytics inside database engine.

In the following we will explain the need for RVFs from the following dimensions: expressive power, execution efficiency, and seamless integration with relational operators in SQL queries.

2.1   Expressive Power

Usually, the set of tuples in a relation represent a set of objects; when an application involves the inter-relationship and global behavior of these objects, coding the application in a UDF with per-tuple input becomes impossible.

2.2   Execution Efficiency

Tuple-wise computation by scalar UDFs often leads to performance penalties incurred from parameter setups of large volume of calls and from repeatedly loading the data commonly used across calls. We explain this using the video image analysis example shown in Fig 2.

In soccer games, a corner kick is awarded to the attacking team when the ball leaves the field of play by crossing the goal line without a score, having been last touched by a defending player. A corner kick may result in a good scoring opportunity such as by a "header". The defending team may choose to form a wall of players in an attempt to force the ball to be played to an area which is more easily defended.

Fig. 2. Images on Corner-Kick

To analyze the strategies in taking and defending a corner statistically, given a collection of sample images of "typical" corner kick scenes, we want to rank them by their popularity through comparing them with a large set of images on corner kicks.

From each corner kick image the SIFT features are extracted which consist of hundreds to thousands of key points, each is a 128-dimensional vector. Since we only compare the similarity of the whole image, we leverage the order-less bag-of-features model [11] to avoid the expensive pairwise point matching. It works as follows: in an image the key points, or SIFT feature vectors are extracted and clustered, and the histogram of key points in clusters is generated as the signature of the image, that is itself a vector, referred to as the *composite feature vector* of the image. Then, for simplicity, the closeness of two images is determined by the similarities of their composite feature vectors using dot product. The computation involves a large table of corner kick images.

CKImages [ID, feature]
and a small table of sample corner kick images, **CKSamples**, with the same schema. An image has several feature vectors but we compose them into a single long vector. A UDT (User Defined Type) is defined for feature vector. Function *sim* computes the similarity of two images based on the similarity of their composite feature vectors.

The following SQL query first derives the closest sample image of each corner kick image (by maximal similarity), then for each sample image s, calculates the number of corner images taking s as the closest sample, and ranks the sample images by that number.

```
[Q1]   SELECT Sid, COUNT(Neighbor) AS n FROM
          (SELECT P.ID AS Neighbor, (SELECT S.ID FROM CKSamples S
            WHERE sim(P.feature, S.feature) = (SELECT MAX(sim(P2.feature, S2.feature))
            FROM CKSamples S2, CKImages P2 WHERE P2.ID = P.ID)) AS Sid
          FROM CKImages P)
          GROUP BY Sid ORDER BY n;
```

where the closest sample image of each corner-kick image, say p, is computed by comparing p with all the sample images. Since the UDF, *sim*, is invoked on the per-tuple basis but unable to receive the whole *CKSamples* relation as input argument, then

− the *CKSamples* relation is not cached but retrieved for each *CKImages* instance p;

− the *CKSamples* relation is also retrieved in a nested query (Query Optimizer turns it to join) for each (tuple) instance p of *CKImages*, for the MAX similarity between p and the sample images.

Our experiment shows quantitatively that such repeated relation retrieval becomes the performance bottleneck. In fact this kind of inefficient computation pattern widely exists in SQL queries due to the lack of RVFs.

2.3 Composed with Relational Operator in Queries

An RVF has at least one input relation, or tuple-set returned from a query; it cannot update its input relations but can generate a new relation as its output, and in this way viewed as a relational operator and a relation data source. An RVF can be defined, for instance, in the way outlined in Fig. 3.

```
DEFINE RVF f (x, y, R₁, R₂)
  RETURN R₃ {
    float a, b;
    Relation R₁ (/*schema1*/);
    Relation R₂ (/*schema2*/);
    Relation R₃ (/*schema3*/);
    PROCEDURE fn(/*dll name*/);
    RETURN MODE SET_MODE;
    INVOCATION PATTERN BLOCK
}
```

– *The relation schemas R_1, R_2 and R_3 denote the "schema" of f, the actual relation instances or query results compliant to those schemas can be bound to f as actual parameters.*
– *The BLOCK input mode means the input relations are passed in at once.*
– *The SET return mode means all the resulting tuples are returned at once.*

Fig. 3. How an RVF is specified

The above nature allows RVFs to be naturally composed with other relational operators or sub-queries in a SQL query; the relation arguments of an RVF can be exressed by queries as well, such as illustrated in the following query.

SELECT * FROM $rvf_1(Q_4, rvf_2(Q_1, Q_2, Q_3))$;

Like other SQL constructs, our notion of RVF is an extension at SQL level which is supported by extending the query processor, rather than by ad-hoc user programs. This also allows us to construct an inter-query dataflow process [5], using RVFs as actors.

3 Invocation Pattern wrt Input Mode and Return Mode

In relational database engines, the argument of a relation operator may be fed in tuple by tuple (e.g. at the probe site of hash-join), or by a set of tuples (e.g. at the build-site of hash-join). If an operator has any tuplewise input, it is called multiple times wrt to that input during execution. In a tuple-wise evaluated query, a parent operator demands its child operator to return the "next" tuple, and recursively the child operator demands its own child operator to return the "next" tuple, ... in the top-down demand driven and bottom-up dataflow fashion. How to deal with input/output relation data constitutes the *invocation patterns*.

The analogy between RVFs and relational operators allows them to be invoked compositionally in a query and allows the notion of invocation pattern to be applied to RVFs. The input mode and return mode of an RVF represents the specific mechanisms for applying the RVF to its input relations and to deliver the resulting relation.

PerTuple Input Mode. The simplest input mode, *PerTuple,* can be defined such that applying *PerTuple* to RVF f with a single input relation R means f is to be invoked for every tuple in R (pipelined). In the query shown in Fig. 4, RVF *per_image_summery* is applied to table CKSamples and plays as the relation data source although it is invoked under the *PerTuple* mode. This is the only input mode under which the use of an RVF and a scalar UDF is interchangeable.

SELECT ID, Summary FROM
per_image_summery_rvf
("SELECT feature FROM CKSamples");

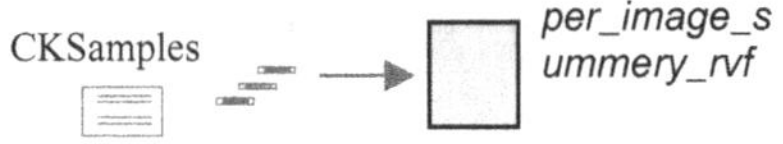

Fig. 4. PerTuple input mode

Block Input Mode. Under the *Block* input mode, as the query shown in Fig 5 for our corner-kick image ranking example, the RVF, *ck_ rvf1*, is called only once in processing a query, with both relations, CKImages and CKSamples, are cached in. The block input mode opens the potential for "in-RVF data parallel computation"; however, when the input relation is sizable, this invocation mode is inappropriate as it may run out memory.

SELECT r.sid, COUNT(r.neighbor) AS n FROM
  ck_ rvf1 ("SELECT * FROM CKImages",
  "SELECT * FROM CKIsamples") r
GROUP BY r.sid ORDER BY n;

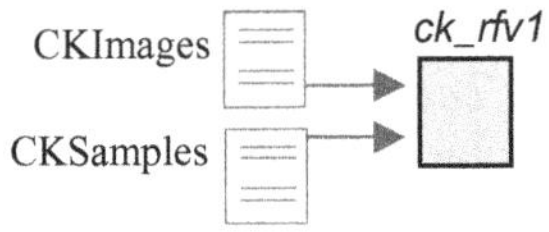

Fig. 5. Block input mode

An input relation can be cached in a RVF as a whole provided that the relation is declared as **static** (by default). An RVF can be treated as a block operation only If all its input relations are static.

PerTuple/Block Input Mode. Under this input mode, apply RVF f to 2 or more input relations where the first argument relation is denoted by R_{left}, means that f is to be invoked for every tuple in R_{left} (pipelined), in combination of the whole tuple-sets of other relations, under the assumption that the instances of the other relations are small enough to be reside in memory. For the above corner-kick image ranking example, the query with RVF under this invocation mode is specified as below; it is executed image by image on the *CKImages* table, but caches in all sample images as initial data (Fig 6).

[Q2] SELECT Sid, COUNT(Neighbor) AS n FROM (
    SELECT P.ID AS Neighbor, ***ck_ rvf2*** (P.ID, P.feature,
      "SELECT * FROM CKIsamples") AS Sid
    FROM CKImages P)
  GROUP BY Sid ORDER BY n;

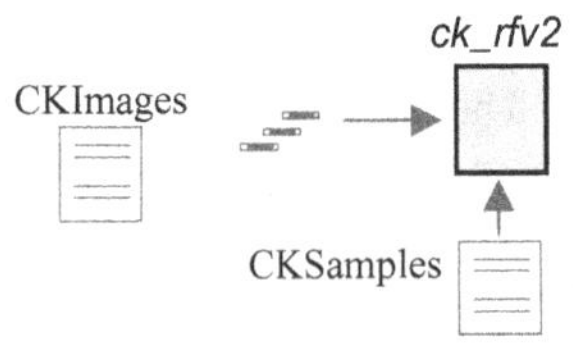

Fig. 6. Block input mode

Tuple Return Mode. An RVF under TUPLE_MODE returns one tuple-per-call in multiple calls when invoked in a query, typically once for each input tuple.

Set Return Mode. An RVF under SET_MODE returns the entire resulting tuple-set in a single call.

Additional batch modes in between the above modes might be added in the future. Confining RVF invocation to designated input and return mode underlies *focused* system support to interact RVF execution with query processing efficiently.

4 Separating RVF Shell and User-Function

Let us consider two parts of an RVF (and in general a UDF): ***RVF shell*** and ***user-function*** where the *RVF shell* deals with the interaction with query processing in parameter passing, data conversion, initial data preparation, memory management, etc. and the *user-function* contains application logic only and is plugged in the shell.

To describe the functionalities of RVF shell, let us provide some background on how functions are executed in query processing. Like other relational operators, a function executed in a query may be called multiple times, one for each returned tuple. Accordingly a function is coded with three cases: FIRST_CALL, NORMAL_CALL and LAST_CALL (also referred to as INIT_CALL, NEXT_CALL and FINAL_CALL). The FIRST_CALL is executed only once in the first time the function is called in the hosting query which provides initial data; the NORMAL_CALL is executed in each call including the first call, for doing the designated application; therefore there would be multiple NORMAL_CALLs if the function is called one tuple at a time, or only a single NORMAL_CALL if it is called only once. LAST_CALL is made after the last normal call for cleanup purpose. The query executor keeps track the number of calls of the UDF during processing a query, and checks the end-of-data condition for determining these cases.

With the above background information, let us describe the functionality of RVF shell in interacting the RVF execution with the host query processing.

– When an RVF is defined, the information about its name, arguments, input mode, return mode, dll entry-point, etc, is registered into the FUNCTIONS meta-table and the FUNCTION_PARAMS meta-table. These tables are to be retrieved by the RVF shell programs.

– When the RVF is invoked, several handle data structures are provided by subclassing the corresponding ones in query executor. Handle of RVF Execution (***hFE***) keeps track, at a minimum, the information about input/output relation argument schemas, input mode, return mode, result set, etc. Handle of RVF Invocation Context (***hFIC***) is used to control the execution of the RVF across calls. hFIC has a pointer to the hFE, and at a minimum keeps track the information on number of calls, end-of-data status, memory context (e.g. life span one or multi-calls), and a pointer to user-provided context known as scratchpad for retaining certain application data between calls. hFIC has a pointer to hARG, a data structure generated from RVF definition for keeping actual argument values across calls.

– During function execution, the RVF container uses several system functions and macros to manipulate the hFIC structure and perform RVF execution. For instance of multi-calls, the steps of RVF invocation include the following.

a) In the first call (only), initialize the hFIC to persist across calls; evaluate each relation argument expressed by a relation name or a query in terms of launching a query evaluation sub-process where the argument query is parsed, planned and executed; convert the complex DBMS internal tuple structures to an array of simple data structures to be passed in the "*user-function*"; initiate other arguments and possibly the scratchpad.

b) In every function call, including the first, set up for using the hFIC and clearing any data left over from the previous pass; get non-static input argument values; invoke **user-function** where the input and returned relations are array of structures defined in the corresponding header files (there is no DBMS internal call within the *user-function* body); convert the data generated by user-function back to DBMS internal data structures, and store them in the result-set pointed by hFE. If the return mode is TUPLE_MODE, return the first resulting tuple to the caller; otherwise if the return mode is SET_MODE, return the entire result-set.

c) Finally, do clean up and end the RVF execution.

In order to ease the development of RVFs, we further investigate the follows:

– separate an RVF into two parts: *RVF shell* and *user-function* under specific input and return modes;
– provide high-level *RVF Shell APIs* for building the shell but shading the DBMS internal details from RVF developers;
– generate RVF shells based on RVF specifications, input and return modes and SROM.

With the above solutions, the major task left to analytics users is to plug in the shell the "user-function" that contains only application logic and free of DBMS internal data structures and calls.

5 Automate RVF Shell Generation

Our further goal is to allow RVF shell provisioning automated, i.e. system generated from RVF specifications under the given input mode and return mode.

5.1 Simple Relation-Object Mapping (SROM)

A common functionality provided by RVF shells is to convert DBMS internal data structures for relation objects into simplified data structures to be manipulated in "user-functions". This is opposite to converting system internal data into string arguments as supported by some DBMSs, which can only deal with simple data and often sacrifices performance. However, coding such conversion in terms of DBMS "system programs" requires the familiarity of system internal details which is in general beyond the expertise of analytical users. To free the RVF development from such burden, we defined the mappings from a relation schema to the data structures of the corresponding tuple (as C-struct) and tuple-set (as array of such C-structs). A language specific *simple relation-object mapping* protocol (SROM) is introduced to underlie the above data mapping. The SROM for C is used to generate C typedefs in a header file based on the given relation schema and the correspondence between SQL types and C-types. User Defined Types (UDTs) with basic components are also covered.

In the above image analysis example, the composite feature vector of an image is represented by a FloatVector object. Our implementation is based on the well-known LIBSVM, which represents sparse vectors with (index:value) pairs to avoid storing too many 0's. For example, (1:32 2:44 4:69 6:89) stands for the vector (32, 44, 0, 69, 0, 89). In Postgres, these types are declared as the following composite UDTs.

```
CREATE TYPE FloatVectorType AS (
    mask BIT VARYING(100),
    floatVector float4 []
);
```

Then we create the following tables using these types.

```
CREATE TABLE CKImages (          CREATE TABLE CKSamples (
    ID INTEGER NOT NULL,             ID INTEGER NOT NULL,
    feature FloatVectorType          feature FloatVectorType
);                               );
```

Based on the above relation declarations, the following header file is generated for the *use-function* (some type reuse intelligence is under development).

```
typedef struct {
    byte * mask;
    float4 * vector;
} FloatVectorType;

typedef struct {                 typedef struct {
    int ID;                          int ID;
    FloatVectorType feature;         FloatVectorType feature;
} CKImage;                       } CKSample;

typedef struct {                 typedef struct {
    CKImage * CKImageArray;          CKSample * CKSampleArray;
    int tuple_num;                   int tuple_num;
} CKImages;                      } CKSamples;
```

Then based on these typedefs the user can provide functions *allocCKImages(n)*, *deallocCKImages()*, etc. These functions are invoked in some API functions and passed in as *pointers* – a *coding trick* allowing us to separate generic APIs and application specific functions. The hARG data structure for holding the initial arguments of this RVF, say, *ck_rvf2_args*, is also generated.

5.2 RVF-Shell APIs

Based on RVF specifications, invocation patterns and SROM, a set of high-level *RVF Shell APIs* are provided for building the shell; these APIs shade the DBMS internal details from RVF developers. Below we use RVF *ck_rvf2* given above as an example to show the use of these APIs.

The RVF is constructed with TUPLE_MODE return for the closest sample image of each given image. The **user-function,** *find_closest_sample()* does not involve any DBMS system internal data structure and function. It takes ID, feature of an image and the array of KCSamples as input, and returns a relation as output. These input/output data are converted from/to the query processing internal objects by the appropriate Shell APIs. We illustrate these APIs (upper-case with RVF_ prefix) by the following pseudo RVF that is specific to the already registered input and return mode. For simplicity we omitted all exception handling and on-error early returns. Note that an FIRST_CALL is also a NORMAL_CALL.

```
SQLUDR_INT32 ck_rvf2(RVF_ARGS) {
  int rv; RVFCallContext *h; ck_rvf2_args *hARGS; CKSamples *samples;
  if (RVF_IS_FIRST_CALL()) {
    h = RVF_FIRST_CALL_BEGIN();
    RVF_ALLOC_ARGS(h->hARGS, &allocCk_rvf2_args);
    h->hARGS->Samples = (Samples *)
        RVF_GET_INPUT_RELATION(RVF_ARG(2), &allocSamples);
    RVF_FIRST_CALL_END(h);
  }
  if (RVF_IS_NORMAL_CALL()) {
    h = RVF_NORMAL_CALL_BEGIN();
    Samples *samples = h->hARG->Samples;
    int ID = (int) RVF_GET_INPUT_TUPLE_FIELD(RVF_ARGS(0));
    FloatVectorType *feature = (FloatVectorType *)
      RVF_GET_INPUT_TUPLE_FIELD(RVF_ARGS(1), &allocFeature());

    /*user-function*/
    int sid = find_closest_sample (ID, feature, samples);

    RVF_RETURN_NEXT(sid);
    RVF_NORMAL_CALL_END(h);
  }
  if (RVF_IS_LAST_CALL()) {
    RVF_FREE (samples, &deallocSamples);
    RVF_FREE (h->hARGS, &deallocCk_rvf2_args);
    RVF_FINALIZE(h);
  }
  return rv;
}
```

Some of the APIs involved above can be explained as below. API RVF_GET_INPUT_RELATION() retrieves the tuple-set of the specified relation or query result and populate the corresponding C-structure objects based on SROM; API RVF_RETURN_NEXT() converts result into the tuple format recognized by the query processor; API RVF_FREE de-allocates memory; using both the DBMS specific memory management utilities and those provided for the data structures used inside the user-function, with the later passed in as function pointer for keeping the generality of the API.

5.3 RVF-Shell Generation

Based on a well defined invocation pattern, it is possible to single out the development of the "user-function", that contains application logic only, from the development of the RVF-shell, and to provide tools for generating the RVF-shell automatically.

The proposed RVF-shell generation mechanism is illustrated in Fig 7, where the responsibilities of system and user are well separated, and the system responsibility for generating RVF shell is automated. With this approach, an RVF is developed in the following way.

– Based on the RVF declaration stored in system meta tables, the system *generates* the header file containing the RVF argument data structure declarations to be used in the "user-function"; the RVF shell skeleton including FIRST_CALL, NORMAL_CALL, LAST_CALL, etc; and the API calls for retrieving argument relations, converting data structures; switching memory contexts, allocating and deallocating memories.

– The user provides the ***user-function*** containing application logic only without DBMS internal system calls and data structures. Optionally the user also deals with other initial data accessible to the *user-function* using the scratchpad.

– A complete RVF is made by plugging the ***user-function*** in the RVF-shell.

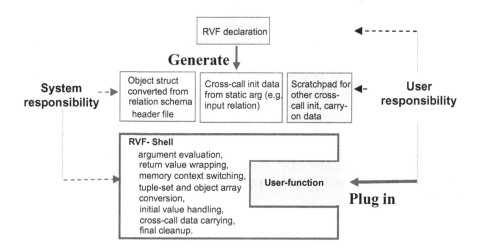

Fig. 7. RVF-shell Generation

For example, given the declaration of RVF, ck_rvf2, which is stored in metatables, it is built with the following mechanisms.
– A header generation utility
```
RVF_RO_META_GEN ("ck_rvf2")
```
is responsible for generating the header files in the way described above.
– A shell generation utility is responsible for generating the RVF shell.
```
RVF_SHELL_GEN ("ck_rvr2")
```

The FIRST_CALL, NORMAL_CALL, LAST_CALL in the generated RVF shell provide the following functionalities in the execution of the RVF.

- The RVF developer ends up with coding a function

    find_closest_sample *(ID, feature, samples);*

The scratchpad for holding other initial data (rather than *CKSamples*) is null in this example; but in general it is an extern pointer to user defined initial values, and the user has the opportunity to add any other data to be carried-on across calls.

- The complete RVF is built by plugging the *user-function* in the shell.

In this way the pure application oriented *user-function* is made independent of platform specific system calls, just like what featured by a MapReduce platform. While the common set of invocation patterns are provided, many applications can be easily pushed down to the DBMS layer as RVFs.

6   Experimental Results

We use the open-sourced Postgres database engine as our prototyping vehicle to test this innovated approach. The experiments are set up on a HP ProLiant DL360 G4 server with 2 x 2.73 Ghz CPUs and 7.74 GB RAM, running Linux 2.6.18-92.1.13.el5 (x86_64). With the goal of studying the computation workload, we used a moderate set of feature vectors extracted from real images, and a large set of features derived from them as the generated test data. We grouped the image SIFT features to up to 2048 clusters but in this experiment only 100 clusters are used, therefore a histogram vector (image signature), or so called composite feature vector, has 100 dimensions. Our experiments show the proven scalability of RVF based computation in data size and in applications, and superior performance compared with using conventional UDF.

We compare running the query Q2 using RVF *ck_rvf2* in PerTuple/Block mode with running Q1 using scalar UDF. We ran these queries with different data load (number of CKImages) with 100 CKSamples. The performance comparison is shown in Fig 8.

	100K	200K	300K	400K	500K	600K	700K	800K	900K	1M
Scalar UDF	1741.2	3482.4	5223.6	6964.8	8700.2	10443.3	12143.3	13856.6	15597.1	17300.3
RVF	12.956	25.032	38.71	52.117	65.491	78.823	91.032	103.996	118.023	131.45

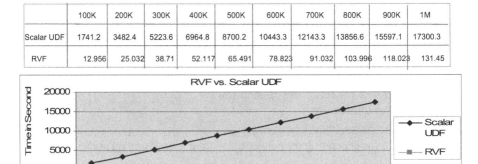

Fig. 8. Using RVF over-performs using client side and scalar UDF

The following is indicated in this experiment. The inability of the scalar UDF to cache-in the CKSamples relation forces the engine to perform multiple joins, to make repeated retrieval of CKSamples and for each sample, repeated MAX similarity calculation (since the MAX state cannot be kept in SQL). Such overhead is added to processing each CKImages instance and is *proportional* to the number of them, therefore becomes serious when the CKImages table is sizable.

The RVF-version, in contrast, is executed with a very streamlined data flow, where the state of the input relation can be traced – the CKSamples can be cached in and the closest sample of an image is calculated only once; all these contribute to the high-performance of the query invoking RVF *ck_rvf2*.

7 Conclusions

In this research we tackled two major limitations found in the existent UDF technology: lack of set input or output which causes the inability of application modeling and the inefficiency of execution, and the difficulty in integrating UDFs with the query engine. These limitations actually prohibit the use of UDFs for complex applications. For solving these problems, we extended the UDF technology in both semantic and system dimensions.

We have implemented the system support to RVFs. For efficiency, we opt to have RVF execution tightly integrated with the query engine. However, we also focused on relieving analytics users from the system internal details. We treat an RVF as two separated parts: the *user-function* containing application logic only, and the *RVF-shell* for executing the RVF in the query processing environment, and provided the system tools for generating RVF-shells automatically. With this approach, the responsibility of an analytics user is limited to plugging the "*user-function*" in the *RVF-shell* without dealing with DBMS internal details.

Prototyped on Postgres, our experience reveals the benefits of the proposed solutions in integrating complex applications to the SQL framework, in significantly enhanced performance, and in easing the UDF development. All these are essential data engineering requirements for making the UDF technology practically usable in the syntheses of data intensive analytics computation and data management.

References

1. Argyros, T.: How Aster In-Database MapReduce Takes UDF's to the next Level (2008), http://www.asterdata.com/
2. Chaiken, R., Jenkins, B., Larson, P.-Å., Ramsey, B., Shakib, D., Weaver, S., Zhou, J.: SCOPE: Easy and Efficient Parallel Processing of Massive Data Sets. In: VLDB (2008)
3. Chen, Q., Hsu, M.: Data-Continuous SQL Process Model. In: Proc. 16th International Conference on Cooperative Information Systems (CoopIS 2008) (2008)
4. Chen, Q., Hsu, M.: Inter-Enterprise Collaborative Business Process Management. In: Proc. of 17th Int'l Conf on Data Engineering (ICDE 2001), Germany (2001)
5. Chen, Q., Hsu, M.: Support Dataflow Applications inside Database Engine. Submitted to ER 2009 (2009)

6. Cooper, B.F., et al.: PNUTS: Yahoo!'s Hosted Data Serving Platform. In: VLDB (2008)
7. Dayal, U., Hsu, M., Ladin, R.: A Transaction Model for Long-Running Activities. In: VLDB 1991 (1991); Received 10 Year Best Paper Award in 2001
8. Dean, J.: Experiences with MapReduce, an abstraction for large-scale computation". In: Int. Conf. on Parallel Architecture and Compilation Techniques. ACM Press, New York (2006)
9. DeWitt, D.J., Paulson, E., Robinson, E., Naughton, J., Royalty, J., Shankar, S., Krioukov, A.: Clustera: An Integrated Computation And Data Management System. In: VLDB (2008)
10. Jaedicke, M., Mitschang, B.: User-Defined Table Operators: Enhancing Extensibility of ORDBMS. In: VLDB (1999)
11. Lowe, D.: Distinctive image features from scale-invariant key points. International Journal of Computer Vision 60(2), 91–110 (2004)
12. Moran, B.: UDFs Endanger Performance, `http://www.sqlmag.com/Article/ArticleID/42139/sql_server_42139.html`
13. Novick, A.: Drilling Down into Performance Problem. In: Transact-SQL User-Defined Functions, ch. 11, pp. 235–244, Wordware Publishing (2004) ISBN 1-55622
14. Ordonez, C., Garcia-Garcia, J.: vector and Matrix Operations Programmed with UDFs in a Relational DBMS. In: CIKM 2006 (2006)

High Performance Analytics with the R^3-Cache

Todd Eavis and Ruhan Sayeed

Concordia University, Montreal, Canada
eavis@cs.concordia.ca, r_sayee@encs.concordia.ca

Abstract. Contemporary data warehouses now represent some of the world's largest databases. As these systems grow in size and complexity, however, it becomes increasingly difficult for brute force query processing approaches to meet the performance demands of end users. Certainly, improved indexing and more selective view materialization are helpful in this regard. Nevertheless, with warehouses moving into the multi-terabyte range, it is clear that the minimization of external memory accesses must be a primary performance objective. In this paper, we describe the R^3-cache, a natively multi-dimensional caching framework designed specifically to support sophisticated warehouse/OLAP environments. R^3-cache is based upon an in-memory version of the R-tree that has been extended to support buffer pages rather than disk blocks. A key strength of the R^3-cache is that it is able to utilize multi-dimensional fragments of previous query results so as to significantly minimize the frequency and scale of disk accesses. Moreover, the new caching model directly accommodates the standard relational storage model and provides mechanisms for pro-active updates that exploit the existence of query "hot spots". The current prototype has been evaluated as a component of the Sidera DBMS, a "shared nothing" parallel OLAP server designed for multi-terabyte analytics. Experimental results demonstrate significant performance improvements relative to simpler alternatives.

1 Introduction

Online Analytical Processing (OLAP) has become one of the cornerstones of contemporary data warehousing systems. By offering an intuitive, easily navigable multi-dimensional perspective of corporate data, OLAP empowers decision makers with the ability to assess and quantify operational trends and patterns that underly the growth of the organization. Moreover, by largely shielding the users from the often overwhelming volume of transactional data in the "raw" warehouse, OLAP systems and interfaces have at least the potential to support a much more interactive form of analysis for non-experts users.

Of course, in order to realize this potential, the underlying systems must not only hide the scale and complexity of the data warehouse, but they must also offer performance in keeping with the requirements of interactive or real time analysis. In the data warehousing context, this has lead to innovations such as selective materialization of OLAP views or group-bys, as well as the design

T.B. Pedersen, M.K. Mohania, and A M. Tjoa (Eds.): DaWaK 2009, LNCS 5691, pp. 271–286, 2009.

of DW-specific indexing structures. Nevertheless, such methods are primarily disk-oriented and, while certainly important in the current context, still place significant processing burdens on the OLAP/DW DBMS.

As is the case with DBMS systems in general, memory-resident caches can and should be used to improve query response. In fact, caches can be even more important to OLAP systems, given the massive processing costs often associated with analytical queries on tera-scale data sources. Unlike traditional DBMS caches, however, the fundamental structure of OLAP queries can be exploited to dramatically improve the capabilities of the cache manager. Specifically, OLAP queries tend to be "cubic" in nature; in other words, the most common pattern is a multi-dimensional range query that defines a contiguous hyper-cubic region in the data space. That being the case, existing buffer pages in OLAP-aware caches can subsume future queries that fall within the hyper-cubic boundary. More powerfully still, new queries that spatially overlap multiple existing pages can be dynamically transformed so as to minimize the number of disk accesses required. Efficiently exploiting this notion of the geometric cache is a crucial performance concern for real world OLAP servers.

In this paper, we present the R^3-cache, a *cube*-oriented caching framework that supports the spatial manipulation of both full and partial query matches. Structurally, R^3-cache is in fact based upon the R-tree, a multi-dimensional disk-based indexing structure often seen in research and industrial settings. In addition to modifications to the tree structure that reduce bounding box overlap, the cache manager is also capable of pro-actively pre-filling targeted query regions. The caching model is, in turn, integrated into the Sidera OLAP DBMS, a fully parallelized "shared nothing" server that seeks to provide robust, high performance analytics for today's massive decision support environments.

The remainder of the paper is organized as follows. In Section 2, we review related work. Section 3 briefly presents the architecture of the larger Sidera DBMS, with the new R^3-cache then presented in Section 4. Experimental results are provided in Section 5, followed by final conclusions in Section 6.

2   Related Work

The interest in OLAP as a research pursuit grew out of the seminal *data cube* paper by Gray et al [8]. Subsequently, researchers focused on fundamental construction algorithms for OLAP cubes, with a particular emphasis on the efficient generation of all 2^d sub-cubes or *group-bys* in the d-dimensional space [1,2]. As it became clear that the computational storage requirements of this model were prohibitive, numerous algorithms for partial cube generation were proposed. Such research focused on both the identification of the optimal subset of group-bys [18], as well as the computation of the selected subset [3]. A common theme of more recent OLAP research has been the design of algorithms and data structures that support hierarchical dimension representation. By this we mean that dimensions often define a multi-level aggregation hierarchy (e.g., product – brand – category) that gives rise to interactive "drill down" and "roll up' analysis. Hierarchy-aware data structures include the Cure Cube [13], map-Graph [7], and the Cube File [10]. We note that the notion of dimension hierarchies is particularly relevant in the current context as caches *must* be amenable to hierarchical processing.

The R-tree, the fundamental indexing model underlying the current research, has had a long and rich academic history since it was first presented by Guttman [9], who described the structure, processing model, and algorithms for node splitting. In the context of data warehousing, it has been used by Roussopoulos et al. to define the cube tree, a *data cube* indexing model based upon the concept of a *packed* R-tree [16]. Eavis and Cueva have also described compression and block organization mechanisms for OLAP-centric R-tree implementations [5].

Finally, a great deal of work has been presented in the general area of database caching. A common theme is the use of materialized views to reduce response time in large database environments. Recent proposals in this vein include the Materialized Query Table of Phan and Li [15], and Luo's "partial materialized views" [12], though neither could be regarded as warehouse-specific. An earlier approach by Kotidis and Roussopoulos [11] that employs a dynamic framework to materialize and re-use previous query fragments does, in fact, target data warehouse settings. Still, we note that materialization-based mechanisms represent a very rudimentary form of caching. With respect to true, data warehouse specific caches, the literature is actually quite thin. Shim et al. describe an in-memory caching model for partial query re-use [17]. However, their approach is limited to subsumption-only processing. To date, the most flexible caching model is that presented by Deshpande and Naughton [4]. This framework supports true partial matches, as well as hierarchical aggregation facilities. Unlike the approach described in this paper, however, it is intended primarily for chunk-based MOLAP servers (as opposed to relational systems), and excels in lower dimensional spaces (e.g, 2–3 dimensions).

3   The Sidera Parallel OLAP DBMS

In the next section, we will present a concise overview of the R^3-cache framework. However, because the system is in fact fully integrated into the larger Sidera OLAP DBMS [6], we begin with a brief description of the structure and function of the OLAP server itself. As previously noted, Sidera has been designed from the ground up as a parallel "shared nothing" platform for the resolution of complex multi-dimensional analytic queries. The current system consists of approximately 70,000 lines of C++ code and runs on a 17-node, 34-processor HP Proliant Linux cluster. Subsystems exist for data cube generation and distribution, as well as multi-dimensional selectivity estimation, OLAP indexing, and the manipulation of dimensional hierarchies. Current projects are extending the server with conceptual modeling facilities, high availability and fault tolerance features, and object oriented query interfaces. Figure 1(a) provides a simple illustration of the architectural framework. Note that Sidera is essentially

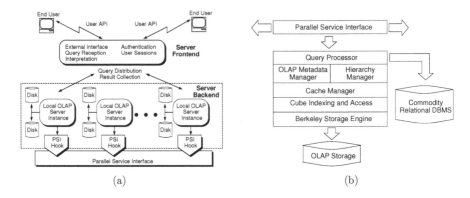

Fig. 1. (a) The Sidera architecture (b) The OLAP stack

constructed as a series of logically independent back end servers that are transparently bound together by a Parallel Service Interface (constructed on top of the Message Passing Interface). In essence, each server knows nothing of the existence of its sibling servers and operates solely on the cube fragments associated with its local resources. Specifically, individual cuboids are striped — as per their position along a Hilbert space filling curve — onto each of the p processors in the cluster. Distributed R-tree indexes then allow the query resolution engine to extract and integrate records encapsulated by common bounding boxes on each node. Ultimately, each server contributes equally to the resolution of every individual query, ensuring full parallelization across arbitrary load levels.

Each of the local servers supports its own OLAP *stack*, of which the caching module is but one component. Figure 1(b) depicts the basic elements of the stack and the relationship between them. Note that the caching subsystem sits below the query processor and hierarchy manager but above the low level indexing and storage components. In fact, both the cache and storage engine are oblivious to the hierarchies themselves and represent data solely at the base level of each dimension. It is the job of the hierarchy manager to transparently *map* final result sets — with the aid of the query processor — between arbitrary levels of the dimension hierarchies (i.e., roll-up and drill-down). This modular approach dramatically simplifies the processing logic of both the caching and indexing subsystems. A full discussion of the Hierarchy Manager is provided in [7]. Finally, we note that Sidera is designed specifically as an analytics server, and does not attempt to function as an "all things to all people" system. As such, detail-level ad hoc querying is (transparently) funneled to a local data warehouse *partner* (i.e, a commodity DBMS).

4   The R^3-Cache

A primary motivation when investigating new caching models for Sidera was the importance of building upon the current framework, so as to exploit both existing software and facilities. With respect to the resolution of multi-dimensional queries, we note that Sidera employs a (parallelized) R-tree model in which group-by fragments are distributed to the p nodes of the cluster and subsequently packed and compressed via a Hilbert space filling curve. Recall that the R-tree is a hierarchical, d-dimensional tree-based index that organizes the query space as a collection of nested, possibly over-lapping hyper-rectangles. In practice, R-trees are one of the few true multi-dimensional indexes that have been consistently utilized in production settings.

As such, we chose to leverage our previous experiences in designing disk based multi-dimensional indexing structures by exploiting the strengths of the R-tree. That being said, it must be noted that a direct implementation of the R-tree would be ill-suited to a caching framework. In particular, the tendency of the tree to produce bounding box overlap — particularly at the wide leaf level — represents a serious constraint on cache responsiveness. For this reason, we propose in this paper a variation on the R-tree that we call the Non-overlapping R-tree (NOR-tree). In the remainder of this section, we describe how the NOR-tree is used to provide efficient OLAP cache support.

4.1   NOR-Tree Search and Query Decomposition

In order to illustrate how the NOR-tree structure differs from that of a conventional R-tree, we will begin with a look at the process of resolving a typical multi-dimensional query. Let us assume that an R-tree style cache currently holds one or more previous query results and that a new query Q has just arrived. If Q does not intersect the boundaries of any existing query in the cache, then we clearly must go to disk for the result. Likewise, if Q is fully subsumed by an existing query, then the cache manager can process the contents of the cache buffer and return the result immediately.

However, if Q intersects an existing query, then we have what we call a *partial match*. The R^3-cache manager resolves such queries by *decomposing* the new query Q into two query sets, A and B. The first set contains $A_i \in A$, where $0 \leq i \leq d$ and $A = N \cap Q$, with N indicating the node to be searched. In effect, the set A consists of one or more queries representing that portion of Q that intersects the cache node N. The second set contains $B_i \in B$, where $0 \leq i \leq d$ and $B = (N \cup Q) - Q$. In this case, B consists of sub-queries whose boundaries lie outside of N. Moreover, the constituent sub-queries of A and B are constructed so as to be *non-overlapping*. In terms of the mechanism for defining the sub-query regions, the search algorithm essentially uses the "corner points" of the subsumed box to define new hyper-rectangles. We will look at this process more closely in Section 4.3.

At this point, results from the query set A can be processed directly from the cache, while the query set B is sent to the backend storage engine. Data returned from the storage engine is inserted into the buffers corresponding to query set B and subsequently combined with A and returned to the user. Algorithm 1 formally describes the process.

Algorithm 1. Basic search: $Search(Q, M, L)$

Input: The d-dimensional query box Q, the NOR-tree M, and a list L to contain newly created NOR-tree boxes.
Output: Result set R consisting of cached records + database records
1: **for** each leaf level bounding boxe m in M that touches Q **do**
2:    **if** $Q \subseteq m$ **then**
3:        Return R directly from box m
4:    **if** $Q \cap m \neq \emptyset$ **then**
5:        Insert $((Q \cup m) - m)$ *into* L
6:        Add cached data in $(Q \cap m)$ to R
7: **if** $L \equiv \emptyset$ **then**
8:    Retrieve Q from database and return as R
9: **else**
10:    Retrieve L from database, add cached data, and return as R

Figure 2 demonstrates how this would work with a simple example. In this case, "Query 1" has already been executed and its contents are now stored in the cache. When "Query 2" arrives, the cache manager determines a partial intersection with the "Query 1" object in the cache. "Query 2" is now decomposed into a pair of hyper-rectangular query sets, A and B. A is processed from the cache, while the two B queries are delivered to the disk backend. When the results for the two B sub-queries are retrieved, they are first inserted into the cache as independent buffer pages. A and B are subsequently combined and returned to the user.

Figure 3 concisely illustrates the difference between a standard R-tree search and the one performed by the R^3-cache. In both models, we begin at the same point. In Stage 1, the R-tree is empty; it has no data points, child nodes, or bounding boxes. When the first query, "Query 1", is executed, the R-tree is searched and no match is found. Subsequently, "Query 1" is sent to the backend data warehouse. Query results returned from the warehouse are then stored in the root node. At this point, the root is also the leaf node. In Stage 2, when "Query 2" is executed, the root node is searched and we find that "Query 2" partially matches the root. The root node expands to hold both "Query 1" and "Query 2". At this stage, our data structure is essentially a standard R-tree in that it houses overlapping leaf nodes. In the final phase, however, overlapping leaf nodes in the R^3-cache are split into multiple leaf nodes (in this case, four). More importantly, none of the cache objects — Q1, Q2, Q3, or Q4 — share any portion of the space.

As a final point, we note that the non-overlapping characteristic of the NOR-tree refers exclusively to the leaf nodes. Parent level bounding boxes are still free to intersect. In practice, this is an advantage, rather than a disadvantage as it permits a much simpler tree maintenance algorithm. Little is sacrificed in terms of performance since, as noted previously, most of the nodes in an R-tree are in fact found at the leaf level.

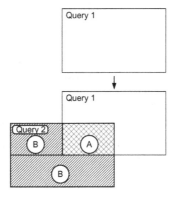

Fig. 2. Query decomposition

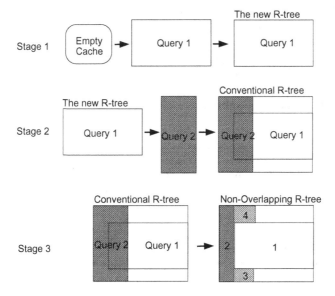

Fig. 3. A partial match: R-tree versus NOR-tree

4.2 Cache Insertion and Partitioning

In general, our insertion mechanism is similar to that of the standard R-tree. In short, we insert data points into leaf nodes. If the leaf node grows larger than its maximum size (i.e., a user-defined multiple of the OS page size), we split it recursively into sub-nodes. It is crucial, of course, to minimize the number of intersections between the leaf nodes and incoming range queries since an increased number of intersections negatively impacts cache performance. In fact, our partitioning approach is based upon the mechanism first proposed by

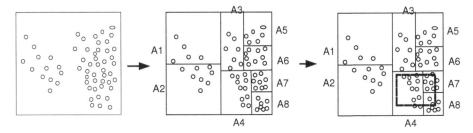

Fig. 4. Node insertion and splitting. The dashed line represents a possible user query.

Muralikrishna and DeWitt [14]. Here, an equi-depth histogram called an *hTree* is used to recursively partition the space by halves. So, for a dimension count d and $D = \{d_1, d_2, ...d_d\}$, where D is the set of dimensions in our data space, the root node A will be split on the medians along the sequence $\{d_1, d_2, ...d_d\}$. As Figure 4 demonstrates, the generation of an equi-depth tree results in B rectangles, such that each rectangle encloses approximately $\frac{R}{B}$ points within it, where R denotes the total number of points in the data space. Note how the box-like subdivisions of the space minimize intersection with the user query. In contrast, naive partitioning algorithms tend to create partitioning "stripes" that can exaggerate the effect of overlap. Algorithm 2 compactly describes the core logic for insertion and node-splitting.

Algorithm 2. Cache insertion: $Insert(D_{new}, N, S_N)$

Input: The set of points to be inserted D_{new}, the targeted leaf node N, and the
      dimension upon which N was previously split S_N. For convenience, we also define
      the maximum node size M.
Output: The updated cache
1: Insert D_{new} into N
2: **if** $sizeof(N) > M$ **then**
3:    Split N on $(S_N + 1)$ to get new leaf-nodes N_1 and N_2
4:    **if** $sizeof(N_1) > M$ **then**
5:        $Insert(\emptyset, N_1, (S_N + 1))$
6:    **if** $sizeof(N_2) > M$ **then**
7:        $Insert(\emptyset, N_2, (S_N + 1))$

4.3   Leaf Node Merging

One of the problems with the query decomposition model — at least as we've described it thus far — is that it has the potential to create a significant number of new cache buffers. In the worst case, when a new query completely subsumes an existing node, it can generate $O(d^3)$ new queries, where d is the number of dimensions. Strictly speaking, the number of new boxes in the worst case is exactly $d^3 - d + 2$. So, for a two-dimensional cache, the worst case scenario

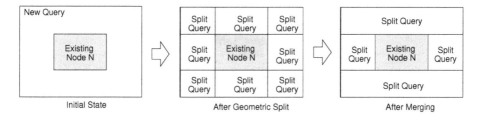

Fig. 5. Query merging

generates $2^3 - 2 + 2 = 8$ new boxes. Likewise, for a three dimensional cache, the worst case scenario would be $3^3 - 3 + 2 = 26$ new boxes. The second box in Figure 5 illustrates the basic problem. Here, a new query in the two-dimensional space subsumes the existing node N. In this case, 8 new back end queries could be generated. These numbers are simply unsustainable in caching scenarios.

To address this problem we utilize a technique whereby adjacent query elements are merged. Let d represent the number of dimensions for queries A and B. Further, let the set of ranges $R_A = (R_{A^1}, R_{A^2}, ...R_{A^d})$ define the range-set for A, while the set of ranges $R_B = (R_{B^1}, R_{B^2}, ...R_{B^d})$ defines the range-set for B. For query A and query B to be merged, we need to satisfy the following criteria. For $j = d - 1$, $R_{A^i} \equiv R_{B^i}$ must hold for the dimension count j, where $1 \leq i \leq d$, and $R_{A^k} \neq R_{B^k}$ must hold true for just one dimension k, which can be any dimension between 1 to d. In addition, we must have either (i) R_{A^k}, $R_{A^{k1}} = R_{B^{k0}} - 1$, where $R_{B^{k0}}$ is the lower dimension for R_{B^k}, or (ii) R_{B^k}, $R_{B^{k1}} = R_{A^{k0}} - 1$, where $R_{A^{k0}}$ is the lower dimension for R_{A^k}. Simply put, application of this logic combines contiguous hyper-rectangles so as to reduce the number of new queries from $O(d^3)$ to $O(2d)$.

In Figure 6, we demonstrate how the technique is used in order to reduce the worst case decomposition of a 3-D space from 26 query boxes to just six by merging adjacent cache buffers. In (b), we have nine adjacent boxes on the front plane. We can therefore combine them to reduce the number of boxes to one. Similarly in (c), we can combine the nine adjacent boxes and reduce them to one. In (d) and (e), we merge the three adjacent boxes on each side plane to reduce six boxes to just two. Finally, (f) and (g) are already single boxes, so there is nothing to merge here. We are therefore left with a total of just six boxes.

We can formally verify the proposition that box merging guarantees a reduction from $O(d^3)$ to $O(2d)$ as follows. We begin with the original decomposition process, which we indicated produced a worst case box count of $d^3 - d + 2$. In the first step, the algorithm splits the box in any one direction. This creates $d - 1$ box sets, with each set consisting of d^2 adjacent boxes. The total is therefore $d^2 \times (d - 1)$ boxes. During the second step, the remaining area of the boxes is split again. But this time the process creates $d - 1$ sets, with each set consisting of d adjacent boxes, for a total of $d \times (d - 1)$ boxes. After the first two steps of splitting, there will always be two boxes remaining. Combining all

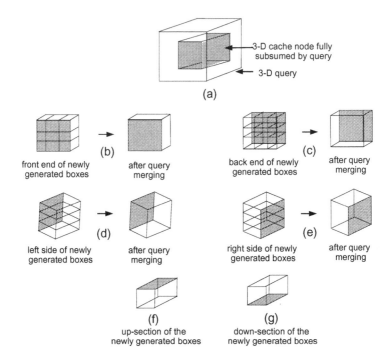

Fig. 6. Query merging in a 3-D cache

individual boxes now gives a total of $d^2 \times (d-1) + d \times (d-1) + 2$. Therefore, $d^2 \times (d-1) + d \times (d-1) + 2 \Rightarrow d^3 - d^2 + d^2 - d + 2 \Rightarrow d^3 - d + 2$. Now, let us apply the query merging solution. Since our query merge combines adjacent boxes into a single box, the previous equation can be rewritten simply as $(1 \times (d-1)) + (1 \times (d-1)) + 2 = 2d$.

As a final point, we note that actually decomposing subsumption queries into the full set of $O(d^3)$ boxes, many of which will subsequently merged, is unnecessary in practice. Instead, the process can be made more efficient by integrating the two mechanisms so that the $O(2d)$ boxes are directly generated. In practice, the overhead of the modified decomposition phase is negligible for the modest dimension counts seen in OLAP environments.

4.4   Hot Spots and Pro-active Insertion

The system, as described thus far, is capable of supporting both full match and partial match queries against a relational backend. Initial testing of the system, however, revealed performance improvements that were less dramatic than expected. A closer examination of the cache logs revealed that while the subsystem was working exactly as expected, much of the potential performance improvement was lost as user queries consistently required a small percentage of records to be retrieved from disk. Because the cost of disk access relative to RAM

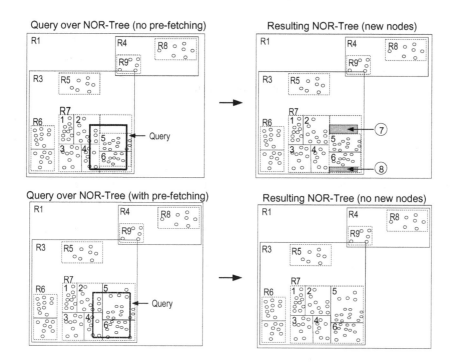

Fig. 7. Comparison of R-tree with pre-fetching and without pre-fetching

access is so extreme, partial match caching — while theoretically attractive — can be surprisingly ineffective in real systems.

We address this shortcoming by building upon a fundamental observation that underlies much of the Sidera design. Specially, in contrast to the "small query/high volume" character of transaction processing systems, data warehousing and OLAP platforms tend to be "big query/low volume" systems. As such, the Sidera philosophy is, as much as possible, to utilize its resources *between* queries to improve later response times. In other words, "make the system wait, not the user." The R^3-cache manager therefore monitors the cache on an ongoing basis in order to identify *hot spots* in the cache; in other words, it locates regions deemed to have high cache value. As it identifies these areas, it "fills gaps" in hot regions in anticipation of future requests. At the same time, it may release pages that it deems to be of little value.

Of course, pre-fetching data is not terribly useful by itself (or might even be counter-productive) if the pre-fetched data has a low probability of being queried. For this reason, it's important to fetch only those data points that have a high probability of eventual query access. As such, each cache node is augmented with the following meta data: *query size* (query result-set size in bytes), *frequency* (number of node accesses versus time in the cache), and *response time* (the time it would take to refresh this node from disk). A *heat estimate* is then generated by normalizing each value into a (0,1) range and calculating the sum. It should be clear that the heat estimate is more that a measure of frequency. It also identifies cache regions that would be expensive to re-compute. While relatively simple, the heat estimate provides a concise picture of the value of the various nodes in the tree. More importantly, heat estimates can be aggregated into parent boxes so as to identify broad regions of interest in the cache.

Let's look at a graphical illustration of the effectiveness of pre-fetching. Figure 7 starts with a "non pre-fetching" NOR-tree containing a number of previous queries. The first image shows a query Q overlapping the nodes R7.2, R7.4, R7.5, R7.6, plus two empty areas beneath. The resulting NOR-tree shows two new nodes R7.7 and R7.8. Not only do these two nodes result in a cumbersome R-tree that has the potential to create many small nodes along the narrow edges, but the associated data had to be fetched at query time from the backend database which, of course, reduces cache efficiency. If we compare this outcome to that of the second image, where the NOR-tree actively pre-fetches data, we see that the cache does not have to create any new nodes or fetch any data from the disk. In effect, this has already been done in the expectation that future queries would hit this general region.

5   Experimental Results

While the caching framework continues to evolve, we have undertaken numerous tests to assess its ability to tangibly improve OLAP query performance. Due to the space limitations of this paper, we will highlight just a handful of the key results here. We note at the outset that although Sidera is an inherently parallel system, the results included in this section are restricted to a single node so as not to complicate their interpretation. With respect to the test parameters themselves, we synthetically generate data sets using a data generator developed for the Sidera system. Using the generator allows us to test the platform under a variety of conditions, including pronounced skew patterns (we use a zipfian skew factor of 1.0 for the tests shown in this section). Query streams are also generated using a system constructed specifically for Sidera. Queries are multi-dimensional in nature (i.e., range queries), with the dimension count and dimensional ranges randomly defined. The generator produces Roll Up and Drill Down queries by iteratively adjusting new queries in several follow-up passes. To encourage the appearance of hot spots, we automatically generate the first 300 queries, then manually replicate and slightly modify subsets of these queries to repeatedly hit specific areas in the space. We note that there is no standard way to evaluate OLAP caches so we believe that this configuration represents a reasonable starting point.

In Figure 8, we provide a baseline evaluation illustrating performance for the R^3-cache versus the same Sidera system without the caching option available. Tests were run for input sets of one million and ten million records (using attribute cardinalities between 10 and 1000) with the cache capacity set at 10% of the size of the input set (note that the fully materialized aggregated cubes would be much larger). At both input sizes, we see a run-time for the cache based

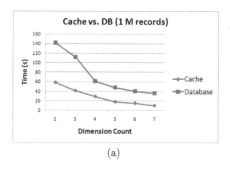
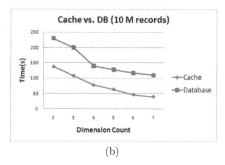

(a)   (b)

Fig. 8. Cache performance relative to standard DBMS indexes (a) 1 M records (b) 10 M records

system that is about half that of the non-cache system at low dimensions, and about one third the cost at higher dimensions. We note that result sets in higher dimensions tend to be smaller due to increased specificity of the queries. However, costs do not necessarily drop proportionally as even small queries require some degree of disk access.

We also looked at the effect of our pro-active insertion model. Figure 9 provides a comparison of pro-active insertion versus a standard "static" cache, using the data sets from the running example. The performance pattern is similar to that seen in the previous test. In fact, this result underscores a very important point. Without pro-active caching, the performance of the standard caching system is essentially equivalent to a well indexed database with no caching system at all! In fact, this should not be so surprising. Specifically, given that a disk access may be several orders of magnitude more expensive that an equivalent, in-memory access, a single miss in the cache effectively destroys much of the benefit of a multi-dimensional cache.

We use the same test sets and parameters to compare the relative performance of the relational R^3-cache and the chunk-style MOLAP cache proposed by

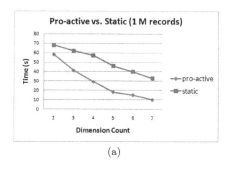
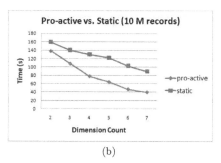

(a)   (b)

Fig. 9. Pro-active caching performance versus static caching (a) 1 M records (b) 10 M records

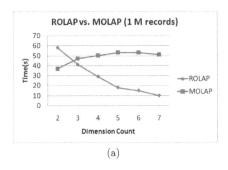
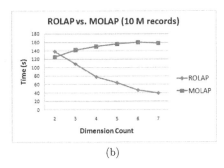

Fig. 10. ROLAP performance versus the MOLAP alternative on (a) 1 M records (b) 10 M records

Deshpande and Naughton [4]. Specifically, we wanted to assess performance beyond the two or three dimensional spaces in which MOLAP sytems typically excel. The results are provided in Figure 10. Note that the chunking cache does indeed provide superior performance in the simple 2-d space. This is not unexpected as a considerable amount of subsumption resolution takes place in two dimensions. At higher dimensions, however, the space becomes much more fragmented and the MOLAP cache has no mechanism to find or fill hot spot regions. The result is a fairly flat performance curve. In fairness, we note that our implementation of the MOLAP cache was relatively simple and could undoubtedly be further optimized. Still, at this juncture, it seems unlikely that the MOLAP model would offer superior performance in higher dimensions.

In the final two graphs, we look more closely at the ROLAP/MOLAP comparison. In Figure 11(a), we depict the cache hit ratio for the two systems, using the test set of Figure 10(a). As we move towards 600 queries, we see the "prefetch" ROLAP solution producing a hit ratio that is approximately double that of the "fetch-less" MOLAP solution. These results are in keeping with those of Figure 9. Figure 11(b), on the other hand, displays the results for the same test

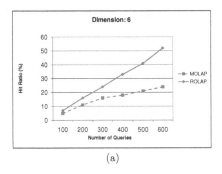
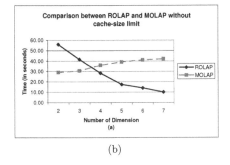

Fig. 11. ROLAP versus the MOLAP alternative (a) hit ratio (b) no cache limit

parameters, but removes all limitations on the size of the cache. Here, we can see a modest improvement for the MOLAP system. In short, this is due to the fact that a larger number of "memory hungry" MOLAP arrays can be materialized in memory (by default, MOLAP arrays must store empty cells). We emphasize, however, that limitless caches are highly unrealistic in practice. The existence of memory constraints in fact strongly supports the implementation of ROLAP caches that only store valid data points.

6 Conclusions

We have described a new relational caching framework that has been integrated into the Sidera parallel OLAP DBMS. The cache — based upon an in-memory variation on the classic R-tree — has shown considerable promise in the low to medium dimensional spaces commonly found in OLAP query environments. In particular, experimental work has demonstrated the importance of incorporating pro-active insertion policies into cubic caching schemes so as to mitigate the effects of extraneous disk accesses. Furthermore, by exploiting the synergies that naturally exist between Sidera's R-tree based disk indexes, the R-tree oriented caching subsystem, and the hierarchy translation facilities that transparently operate on results returned from *either* the cache or the disk, we believe that the current system represents a concrete blueprint for the design of practical high performance OLAP servers.

References

1. Agarwal, S., Agrawal, R., Deshpande, P., Gupta, A., Naughton, J.F., Ramakrishnan, R., Sarawagi, S.: On the computation of multidimensional aggregates. In: VLDB 1996: Proceedings of the 22th International Conference on Very Large Data Bases, pp. 506–521 (1996)
2. Beyer, K., Ramakrishnan, R.: Bottom-up computation of sparse and iceberg cubes. In: ACM SIGMOD, pp. 359–370 (1999)
3. Dehne, F., Eavis, T., Rau-Chaplin, A.: Efficient computation of view subsets. In: DOLAP 2007: Proceedings of the ACM tenth international workshop on Data warehousing and OLAP, pp. 65–72 (2007)
4. Deshpande, P., Naughton, J.F.: Aggregate aware caching for multi-dimensional queries. In: EDBT 2000: Proceedings of the 7th International Conference on Extending Database Technology, pp. 167–182 (2000)
5. Eavis, T., Cueva, D.: The lbf r-tree: Efficient multidimensional indexing with graceful degradation. In: International Database Engineering and Applications Symposium (IDEAS), pp. 241–250 (2007)
6. Eavis, T., Dimitrov, G., Dimitrov, I., Cueva, D., Lopez, A., Taleb, A.: Sidera: A cluster-based server for online analytical processing. In: Grid Computing, High-Performance and Distributed Applications (GADA), pp. 1453–1472 (2007)
7. Eavis, T., Taleb, A.: Mapgraph: efficient methods for complex olap hierarchies. In: CIKM 2007: Proceedings of the sixteenth ACM conference on Conference on information and knowledge management, pp. 465–474 (2007)

8. Gray, J., Bosworth, A., Layman, A., Pirahesh, H.: Data cube: A relational aggregation operator generalizing group-by, cross-tab, and sub-totals. In: ICDE, pp. 152–159 (1996)
9. Guttman, A.: R-trees: A dynamic index structure for spatial searching. In: ACM SIGMOD, pp. 47–57 (1984)
10. Karayannidis, N., Sellis, T.: Hierarchical clustering for olap: the cube file approach. The VLDB Journal 17(4), 621–655 (2008)
11. Kotidis, Y., Roussopoulos, N.: A case for dynamic view management. ACM Transactions On Database Systems 26(4), 388–423 (2001)
12. Luo, G.: Partial materialized views. In: ICDE, pp. 756–765 (2007)
13. Morfonios, K., Ioannidis, Y.: Cure for cubes: cubing using a rolap engine. In: VLDB 2006: Proceedings of the 32nd international conference on Very large data bases, pp. 379–390 (2006)
14. Muralikrishna, M., DeWitt, D.J.: Equi-depth multidimensional histograms. SIGMOD Rec. 17(3), 28–36 (1988)
15. Phan, T., Li, W.-S.: Dynamic materialization of query views for data warehouse workloads. In: ICDE, pp. 436–445 (2008)
16. Roussopoulos, N., Kotidis, Y., Roussopoulos, M.: Cubetree: Organization of the bulk incremental updates on the data cube. In: ACM SIGMOD, pp. 89–99 (1997)
17. Shim, J., Scheuermann, P., Vingralek, R.: Dynamic caching of query results for decision support systems. In: International Conference on Scientific on Scientific and Statistical Database Management (SSDBM), pp. 254–263 (1999)
18. Shukla, A., Deshpande, P., Naughton, J.F.: Materialized view selection for multidimensional datasets. In: VLDB 1998: Proceedings of the 24rd International Conference on Very Large Data Bases, pp. 488–499 (1998)

Open Source BI Platforms: A Functional and Architectural Comparison

Matteo Golfarelli

DEIS, University of Bologna, Viale Risorgimento 2, Bologna, Italy
matteo.golfarelli@unibo.it

Abstract. While in the past the BI market was strictly dominated by closed source and commercial tools, the last few years were characterized by the birth of open source solutions: first as single BI tools, and later as complete BI platforms. An Open Source BI platform provides a full spectrum of BI capabilities within a unified system that reduces the overhead for the development and management of each application, and lets the user feel like he/she was using a single BI solution. This paper proposes a comparative evaluation of three different Open Source BI platforms (namely JasperSoft, Pentaho and SpagoBI) aimed at understanding their current features, their future potentialities and their limits when adopted in real projects as well as a basis for research prototyping. Overall we try to understand if the open source phenomenon will be able to become a valid alternative to commercial platforms within the BI context.

1   Introduction

While in the past the BI market was strictly dominated by closed source and commercial tools (see for example [1] for different vendors' market shares), the last few years were characterized by the birth of open source (OS) solutions. At first OS BI tools covered isolated portions of the DW process with a limited set of functionalities that made them appear as toys if compared to large commercial BI platforms. Consider for example the initial releases for Octopus as to ETL, Mondrian as to OLAP servers, and JPivot as to OLAP clients (see [2] for a complete listing). While single tools still keep evolving with an increasing number of features and a higher level of reliability, the turning point in OS BI was the birth of OS BI platforms. An *OS BI platform* provides a full spectrum of BI capabilities within a unified system that reduces the overhead for the development and management of each application, and lets the user feel like he/she was using a single BI solution.

Commercial platforms are commonly considered superior to OS ones. Nevertheless, we believe that OS BI platforms will evolve much faster than commercial ones since they are not constrained by compatibility problems and rigid (or even obsolete) architectures. Furthermore, OS solutions can exploit the contributions of the OS development community, that relies on hundreds of programmers and designers as well as on the direct involvement of researchers.

This paper presents a comparative evaluation of three different OS BI platforms (namely JasperSoft, Pentaho and SpagoBI) aimed at understanding their current

T.B. Pedersen, M.K. Mohania, and A M. Tjoa (Eds.): DaWaK 2009, LNCS 5691, pp. 287–297, 2009.

features, their future potentialities and their limits when adopted in real projects. Overall we try to understand if the open source phenomenon will be able to become a valid alternative to commercial platforms within the BI context. OS BI platforms are not only attracting practitioners but also researchers since the availability of the source code makes them a perfect framework for prototyping and testing research findings. Furthermore both the European Community [3] and the United States government, as well as many other countries [4] are urging for the adoption of open source solutions in their research programs and more in general in the ICT area as a lever for increasing competitiveness [5]. Nowadays, in several areas such as e-health and e-government, funding calls suggest (or occasionally require) the use of open source.

The diffusion of OS BI technologies is also supported by private companies and consortiums. For example, BI Initiative [6] is an interesting OW2 project aimed at the diffusion of OS BI technologies. In particular BI Initiative is aimed at improving the coordination effort in the OS BI context, increasing the use of OS BI solutions at enterprise level, strengthening connections between integrators, vendors, users and the research communities and finally attracting more attention from the research activities to foster innovative BI solutions and practices.

The only scientific paper focusing on OS BI is the one proposed by Thomsen and Pedersen [2]: this interesting survey focuses on functionalities available in single tools but it does not consider BI platforms. A large number of comparative analyses are periodically published by software vendors, that obviously report a biased point of view, as well as independent groups. These reports (see for example [7,8]) are typically tailored on practitioners' needs and focus on technical aspects rather than studying the overall characteristics of the suite. The quality of OS software has been studied in three projects funded by the European Union, namely Flossmetric – Free/Libre Open Source Software Metrics - [9], Qualoss - Quality in Open Source Software - [10], and SQOOS - Software Quality Observatory for Open Source Software - [11]. The three projects converged to a unique initiative, named flossquality, aimed at developing a high level methodology to benchmark the quality of OS software and to apply it to a large number of OS projects. None of the platforms considered have currently been analyzed.

Our paper is thus the first one studying the added value of OS BI platforms; it evaluates comparatively the philosophy of the different platforms as well as their architecture, functionalities and usability. We will not consider efficiency aspects since they are strictly determined by the single BI tools which are often shared by the different platforms.

The paper is structured as follows. Section 2 briefly describes how the comparison has been conducted and introduces the key aspects that have been analyzed. Section 3 describes the platforms from different points of view, while in section 4 the results of the comparison are reported and discussed.

2   Method of Conducting the Comparison

This work comes from the interest in exploring the OS BI platforms shared by our research group and three Italian consulting firms that intend to propose OS based applications to their customers. The outcome of the analysis is the fusion of our independent analysis and their work on field testing and evaluation. All the consulting firms[1] involved are specialized in BI projects and they usually develop their applications using commercial BI suites.

We initially defined an evaluation grid describing in details the aspects to be investigated. The evaluation criteria were derived from the models available in the literature for general purpose software [12,13] and they were specialized to fit BI software specificities. The resulting grid was shared with the consultant firms and further discussed and integrated. Each consultant firm carried out one or more porting of real projects previously implemented through commercial BI suites. The compiled grids were finally shared and discussed with the other participants. In the current work we only report and summarize the evaluations concerning the platforms while we do not study in depth the features of each single BItool. The comparison hinges on the following key aspects:

- Non-technical: platform philosophy, type of licensing and availability of enterprise editions.
- Architectural: in terms of the global framework, modules and their relationships, programming languages and supported operational systems.
- Functional: in terms of functionalities provided natively by the platforms or made available to the users through the integrated BI tools.
- Meta-data: in terms of expressiveness, completeness, standardization and level of reusability.
- Security: in terms of functionalities provided for authentication and profiling of the users, interfaces to external authentication systems and secure data transmission.
- Usability: both from the user viewpoint, in terms of level of transparency in using the different tools, and from the developers' and system administrators' viewpoint in terms of complexity of installation and administration as well as development of applications, quality of manuals and forums.

3 Platforms Description

The platforms we considered are JasperSoft BI Suite [14], Pentaho BI Suite [15] and SpagoBI [16] and the versions considered are those released by December 31 2008. In the following will refer to them with the names Jasper, Pentaho and SpagoBI, respectively. Please note that in many cases there is a gap between the functionalities that are actually available to the users and those expected by the project road map for a given release. We will adopt a strict policy and we will disregard those features that have been only sketched.

3.1 Non-technical Aspects

The three platforms adopted two different open source models:

---

[1] We do not report the company names since they required to remain anonymous in order to avoid marketing activities by both open source and commercial software producers.

- Commercial open source: this model provides for an open source product that meets the user's basic needs (i.e. community edition); an enterprise edition of the product can be purchased and it usually includes enhanced features as well as support and training services. Jasper and Pentaho fit into this model. Their community editions are covered by the GNU General Public License (GPL) and Mozilla Public Licence (MPL) respectively while commercial agreements are needed for the enterprise releases.
- Free and Open Source Software (FOSS): the product is completely free, no enterprise solution is available, thus all the functionalities are available to the community for free. SpagoBI fits into this model. It is distributed under the GNU LGPL license.

Without entering into details, the right to freely use, modify, and redistribute software is fundamental to the GNU GPL agreements [17] and if you release a modified version of a software, you may be obliged to contribute your entire work to the open source community under the same type of agreement. On the other hand, a commercial agreement typically allows you to use but not to distribute the software. Usually, a different type of agreement (OEM license) is needed for profit developers who want to include BI capabilities in their applications.

According to the OS philosophy, platform functionalities can either be developed internally by the software house that owns the platform (e.g. JasperReport was born within JasperSoft Corporation) or, more frequently, they can be achieved by plugging a module implemented in a different OS project. Module plugging can be obtained by:

- Integration: a software interface is defined in order to control and to exploit module functionalities directly and transparently through the platform. The intellectual property of the software does not change, and the original developers remain in charge of maintaining and evolving the module.
- Acquisition: the intellectual property of the software is acquired and the original project terminated. The buyer will be in charge of maintaining and evolving the module.
- Technological partnership: stands in the middle between integration and acquisition. The original project remains alive and it is maintained by the original developers. The partner that incorporates the module influences its evolution and collaborates to its maintenance. The module usually appears with a different name in the new platform.

The policy adopted changes depending on the complexity of modules and on its relevance to the platform. Pentaho often has recourse to acquisition (e.g. Pentaho ETL comes from the Kettle project) while SpagoBI is strictly based on integration; finally Jasper mainly exploits partnerships (e.g. JasperETL was developed through a partnership with Talend that still maintains Talend Open Studio that is also integrated in SpagoBI).

Platforms that acquire the BI engines or have strong partnerships with their original developers can steer and control the engine evolutions and ensure a higher level of quality; on the other hand, platforms that integrate third-party modules can lean on wider developer communities and can more easily include new BI projects.

3.2 Architectural Aspects

An OS BI platform provides a full spectrum of BI capabilities within a unified system that reduces the overhead for the development and management of each application, and lets the user feel like he/she was using a single BI solution.

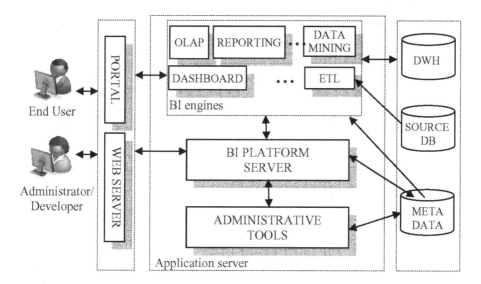

Fig. 1. Reference architecture for BI OS platforms. Arrows entering group of modules mean that communication concerns all the modules.

OS BI platforms are developed using Java since the modules they rely on are based on this technology. They typically require an application server and the users, as well as system administrators and developers, access them through a web browser. The platforms adopted the same architecture that is sketched in Figure 1: the platform core is a web application that stands in the middle between BI engines that implement each single BI functionality and the databases that store the required information. The users access the system through a web client that can be connected either to a portal or directly to a web server. A meta knowledge layer completes the picture and is crucial to provide the platforms with the necessary "intelligence". A typical user-platform interaction includes the following steps: (1) the user requiring a given document logs into the portal or directly into the platform server; (2) the platform server verifies if the user profile allows him/her to access the document requested; (3) the platform server opens the connection to the data source; (4) the platform server also activates the BI engine involved and passes it the user credentials, the necessary meta-information as well as the connection to the data source; (5) the BI engine produces the document and makes it available to the user through the web server or the portal.

Table 1. Modules building up the considered BI OS platforms, alternative configurations are possible

Modules	JasperSoft	Pentaho	SpagoBI
Application Server	JBoss	JBoss	JBoss
Authentication and user profiling	Acegi	Acegi	Integrated in eXo Portal
Collaboration	-	-	Dossier
Dashboard	JFreeChart	JFreeChart	Openlaszlo
Data Mining	-	Weka	Weka
DBMS	MySQL, Oracle, SQL Server, PostgreSQL, etc.	MySQL, Oracle, SQL Server, PostgreSQL, etc.	MySQL, Oracle, SQL Server, PostgreSQL, etc.
ETL	JasperETL	Pentaho Data Integration	Talend Open Studio
Geo-referencing	Google Maps	Google Maps	GEO
Job Scheduler	Quartz	Quartz	Quartz
OLAP	Mondrian&Jpivot	Mondrian&Jpivot	Mondrian&Jpivot
Portal	Liferay	JBoss Portal	ExoPortal, Liferay
Query by Example	-	-	Hibernate
Reporting	JasperReport	Pentaho Report Designer, JasperReport, BIRT	JasperReport, BIRT
Single sign on	Acegi	CAS	CAS
Web Server	Tomcat	Tomcat	Tomcat

Beside front-end functionalities the platforms include back-end ones as for example ETL services and scheduling services necessary to automate report updates. In all these cases engines are activated directly by the platform or by the system administrator.

Table 1 shows the main modules building up the platforms considered. Many of the modules are shared, some of them are evolutions of a different open source project (e.g. Jasper ETL comes from Talend Open Studio), others have been developed internally and belong to the same software house that is charge of the platform

(JasperReport is the most widespread modules for BI reporting, while GEO is the module developed by SpagoBI team for geo-referenced analysis) - reusing and sharing underlie OS software development. Table 1 also shows that some modules are standard de facto within BI OS: in particular the Mondrian OLAP engine and the JPivot graphical interface are the standard solutions for OLAP, while Weka is the standard data mining module.

3.3 Metadata

Within a BI platform, metadata largely determine the behavior it can exhibit, and the expressivity of reports and OLAP analyses. Metadata store the structure of data sources and multidimensional cubes, the content of the reports and the actions to be executed within an ETL process. Metadata also store user profiles as well as information related to scheduling and auditing.

We distinguish between platform metadata and BI engine metadata. In fact, metadata necessary to specific BI functionalities are usually created outside the platforms by editing an XML file or by exploiting simple graphical tools. Only afterwards can they be imported in the BI platform. Although, they model the same information, metadata belonging to different engines are differently coded and cannot be reused. This obviously affects development and maintenance negatively. For example, the multidimensional structure of a cube must be defined repeatedly if the cube is involved in an OLAP analysis, in a report or in a ETL process. We believe that this is the main shortcoming of OS BI platforms compared to commercial ones that are typically based on a unique and integrated metadata repository. Within community editions metadata are stored in XML files, while the two enterprise editions provide for a DBMS based metadata repository. Although all three platforms declare that their metadata are CWM-compliant [18] no interoperability tools have been released yet.

3.4 Functional Aspects

Table 2 reports the main functionalities made available by the platforms. If we consider the completely free version of the suites (i.e. community editions) SpagoBI overcomes Pentaho and Jasper that make available many of the advanced features only in the enterprise editions. We will not discuss in detail each single item in the table since most of them are self explaining, we will briefly describe the infrequent terms instead. The term *Query by Example* refers to the capability of running free inquiring over a database schema using a graphical interface that does not require the user to be an SQL expert, while *Ad-hoc reporting* refers to the availability of a graphical interface that allows each user to create his own reports directly from a web interface. The term *collaborative BI* refers to functionalities that allow BI results to be shared between managers in order to reach a concerted decision. Finally, *report validation workflow* stands for the possibility of defining a set of states and approval steps a report and its data must pass through before being finally published.

Table 2. Main functionalities made available by the platforms; community and enterprise releases are distinguished

Functionalities	SpagoBI	Pentaho	Pentaho Ent. Ed.	Jasper	Jasper Ent. Ed.
Activities scheduling	√	×	√	×	√
Ad-hoc reporting	×	×	√	×	√
Auditing	√	×	√	√	√
Collaborative BI	√	×	×	×	×
Data Mining	√	√	√	×	×
Dashboard	√	√	√	×	√
Document export	√	√	√	√	√
ETL	√	√	√	√	√
Geo-referenced analysis	√	√	√	×	√
OLAP	√	√	√	√	√
Query by Example	√	×	×	×	×
Report validation workflow	√	×	√	×	×
Reporting	√	√	√	√	√
User profiling	√	×	√	×	√

Security issues are particularly relevant in data warehousing. All the platforms allow secure data transmission as well as user authentication, while they offer pretty different functionalities for user profiling. Typically DBMSs are not suitable for defining the security policies relevant in a BI application, thus BI platforms are in charge of their definition. In advanced commercial solutions profiling is based on security models that govern, in a centralized fashion, three fundamental areas of every BI application: (1) *objects* (e.g. a specific report or an OLAP analysis) each user can use, (2) *cell-level data* each user can access; and (3) *BI functionalities* each user gets (i.e. choosing the types of actions users may perform in the system such as printing, saving, exporting, drilling, pivoting, sorting, formatting and creating reports). As to OS BI platforms only two of these areas are covered; in fact user profiles can grant or deny access to different objects and allow filters to be applied on data retrieval but they cannot restrict the set of BI functionalities a user can run on a given document. More in detail user profiling is made available for free by SpagoBI while Jasper and Pentaho offer this feature only in their enterprise editions: Pentaho community edition only provides user authentication, while Jasper community edition provides a simplified profiling where the access is granted/denied for an entire directory usually containing mode reports or analysis.

As concerns the comparison between community - including SpagoBI - and enterprise editions, differences are not only in terms of functionalities available to the users (see Table 2) but also in terms of utilities for administrators and developers. The main improvements we identified in enterprise editions are:

- Improved administration consoles: the improvement is particularly relevant in Pentaho where the Enterprise console fills the gap with Jasper as concerns usability and functionalities.
- Wizard based configurations: most configuration activities are based on wizards and do not require a manual access to configuration files or multiple access to menus.
- Process monitoring: front-end (e.g. query execution) as well as back-end (e.g. ETL) processes can be monitored and analyzed in order to optimize their execution.
- ETL debugging environment: it is available and determines a strong reduction of the development effort.

Administrators and developers are further supported through a wider documentation, a knowledge base as well as consultant and training services. Obviously such enhancements, together with warranties and certification of the software on a larger number of operating systems, applications servers, DBMSs, etc., become more and more relevant when you are developing a mission-critical application or when you are planning to adopt the platform in a large and complex organization.

3.5 Usability

Usability enables the users to easily access BI functionalities and it ensures developers and administrators a high productivity.

From the user point of view platforms usability is largely determined by the BI engines composing them. We consider the usability of those engines qualitatively satisfactory. Although they do not reach the level of refinement of the commercial suites, their graphical features give the developed applications an appreciable look-and-feel. OS BI platforms also succeed in hiding the access to different tools.

From the administrators' point of view usability is determined by the easiness in administering the platform and adding new functionalities, in particular:

- Complexity of the installing and configuring process: installing procedures are in general quite easy. This is particularly true for Pentaho and JasperSoft whose installation procedures completely rely on a wizard that also includes the installation of the BI engines. SpagoBI installation requires manually modifying eXoPortal configuration files and it does not include BI engines that must be installed separately.
- Administration complexity: the different usability is well perceived when you register a new report or analysis. As described in Section 3.2 functionalities (e.g. a report, an OLAP analysis, an ETL process) are usually developed outside the platform and then imported before making them available. In SpagoBI and even more in Jasper we appreciated the easiness of the form-based procedure. Much effort is needed in Pentaho where functionalities registration is based on Action Sequences: an Eclipse procedure that may become quite complex since it is not adequately

supported by appropriate debug information and documentation. This problem is partially solved in the enterprise edition that includes a debug tool for Action Sequences.

- Problem solving and training effort: manuals have a good quality and they allow most of the problems to be solved. Besides, in line with OS philosophy, several practitioners' forums make available a high number of technical tips. The quality of information and the activeness of the forums are strictly related to the number of the platform users. During our analysis, the richness and most active forum was the one from Pentaho (more than 20,000 registered users). The Jasper community is even larger (about 90,000 registered users) but we experienced in many cases longer response time (about 2-3 days for receiving an answer). SpagoBI community is definitely smaller and so the activeness of its forum (the number of registered users is unavailable, but only six thousands posts have been submitted since 2006). Finally, the adoption of standard and well-known programming languages does not require programmers and administrators to have any particular skill.

4 Discussion and Conclusions

Our analysis shows that OS BI platforms determine an added value with respect to single BI tools since they allow several functionalities to be accessed transparently and a set of processes to be centralized and simplified thus reducing the administration and development effort. We believe that the main shortcoming of the platform is the absence of a fully centralized and unified metadata layer, as this reduces reusability and integration. The capabilities of the administrative tools could also be improved in the community editions – this concerns in particular Pentaho.

SpagoBI makes available a remarkable number of BI functionalities even if it adopts a free open source model. As concerns the functionalities offered to the users SpagoBI is comparable to the enterprise editions by Jasper and Pentaho. From that observation we can infer that integration (instead of acquisition) allows an easier plug of new modules and gives the original developers the possibility to improve them. On the other hand, acquisition ensures a higher quality of the modules and a road map compatible with the owner's one. These are mandatory needs for distributing certified editions.

Although OS BI platforms are still not as sophisticated as commercial ones we can state that they got a sufficient level of reliability and must be considered a valid alternative to commercial suites. This is particularly true in small and medium-sized enterprises where the quantity of data and the workload are not critical points. Several companies are evaluating the use of OS BI in pilot projects where budget constraints are typically very tight. The main risks related to an investment in OS technology come from unexpected termination of the project that will no longer be maintained and evolved or, even worse, from the adoption of a more restrictive licensing of the new releases that prevents using or distributing them. Finally, due to the short history of such products, it is impossible to predict if, apart from the initial investment, the companies that are in charge of the platforms will earn enough from services and application developments to stay on the market.

According to their road maps and evolution trends OS BI platforms will equal commercial ones in a few years. In order to really do better than commercial solutions, we argue, OS BI platforms should not only replicate commercial functionalities with lower costs for the final users, but should also propose innovative functionalities according to the most sophisticated requirements of business users. Coupling twenty years of experience in building BI software with the more recent results on BI research can really make the difference.

References

[1] OLAP Report: OLAP Market Share Analysis Retrieved March 12 (2009)The Olap Report web site, http://www.olapreport.com/market.htm

[2] Thomsen, C., Pedersen, T.B.: A survey of open source tools for business intelligence. In: Tjoa, A.M., Trujillo, J. (eds.) DaWaK 2005. LNCS, vol. 3589, pp. 74–84. Springer, Heidelberg (2005)

[3] DG Information Society and Media: Towards a European Software Strategy – Report of an Industry Expert Group, The European Commission web site, http://www.nessi-europe.com/ (Retrieved March 12, 2009)

[4] Lewis, J.A.: Government Open Source Policies. Center for Strategic and International Studies (2007)

[5] Ghosh, R.A.: Study on the: Economic impact of open source software on innovation and the competitiveness of the Information and Communication Technologies (ICT) sector in the EU, The European Commission web site, http://ec.europa.eu/enterprise/ict/studies/publications_en.htm (Retrieved, March 12, 2009)

[6] BI Initiative web site, http://www.ow2.org/view/BusinessIntelligence/

[7] Optaros. Open Source Catalog, Optaros white paper (2009)

[8] Smile. Décisionnel Solutions open source (2009)

[9] Flossmetric Project web site, http://www.flossmetrics.org/

[10] Qualoss Project web site, http://www.flossmetrics.org/

[11] SQOOS Project web site, http://www.sqo-oss.eu/

[12] Samoladas, I., Gousios, G., Spinellis, D., Stamelos, I.: The SQO-OSS Quality Model: Measurement Based Open Source Software Evaluation. Open Source Development, Communities and Quality 275, 237–248 (2008)

[13] Ortega, M., Perez, M., Royas, T.: Construction of a Systemic Quality Model for Evaluating a Software Product. Software Quality Journal 11, 219–242 (2003)

[14] JasperSoft web site, http://www.jaspersoft.com/

[15] Pentaho web site, http://www.pentaho.com/

[16] SpagoBI web site, http://spagobi.eng.it/

[17] GNU General Public License web site, http://www.gnu.org/copyleft/gpl.html

[18] CWM: Common Warehouse Metamodel Specification - Version 1.1, OMG Inc. (2003)

Ontology-Based Exchange and Immediate Application of Business Calculation Definitions for Online Analytical Processing

Matthias Kehlenbeck and Michael H. Breitner

Institut für Wirtschaftsinformatik, Leibniz Universität Hannover
{kehlenbeck,breitner}@iwi.uni-hannover.de

Abstract. Business users define calculated facts based on the dimensions and facts contained in a data warehouse. These business calculation definitions contain necessary knowledge regarding quantitative relations for deep analyses and for the production of meaningful reports. The business calculation definitions are implementation and widely organization independent. But no automated procedures facilitating their exchange across organization and implementation boundaries exist. Separately each organization currently has to map its own business calculations to analysis and reporting tools. This paper presents an innovative approach based on standard Semantic Web technologies. This approach facilitates the exchange of business calculation definitions and allows for their automatic linking to specific data warehouses through semantic reasoning. A novel standard proxy server which enables the immediate application of exchanged definitions is introduced. Benefits of the approach are shown in a comprehensive case study.

1 Introduction

For decision support business users have to perform analyses and create reports based on large data sets from heterogeneous sources. Data warehouses (DW) facilitate this decision support by integrating data from different systems and providing them in a consistent multidimensional (MD) model to analysis and reporting tools [1][2]. With the help of these tools business users build queries using the dimensions and facts contained in the MD model and define additional calculated facts based on them. These business calculation definitions contain necessary knowledge regarding quantitative relations for the performance of deep analyses and the production of meaningful reports. Although they are implementation and, to a large extend, organization independent, no automated procedures facilitating their exchange across organization and implementation boundaries exist. Therefore, each organization currently has to map its own business calculations to online analytical processing (OLAP) tools separately.

A major obstacle for the exchange of business calculation definitions is its missing division into organization independent and organization specific parts as well as its missing abstraction from implementation specific details. Moreover,

T.B. Pedersen, M.K. Mohania, and A M. Tjoa (Eds.): DaWaK 2009, LNCS 5691, pp. 298–311, 2009.
© Springer-Verlag Berlin Heidelberg 2009

organizations and implementations use different names for entities with the same meaning and describe the relations between these entities in different languages.

In order to overcome these obstacles, this paper presents an innovative approach which structures business calculation definitions for OLAP into distinct layers of ontologies and enables business users to exchange definitions using standard technologies, e. g. by means of office documents or web pages. Exchanged definitions are automatically linked to specific data warehouses and immediately provided for OLAP.

The benefits of this proposal are demonstrated in a comprehensive case study in which business calculation definitions are created, exchanged and consolidated as well as automatically linked to a specific data warehouse and used as a semantic middleware layer while querying Microsoft Analysis Services (MSAS) [3], Penhao Analysis Services (PAS) [4] and SAP NetWeaver Business Intelligence (SAPBI) [5].

The remainder of this paper is structured as follows: Section 2 presents an overview of related work. Section 3 describes the approach for the ontology-based exchange and immediate application of business calculation definitions for OLAP. A comprehensive case study is provided in section 4. Finally, section 5 points out the conclusions.

2   Related Work

In the last few years, there has been a growing interest in metamodels and ontologies [6]. As these terms are closely related [7], the most relevant approaches are briefly described for both of them.

The Common Warehouse Metamodel (CWM) is a standardized metamodel for data warehouse metadata [8] and can be used in conjunction with the CWM Metadata Interchange Patterns (CWM MIP) [9] and XML Metadata Interchange (XMI) [10] for exchange. Januszewski and Pankowski use the Behavioral Metamodel part of the CWM to create an implementation independent description of a calculation function for a business quantity [11]. This description is primarily intended for the subsequent implementation of the function on a data warehouse platform and does not distinguish between business entities and their data warehouse representations. Futhermore, the description does not define the meaning of the contained terms. The authors suggest to use the Business Nomenclature part of the CWM to define those terms. However, this part is far less expressive than the available ontology languages.

The Model-Driven Architecture (MDA) is an approach which emphasizes the use of models in software development [12]. The requirements and the vocabulary of business users are described in the Computation Independent Model (CIM), which forms the basis to create the Platform Independent Model (PIM). Finally, the PIM is used to derive a Platform Specific Model (PSM). These models can be described using the Unified Modeling Language (UML) [13], transformed using Query/View/Transformations (QVT) [14] and exchanged using XMI. Mazón and Trujillo describe a comprehensive MDA approach for the development of data warehouses [15]. They use UML profiles to create a PIM for the multidimensional model and derive a corresponding PSM in the CWM using QVT transformations. A similar approach for queries is presented by Pardillo et al. [16]. None of these approaches deals with the modeling of calculated facts.

The Web Ontology Language (OWL) is a machine interpretable knowledge representation language [17]. OWL provides its three sublanguages OWL Lite, OWL DL and OWL Full. As OWL Lite and DL are both based on description logics (DL), they can be used in conjunction with available semantic reasoners. Xie et al. use an extended OWL DL to represent a conceptual enterprise data model and a conceptual multidimensional model [18]. Based on these models, business users define analysis requirements. Afterwards IT specialists map new entities to the data warehouse model and use a deployment engine to create a dedicated data mart. As an exchange of entities and their relations is not intended, the approach does not distinguish between organization independent and organization specific parts and provides for the definition of calculated facts in a proprietary language.

Diamantini and Potena propose to annotate data cubes by describing the contained facts with a business and a mathematic ontology [19]. The business ontology is described using OWL while the mathematic ontology is described using MathML [20] and OpenMath [21]. Both ontologies are maintained exclusively by IT specialists after they made changes to data cubes. The exchange of entities and their relations is not discussed.

This contribution is the first approach which facilitates the exchange of business calculation definitions between business users across organization and implementation boundaries and enables their immediate application to specific data warehouses. Moreover, the approach exclusively uses standard technologies and is therefore easy to implement and maintain.

3 Ontology-Based Exchange and Immediate Application of Business Calculation Definitions for OLAP

As business and data warehousing are different domains, their entities and relations are mapped to distinct ontologies. Business entities and their relations are only contained in the business ontology while data warehouse entities and their relations are only contained in the data warehouse ontology. The business ontology is divided into an organization independent and an organization specific part to facilitate the exchange of business calculation definitions. Likewise, the data warehouse ontology is divided into an implementation independent and an implementation specific part to increase its reusability. Business and data warehouse ontology are combined with each other by the mapping ontology. The mapping ontology consists of a manually created and an automatically inferred part. All ontologies are defined in OWL DL and are used in conjunction for OLAP. This is illustrated in Figure 1 and concisely described in the following subsections.

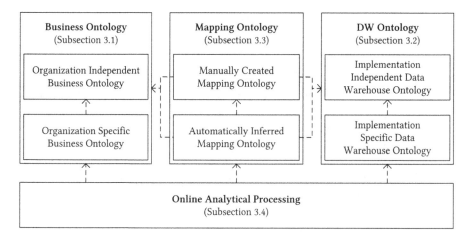

Fig. 1. Business calculation definitions and DW models are structured into distinct ontologies. They are solely combined with each other by the mapping ontology. All ontologies are used in conjunction for OLAP. Arrows (- →) correspond to dependencies.

3.1   Business Ontology

The business ontology is based on the concepts function, term, operator, quantity, object and set. Functions have exactly one calculation term and return a quantity with its result. Additionally, functions may have one restriction term which constrains the domain of application. Terms are either single quantities, objects or sets, or they have exactly one operator and at least one other term as an operand. Operands may take on different roles, e. g. the role of a dividend or divisor, depending on the used operator. Operators, quantities, objects and sets are uniquely identified by their name. Individuals are sorted into the organization independent respectively specific ontology according to their nature. Figure 2 illustrates the concepts and relations of the business ontology.

Provided that an ontology is defined in a prevalent language, like OWL, it is well suited for its exchange and consolidation with other ontologies [22]. However, it is also desirable to facilitate the exchange of single functions. Prevalent languages for the description of mathematical functions are MathML and Open-Math. MathML provides its sublanguages MathML Presentation and MathML Content. MathML Presentation focuses on the display of expressions while MathML Content and OpenMath focus on their semantic meaning. Functions of the business ontology can be transformed to MathML Content or OpenMath expressions and vice versa. Therefore, it is also possible to use business calculation definitions which were originally created for other purposes than OLAP, e. g. for their description on wiki pages. As MathML and OpenMath provide for the definition of symbols in Content Dictionaries (CD), operators may refer to a corresponding CD and CD base. Additionally, terms may possess a corresponding MathML Content, MathML Presentation and/or OpenMath expression.

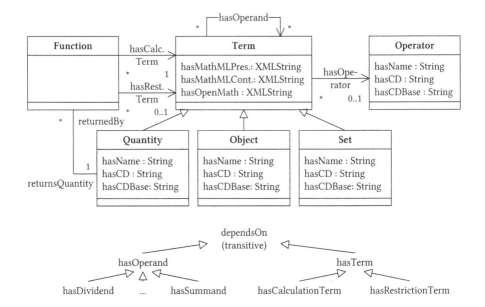

Fig. 2. Concepts, associations (→) and generalizations (⇢) of the business ontology. OWL supports many relation types, e. g. generalized as well as transitive relations.

As MathML and OpenMath expressions can be embedded in any Extensible Markup Language (XML) [23] document, e. g. office documents [24] or web pages [26], their technical exchange is simple. However, their businesslike exchange requires the consolidation of external and internal terms. Therefore, business users have to decide for every external quantity, object or set referred by an exchanged expression whether it has to be replaced by an already existing or taken over as a new individual. To allow for the globally unique identification of these individuals, they may refer to a corresponding CD and CD base as well.

Although functions are stored in a tree structure, the quantities, objects and/or sets required for their evaluation can be inferred using a transitive relation. The business ontology contains an object property `dependsOn` which is transitive and is a super property of the object properties `hasOperand` and `hasTerm`. The inverse object property of `dependsOn` is `requiredBy`. Using these relations, the set of quantities required for the evaluation of on individual function F can be defined in Manchester OWL Syntax [25] by the class expression `Quantity and requiredBy value F`.

3.2 Data Warehouse Ontology

The data warehouse ontology is based on the concepts cube, fact and dimension. Cubes possess an arbitrary number of facts and dimensions. However, each of them may only belong to one cube. All individuals possess a unique identifier.

Individuals can be generated from data warehouse models, e. g. based on CWM or UML, or using data warehouse application programming interfaces

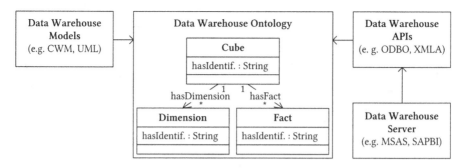

Fig. 3. Alternative information flows ($\rightarrow$) for the generation of the DW ontology

(APIs), e. g. OLE DB for OLAP (ODBO) [27] or XML for Analysis (XMLA) [28], with an OWL API [29] and sorted into the implementation independent respectively specific ontology according to their nature. Figure 3 illustrates the generation of the data warehouse ontology.

3.3 Mapping Ontology

The manually created mapping ontology solely consists of the symmetric object property isMappedTo which maps objects to dimensions respectively quantities to facts. These mappings can be either created manually by IT specialists or, if possible, inferred from Semantic Web Rule Language (SWRL) [30] rules. E. g., the SWRL rule Quantity(?q) $\wedge$ hasName(?q, ?n) $\wedge$ Fact(?f) $\wedge$ hasIdent(?f, ?i) $\wedge$ equal(?n, ?i) $\rightarrow$ isMappedTo(?q, ?f) maps a quantity to a fact, if quantity name and fact identifier equal.

Based on the other ontologies, a semantic reasoner creates the automatically inferred mapping ontology. In particular, it infers which quantities can be provided by which cubes using which functions and mappings. A cube is able to provide an object if it is mapped to one of its dimensions and is able to provide a quantity if it is mapped to one of its facts and/or returned by one of its supported functions. This can be defined by the property chains hasDimension ∘ isMappedTo → isAbleToProvideObject, hasFact ∘ isMappedTo→ isAbleToProvideQuantity and supportsFunction ∘ returnsQuantity → isAbleToProvideQuantity. A cube supports a function if it only uses quantities and objects which the cube is able to provide. The class of functions supported by a cube C can be defined by the expression Function and dependsOn only ((not Object and not Quantity) or (isObjectAvailableForCube value C) or (isQuantityAvailableForCube value C)) and is a subclass of Function and isSupportedByCube value C, where isObjectAvailableForCube is the inverse of isAbleToProvideObject, isQuantityAvailableForCube the inverse of isAbleToProvideQuantity and isSupportedByCube the inverse of supportsFunction.

As these definitions use universal quantifiers, a semantic reasoner which makes the open world assumption requires closure axioms in order to infer the supported functions. These closure axioms can be created automatically by a

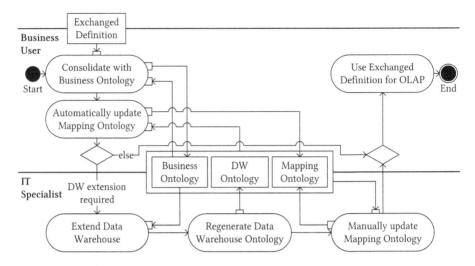

Fig. 4. Workflow from the exchange of a business calculation definition to its utilization for OLAP. Business users predominantly exchange definitions without IT specialists.

tool which enumerates all objects, quantities and sets and complements the object property assertions based on **dependsOn** for all functions with corresponding negative object property assertions. Likewise, the class of supported functions can be defined automatically for each cube.

3.4 Online Analytical Processing

A workflow which describes the activities from the exchange of a definition to its utilization for OLAP is illustrated in Figure 4. It allows for a predominant exchange of business calculation definitions between business users without the participation of IT specialists. As the aforementioned ontologies contain all required information, a defined quantity can be provided automatically, if a data warehouse supports the corresponding function. This provision may take place on the client side, on the server side or in between. A provision on the client or server side would probably require as many different implementations as platforms. However, a provision in between may achieve platform independence by using a data warehouse API. Prevalent data warehouse APIs are ODBO and XMLA. ODBO is based on the proprietary Component Object Model (COM) [31] while XMLA is based on the platform independent SOAP [32] standard. SOAP web services can be described using the Web Services Definition Language (WSDL) [33]. As WSDL documents are available for several data warehouse servers, like MSAS and SAPBI, corresponding client and server side interfaces can be created automatically using a web service framework, e. g. Apache CXF [34]. These interfaces enable the implementation of an XMLA proxy which provides the defined quantities by modifying the communication between client and server. In particular, the responses of XMLA discovery methods and the

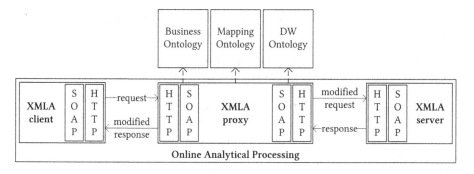

Fig. 5. Immediate application of definitions via XMLA compliant proxy server

requests to XMLA execute methods can be modified. Figure 5 illustrates this approach. As a bridge driver which allows to use an ODBO client in conjunction with an XMLA server exists [35], a wide variety of clients can be used.

4 Case Study

This section contains a case study in which the definitions for two business quantities are produced using a MathML editor, exchanged as an XML document, consolidated with an existing ontology and immediately made available by means of semantic reasoning and an XMLA proxy.

A business user defines the quantities *EBIT Margin* and *EBIT* using a MathML editor, e. g. Integre MathML Equation Editor [36].

$$EBIT\ Margin = \frac{EBIT}{Net\ Sales}\ . \tag{1}$$

$$EBIT = Net\ Income - (Interest\ Income + Tax\ Income)\ . \tag{2}$$

They are saved to an XML document. E. g., the quantity *EBIT Margin* may be represented by the following MathML Content fragment:

```
<mml:apply>
  <mml:csymbol cd="relation1">eq</mml:csymbol>
  <mml:ci>EBIT Margin</mml:ci>
  <mml:apply>
    <mml:csymbol cd="arith1">divide</mml:csymbol>
    <mml:ci>EBIT</mml:ci>
    <mml:ci>Net Sales</mml:ci>
  </mml:apply>
</mml:apply>
```

A different business user adopts these definitions. The quantities as well as their corresponding functions and terms are created in the organization independent business ontology. E. g., the quantity *EBIT Margin* may be represented by the following OWL DL fragment:

```
<Quantity rdf:about="#EBITMarginQuantity">
  <hasName>EBIT Margin</hasName>
</Quantity>
<Function rdf:about="#EBITMarginFunction">
  <hasCalculationTerm rdf:resource="#EBITMarginCalculationTerm"/>
  <returnsQuantity rdf:resource="#EBITMarginQuantity"/>
</Function>
<Term rdf:about="#EBITMarginCalculationTerm">
  <hasOperator rdf:resource="#DivisionOperator"/>
  <hasDividend rdf:resource="#EBITQuantity"/>
  <hasDivisor rdf:resource="#NetSalesQuantity"/>
</Term>
```

In this case study, the referred quantity *Interest Income* is already defined in the organization independent business ontology by the calculation term *Interest Revenue − Interest Expense*. Likewise, the referred quantities *Net Income*, *Net Sales* and *Tax Income* as well as the required quantities *Interest Revenue* and *Interest Expense* are already defined in the organization specific business ontology by calculation terms based on the quantity *Account Balance* and restriction terms based on the object *Account* and a corresponding set. E. g., the quantity *Interest Expense* is defined by the calculation term *− Account Balance* and the restriction term *Account ∈ Interest Expense Accounts*. Therefore, the quantities *EBIT Margin* and *EBIT* ultimately only depend on the quantity *Account Balance*, the object *Account* and the sets *Net Income Accounts*, *Net Sales Accounts*, *Tax Income Accounts*, *Interest Revenue Accounts* and *Interest Expense Accounts*. The quantity *Account Balance* is mapped to a corresponding fact and the object *Account* is mapped to a corresponding dimension of the cube `AdventureWorksCube`. Therefore, a semantic reasoner, e. g. Pellet [37], infers that `AdventureWorksCube supportsFunction EBITMarginFunction` as well as `AdventureWorksCube supportsFunction EBITFunction` and therefore `AdventureWorksCube isAbleToProvideQuantity EBITMarginQuantity` as well as `AdventureWorksCube isAbleToProvideQuantity EBITQuantity`. These axioms are saved to the automatically inferred mapping ontology.

An XMLA proxy makes the quantities immediately available for OLAP. It was implemented as a web service based on the Java API for XML Web Services (JAX-WS) [38] and is currently able to access MSAS, PAS and SAPBI as a web service consumer. XMLA defines two methods: discover and execute. Requests to the discover method of the XMLA proxy are passed unmodified to the responsible server. However, the proxy modifies the received results for the request types `MDSCHEMA_MEASUREGROUPS` and `MDSCHEMA_MEASURES` before passing them on to the client. Results for the request type `MDSCHEMA_MEASUREGROUPS` are complemented by an XMLA fragment like

```
<MEASUREGROUP_NAME>XMLA Proxy</MEASUREGROUP_NAME>
<MEASUREGROUP_CAPTION>XMLA Proxy</MEASUREGROUP_CAPTION>
```

in order to create an additional measure group. Similarly, results for the request type MDSCHEMA_MEASURES are complemented by XMLA fragments like

```
<MEASURE_NAME>EBIT Margin</MEASURE_NAME>
<MEASURE_UNIQUE_NAME>[Measures].[EBIT Marg.]</MEASURE_UNIQUE_NAME>
<MEASURE_CAPTION>EBIT Margin</MEASURE_CAPTION>
<MEASURE_AGGREGATOR>127</MEASURE_AGGREGATOR>
<MEASUREGROUP_NAME>XMLA Proxy</MEASUREGROUP_NAME>
```

which correspond to the facts that can be additionally provided based on the information contained in the ontologies. This information is used to create expression trees, as illustrated in Figure 6, which are subsequently supplemented with the original results for the request type MDSCHEMA_MEASURES to determine the corresponding values of the MEASURE_AGGREGATOR. The latter indicates whether a measure was derived using a single aggregation function (e.g. SUM), a combination of aggregation functions, or using a more complex function. Due to the complementary XMLA fragments, the OLAP client regards the additional facts as available on the data warehouse server.

Requests to the execute method of the XMLA proxy contain commands which can contain queries defined using Multidimensional Expressions (MDX) [39]. The

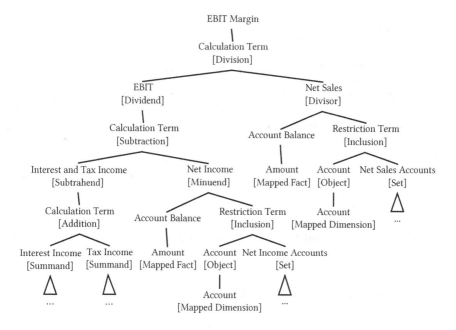

Fig. 6. Expression tree for the exchanged quantity *EBIT Margin*. Triangles represent parts which are not shown in full details due to space restrictions.

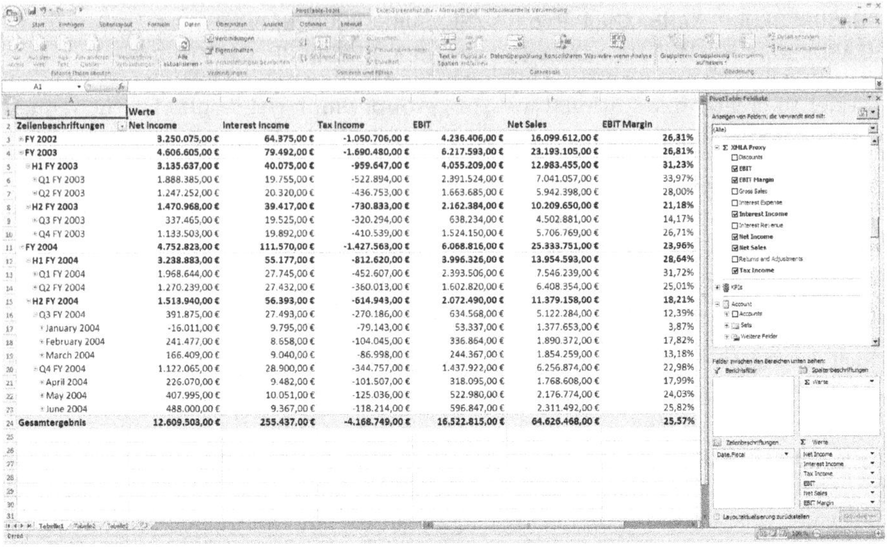

Zeilenbeschriftungen	Net Income	Interest Income	Tax Income	EBIT	Net Sales	EBIT Margin
FY 2002	3.250.075,00 €	64.375,00 €	-1.050.706,00 €	4.236.406,00 €	16.099.612,00 €	26,31%
FY 2003	4.606.605,00 €	79.492,00 €	-1.690.480,00 €	6.217.593,00 €	23.193.105,00 €	26,81%
H1 FY 2003	3.135.637,00 €	40.075,00 €	-959.647,00 €	4.055.209,00 €	12.983.455,00 €	31,23%
Q1 FY 2003	1.888.385,00 €	19.755,00 €	-522.894,00 €	2.391.524,00 €	7.041.057,00 €	33,97%
Q2 FY 2003	1.247.252,00 €	20.320,00 €	-436.753,00 €	1.663.685,00 €	5.942.398,00 €	28,00%
H2 FY 2003	1.470.968,00 €	39.417,00 €	-730.833,00 €	2.162.384,00 €	10.209.650,00 €	21,18%
Q3 FY 2003	337.465,00 €	19.525,00 €	-320.294,00 €	638.234,00 €	4.502.881,00 €	14,17%
Q4 FY 2003	1.133.503,00 €	19.892,00 €	-410.539,00 €	1.524.150,00 €	5.706.769,00 €	26,71%
FY 2004	4.752.823,00 €	111.570,00 €	-1.427.563,00 €	6.068.816,00 €	25.333.751,00 €	23,96%
H1 FY 2004	3.238.883,00 €	55.177,00 €	-812.620,00 €	3.996.326,00 €	13.954.593,00 €	28,64%
Q1 FY 2004	1.968.644,00 €	27.745,00 €	-452.607,00 €	2.393.506,00 €	7.546.239,00 €	31,72%
Q2 FY 2004	1.270.239,00 €	27.432,00 €	-360.013,00 €	1.602.820,00 €	6.408.354,00 €	25,01%
H2 FY 2004	1.513.940,00 €	56.393,00 €	-614.943,00 €	2.072.490,00 €	11.379.158,00 €	18,21%
Q3 FY 2004	391.875,00 €	27.493,00 €	-270.186,00 €	634.568,00 €	5.122.284,00 €	12,39%
January 2004	-16.011,00 €	9.795,00 €	-79.143,00 €	53.337,00 €	1.377.653,00 €	3,87%
February 2004	241.477,00 €	8.658,00 €	-104.045,00 €	336.864,00 €	1.890.372,00 €	17,82%
March 2004	166.409,00 €	9.040,00 €	-86.998,00 €	244.367,00 €	1.854.259,00 €	13,18%
Q4 FY 2004	1.122.065,00 €	28.900,00 €	-344.757,00 €	1.437.922,00 €	6.256.874,00 €	22,98%
April 2004	226.070,00 €	9.482,00 €	-101.507,00 €	318.095,00 €	1.768.608,00 €	17,99%
May 2004	407.995,00 €	10.051,00 €	-125.036,00 €	522.980,00 €	2.176.774,00 €	24,03%
June 2004	488.000,00 €	9.367,00 €	-118.214,00 €	596.847,00 €	2.311.492,00 €	25,82%
Gesamtergebnis	12.609.503,00 €	255.437,00 €	-4.168.749,00 €	16.522.815,00 €	64.626.468,00 €	25,57%

Fig. 7. Microsoft Excel connected to MSAS through the XMLA proxy and an ODBO bridge. Queries may use facts that can be additionally provided based on the information contained in the ontologies just like the original facts from a data warehouse.

XMLA proxy modifies the queries before passing them on to the responsible data warehouse server by complementing them with MDX fragments like

```
MEMBER [Measures].[EBIT Marg.] AS
  [Measures].[EBIT] / [Measures].[Net Sales]
```

which result from transformations of the supplemented expression trees. The received results are passed on unmodified to the client. Due to the MDX fragments, queries can use the additional facts from the ontologies just like the original facts from the data warehouse.

The proxy has been successfully tested with the MSAS, PAS and SAPBI servers as well as the IBM DataQuant for Workstation XMLA [40] and Microsoft Excel [41] ODBO clients. A positive side effect is that the proxy increases compatibility by unifying web service definitions, e. g. enabling to use IBM DataQuant for Workstation with SAPBI. Figure 7 contains a screenshot of Microsoft Excel connected to MSAS through the XMLA proxy and an ODBO bridge.

5 Conclusions and Outlook

This paper focuses on the problem of exchanging business calculation definitions across organization and implementation boundaries. An innovative approach based on Semantic Web technologies which facilitates the exchange of business calculation definitions is presented. This approach also supports the utilization of definitions which were created for other purposes than OLAP. Automatic linking of business calculation definitions to specific data warehouse models through semantic reasoning is enabled. A novel standard proxy server between analysis and reporting clients as well as data warehouse servers is introduced. This proxy server immediately provides exchanged definitions and enables business users to exchange their definitions independently from IT specialists. The benefits of the approach have been outlined in a comprehensive case study.

Research is now dedicated to the design, implementation and evaluation of more advanced analysis and reporting tools with direct support for Semantic Web technologies. In particular, the presented approach will be extended to facilitate the exchange of entire queries. This will require the inclusion of further parts of the multidimensional model.

References

1. Inmon, W.H.: Building the Data Warehouse. Wiley, New York (2005)
2. Kimball, R.: The Data Warehouse Lifecycle Toolkit. Wiley, New York (2008)
3. Microsoft Corporation: Microsoft Analysis Services,
   http://www.microsoft.com/sql/technologies/analysis/default.mspx
4. Pentaho Corporation: Pentaho Analysis Services, http://mondrian.pentaho.org
5. SAP AG: SAP NetWeaver Business Intelligence http://www.sap.com/germany/plattform/netweaver/components/businessintelligence/index.epx
6. Guizzardi, G.: On Ontology, ontologies, Conceptualizations, Modeling Languages, and (Meta)Models. Frontiers in Artificial Intelligence and Applications 155, 18–35 (2007)
7. Object Management Group (OMG): Ontology Definition Metamodel (ODM) Beta 2 (2007), http://www.omg.org/cgi-bin/doc?ptc/07-09-09
8. Object Management Group (OMG): Common Warehouse Metamodel (CWM) 1.1 (2003), http://www.omg.org/spec/CWM/1.1/PDF/
9. Object Management Group (OMG): CWM Metadata Interchange Patterns (2004), http://www.omg.org/cgi-bin/doc?formal/04-03-25
10. Object Management Group (OMG): XML Metadata Interchange 2.1.1 (2007), http://www.omg.org/docs/formal/07-12-01.pdf
11. Januszewski, A., Pankowski, T.: Modeling Analytical Indicators Using Data Warehouse Metamodel. In: Bressan, S., Küng, J., Wagner, R. (eds.) DEXA 2006. LNCS, vol. 4080, pp. 642–646. Springer, Heidelberg (2006)
12. Object Management Group (OMG): MDA Guide 1.0.1 (2003), http://www.omg.org/docs/omg/03-06-01.pdf
13. Object Management Group (OMG): Unified Modeling Language 2.2 Beta 1 (2008), http://www.omg.org/docs/pct/08-04-04.pdf
14. Object Management Group (OMG): MOF Query / View / Transformations 1.0 (2008), http://www.omg.org/docs/formal/08-04-03.pdf
15. Mazón, J.-N., Trujillo, J.: An MDA approach for the development of data warehouses. Decision Support Systems 45, 41–58 (2008)

16. Pardillo, J., Mazón, J.-N., Trujillo, J.: Bridging the Semantic Gap in OLAP Models: Platform-independent Queries. In: Proceedings of the 11th International Workshop on Data Warehousing and OLAP (DOLAP 2008), pp. 89–96. ACM Press, New York (2008)
17. World Wide Web Consortium (W3C): OWL Web Ontology Language, http://www.w3.org/TR/2004/REC-owl-features-20040210/
18. Xie, G., Yang, Y., Liu, S., Qiu, Z., Pan, Y., Zhou, X.: EIAW: Towards a Business-friendly Data Warehouse Using Semantic Web Technologies. In: ASWC 2007 and ISWC 2007, pp. 857–870. IEEE, Los Alamitos (2007)
19. Diamantini, C., Potena, D.: Semantic Enrichtment of Strategic Datacubes. In: Proceedings of the 11th International Workshop on Data Warehousing and OLAP (DOLAP 2008), pp. 81–88. ACM Press, New York (2008)
20. World Wide Web Consortium (W3C): Mathematical Markup Language (MathML) 3.0, http://www.w3.org/TR/2008/WD-MathML3-20081117/
21. The OpenMath Standard 2.0, http://www.openmath.org/standard/om20-2004-06-30/omstd20.pdf
22. de Laborda, C.P., Conrad, S.: Relational.OWL - A Data and Schema Representation Format Based on OWL. In: Proceedings of the 2nd Asia-Pacifc Conference on Conceptual Modelling (APCCM 2005) (2005)
23. World Wide Web Consortium (W3C): Extensible Markup Language(XML) 1.0, http://www.w3.org/TR/2008/REC-xml-20081126/
24. Organization for the Advancement of Structured Information Standards (OA-SIS): The OpenDocument v1.0 Specification, http://www.oasis-open.org/committees/download.php/12572/OpenDocument-v1.0-os.pdf
25. University of Manchester: The Manchester OWL Syntax, http://www.co-ode.org/resources/reference/manchester_syntax/
26. World Wide Web Consortium (W3C): XHTML 1.0: The Extensible HyperText Markup Language, http://www.w3.org/TR/2000/REC-xhtml1-20000126/
27. Microsoft Corporation: OLE DB for Online Analytical Processing (OLAP), http://msdn.microsoft.com/en-us/library/ms717005(VS.85).aspx
28. Microsoft Corporation, Hyperion Solutions Corporation: XML for Analysis Speci-fication 1.1, http://www.xmlforanalysis.com/xmla1.1.doc
29. Horridge, M., Bechhofer, S., Noppens, O.: Igniting the OWL 1.1 Touch Paper: The OWL API. Proceedings of the OWLED 2007, Workshop on OWL: Experiences and Directions. CEUR-WS.org (2007)
30. World Wide Web Consortium (W3C): SWRL: A Semantic Web Rule Language Combining OWL and RuleML, http://www.w3.org/Submissions/2004/SUBM-SWRL-20040521/
31. Microsoft Corporation: COM: Component Object Model Technologies, http://www.microsoft.com/com/default.mspx
32. World Wide Web Consortium (W3C): SOAP 1.2, http://www.w3.org/TR/2007/REC-soap12-part1-20070427/
33. World Wide Web Consortium (W3C): Web Services Definition Language (WSDL) 2.0, http://www.w3.org/TR/2007/REC-wsdl20-primer-20070626/
34. The Apache Software Foundation: Apache CXF: An Open Source Service Frame-work, http://cxf.apache.org/
35. Simba Technologies: SimbaO2X, http://www.simba.com/odbo-to-xmla.htm

36. Integre Technical Publishing Co., Inc.: MathML Equation Editor, http://www.integretechpub.com/zed/
37. Clark & Parsia, LLC: Pellet: The Open Source OWL DL Reasoner, http://clarkparsia.com/pellet/
38. Sun Microsystems, Inc: JSR 224: Java API for XML-Based Web Services (JAX-WS) 2.0, http://jcp.org/en/jsr/detail?id=224
39. Microsoft Corporation: Multidimensional Expressions (MDX) Reference, http://msdn.microsoft.com/en-us/library/ms145506.aspx
40. International Business Machines Corp. (IBM): Dataquant, http://www-01.ibm.com/software/data/db2imstools/db2tools/dataquant/index.html
41. Microsoft Corporation: Microsoft Office Excel, http://office.microsoft.com/en-us/excel/FX100487621033.aspx

Skyline View: Efficient Distributed Subspace Skyline Computation

Jinhan Kim, Jongwuk Lee, and Seung-won Hwang

Department of Computer Science and Engineering
POSTECH, Republic of Korea
{wlsgks08,julee,swhwang}@postech.ac.kr

Abstract. Skyline queries have gained much attention as alternative query semantics with pros (e.g.low query formulation overhead) and cons (e.g.large control over result size). To overcome the cons, subspace skyline queries have been recently studied, where users iteratively specify relevant feature subspaces on search space. However, existing works mainly focuss on centralized databases. This paper aims to extend subspace skyline computation to distributed environments such as the Web, where the most important issue is to minimize the cost of accessing vertically distributed objects. Toward this goal, we exploit prior skylines that have overlapped subspaces to the given subspace. In particular, we develop algorithms for three scenarios– when the subspace of prior skylines is superspace, subspace, or the rest. Our experimental results validate that our proposed algorithm shows significantly better performance than the state-of-the-art algorithms.

1 Introduction

Skyline queries have gained much attention as alternative query semantics with pros (low query formulation overhead) and cons (large control over result size). Example 1 describes a skyline query finding "interesting" hotels.

Example 1 (Skyline query). Consider a user looking for hotels that are close to the beach and reasonably priced. Among 10 hotels plotted in Fig. 1, we say that A is more interesting than I, i.e., A *dominates* I, since A is closer and cheaper than I. After eliminating all "dominated" hotels, remaining hotels are A, B, and E that are viewed as interesting ones regardless of a user-specific preference. They are called *skyline* which is a set of objects not dominated by any other objects.

While the skyline query identifies a subset of interesting objects, it is hard to control the skyline size. In particular, as the number of attributes increases, e.g., to consider additional attributes such as star rating, the skyline size increases exponentially. To address this drawback, subspace skyline queries [14,16,19,20] have been recently studied to control the skyline size with respect to user-specific needs. However, these solutions focus on *centralized* environments and

T.B. Pedersen, M.K. Mohania, and A M. Tjoa (Eds.): DaWaK 2009, LNCS 5691, pp. 312–324, 2009.

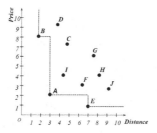

Fig. 1. Illustration of a skyline query

pre-materialize skyline over all possible subspaces, it is non-trivial to extend them to support *distributed* environments such as the Web.

This paper studies the subspace skyline problem in distributed environments such as the Web [4,5,7]. Specifically, datasets can be distributed over several databases provided by different providers. A key challenge is thus to optimize the cost of accessing objects vertically distributed in different sources. Skyline processing in such environments has been only considered for full feature space [2,12]. In a clear contrast, we study how to efficiently support multiple skyline queries defined in diverse subspaces.

In summary, we believe that our paper has the following contributions:

– When there exist no prior results, we improve basic algorithms to exploit data-driven cost estimations and significantly reduce the overall cost.
– Otherwise, we can further reduce cost by reusing previously identified skylines, i.e., *skyline views*. Specifically, to compute the skyline results for a given feature subspace U, we focus on leveraging the subspace skylines on subspace U' that are not disjointed to U, by reusing prior query results. To significantly reduce the cost, we develop algorithms for three scenarios—when U' is the super-space, the subspace, or the rest.

The remainder of this paper is organized as follows. Section 2 states recent studies related to our framework. Section 3 discusses preliminaries to help our framework. Section 4 and Section 5 propose our baseline algorithms with no skyline views and view-based skyline algorithms respectively. Section 6 reports our experimental results, and Section 7 concludes this paper.

2   Related Work

This section first provides an overview of skyline computation in centralized database environments. We then discuss existing skyline algorithms over the distributed environments and how our work distinguishes itself.

Skyline computation in centralized environments: Börzsönyi el at. [3] introduced skyline queries in the database community. Next, Tan et al. [15] proposed a progressive skyline algorithm using bitmap and B-tree structures. Later,

Kossmann et al. [10] and Papadias et al. [13] proposed efficient skyline algorithms based on R-tree. Recently, Lee et al. [11] proposed skyline computation using ZBtree, storing data based on a Z-order curve. The algorithms, however, only focused on full space skyline computations.

Meanwhile, there is recent work for effective subspace skyline computation. Yuan et al. [20] and Xia et al. [19] proposed a *skycube* structure that computes the skylines of all possible subspaces in a batch, and Pei et al. [14] studied the semantic structure between subspace skylines, i.e., *decisive subspace*. Tao et al. [16] proposed a *2D index* supporting subspace skyline computation. However, since these algorithms assume centralized database environments, they do not consider the cost model of accessing objects in distributed environments, i.e., *sorted access* and *random access* [4,5,7]. In a clear contrast, our work aims to identify subspace skylines with minimal cost in distributed scenarios, by reusing prior skyline results.

Skyline computation in distributed environments: The above skyline query processing algorithms have only been studied in centralized environments. Recently there has been growing interest in distributed skyline computations, where objects are distributed vertically or horizontally over data sources.

Vertical model: In the Web scenarios, data are often vertically distributed, e.g., *price* information residing in hotels.com and *distance* in maps.com, as typically assumed in [2,12]. Balke et al. [2] first proposed a distributed skyline computation. Further, Lo et al. [12] proposed a progressive skyline computation under the same model and studied how to find a good "terminating object" using a *linear regression* technique for early termination. Our paper is also based on the same distributed data model, but our work distinguishes itself from these efforts, by (a) devising a more desirable cost estimation model for finding a terminating object, and (b) reusing prior query results for subspace skyline computation.

Horizontal model: In peer to peer scenarios, data are often horizontally distributed, as studied in [6,17,18]. Specifically, Hose et al. [6] focused on Peer Data Management Systems (PDMS) and proposed an algorithm identifying relaxed skyline results. Vlachou et al. [17] addressed skyline computation in a super-peer network, where each peer stores their own local data and computes local skylines for themselves. To support an index structure for skyline computation on the peer-to-peer network, Wang et al. [18] proposed a tree structure like BATON [8], which exploits Z-curve traversal to map multi-dimensional data into one-dimensional space.

3 Preliminaries

This section first presents notations and preference notions to define the skyline problem. Table 1 summarizes notations used in this paper. (Throughout this paper, we use attribute and dimension interchangeably.) For simplicity, suppose that user preference follows a ***strict total order*** on each attribute d_i, denoted as $\succ_{d_i}$. (In Section 6, we will extend the preference into general order as $\succeq_{d_i}$.)

Table 1. Notations used in this paper

Notation	Definition
m	Number of attributes
n	Number of objects
D	Dimension set $(d_1, ..., d_m)$
O	Dataset $(o_1, ..., o_n)$
$S(U)$	Subspace skyline on U
SA.next(d_i)	SA for next preferred objects on d_i
RA(o, D')	RA for o's unknown values on D'

Based on this notion, we formally state dominance and skyline on user-specified subspace U. (These definitions are consistent with the preceding work.)

Definition 1 (Dominance on U). *An object o dominates another object o' on U if and only if $\forall\, d_i \in U\colon o \succeq_{d_i} o'$ and $\exists\, d_j \in U\colon o \succ_{d_j} o'$.*

Definition 2 (Skyline on U). *An object o belongs to skyline on U, denoted as $S(U)$, if and only if $\forall o'(\neq o) \in O$ does not dominate o on U.*

We adopt a vertically distributed data model, as the environment to compute the skyline. This stores different attribute values in multiple distributed data sources, as widely adopted in [4,5,7]. In this model, unknown attribute values are retrieved from such sources through one of the following two access modes:

- Sorted access (SA): Each sorted access on attribute d_i, denoted as SA.next(d_i), retrieves the next object in the descending order of user preference on d_i.
- Random access (RA): For an object o, each random access on attribute set D' retrieves its unknown values.

Full space skyline computation to optimize such access cost in the distributed environments has been proposed [2,12].

In a clear contrast, we aim to support "subspace" skyline processing, where users specify arbitrary subset U of D that represents their information needs. Our goal is, in finding subspace skyline results $S(U)$ for user-specified attribute space U, to reuse the prior query results $S(U')$. More formally, we study how to exploit prior query results $S(U')$ for finding $S(U)$, which falls into one of the following three cases. (For the sake of representation, each case is described with a *tree* structure in which full space is represented as a root node.)

1. **Child skyline view:** U' is a child of U (i.e., $U' \subset U$).
2. **Parent skyline view:** U' is a parent of U (i.e., $U' \supset U$).
3. **Sibling skyline view:** U' is a sibling of U (i.e., $U' - U \neq \phi$, $U - U' \neq \phi$, and $U' \cap U \neq \phi$).

4   Baseline Skyline Algorithm

This section presents our first proposed algorithm, Baseline Skyline Algorithm (BSA), when no prior query results can be leveraged. While IDS [2] and PDS

Table 2. Toy dataset with 3-dimensional 10 objects

rank	1	2	3	4	5	6	7	8	9	10
d_1	a	g	c	j	d	i	f	b	e	h
d_2	b	d	h	f	i	e	a	c	j	g
d_3	c	j	e	b	f	h	d	a	i	g

[12] have similar limitation, our work significantly improves these algorithms by exploiting systematic cost estimation based on data statistics, e.g., *histograms* and *rank index* [1]. BSA will later serve as a baseline as well, when evaluating other proposed algorithms reusing prior results.

As the first step to build a BSA, we observe the commonality of the two algorithms IDS and PDS, which is to perform sorted accesses $SA(d_i)$ on $\forall d_i \in D$, until we access a common object o_x from all d_is. (Due to the limitation of space, all proofs are presented in our extended paper [9].)

Lemma 1 (Termination condition [2]). *At any point, if an object o has been accessed from every sorted access on $d_i \in U$, any objects yet to be accessed cannot be a skyline object.*

From a toy dataset (Table 2), the termination condition holds after one access on d_1 (accessing a), and 7 and 8 accesses each on d_2 and d_3 respectively. Alternatively, a smarter algorithm can satisfy such condition with much less sorted accesses, when c is accessed from all d_is, after 2, 6, and 1 accesses on d_1, d_2, and d_3, respectively. Based on this observation, existing algorithms aimed at finding the "terminating object" causing the minimal sorted access cost C [2,12]. In this paper, we call such a cost-minimal object a *watermark object*, where its cost can be represented as the following function.

$$o_w = \arg\min_i C(o_i)$$

We observe how existing algorithms find such object. IDS [2] assumes that the dataset has an uniform distribution so IDS estimates the number of sorted accesses, using the attribute value itself. PDS [12] enhances such estimation using a linear regression approach of iteratively updating the prediction based on the costs incurred for previously accessed objects. However, experimental result shows that the estimation of PDS also works well only when the dataset has an uniform distribution. In contrast, we develop more enhanced cost estimation for finding a better watermark object. Specifically, our proposed method is categorized into (a) histogram-based estimation and (b) exact evaluation. (They are illustrated as Algorithm 2.)

- **BSA(histogram):** Using histograms on attribute values, we can estimate the rank of the given object. Unlike the estimation schemes of IDS and PDS that work best over uniform distribution, this estimation scheme can adapt to various distributions, as we report in Section 6.
- **BSA(exact):** Using rank based index [1], we can retrieve the rank of each object in each list, from a random access, based on which we can exactly compute the sorted access cost C of the given object.

Algorithm 1. BSA(O, U)

1: $d_c = d_1$, $o_w = \phi$.
2: SKY$^{d_i} = \phi$. {Store final skyline on d_i}, $P = \phi$. {Store pruned objects.}
3: **while** o_w is not seen on any attribute on U **do**
4:    $o_c \leftarrow$ SA.next(d_c).
5:    **if** o_c has a missing value **then**
6:       RA$(o_c, U - \{d_c\})$.
7:    **end if**
8:    **if** $o_c \in P$ or $o_c \prec \exists o_x \{o_x \in$ SKY$^{d_c}\}$ **then**
9:       Push o_c into P.
10:   **else**
11:      Push o_c into SKY^{d_c}, $(o_w, d_c) = $ Cost(o_c, o_w, d_c), Output o_c if not found yet.
12:   **end if**
13: **end while**

Algorithm 2. Cost(o_c, o_w, d_c)

1: RA$(o_c$, a set of attributes with missing values).
2: Update and Estimate $C(o_w), C(o_c)$. {When we use a histogram-based estimation.}
3: **if** $C(o_c) < C(o_w)$ or $o_c \equiv o_w$ **then**
4:    $o_w \leftarrow o_c$.
5:    $d_c \leftarrow$ One of attributes on U that o_w is not seen.
6: **end if**
7: **return** (o_w, d_c).

We now illustrate BSA (Algorithm 1) using the cost estimation above. First, using a sorted access on some attribute d_i, BSA finds a skyline candidate o_c (Line 4-7). Unlike IDS waiting until the termination to decide whether it actually becomes a skyline object or not, BSA "progressively" outputs o_c, if determined as a skyline object. Next, once o_c becomes a skyline object, we update the watermark object o_w if it has a lower cost than the current watermark, i.e., $C(o_c) < C(o_w)$. When the new o_w is determined, we update d_c as some attribute yet to be evaluated for the new watermark object in the *Cost* module (Algorithm 2), and then perform SA on d_c at the next iteration. Such iteration continues until the current watermark object is accessed in all $SA(d_i)$, which satisfies the termination condition (Lemma 1).

To illustrate BSA, we describe an example using a toy (Table 2). First, we perform $SA(d_1)$ to access the top object a. BSA then performs RA on a to evaluate missing values of a. As SKY$^{d_1} = \phi$, no object dominates a so a becomes a skyline object. At this point, a can be returned progressively and pushed into SKY^{d_1}. Meanwhile, object a also becomes the watermark object o_w and BSA changes d_c to some attribute yet to be evaluated on object a. For simplicity, we change the attributes in a round robin manner. In the next iteration, we perform $SA(d_2)$ to access b and b becomes a skyline object. Furthermore, as $C(b) = 13 < C(a) = 16$, o_w is updated to b, and d_c is changed. From $SA(d_3)$, we access c and c becomes a new skyline object and o_w. d_c is changed back to d_1 and $SA(d_1)$ accesses g, which is dominated by a in SKY^{d_1}. BSA thus pushes g into P, and continues to access c, and pushes c into SKY^{d_1} because c is already identified as a skyline object. BSA then changes $d_c = d_2$ because $o_c \equiv o_w$. From $SA(d_2)$, d is accessed, which is not dominated by any object in SKY$^{d_2} = \{b\}$.

Table 3. Toy dataset with 4-dimensional 10 objects extended from Table 2

rank	1	2	3	4	5	6	7	8	9	10
d_1	a	g	c	j	d	i	f	b	e	h
d_2	b	d	h	f	i	e	a	c	j	g
d_3	c	j	e	b	f	h	d	a	i	g
d_4	h	b	d	f	e	i	a	g	c	j

d thus becomes a skyline object, BSA pushes d into SKY^{d_2} but o_w does not change as $C(d) = 14 > C(c) = 12$. After few more $SA(d_2)$, $o_w = c$ is finally accessed, which makes c accessable from all sorted accesses and terminates BSA.

5   Reusing Skyline Algorithm

This section studies how we can enhance BSA using prior query results. Specifically, we consider two main properties as follows:

- **Pruning rules:** Based on prior results, we can identify some objects that can never become skyline. In Algorithm 1, we can initialize P as such objects to avoid any access. Alternatively, instead of evaluating all the missing values $RA(o_c, U - \{d_c\})$ before deciding whether to prune out, we can save RA cost by performing "selective RA" using P.
- **Pre-qualification:** From the prior results, we can identify some objects that are guaranteed to qualify as skylines, before performing any access. Such objects can be returned to users in a progressive manner.

Based on the properties, we propose algorithms that reuse prior results $S(U')$ for computing $S(U)$, for three scenarios when U' is a parent, a child, or a sibling of U. Assume that we only have the prior skyline object IDs not the real values. In the extended version of this paper [9], we discuss how to reuse sibling skyline view by combining the advantages of parent and child views in detail.

5.1   Parent Skyline View (PSV)

We develop how to find $S(U)$ using prior results $S(U')$ when U' is the superset of the U, i.e., $U' \supset U$. Specifically, we develop the following pruning condition for computing $S(U)$, using $S(U')$, as we formally state below.

Lemma 2 (Pruning Condition). *If $o_c \notin S(U')$ and o_c has distinct value, o_c cannot be a new skyline object of U when $U \subset U'$.*

We develop Algorithm PSV using Lemma 2. This identifies the non-skyline objects early on using $S(U')$ to eliminate the unnecessary access cost or dominance checks on such objects. This pruning condition can be easily added to BSA, by populating pruned object set P (Line 2) in Algorithm 1 ,i.e., $P = O - S(U')$.

We now illustrate how PSV works with our toy example (Table 3), when $U = \{d_1, d_2, d_3\}$ and $U' = \{d_1, d_2, d_3, d_4\}$. Using prior query results $S(U') = \{a, b, c, d, e, f, h\}$, we can initialize P as $\{g, i, j\}$ using Lemma 2. First, from

$SA(d_1)$, we access a, which is not dominated by any other object at d_1 and thus becomes a SKY^{d_1}. PSV then performs $SA(d_2)$ to access b and pushes it to SKY^{d_2} and updates b as a new o_w, as its SA cost is lower than a. PSV performs $SA(d_3)$ to access c, which becomes a SKY^{d_3} and new o_w. Next object g from $SA(d_1)$ is in P. PSV can thus save the cost by skipping g. As the next object c is already selected, PSV pushes c into SKY^{d_1} and selects d_2. PSV continues $SA(d_2)$ to access until $o_w = c$ is discovered at d_2 (and all other d_is). S(U) is thus finalized as $\cup_{i=1}^{3} SKY^{d_i} = \{a, b, c, d, e, f\}$.

5.2   Child Skyline View

We move on to the Child Skyline View (CSV) problem, of computing S(U) using S(U') when U' is subset of the attribute U, i.e., $U' \subset U$, based on the property below.

Lemma 3 (Pre-qualification Rule). *Let $U' \subset U$. Object, $o_c \in$ S(U'), can become the skyline of U if o_c has distinct value.*

One naive solution is to simply adopt Lemma 3, to "eagerly" populate $\forall i$: $SKY^{d_i} = S(U')$. With this eager population, a candidate object o_c can be pruned out early on, if it is dominated by some object in S(U'). However, this naive adoption leads to unnecessary comparisons between objects. For example, when some object o_x ranks the second highest with respect to d_i, o_x can only be dominated by the top object on d_i and dominance checking with any other object is unnecessary. However, with eager population, o_x has to be compared with all objects in S(U').

Because of this problem, we study "lazy population". We divide skyline candidates SKY^{d_i} into $OSKY^{d_i}$ (for original skyline for U') and $NSKY^{d_i}$ (for new skyline for U), to store the skyline results that belong to S(U') and that not belong to S(U') respectively, i.e., $SKY^{d_i} = OSKY^{d_i} \cup NSKY^{d_i}$, and delay the population of $OSKY^{d_i}$ as much as possible. Specifically, the object accessed o_c is added to $OSKY^{d_i}$ iteratively, which enables to restrict dominance checks only to those objects that are already accessed from $SA(d_i)$.

We also develop another pruning rule, which enables us to prune out a candidate object with only "selective RA".

Lemma 4 (Pruning Rule). *Let o_c be the current retrieved object on d_c, $d_c \in U'$, by doing SA and $o_c \notin$ S(U'). If $o_c \prec_{U-U'} \forall o_x$, $o_x \in$ S(U') that satisfies $rank_{d_c}(o_x) < rank_{d_c}(o_c)$, then o_c cannot be a new skyline object.*

Using this lemma, instead of performing RAs for all the unknown attributes for each candidate object, we can restrict it to "selective RA" on $RA(o_c, U - U')$ for some skyline objects (lines 12-13). More specifically, CSV algorithm works as follows: If $o_c \in$ S(U'), CSV pushes o_c into $OSKY^{d_c}$. If not, CSV performs RA for $U - U'$ values first. If o_c is dominated by all objects in $OSKY^{d_c}$ with respect to $U - U'$, o_c can be pruned out immediately (and thus saves RA cost). Only for the remaining cases, we fully evaluate all the unknown values.

Algorithm 3. $\mathrm{CSV}(O, U, U', \mathrm{S}(U'))$

1: $d_c = $ One of attributes on U', $o_w = \phi$, $P = \phi$. {Store pruned objects.}
2: $\mathrm{OSKY}^{d_i} = \phi$, $\mathrm{NSKY}^{d_i} = \phi$. {Store original and new skyline of attribute on d_i.}
3: **while** o_w is not seen on any attribute on U **do**
4:       $o_c \leftarrow \mathrm{SA.next}(d_c)$.
5:       **if** $o_c \in P$ **then**
6:             Continue.
7:       **else if** $o_c \in \mathrm{S}(U')$ **then**
8:             Push o_c into OSKY^{d_c}, $(o_w, d_c) = \mathrm{Cost}(o_c, o_w, d_c)$.
9:             Output o_c if not found yet.
10:      **else**
11:            $\mathrm{RA}(o_c, U - U')$
12:            **if** $o_c \prec_{U-U'} \forall o_x, o_x \in \mathrm{OSKY}^{d_c}$ and $d_c \in U'$ **then**
13:                  Push o_c into P.
14:            **else**
15:                  $\mathrm{RA}(o_c, U')$
16:                  **if** $o_c \prec_U \exists o_x, o_x \in \mathrm{OSKY}^{d_c} \cup \mathrm{NSKY}^{d_c}$ **then**
17:                        Push o_c into P.
18:                  **else**
19:                        Push o_c into NSKY^{d_c}, $(o_w, d_c) = \mathrm{Cost}(o_c, o_w, d_c)$.
20:                        Output o_c if not found yet.
21:                  **end if**
22:            **end if**
23:      **end if**
24: **end while**

We now illustrate how CSV works with our toy example (Table 2) when $U = \{d_1, d_2, d_3\}$ and $U' = \{d_1, d_2\}$. CSV first performs SA on d_1 and accesses the top object a. Because $a \in \mathrm{S}(U') = \{a, b, d\}$, CSV pushes a into OSKY^{d_1} and a becomes the first o_w. CSV selects the next attribute as d_2 and pushes the top object b into OSKY^{d_2}. As $C(b) = 13 < C(a) = 16$, o_w is updated to b. Next, c is accessed from $SA(d_3)$. As c is a skyline object and $c \notin \mathrm{S}(U')$, CSV pushes c into NSKY^{d_3} and changes o_w again to c as $C(c) < C(b)$. Going back to $SA(d_1)$, we access $g \notin \mathrm{S}(U')$. CSV thus performs "selective RA" on $U - U' = \{d_3\}$. As $g \prec_{d_3} \forall o_x \in \mathrm{OSKY}^{d_1} = \{a\}$, g can be pruned out without additional RAs. Since the next object c is already found as a skyline object, CSV selects the next attribute d_2. CSV continues sorted accesses until it reaches h and performs selective $\mathrm{RA}(h, \{d_3\})$. As h is is partially dominated by b in $\mathrm{OSKY}^{d_2} = \{b, d\}$, CSV fully evaluates h, which is still dominated by b. Thus, h is pruned. After a few more sorted accesses, CSV reaches $o_w = c$ which is discovered at all d_is and the final result $\mathrm{S}(U)$ is $\sum \mathrm{OSKY}^{d_i} \cup \mathrm{NSKY}^{d_i} = \{a, b, c, d, e, f\}$.

6   Experiments

This section validates the efficiency of our algorithms using synthetic datasets. Specifically, the parameters used and their default values are described in Table 4. (For brevity, we have omitted the results for correlated and anti-correlated datasets, which were consistent with the results reported in this section).

As for performance metrics, we use the following two measures: (1) the number of objects accessed, and (2) the number of random accesses, which directly affect the overall cost in distributed environments. (We do not report the number

Table 4. Parameters for experimental settings

Parameter	Value : Default
Data size n	$[100K, 500K]$: $100K$
Dimensionality m	$[3,7]$: 5
Data distribution	Uniform, Gaussian, Zipfian : Gaussian
Correlation	Correlated, Independent, Anti-Correlated : Independent

of sorted accesses, as it is proportional to the number of objects accessed.) Experiments were carried out on an Intel(R) Pentium(4) with 3.00 GHz processor and 1GB RAM running Windows XP.

We implemented the following five algorithms with C$^\#$ language as follows:

- Improved Distributed Algorithm (IDS) [2]
- Progressive Distributed Skyline Algorithm (PDS) [12].
- BSA: Our baseline algorithm when there are no prior skyline results. We implement it in two versions:
  - BSA(histo): estimating the cost of accessing an object using attribute value histograms.
  - BSA(exact): exactly computing the object access cost using rank index.
- Our framework: Our skyline view algorithms, PSV, CSV, and SSV are combined into a single framework. Given a skyline query, we randomly select a skyline view, and compute the skyline.

While we assume a strict total order for user preference on attribute values, it can be easily extended to support a general total order.

6.1 Experiments on No Skyline Views

We validated the efficiency of BSA(histo) and BSA(exact) against previous distributed skyline algorithms. For BSA(histo), we set BucketSize = $\lfloor n \times 0.01 \rfloor$. All objects within the same bucket tie as the same rank, i.e., HistRank(o_i) = $\lfloor$ RealRank(o_i) / BucketSize $\rfloor \times$ BucketSize + 1, which is synthetically generated. The histogram is pre-constructed by scanning an entire dataset once.

Fig. 2 reports our comparison results of BSA with IDS and PDS. Significantly lower costs of BSA suggest that the rank estimation of our proposed algorithm is more accurate than that of IDS and PDS. IDS, assuming uniform distribution, performs comparably with BSA in the uniformly distributed data, but its performance deteriorates for other distributions, Gaussian and Zipfian. Meanwhile, PDS, leveraging linear regression for rank estimation, incurs high cost, due to the following reasons: First, linear regression assumes linear relationship between values and rankings which does not hold for Gaussian and Zipfian distributions. Second, even for distributions where such relationship holds, the accuracy depends on the quality of samples used for regression. Meanwhile, samples used in PDS are biased to highly ranked objects, which negatively affects the estimation quality.

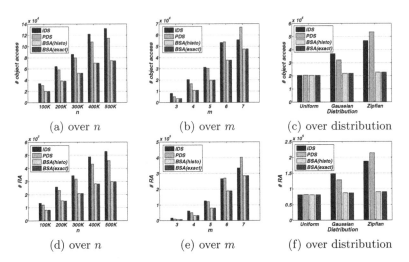

(a) over n (b) over m (c) over distribution

(d) over n (e) over m (f) over distribution

Fig. 2. Comparisons of distributed skyline algorithms without skyline views

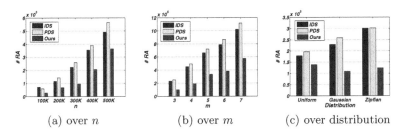

(a) over n (b) over m (c) over distribution

Fig. 3. Comparison between existing algorithms and our reusing skyline views

6.2 Experiments on Skyline Views

We validated the effectiveness of our framework reusing prior skyline views against IDS and PDS. Given a skyline query, we first generated skyline views with $m \pm 1$ overlapped dimensions with respect to the skyline query with m-dimensional space– The subspace size of a child view and a parent view is $m - 1$ and $m + 1$, where the overlapped subspace is $m - 1$ and m, respectively. Similarly, the size of a sibling view is m in which its overlapped subspace is $m - 1$. We then randomly selected one of the three algorithms, and identified the skyline.

We empirically confirmed that our framework effectively reuses prior skyline results, which contributes to saving overall computational costs. Figure 3 describes that our framework outperforms IDS and PDS with various parameter settings. Due to the limitation of space, we do not report the number of object accesses. However, based on the consistent behaviors of object access costs and RA costs in Figure 3, we can infer similar behaviors.

7   Conclusion

This paper studies the subspace skyline query processing problem in distributed environments. In particular, we first design a baseline distributed skyline computation with an improved cost estimation method. We then study how to reuse prior skyline results for the space that is a superset or a subset of the given subspace. Our experiments validate that our proposed framework can significantly save overall costs by reusing prior results.

Acknowledgement

This work was supported by Engineering Research Center of Excellence Program of Korea Ministry of Education, Science and Technology (MEST) / Korea Science and Engineering Foundation (KOSEF), grant number R11-2008-007-03003-0.

References

1. Akbarinia, R., Pacitti, E., Valduriez, P.: Best position algorithms for top-k queries. In: VLDB (2007)
2. Balke, W.-T., Güntzer, U., Zheng, J.X.: Efficient distributed skylining for web information systems. In: Bertino, E., Christodoulakis, S., Plexousakis, D., Christophides, V., Koubarakis, M., Böhm, K., Ferrari, E. (eds.) EDBT 2004. LNCS, vol. 2992, pp. 256–273. Springer, Heidelberg (2004)
3. Börzsönyi, S., Kossmann, D., Stocker, K.: The skyline operator. In: ICDE (2001)
4. Chang, K.C., Hwang, S.: Minimal probing: Supporting expensive predicates for top-k queries. In: SIGMOD (2002)
5. Fagin, R., Lote, A., Naor, M.: Optimal aggregation algorithms for middleware. In: PODS (2001)
6. Hose, K., Lemke, C., Sattler, K.: Processing relaxed skylines in pdms using distributed data summaries. In: CIKM (2006)
7. Hwang, S., Chang, K.C.: Optimizing access cost for top-k queries over web sources. In: ICDE (2005)
8. Jagadish, H., Ooi, B.C., Vu, Q.H.: Baton: A balanced tree structure for peer-to-peer networks. In: VLDB (2005)
9. Kim, J., Lee, J., Hwang, S.: Skyline view: Efficient distributed subspace skyline computation (2009), http://ids.postech.ac.kr/skylineview.pdf
10. Kossmann, D., Ramsak, F., Rost, S.: Shooting stars in the sky: An online algorithm for skyline queries. In: VLDB (2002)
11. Lee, K.C.K., Zheng, B., Li, H., Lee, W.-C.: Approaching the skyline in z order. In: VLDB (2007)
12. Lo, E., Yip, K.Y., Lin, K.-I., Cheung, D.W.: Progressive skylining over web-accessible database (2006)
13. Papadias, D., Tao, Y., Fu, G., Seeger, B.: An optimal and progessive algorithm for skyline queries. In: SIGMOD (2003)
14. Pei, J., Jin, W., Ester, M., Tao, Y.: Catching the best views of skyline: A semantic approach based on decisive subspaces. In: VLDB (2005)

15. Tan, K., Eng, P., Ooi, B.C.: Efficient progressive skyline computation. In: VLDB (2001)
16. Tao, Y., Xian, X., Pei, J.: SUBSKY: Efficient computation of skylines in subspaces. In: ICDE (2006)
17. Vlachou, A., Doulkeridis, C., Kotidis, Y., Vazirgiannis, M.: Skypeer: Efficient subspace skyline computation over distributed data. In: ICDE (2007)
18. Wang, S., Ooi, B.C., Tung, A.K.H., Xu, L.: Efficient skyline query processing on peer-to-peer networks. In: ICDE (2007)
19. Xia, T., Zhang, D.: Refreshing the sky: The compressing skycube with efficient support for frequent updates. In: SIGMOD (2006)
20. Yuan, Y., Lin, X., Liu, Q., Wang, W., Yu, J.X., Zhang, Q.: Efficient computation of the skyline cube. In: VLDB (2005)

HDB-Subdue: A Scalable Approach to Graph Mining*

Srihari Padmanabhan and Sharma Chakravarthy

IT Laboratory & Department of Computer Science and Engineering
The University of Texas at Arlington, Arlington, TX 76019
`sharma@cse.uta.edu`

Abstract. Transactional data mining (association rules, decision trees etc.) has been effectively used to find non-trivial patterns in categorical and unstructured data. For applications that have an inherent structure (e.g., social networks, proteins), graph mining is useful since mapping the structured data into a transactional representation will lead to loss of information. Graph mining is used for identifying interesting or frequent subgraphs. Database mining uses SQL and relational representation to overcome limitations of main memory algorithms and to achieve scalability.

This paper presents a scalable, SQL-based approach to graph mining – specifically, interesting substructure discovery. The most general form of graphs including directed edges, multiple edges between nodes, and cycles are handled by our approach. Our primary goal in this work has been to address scalability, and map difficult and computationally expensive problems such as pseudo duplicate elimination, canonical labeling, and isomorphism checking into SQL-based counterparts. The notion of minimum description length (MDL) has been cast into corresponding metric for relational representation. Our experimental analysis shows that graphs with Millions of nodes and edges can be handled by the algorithm and the approach presented in this paper.

1 Introduction

Transactional data mining is widely used in detecting interesting patterns from unstructured data using techniques such as association rule mining. Graph mining is appropriate for mining data with inherent complex structure. Although main memory based data mining algorithms exist, they typically face two problems with respect to scalability: i) storing the entire graph (or its adjacency matrix) in main memory may not be possible, and ii) the computational space requirements of the algorithm may be more than what is available.

Database mining has been an answer to scalability without having to incorporate complex buffer management strategies in mining algorithms. Database mining uses SQL and relational representation to overcome limitations of main

---

* This work was supported, in part, by NSF grant IIS 0534611.

T.B. Pedersen, M.K. Mohania, and A M. Tjoa (Eds.): DaWaK 2009, LNCS 5691, pp. 325–338, 2009.

memory algorithms – buffer management and scalability. Several database mining algorithms [5] have been proposed for transactional data mining demonstrating portability of code, scalability, and the ability to mine directly over relations where the data is collected in the first place. Simply put, database mining brings algorithms to data instead of taking data to algorithms as is the case for conventional mining. This paper presents an SQL-based approach to graph mining (specifically, interesting substructure discovery. The most general form of graphs including directed edges, multiple edges between nodes, and cycles are handled by our approach. Our primary goal in this work has been to address scalability, and map difficult and computationally expensive problems such as pseudo duplicate elimination, canonical labeling, and isomorphism checking into SQL-based counterparts. The notion of minimum description length (MDL) has been cast into corresponding metric using relational representation. Our experimental analysis shows that graphs with millions of nodes and edges can be handled by the proposed algorithm. Hierarchical reduction used for inferring second order and meta concepts is also handled by our approach.

The rest of the paper is structured as follows: Section 2 presents the HDB-Subdue algorithm along with graph representation and candidate generation. section 3 details the key steps – pseudo duplicate elimination, canonical labeling, and counting using SQL on the relational representation. Section 4 illustrates the experimental analysis and comparison with Subdue and other relational approaches. Section 6 briefly places our work in the context of graph mining. Section 7 contains conclusions.

2   Relational Approach to Graph Mining

HDB-Subdue [6] closely follows the Subdue algorithm in candidate generation (expanding substructure of size n to size n+1) using joins. However pseudo duplicate elimination and canonical ordering are the difficult components to map to SQL. Algorithm 1 outlines the steps in HDB-Subdue

---

Algorithm 1. HDB-Subdue Steps

Bulk load vertices and edges from flat file into vertex and edge table
for $i = 1$ to $numOfCompressions$ do
   Create oneedge table from vertex and edge table
   for $j = 1$ to $MaxSize$ do
      Join instance_j table and oneedge table to generate instance_j+1
      Canonically order instance_j+1 table on vertex numbers
      Eliminate pseudo duplicates from instance_j+1
      Canonically order instance_j+1 on vertex labels
      Group By vertex and edge labels to obtain instance count and insert into sub_fold_j+1
      Calculate DMDL Value for substructures in sub_fold_j+1
      Apply Beam on sub_fold_j+1 to pick top 'Beam' number of substructures
      Retain only the instances of substructures in current sub_fold_j+1
   end for
   Pick the best substructure based on highest DMDL Value
   Compress the graph by removing the vertices and edges that are in the best substructure
   if $achievedCompression = ZERO$ **then**
      Break
   end if
end for

Graph Representation and Candidate Generation: We first describe how graphs are represented in a database and expanded to generate subgraphs of different sizes. Since databases only support relations, we need to represent a graph as a tuple in a relation. The vertices in the graph are inserted into a relation called Vertices, and the edges are inserted into a relation called Edges. For the graph shown in Figure 1, the corresponding vertices and the edges relations are shown in Table 1 and Table 2.

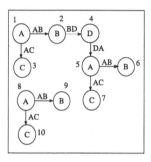

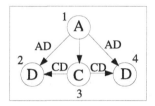

Fig. 1. Example Graph 1   **Fig. 2.** Example graph 2

From the vertices and edges relations, one edge substructures are created by joining the *vertices* and *edges* relations into a relation called *oneedge* (shown in Table 5). This table contains all substructures of size one. For a one edge substructure, the edge direction is always from the first vertex to the second vertex. In order to represent a graph completely, it is necessary to represent all the edge details including their directionality. Merely representing the directionality does not allow one to represent multiple edges between two vertices with the same edge label. In order to support multiple edges, we assign a unique edge number for each edge in the *edges* table. In general, substructures of size i are generated by joining instance_(i-1) relation with oneedge relation. In case of substructures that have 2 or more edges, we would need attributes to denote the direction of the edges. The From and To (F and T for short) attributes in the instance_n table serve this purpose. An n edge substructure is represented by n+1 vertex numbers, n+1 vertex labels, n edge numbers, n edge labels and n From and To pairs. In general, 6n+2 attributes are needed to represent an n-edge substructure. Note that the edge numbers are not part of the input. Edge numbers are assigned by the system to distinguish between edges between the same vertices and have the same edge label. Though edge

Table 1. Vertices

Vert No	Vert Name
1	A
2	B
3	C
4	D
.	.
10	C

Table 2. Edges

Vert 1	Vert 2	Edge
1	2	AB
1	3	AC
2	4	BD
.	.	.
8	10	AC

Table 3. Instance_2

V1	V2	V3	VL1	VL2	VL3	E1	E2	EL1	EL2	F1	T1	F2	T2
1	2	4	A	B	D	1	3	AB	BD	1	2	2	3
1	2	3	A	B	C	1	2	AB	AC	1	2	1	3
2	4	5	B	D	A	3	4	BD	DA	1	2	2	3
4	5	6	D	A	B	4	5	DA	AB	1	2	2	3
5	6	7	A	B	C	5	6	AB	AC	1	2	1	3
4	5	7	D	A	C	4	6	DA	AC	1	2	2	3
8	9	10	A	B	C	7	8	AB	AC	1	2	1	3

Table 4. Sub_Fold_2

VL1	VL2	VL3	EL1	EL2	F1	T1	F2	T2	COUNT	DMDL
A	B	C	AB	AC	1	2	1	3	3	1.8
A	B	D	AB	BD	1	2	2	3	1	0.9
B	D	A	BD	DA	1	2	2	3	1	0.9
D	A	B	DA	AB	1	2	2	3	1	0.9
D	A	C	DA	AC	1	2	2	3	1	0.9

Table 5. Oneedge table

V1	V2	EdgeNo	EdgeLabel	V1 Name	V2 Name
1	2	1	AB	A	B
1	3	2	AC	A	C
2	4	3	BD	B	D
4	5	4	DA	D	A
5	6	5	AB	A	B
5	7	6	AC	A	C
8	9	7	AB	A	B
8	10	8	AC	A	C

Table 6. Updated Sub_Fold_2

VL1	VL2	VL3	EL1	EL2	F1	T1	F2	T2	COUNT	DMDL
A	B	C	AB	AC	1	2	1	3	3	1.8

numbers are part of every instance_n table, owing to the space constraint, we will be showing it only in sections where they are necessary. Instance_2 relation for the graph in Figure 1 is shown in Table 3. The Fi and Ti are relative to the substructure. That is, the values of the attributes Fi and Ti do not indicate the actual vertex numbers but the attributes whose value correspond to the vertex numbers. Once substructure instances of a particular size has been generated, we need to find isomorphic substructures and count them. If everything is in order, projecting on the vertex labels, edge labels and connectivity attributes in the instance_n table (i.e., not including the vertex number and edge number) and doing a GROUP BY on the same attributes will produce the count of each substructure. Using this count we calculate the DMDL [1] value of each substructure and insert it into a table called Sub_Fold_n. Sub_Fold_2 is shown in Table 4. Substructures of count 1 are eliminated from the Sub_Fold_2 as they do not contribute to the repeated higher edge substructures (using the principle of subsumption which is valid for subgraphs) The updated Sub_Fold_2 is shown in Table 6. Removal of single instance edges (based on the subsumption property) is applicable to exact match. Often we may be interested only in expanding k best substructures. To achieve this we sort the Sub_Fold_n in the descending order of the evaluation metric (DMDL or Count) and retain only the top k substructures (k corresponds to the beam parameter value in Subdue). In order to get at the instances corresponding to the substructures in Sub_Fold_n, we join the updated Sub_Fold_n with Instance_n and insert the resulting instances into another table called InstanceIter_n. Only the instances present in InstanceIter_n participate in the next level of expansion.

3   Details of HDB-Subdue

Although a single edge substructure (substructure of size 1) can be readily represented using a few attributes, for larger substructures, edge connectivity, directions of the edge are very important to check for duplicates and isomorphism. Further, there is a need to manipulate them in various ways to compensate for

Table 7. HDB-Subdue instances

V1	V2	V3	V4	VL1	VL2	VL3	VL4	EL1	EL2	EL3	F1	T1	F2	T2	F3	T3
1	2	3	-	A	D	C	-	AD	AC	CD	1	2	1	3	3	2
1	4	3	-	A	D	C	-	AD	AC	CD	1	2	1	3	3	2
1	4	3	2	A	D	C	D	AD	AC	CD	1	2	1	3	3	4

unconstrained expansion. Based on these requirements, the following representation has been used[1].

3.1 Connectivity Map

It is necessary and sufficient to have both originating and terminating vertex numbers for each edge in the substructure. Consider the graph shown in Figure 2.

Table 7 shows the connectivity map where all its instances of size 3 are properly represented. When we project on the vertex and edge labels we will get the correct count of instances if there are any duplicates (*assuming that the subgraphs have been ordered canonically*).

The number referred by the connectivity attributes correspond to the column position where the vertex occurs and does not refer to the actual vertex number. Also in case of cycles or multiple edges – when the vertices repeat – we set the repeating vertex's vertex number and vertex label to vertex invariant markers, '0' and '-' respectively. In the connectivity map, when referring to a vertex index, we always use the first occurrence of that vertex and never point to the attributes containing the vertex invariant markers.

3.2 Iterative Expansion/Duplicate Elimination

The substructure expansion in HDB-Subdue is not constrained[2] to make sure that all possible substructures of size i+1 are generated from substructures of size i. This leads to the possibility of the same i+1 size substructure being generated in more than one way. For example, let us consider a two edge expansion for the graph in Figure 3. The edge 'AB' can grow into 'AB, CA' or the edge 'CA' can grow into 'CA, AB' as shown in Table 8. These two substructures are essentially the same substructures but has been grown in two different ways. We term these as pseudo duplicates as they are indeed duplicates once the vertex and connectivity are rearranged in the table without changing the graph structure. Similarly 'AD' can grow into 'AD, CA' or 'CA' can grow into 'CA, AD'. In general an n edge substructure (with n+1 vertices) can grow into an n+1 edge substructure in n+1 (one way from each vertex) ways. If we do not eliminate the pseudo duplicates, the same substructure will come out as different ones. Eventually, this will result in some genuine substructures being pruned when we apply the beam (choose the top beam structures to expand in the next iteration as is done in Subdue).

---

[1] Our earlier representation in EDB-Subdue had less number of attributes, but made assumptions on connectivity which failed to identify all duplicates.

[2] If the expansion is constrained in any way, there is no guarantee that all substructures will be generated for an arbitrary graph input.

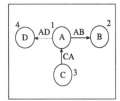

Fig. 3. Pseudo duplicates

Table 8. Instance table - Before canonical ordering

Id	V1	V2	V3	VL1	VL2	VL3	EL1	EL2	F1	T1	F2	T2
1	1	2	3	A	B	C	AB	CA	1	2	3	1
2	3	1	2	C	A	B	CA	AB	1	2	2	3
3	1	3	4	A	C	D	AD	CA	1	3	2	1
4	3	1	4	C	A	D	CA	AD	1	2	2	3

In order to identify two or more substructure instances as pseudo duplicates of each other, vertex numbers and the connectivity attributes are necessary. If two instances have the same vertex numbers and edge directions then we can identify them as pseudo duplicates. In SQL we can identify the pseudo duplicates only if the vertex numbers and connectivity map of all the instances are canonically ordered. Since databases do not allow rearrangement of columns, to obtain canonical ordering, we have to transpose the rows of each substructure into columns, sort them and reconstruct them to get the canonical order.

Owing to the table space constraints, canonical ordering of only the second and fourth instance of Table 8 are shown below, as the first and third instances are already sorted on vertex numbers. We project the vertex numbers and vertex names from the instance table and insert them row wise into a relation called unsorted as shown in Table 9. We also include the position in which the vertex occurs in the original instance. To differentiate between the vertices of different instances we carry the primary key Id from the instance table onto the unsorted table. Next we sort the table on *Id and vertex number* and insert it into a table called Sorted as shown in Table 10 with its *New* attribute pointing to the new position of the vertex within its instance (identified by *Id*) and the attribute *Old* pointing to the old position of the vertex.

Similarly the connectivity attributes are also transposed into a table called Old_Ext as shown in Table 11. Since the sorting on vertex numbers has changed its position we need to update the connectivity attributes to reflect this change. Therefore we do a 3-way join of two copies of Sorted and one copy of Old_Ext tables on the Old attribute of the Sorted table to get the updated connectivity attributes which we call New_Ext as in Table 12. Next we sort the New_Ext table on Id and the attributes F (From vertex) and T (Terminating vertex) to create Sorted_Ext as in Table 13.

Now that we have the ordered vertex as well as connectivity map tables, we can do a $2n+1$ way join (where n is the current substructure size) of $n+1$ Sorted tables and n Sorted_Ext tables to reconstruct the original instance in the

Table 9. Unsorted

Id	V	VL	Pos
2	3	C	1
4	3	C	1
2	1	A	2
4	1	A	2
2	2	B	3
4	4	D	3

Table 10. Sorted

Id	V	VL	Old	New
2	1	A	2	1
2	2	B	3	2
2	3	C	1	3
4	1	A	2	1
4	3	C	1	2
4	4	D	3	3

Table 11. Old_Ext

Id	EL	F	T
2	CA	1	2
2	AB	2	3
4	CA	1	2
4	AD	2	3

Table 12. New_Ext

Id	EL	F	T
2	CA	3	1
2	AB	1	2
4	CA	2	1
4	AD	1	3

Table 13. Sorted_Ext

Id	EL	F	T
2	AB	1	2
2	CA	3	1
4	AD	1	3
4	CA	2	1

Table 14. Instance table - After canonical ordering

Id	V1	V2	V3	VL1	VL2	VL3	EL1	EL2	F1	T1	F2	T2
1	1	2	3	A	B	C	AB	CA	1	2	3	1
2	1	2	3	A	B	C	AB	CA	1	2	3	1
3	1	3	4	A	C	D	AD	CA	1	3	2	1
4	1	3	4	A	C	D	AD	CA	1	3	2	1

canonical order. Table 14 shows the substructures after canonically ordering the vertex numbers and the connectivity attributes. After having the instances ordered, a GROUP BY on the vertex numbers and the connectivity attributes will bring all the pseudo duplicates together and we can retain the instance with highest Id value and eliminate the rest. The choice of highest Id value is arbitrary and one could have chosen smallest Id value too.

The pseudo duplicate identification and removal essentially eliminates identical graphs represented by two different tuples as they were expanded in different ways. This turns out to be computationally expensive (2n+1 joins and group by). We are currently investigating an efficient alternative.

3.3   Canonical Label Ordering

Since the substructure expansion is unconstrained it is likely that two *instances* of the *same structure* start from different initial edge and grow in a different order. Although these instances are similar they may not group together when counting for instances. For example consider a two edge substructure 'AB, AC' in Figure 1. There are three instances of this substructure as shown in Table 15. When we project by the vertex and edge labels to count the number of instances of the substructure the second and third instance will group together resulting in a count of two instead of three. This is because the first instance started with the edge AB and expanded to 'AB, AC' and other two instances started with AC to grow to 'AC, AB'.

Table 15. Before ordering

V1	V2	V3	VL1	VL2	VL3	E1	E2	F1	T1	F2	T2
1	2	3	A	B	C	AB	AC	1	2	1	3
5	7	6	A	C	B	AC	AB	1	2	1	3
8	10	9	A	C	B	AC	AB	1	2	1	3

Table 16. After ordering

V1	V2	V3	VL1	VL2	VL3	EL1	EL2	F1	T1	F2	T2
1	2	3	A	B	C	AB	AC	1	2	1	3
5	6	7	A	B	C	AB	AC	1	2	1	3
8	9	10	A	B	C	AB	AC	1	2	1	3

To make the count independent of the order of substructure growth, we rearrange the vertex and edge labels (remember that vertex numbers and connectivity were rearranged for pseudo duplicate elimination) in lexicographic order so that all instances of same substructure have their vertex and edge labels occurring in the same order. The sorting is done in the same way as it was done in pseudo-duplicate elimination. Canonical label ordering, although similar to vertex and connectivity ordering, serves a different purpose. Here we are ordering the instances, so that the instances of the *same substructure*, even though grown in different order, would group together when we do a GROUP BY on the vertex and edge labels and connectivity attributes. In contrast, canonical vertex and connectivity ordering helped us identify *duplicate structures* that were due to the growing of the *same structure in different ways*.

3.4   Multiple Edges

Most often one can find double or triple bonds occurring in chemical compounds. The atoms (such as carbon, hydrogen etc.) can be treated as vertices and the double and triple bonds can be viewed as multiple edges between the vertices as shown in Figure 4. Social networks are also gaining popularity in the applications of graph mining and often one can find multiple relationships existing between two people. These examples substantiate the need for handling multiple edges correctly during graph mining..

The multiple edges between the same pair of vertices have the same starting and ending vertex number, same connectivity map and may have the same edge label. Hence, to distinguish between the multiple edges we use *edge numbers* which are unique across all the edges. From the edge number one can differentiate two substructures based on the edge numbers associated with the edges. One-edge table in Table 17 shows the *Edge No* attribute with unique edge numbers for multiple edges as well as other edges.

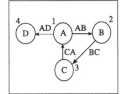

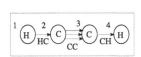

Fig. 4. Acetylene   Fig. 5. Cycle

Table 17. Oneedge table with multiple edges

Vertex1	Vertex2	EdgeNo	EdgeLabel	Vertex1Name	Vertex2Name
1	2	1	HC	H	C
2	3	2	CC	C	C
2	3	3	CC	C	C
2	3	4	CC	C	C
3	4	5	CH	C	H

Table 18. Instance table without Vertex invariants

V1	V2	V3	V4	VL1	VL2	VL3	VL4	EL1	EL2	EL3	F1	T1	F2	T2	F3	T3
1	2	3	1	A	B	C	A	AB	BC	CA	1	2	2	3	3	4
2	3	1	2	B	C	A	B	BC	CA	AB	1	2	2	3	3	4

Table 19. Instance table with Vertex invariants

V1	V2	V3	V4	VL1	VL2	VL3	VL4	EL1	EL2	EL3	F1	T1	F2	T2	F3	T3
1	2	3	0	A	B	C	-	AB	BC	CA	1	2	2	3	3	1
2	3	1	0	B	C	A	-	BC	CA	AB	1	2	2	3	3	1

Table 20. Instance table after canonical ordering

V1	V2	V3	V4	VL1	VL2	VL3	VL4	EL1	EL2	EL3	F1	T1	F2	T2	F3	T3
0	1	2	3	-	A	B	C	AB	BC	CA	2	3	3	4	4	2
0	1	2	3	-	A	B	C	AB	BC	CA	2	3	3	4	4	2

3.5 Handling Cycles in Input Graph

A figure similar to one shown in Pseudo duplicate elimination section, but with
a cycle is shown in Figure 5. From this figure two valid 3-edge expansions are
possible. One starting with vertex 1, adding vertex 2 (edge AB), adding vertex 3
(edge BC) and then terminating at vertex 1 (edge CA). Another expansion can
start with vertex 2, adding vertex 3 (edge BC), adding vertex 1 (edge CA) and
then terminating at vertex 2 (edge AB). Both the expansions occur because of
HDB-Subdue's unconstrained expansion.

The instance table with these expansions is shown in Table 18. Both these
instances are essentially the same, the second being the duplicate of the first. To
identify the duplicate we would order the vertex numbers and the connectivity
attributes canonically. Even when we order the instances canonically, we will
not be able to identify the pseudo duplicate because the first instance would
have the vertex numbers ordered as '1 1 2 3' and the second instance would
have the vertex numbers ordered as '1 2 2 3'. So a GROUP BY query on the
vertex numbers and connectivity attributes will not group these two instances
together.

The solution to this problem is to mark each repetition of the vertex by vertex
invariants, '0' for a repeating vertex number and a '-' for the corresponding vertex
label. For this example, in the first instance the second occurrence of 1 is marked
with '0' and its vertex label marked with '-'. In the second instance the second
occurrence of 2 is marked with '0' and its vertex label marked with '-'. The
instance table with vertex invariant markers is shown in Table 19. Now if we
canonically order the instances by vertex numbers and connectivity attributes
the vertex numbers of the first instance, '1 2 3 0' will be ordered as '0 1 2 3',
and the vertex numbers of the second instance '2 3 1 0' will be ordered as '0
1 2 3' as shown in Table 20. Now when we group by the vertex numbers and
the connectivity attributes we can easily identify that the second instance is a
duplicate of the first and eliminate the second. Observe that the connectivity
attributes do not refer to the columns marked with vertex invariants.

4   Experimental Analysis

This section gives the experimental results of HDB-Subdue and compares it with that of Subdue (and EDB-Subdue). A configuration file is used to input parameters for one or more experiments and the results are written to a log file. Graphs and timing analysis is done using the log file. Experiments were conducted on several data sets of different sizes – from very small to very large to verify the trend of computation time as the data size increases. For the purposes of comparison with Subdue and EDB-subdue, graphs with and without multiple edges, with and without cycles were used. All the experiments were performed on a Linux machine using Oracle9i Release 9.2.0.1.0. The machine was running on dual processors with 2 GBytes of memory. All experiments were performed 4 times and the last 3 were used for computing the average time. The cold start case was discarded.

Without cycles and multiple edges: The substructures that were embedded are shown in Figure 6. The number of instances embedded is proportional to the size of the data set. We embedded roughly two times more substructures when the data set doubles. The graphs were generated by a synthetic graph generator.

The same values of parameters (Maxsize (=5), beam (=4), and iterations (=1)) were used. Additionally, we disabled the ties while using beam to limit substructures, and the substructure evaluation metric was set to DMDL. Subdue, EDB-Subdue and HDB-Subdue discovered the same substructures with all the embedded instances. The performance comparison is shown in Figure 7. The X-axis shows the data set size and the Y-axis shows the running time in seconds. From the graph we can observe that Subdue performs well for very small data sets, but slows down when the data set grows more than 2500 vertices and 5000 edges. From this point HDB-Subdue performs better than Subdue. Also we can notice that EDB-Subdue performs slightly better than HDB-Subdue because it does not perform pseudo-elimination, canonical ordering and other functions that HDB-Subdue does. A log scale is used to keep the presentation meaningful to understand.

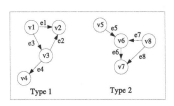

Type 1   **Type 2**

Fig. 6. No cycles, multiple edges

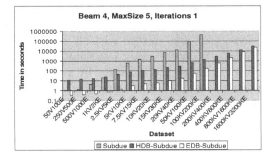

Fig. 7. No cycles, multiple edges

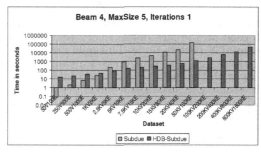

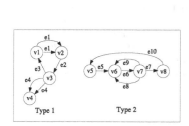

Fig. 8. With cycles, multiple edges   **Fig. 9.** Graph with cycles and multiple edges

Multiple Edges and Cycles: We present the experiments performed on Subdue and HDB-Subdue for data sets in which cycles and multiple edges are present. EDB-Subdue is not compared because it does not handle multiple edges in the input. The embedded substructures are shown in Figure 8. Same parameter settings as in the previous experiments are used.

Subdue and HDB-Subdue discovered the same substructure and all the embedded instances. From the comparison chart shown in Figure 9, we can clearly see that HDB-Subdue starts to outperform Subdue when the input graph size crosses 2500 vertices and 5000 edges as before. We can also observe that, in the presence of multiple edges, a data set of same size takes more time to complete than the data set without multiple edges. This is due to the fact that the number of tuples generated and retained in each iteration is more in the presence of multiple edges and cycles.

5   Observations

Apart from conducting experiments on various data sets, we also analyzed the performance of different types of SQL queries and their alternatives. We also explored the use of index on frequently used tables to study the performance improvement of HDB-Subdue.

In-place Deletes: The canonical table contains the instances and their duplicates ordered by the vertex numbers and the connectivity attributes. Each instance is identified by a unique identifier attribute called Id. To identify the duplicate instances we do a GROUP BY on the Id, vertex number and canonical attributes and select the instance with the maximum Id from each group

Table 21. In-place delete Vs Join query (Time in Secs)

Module	200KE400KE In-place	200KE400KE New Table	400KE800KE In-place	400KE800KE New Table
Pseudo_2	9474.58	1103.37	36028	2229.14
Pseudo_3	676.7	284.51	2119.24	590.29
Pseudo_4	252.31	185.01	641.19	383.53
Pseudo_5	2.72	3.05	4.97	4.24

and insert it in a table called instance_N_pseudo (The choice of max Id is arbitrary. One could choose min Id too.) To delete the duplicate instance, earlier we used an in-place delete query that deletes instances whose Id is not in the instance_N_pseudo table. This query took a longer time to complete. So we tried an alternate query that creates a canonicaltemp table by joining the canonical table with the instance_N_pseudo table on the Id attribute. Since this join produces only the instances we indented to retain, we can drop the canonical table and rename the canonicaltemp table as the canonical table. The in-place delete and join query are shown below.

```
IN-PLACE DELETE: DELETE FROM canonical_N
                 WHERE Id NOT IN
                           (SELECT Id FROM instance_N_pseudo)

ALTERNATIVE: Create a new table using a join query
INSERT INTO canonicaltemp (
  SELECT c.v1, .. c.vN+1, c.v1L, .. c.vN+1L,
         c.e1 .. c.eN, c.e1L, .. eN+1L, c.f_1,
         c.t_1, ..., c.f_N, c.t_N, c.Id
  FROM canonical_N c , instance_N_pseudo i
  WHERE c.Id = i.Id)

DROP TABLE canonical_N
CREATE TABLE canonical_N
   AS SELECT * FROM canonicaltemp
```

The comparison of running time of HDB-Subdue using the In-place query versus creating a new table using join query is show in Table 21. The table shows the results for two data sets, one with 200K vertices and 400K edges and the other with 400K vertices and 800K edges. We can observe from the table that the time taken by the pseudo duplicate elimination module in each iteration (the entries Pseudo_1, Pseudo_2, etc..) for join query is much lesser than that of the in-place delete query, corresponding to 85-90% improvement in iteration 2 and 50-60% on an average in all iterations.

Correlated Queries: A correlated subquery is a subquery that contains a reference to a table that appears in the outer query. During earlier phase of implementation of this algorithm, we used a correlated subquery to delete edges that appear only once, from the one-edge relation. The beamdel relation contains the edges that appear only once. So a delete query identifies those substructures of one-edge that appear in beamdel table and deletes them. The query for this delete is given below.

```
DELETE FROM oneedge o
WHERE EXISTS (
   SELECT *
   FROM beamdel b where b.v1L = o.v1L and
        b.v2L=o.v2L and b.e1L=o.Label)

DELETE FROM oneedge o
WHERE edgeno IN (SELECT edgeno
                 FROM beamdel)
```

Table 22. Performance of Correlated queries (TimeUnit: Sec)

	15KV30KE	15KV30KE	50KV100KE	50KV100KE
Module	Non-Correlated	Correlated	Non-Correlated	Correlated
Iter1_1	19.25	77.94	57.38	694.36

Table 23. Performance using Indexes (Time Unit: Seconds)

	15KE30KE	15KE30KE	50KE100KE	50KE100KE
Module	Without	With	Without	With
Total	311.41	319.62	1844.1	1885.72

The second query shown above is a non-correlated alternative for the correlated query to do the same function. The time taken in seconds for the correlated versus the non-correlated query with the percentage improvement is shown in Table 22. From this we can conclude that it is useful to avoid correlated queries and adopt an alternate form without correlation (a join query) where possible.

Using Indexes: To speed up the execution of queries, one option is to create an index on the tables that are repeatedly used. One of the tables on which an index was created is the one-edge table, because it is joined with instance_i during each expansion. We also created indexes on Sorted and the Sorted_Ext tables because we perform a 2n+1 way join during instance_i table reconstruction in pseudo-duplicate elimination step. Contrary to our expectations the performance of HDB-Subdue did not improve. We used the data set without multiple edges and cycles shown in Figure 6 with a beam of 4 and for one iteration. The running times of the individual modules for 15KV30KE data set and 50KV100KE data set without and with indexing is shown in Table 23. From the total time, we can see that the performance dropped down by little more than 2%. Usage of index needs to be carefully analyzed and cannot be taken for granted.

6 Related Work

During the past decade, the field of graph mining has emerged as a novel field of research, investigating interesting research issues and developing algorithms for challenging real-life applications [8]. Seminal work in this area include Subdue [2], FSG [4], gSpan [9], AGM [3,8], and PrefixScan [7]. Subdue [2] identifies the best substructure in a graph (or a forest) that compresses the graph using the minimum description length (or MDL). MDL has been shown to be domain independent and uses the information theoretic measure of the number of bits used to represent the graph. The downside in Subdue main memory algorithm is its inability to scale to large problems. This motivated the development of a preliminary version EDB-Subdue [1], a graph mining algorithm using SQL and stored procedures which did not address cycles, duplicate elimination, and canonical label ordering. FSG [4] (Frequent SubGraphs) mines frequent subgraphs from a set of graphs using user-supplied support. FSG is analogous to frequent-itemset mining except that it is applied to graphs instead of transasctions. gSpan [9] (graph based Substructure pattern mining) is an alternative approach to perform frequent graph mining. It overcomes some of the challenges faced by Apriori-like algorithms. AGM [3] (Apriori based Graph Mining) approach mines induced frequent subgraphs in graph structured transactions. Similar to market-basket Analysis, the interestingness of a subgraph is defined along the support and the confidence of the Apriori algorithm.

7   Conclusions

In this paper, we have presented an scalable, SQL-based graph mining approach. We have shown how some of the complex aspects (pseudo duplicate elimination and canonical labeling) can be accomplished efficiently in SQL. Our observations shed light on optimization aspects.

References

1. Balachandran, R., Padmanabhan, S., Chakravarthy, S.: Enhanced DB-subdue: Supporting subtle aspects of graph mining using a relational approach. In: Ng, W.-K., Kitsuregawa, M., Li, J., Chang, K. (eds.) PAKDD 2006. LNCS, vol. 3918, pp. 673–678. Springer, Heidelberg (2006)
2. Cook, D.J., Holder, L.B.: Graph-based data mining. IEEE Intelligent Systems 15(2), 32–41 (2000)
3. Inokuchi, A., Washio, T., Motoda, H.: Complete mining of frequent patterns from graphs: Mining graph data. Mach. Learn. 50(3), 321–354 (2003)
4. Kuramochi, M., Karypis, G.: Frequent subgraph discovery. In: ICDM 2001: Proceedings of the 2001 IEEE International Conference on Data Mining, Washington, DC, USA, pp. 313–320. IEEE Computer Society Press, Los Alamitos (2001)
5. Mishra, P., Chakravarthy, S.: Performance evaluation and analysis of k-way join variants for association rule mining. In: BNCOD, pp. 95–114 (2003)
6. Padmanabhan, S.: HDB-Subdue: A relational database approach to graph mining and hierarchical reduction. Master's thesis, Department of Computer Science and Engineering, University of Texas at Arlington /Students/sharma/theses/Pad05MS.pdf (December 2005), http://itlab.uta.edu/ITLABWEB
7. Pei, J., Han, J., Mortazavi-Asl, B., Pinto, H., Chen, Q., Dayal, U., Hsu, M.: Prefixspan: Mining sequential patterns by prefix-projected growth. In: Proceedings of the 17th International Conference on Data Engineering, Washington, DC, USA, pp. 215–224. IEEE Computer Society Press, Los Alamitos (2001)
8. Washio, T., Motoda, H.: State of the art of graph-based data mining. SIGKDD Explor. Newsl. 5(1), 59–68 (2003)
9. Yan, X., Han, J.: gspan: Graph-based substructure pattern mining. In: ICDM 2002: Proceedings of the 2002 IEEE International Conference on Data Mining (ICDM 2002), Washington, DC, USA, p. 721. IEEE Computer Society Press, Los Alamitos (2002)

Mining Violations to Relax Relational Database Constraints[*]

Mirjana Mazuran, Elisa Quintarelli, Rosalba Rossato, and Letizia Tanca

Dipartimento di Elettronica e Informazione – Politecnico di Milano
{mazuran,quintare,tanca}@elet.polimi.it, rosalba.rossato@gmail.com

Abstract. Frequent constraint violations on the data stored in a database may suggest that the represented reality is changing, and thus the database does not reflect it anymore. It is thus desirable to devise methods and tools to support (semi-)automatic schema changes, in order for the schema to mirror the new situation. In this work we propose a methodology and the RELACS[1] tool, based on data mining, to maintain the domain and tuple integrity constraints specified at design time, in order to adjust them to the evolutions of the modeled reality that may occur during the database life. The approach we propose allows to isolate frequent and meaningful constraint violations and, consequently, to extract novel rules that can be used to update or relax the no longer up-to-date integrity constraints.

1 Introduction

The information related to a specific reality is represented in a database by means of *data*; the correct interpretation and correlation of the stored data allows one to capture information about the represented reality. For these reasons, data stored in a database have to satisfy certain semantic conditions that are usually observed in the underlying application context as well. The real-world semantics of a database instance is expressed by the schema plus other properties, called *integrity constraints*. Typically, a database management system checks the satisfaction of integrity constraints and rejects those updates that violate them. However, there exist several database applications where data are not necessarily always consistent[13,19].

A first and important example of a possible data-inconsistent scenario is the case of data integration: it is possible that, while each data-source satisfies the constraints when considered separately from other data-sources, after the integration step some constraints do not hold any more on the integrated scenario. Another typical case is the modification produced in the reality by a change in government policies: for example, in Italy a recent change in the school regulations allowed 5-year-old children to access primary school, which was forbidden

---

[*] This research is partially supported by the Italian MIUR project ARTDECO and by the European Commission, Programme IDEAS-ERC, Project 227977-SMScom.
[1] RElational vioLation Analysis for Constraint Satisfaction.

T.B. Pedersen, M.K. Mohania, and A M. Tjoa (Eds.): DaWaK 2009, LNCS 5691, pp. 339–353, 2009.

before. Accordingly, the constraint saying that only 6-year-old pupils can be enrolled must be relaxed.

One of the main related problems that have been investigated is query answering in the presence of inconsistency – e.g., when a double value is inserted for the same object [6,11]. Its solution may be complex and costly, and the related information loss undesirable. Several works, proposed in the database literature, try to solve or reduce inconsistency and define different notions of consistent query answering [14,15,16,20,9]; thoroughly discussed in Section 5. Differently from the above cited works, this paper's main goal is to propose a data mining methodology to modify integrity constraints, in order to adjust them to the evolutions of the modeled reality that may occur during database life. In particular, we focus our attention on *frequent* attempts at constraint violations; in our opinion frequent violations of the same constraints may suggest that the semantics of the represented reality is changing with respect to what has been statically modeled at design time. Our idea is to analyze such frequent violations, and learn from them in order to produce "relaxed" constraints that they satisfy. Data mining algorithms allow us to automatically discover frequent correlations of values in the data, thus represent an efficient and effective alternative to querying the data sources for discovering constraint violations – i.e., facts satisfying the denial form of the constraint [9].

Contributions

We propose a strategy and a tool, RELACS, to keep trace of attempted constraint violations, in order to compare the legal information (contained in the current database) with the violating one –the so-called *anomalies*– and to propose to the user modifications of the integrity constraints which are frequently violated. The main result is the application of data mining algorithms (i) to isolate the frequent and significant anomalies, which represent constraint violations that have to be considered, (ii) to extract itemsets that can be used to update/relax tuple and domain integrity constraints and, consequently, (iii) to restore into the database the formerly violating tuples which now do not violate the (new) constraints any more.

In this work we focus on tuple constraints, and suppose that violations be injected into the database by means of tuple insertions or modifications: a violation that is caused by an update operation is considered as the deletion of the original tuple followed by the insertion of a new, violating one. A prototype of RELACS has been implemented, to test the effectiveness of our approach.

Structure of the paper

Section 2 sets the background on the relational database model, integrity constraints and data mining concepts. Section 3 describes how, by means of data mining techniques, we are able to extract frequent constraint violations. The relaxing phase of the analyzed constraints and the main properties of our approach are presented in Section 4. Our contribution with respect to the related studies is presented in Section 5. Section 6 presents the RELACS prototype and some experiments performed on a real relational dataset. Future work and concluding remarks are reported in the last section.

2   Background and Notation

In this section we summarize some background definitions and notations that are commonly accepted in database theory [11,17].

Let $\mathcal{U}$ be a finite set of attribute names; we denote attributes by capital letters from the beginning of the alphabet (e.g., A, B, C, A_1, etc.), while capital letters from the end of the alphabet (e.g., U, X, Y, Z, X_1, etc.) are used to denote sets of attributes. Let $\mathcal{D}$ be a finite set of domains, each containing atomic values for the attributes; the domain D_j contains the possible values for the attribute A_j. Without loss of generality, in the following each D_i may be either the domain $\mathcal{C}$ of uninterpreted constants or the domain $\mathcal{N}$ of numbers. Moreover, assume that domain elements with different names be different. The set $\mathcal{OP} = \{=, \neq, <, >, \leq, \geq\}$ contains the numerical predicates, while the domain $\mathcal{C}$ has only equality ($=$) and inequality ($\neq$) as built-in predicates. The special value null, considered to be included both in $\mathcal{C}$ and $\mathcal{N}$, is used to represent the null value. A relation (or relation state or instance) r of the relation schema $R(A_1, A_2, \ldots, A_n)$, is a finite set of tuples of the form $t = (v_1, v_2, ..., v_n)$, where each value v_k, $1 \leq k \leq n$, is an element of D_k. We recall that $t[A_i]$ denotes the value assumed by the attribute A_i in the tuple t (i.e., v_i), and, given an instance r on $R(U)$ and a subset X of U, $R[X]$ is the set of tuples of r obtained by considering only the values assumed by the attributes X.

Given a set X of attributes, let $\bar{x}$ represent the associated variables; we deal with the following *intra-relation* integrity constraints:

- *Domain constraints*: Let A_j be an attribute of the relational schema $R(A_1, \ldots, A_n)$, and γ_j a formula describing a simple constraint of the form $\gamma_j \overset{\text{def}}{=} x_j \theta_k v_k$ where $v_k \in D_j$ and $\theta_k \in \mathcal{OP}$. Let Γ_j be an n-ary conjunction of simple basic constraints on the possible values assumed by the attribute A_j: $\Gamma_j \overset{\text{def}}{=} \bigwedge_{h=1}^{n} \gamma_{j_h}$
  The general $\mathcal{L}_\mathcal{R}$-formula for the description of a domain constraint on the attribute A_j is a (m-ary) disjunction of non-contradictory conjunctions [2] of basic formulae γ_j: $\forall x_1, \ldots, x_j, \ldots, x_n.\ R(x_1, \ldots, x_j, \ldots, x_n) \rightarrow \bigvee_{k=1}^{m} \Gamma_{j_k}$
  A special case of domain constraint is the *Not-Null constraint*, where the right-hand side part of the implication γ_j is composed by one basic constraint of the form $x_j \neq$ null.
- *Tuple constraints* are a generalization of domain constraints, involving more than one attribute of the same tuple. The right hand side of the implication (1) is a disjunction of conjunctions of γ formulae, containing comparisons between attributes of R or between an attribute and a value of the corresponding domain[3]. In the rest of the paper we will use the term "tuple constraint"

---

[2] A conjunction of Γ-formulae on the attribute A_j is contradictory when it reduces the set of admissible values for A_j to the empty set.

[3] We do not deal with tuple constraints imposing that the value of an attribute be obtained as an (arithmetic) expression on the values assumed by other attributes (e.g, in a database storing information about goods, the net weight of a good is obtained as gross weight minus tare). This limitation is due to the fact that data mining algorithms extract values but do not perform computations on them.

for both domain and tuple constraints, and denote by $\forall \overline{x}.R(\overline{x}) \to \Gamma(\overline{y})$ with $\overline{y} \subseteq \overline{x}$, their general form.

Our proposal to relax constraints makes use of itemsets extracted from datasets. Itemsets describe the co-occurrence of data items in a data collection [2]. In our framework, a data item is a pair *(attribute,value)*. An itemset is a set of correlated data items that occur together, and each itemset is characterized by the frequency of the co-occurrence in the dataset, which is called *support*. An itemset formed by k co-occurring data items is called k-itemset, and we call k the *itemset dimension*.

Example 1. Suppose we have the following relational schema $\mathcal{R}$ of a school:

Pupil(P_ID, Pname, Psurname, Pbirthyear, sex, citizenship, Paddress, Pcity, Pprovince)
Teacher(T_ID,Tname,Tsurname,Tbirthyear,Taddress,Tcity,Tprovince,role)
Enrolment(P_ID,Penrolyear,class,section)
Course(T_ID,year,subject,class,section)
SchoolReport(P_ID,Pschoolyear,subject,mark)
Subject(S_ID,description,program,area)

Besides the primary key and the intuitive foreign key constraints, assume that the following additional integrity constraints be defined on this schema:

$$\forall s, d, p, a.Subject(s, d, p, a) \to (p \neq \texttt{null}) \tag{C_1}$$

$$\forall s, d, p, a.Subject(s, d, p, a) \to (a = math \vee a = art \vee a = sport) \tag{C_2}$$

$$\forall p, y, s, m.SchoolReport(p, y, s, m) \to (m = A \vee m = B \vee m = C \vee m = D) \tag{C_3}$$

That is, constraint C_1 requires that the program of a subject cannot be null, constraint C_2 establishes the admissible values for the area of a subject, and constraint C_3 states that the mark assigned to a pupil, for a given subject, must belong to the set $\{A, B, C, D\}$. $\diamond$

3   Mining Constraint Violations

In this section we describe our proposal to extract – by means of data mining techniques – frequent constraint violations, and then analyze interesting anomalies that will be used to relax constraints that are no longer up-to-date.

3.1   Storing Constraint Violations

In general, when we access a database by using one of the common DBMSs, user operations that violate the design-time defined constraints are rejected. Our strategy keeps trace of the rejected insertions and updates. For each such operation, we keep the *violating tuple*, i.e., in case of INSERT operation, the tuple the user has attempted to insert, in case of UPDATE, the tuple which would have resulted after the update. The set of violating tuples is analyzed by using mining techniques to find out frequent anomalies.

We assume the user is "not intentionally cheating"; that is, s/he does not try to correct an operation that has been previously aborted by the DBMS, for instance by inserting new attribute values with a different semantics w.r.t. those causing the violation. This assumption is needed because in the mining phase we use both the violating tuples and the tuples of the database, thus for the approach to be effective the database must adhere to the constraints' intended semantics. Consider the case of a user who tries to insert, into the *Subject* table, information about a course having a null program, because the program for that course does not exist. We do not want to consider the possibility that, in this case, after the violation notification, the user tries to insert a new tuple with a fictitious program (e.g., "Program XYZ") because our aim is to analyze the semantics of violations in order to adjust constraints that are too strict (e.g., the outcome of the process is to admit null programs only for certain courses).

According to our technique, for each relation R an associated relation with the same schema, R^v, is created, containing the violating tuples. By storing violations, it is possible to compare the information contained into the original instance r with the violating one, represented by the tuples of r^v.

Let the pair $(r, \mathbb{IC})$ denote the instance r with the associated set of (intra and inter) integrity constraints $\mathbb{IC}$. A tuple s is in r^v if it contains a violation to some constraint(s) of $\mathbb{IC}$. For each constraint $C \in \mathbb{IC}$, let r^v_C be the set of tuples of r^v violating C.

Our proposal can be sketched as follows.

- For each constraint $C \in \mathbb{IC}$, we isolate frequent and significant anomalies, i.e., those tuples that appear in r^v with a support greater than a fixed threshold sup_{min}. Consequently, we extract itemsets $\mathfrak{I}$ that can be used to relax the original set of integrity constraints $\mathbb{IC}$.
- Once the extraction phase is completed, a relaxation phase is applied, and each constraint C which has been relaxed into a new constraint, denoted by C^*, is submitted to the designer for approval.
- For each constraint $C \in \mathbb{IC}$, if C^* has been approved by the designer, the tuples of r^v used to mine the useful itemsets $\mathfrak{I}$, denoted by $r^v_{\mathfrak{I}}$, do not violate the updated constraint C^* any more, thus if they do not violate another constraint as well, they can be safely integrated within r.

It is important to highlight that $r^v_{\mathfrak{I}} \subseteq r^v_C$ because our data mining approach considers only frequent anomalies w.r.t. a constraint C, i.e., it considers only those itemsets having a support greater than the fixed threshold value.

3.2   Mining Violations for Tuple Constraints

As described in Section 2, a tuple constraint C can be written as an implication of the form $\forall \overline{x}.R(\overline{x}) \rightarrow \Gamma(\overline{y})$, with $\overline{y} \subseteq \overline{x}$.

We call **length** of C (denoted as $len(C)$), the number of variables in $\overline{y}$; constraint (C_1) of Example 1 is a constraint of length 1.

Definition 1. *Given a tuple constraint*

$$C \stackrel{def}{=} \forall \overline{x}.R(\overline{x}) \rightarrow \Gamma(\overline{y})$$

with $\overline{y} \subseteq \overline{x}$, *an itemset* $I = \{\langle A_1, v_1 \rangle, \ldots, \langle A_k, v_k \rangle\}$, *formed by* k *values for the attributes* $\{A_1, \ldots, A_k\}$ *is* C**-dependent** *iff for each* $y_i \in \overline{y}$, $\langle A_i, v_i \rangle \in I$, *with* $A_i = att(y_i)$ *and* $v_i \in D_i$.

An itemset I is C-dependent when it describes the correlation among (at least) the values assumed by all the attributes involved in $\Gamma(\overline{y})$ (i.e., the set of attributes Y), thus, I contains at least the data items $\langle A_i, v_i \rangle$ for each attribute $A_i \in Y$.

Given a tuple constraint C, we extract from the table r^v the C-dependent itemsets containing the data items which cause $\Gamma(\overline{y})$ to be false, and their correlation with other attributes values. This is because we want to see whether the anomaly is regularly associated to some specific value(s) of other attributes.

For example, for the not-null constraint $C_1 = \forall a, s, d, p$ $Subject(s, d, p, a) \rightarrow$ $(p \neq \text{null})$ we have to mine from the table $Subject^v$ the frequent itemsets that are C_1-dependent, i.e., they contain the data item $\langle Subject^v.program, \text{null} \rangle$ and highlight correlations of the attribute *program* with other attributes. In other words, we see whether the attempts to insert a null program for some subject are caused by the nature of that subject, which is expressed by its properties (the other attributes of the relation *Subject*). Suppose for instance that we mine the itemset $I_1 = \{\langle program, \text{null} \rangle, \langle description, lab \rangle\}$ with support 0.95. This might be an indication that lab courses do not provide a program, thus the constraint has to be relaxed in order to accommodate this exception.

Suppose that, for the aforementioned not-null constraint C_1, we mine also the itemset $I_2 = \{\langle program, \text{null} \rangle, \langle description, lab \rangle, \langle area, math \rangle\}$ with support 0.85. We have to understand which one, between I_1 and I_2, is a "good-candidate" for relaxing C_1, i.e., it is not too generic. In fact, even if in I_1 the *program* attribute of the *Subject* table assumes the value **null** when its *description* is *lab*, it could still be that some labs – not in the math area – have a non-null *program* in the *Subject* table. However, if this were the case, we should find some lab tuples with non-null program in the real database, which would tell us that having a null-program is not a peculiarity of all the lab subjects. Thus, further analysis of the exceptions is in order, and we discover, by mining I_2, that a particular area –the math one– co-occurs with the lab subject in the violations of the constraint C_1.

Once we choose the candidate itemset, we have to check this itemset against the database, and may discover that it is more appropriate to relax C_1 using I_2 instead of I_1. Note that I_2 has an additional condition, thus is more specific than the previous one.

We say that a C-dependent k-itemset I, with support s, is a "good-candidate" for relaxing a constraint C on R, if I is "significantly frequent", and C is violated by those tuples of r^v which contain the values specified in the items of I.

The concept is formalized by the following definition.

Definition 2. *Given a tuple constraint*

$$C \stackrel{def}{=} \forall \overline{x}.R(\overline{x}) \rightarrow \Gamma(\overline{y})$$

on R, with $\overline{y} \subseteq \overline{x}$, we say that an itemset $I = \left\{ \langle \widetilde{A_1}, \widetilde{v_1} \rangle, \dots, \langle \widetilde{A_k}, \widetilde{v_k} \rangle \right\}$ with support s_i, is a C-good candidate of dimension k w.r.t. an a-priori established minimal support sup_{min} iff

1. *I is a C-dependent k-itemset, with $k > len(C)$;*
2. *$s_i \geq sup_{min}$;*
3. *$Q \not\models C$, where $Q = \{t | t \in r^v, \forall i \in \{1, \dots, k\} . t[\widetilde{A_i}] = \widetilde{v_i}\}$.*

Note that point 1 requires that R be strictly greater than $len(C)$, because we are looking for correlations between violations and at least one more attribute. To choose the C-good candidates to relax a tuple constraint C on R, we propose the algorithm `ItemExtractor`, reported in Algorithm 1.

Algorithm 1. The pseudocode of `ItemExtractor` algorithm.

Input: a tuple constraint C on R with $len(C) = l$;
   r and r^v denoting the consistent and the inconsistent instances of R;
   k_{max} the maximal dimension of mined itemsets;
   sup_{min} minimal support value;
Output: $\mathfrak{I}$ a minimal set of itemsets representing interesting anomalies or the empty set;

1: $\mathfrak{I} = \emptyset, \Upsilon = \emptyset$
2: $k = l + 1$
3: Let $\mathtt{I_k} = \left\{ I_i \mid I_i \text{ is a C-good candidate with dimension k w.r.t. } sup_{min} \right\}$
4: **while** $k \leq k_{max}$ and $\mathtt{I_k} \neq \emptyset$ **do**
5:   $\mathtt{I_{k+1}} = \emptyset$
6:   **for all** $I_i = \{\langle A_1, v_1 \rangle, \dots, \langle A_n, v_n \rangle, \langle \widetilde{A_1}, \widetilde{v_1} \rangle, \dots, \langle \widetilde{A_l}, \widetilde{v_l} \rangle\} \in \mathtt{I_k}, \quad n + l = k,$
      $\forall y_i \in \overline{y} \, att(y_i) = \widetilde{A_i}$ **do**
7:     Let $q^*(I_i) = \left\{ t \in r \, \middle| \, \begin{array}{l} \forall j \in \{1..n\} . t[A_j] = v_j \wedge \\ \forall i \in \{1..l\} . t[\widetilde{A_i}] = w_i, \ \Gamma_{[\{y_1, \dots, y_l\} \leftarrow \{w_1, \dots, w_l\}]} \text{ is true} \end{array} \right\}$
8:     **if** $q^*(I_i) = \emptyset$ **then**
9:       $\mathfrak{I} = \mathfrak{I} \cup \{I_i\}$
10:    **else**
11:      $\Upsilon = \Upsilon \cup \{I_i\}$
12:    **end if**
13: **end for**
14: **end while**
15: **for all** $I_i \in \Upsilon$ **do**
16:   $\mathtt{I_{k+1}} = \mathtt{I_{k+1}} \cup \left\{ I_j \, \middle| \, \begin{array}{l} I_j \text{ is a C-good candidate with dimension k+1 w.r.t. } sup_{min}, \\ I_j \supseteq I_i, \forall I_x \in \mathfrak{I}. \ I_j \not\supseteq I_x \end{array} \right\}$
17: **end for**
18: $k = k + 1$

Observe that, given a constraint C, among the different itemsets satisfying $\neg C$, it is convenient to choose *minimal* itemsets to relax C. In fact, our purpose is to characterize classes of frequent violations that are as general as possible, and moreover, by considering minimal itemsets we also limit the complexity of the mining process. Thus we have a trade-off between generality – i.e., small, possibly too generic itemsets – and precision – i.e., large, but inefficient and less "inclusive" itemsets. Intuitively, in the extreme case, each itemset coincides with a unique violating tuple, and the system would propose as new constraint that includes a disjunction of the violating tuples, which of course is not our objective.

The goal of the algorithm `ItemExtractor` is thus to produce a set $\mathfrak{J}$ of good candidates that will be used to relax a constraint C. We remark that good candidates are mined *having fixed a support threshold*; thus, if no frequent itemsets representing anomalies for a constraint are found, the result of the algorithm is the empty set and the constraint cannot be relaxed.

According to Definition 2, the first step (line 3) produces the mined set I_k of frequent k-itemsets (with $k = len(C)+1$) containing the data items violating C, i.e., $\left\{\langle \widetilde{A_1}, \widetilde{v_1}\rangle, \ldots, \langle \widetilde{A_l}, \widetilde{v_l}\rangle\right\}$. For each k-itemset in I_k (cyclic construct of line 6), we check whether it is a *minimal* "good-candidate" to relax C. For this purpose, we apply the query q^* (line 7) to check if in the instance r there is any tuple t that is C-consistent, it assumes values different from $\widetilde{v_i}$ and *does not violate* the considered constraint C, but for the other attributes $A_1 \ldots, A_n$ present in the current itemset it assumes exactly the mined values $\widetilde{v_1}, \ldots, \widetilde{v_l}$. If q^* has an empty result, then the current itemset I_k is a minimal "good-candidate" and is added to the set $\mathfrak{J}$ (line 9). If not, we add I_k to an auxiliary set Υ. After considering all the k-itemset, we have to look for frequent $(k + 1)$-itemsets, extending the I_k itemsets in Υ, but not using attributes that are in k-itemsets that have been previously included in $\mathfrak{J}$. Example 2 will clarify the rationale behind this.

In order to guarantee determinism, all the mined itemsets with a given dimension k are first analyzed to check whether they are good candidates, and only after this phase non-good candidates are extended.

Example 2. Suppose that for constraint C_1 we have mined the set $S = \{I_1, I_2\} = \{\{\langle program, \text{null}\rangle, \langle description, \text{gym}\rangle\}, \{\langle program, \text{null}\rangle, \langle description, \text{lab}\rangle\}\}$

For the 2-itemset I_1, we verify that in the *Subject* table there is no tuple with **gym** value for the *description* and a non-null value for the *program*, i.e., that $q^*(I_1) = \emptyset$. This means that the relation *Subject* does not contain any gym course with a non-null program, thus, probably, a null program is allowed for all gym courses. We conclude that I_1 is a "good-candidate" that should be added to $\mathfrak{J}$.

For the 2-itemset I_2, since $q^*(I_2) \neq \emptyset$, we find a tuple t in the *Subject* table with $t[description] = \text{lab}$ and $t[program] \neq \text{null}$. Thus, we have to conclude that I_2 expresses a too generic condition, that is, there might be some further property that, associated to the description "lab", accounts for the empty program. Thus, we try to extend I_2 by mining 3-itemsets. Suppose now that we mine with a sufficient support only the 3-itemset $I_3 = \{\langle program, \text{null}\rangle, \langle description,$

$lab\rangle, \langle area, math\rangle\}$ which extends I_2. If we do not find any tuple in $Subject$ with lab value for $description$, $math$ value for the $area$ attribute, and a non-null value for $program$ (i.e., $q^*(I_3) = \emptyset$), then we add I_3 to $\Im$ and the algorithm stops, returning $\Im = \{I_1, I_3\}$.   ◇

Note that, for the previous example, line 12 specifies that I_3-itemsets must extend I_2, because I_2 is not a good candidate, but without including the 2-itemset I_1, that has already been added to $\Im$. The reason is that the condition $program = $ null $\wedge\ description = gym$ derived from I_1 will be used to relax C_1, and thus using it for further extending itemsets would be redundant.

Once again, note that line 4 of `ItemExtractor` limits the itemset we check to be below a maximum length, because the computation time increases with the itemset dimension.

Consider now a generic tuple constraint. In this case, the right part of the implication is not an atomic inequality, but a more complex combination of boolean comparisons applied to attributes of a relation R. Thus, the cause of inconsistency is not as easily recognizable as it was for the non-null case.

For example, considering constraint C_3, in line 3 of Algorithm 3.2, where C_2-good candidates are defined, I_2 is the set of 2-itemsets which contain a data item $\langle mark, m\rangle$, where m is any value that differs from A, or B, or C, or D. The same consideration holds for line 12, where the $(k+1)$-itemsets are good-candidates that must violate C_2.

A frequent 2-itemset we could obtain from the algorithm `ItemExtractor` for this constraint is $I = \{\langle mark, $ passed$\rangle, \langle subject, $ A0010$\rangle\}$ stating that the mark "passed" is frequently associated with the Subject with code "A0010". By using I to relax C_2 we obtain the constraint $C_2^* = \forall p, y, z, m.\ SchoolReport(p, y, z, m) \rightarrow (m = A \vee m = B \vee m = C \vee m = D) \vee (m = passed \wedge z = A0010)$

Theorem 1. *Soundness and Completeness. Given a pair $(r, \mathbb{IC})$ and a constraint $C \in \mathbb{IC}$,* `ItemExtractor` *produces as output the minimal set $\Im$ containing all and only the minimal C-good candidates I_i having support greater than sup_{min} and such that $q^*(I_i) = \emptyset$.*

Proof. The proof can be found in [18].

4   Relaxing Tuple Constraints

The procedure to relax violated constraints is based on the previously mined anomalies. For each tuple constraint $C \stackrel{def}{=} \forall \overline{x}.R(\overline{x}) \rightarrow \Gamma(\overline{y})$ with $\overline{y} \subseteq \overline{x}$, we consider the set of itemsets $\Im = \{I_1, \ldots, I_n\}$ produced by the algorithm `ItemExtractor`, with $I_i = \{\langle A_1, v_1\rangle, \ldots, \langle A_{m_i}, v_{m_i}\rangle\}\ \forall i \in \{1..n\}$ and modify the constraint C as follows:

$$C^* = \forall \overline{x}.R(\overline{x}) \rightarrow \left(\Gamma(\overline{y}) \bigvee_{i\in\{1,\ldots,n\}} (\bigwedge_{j\in\{1,\ldots,m_i\}} y_j = v_j)\right) \tag{1}$$

After the violated constraints C_i, with $i = 1, \ldots, n$, have been successfully relaxed by producing the new constraints C_i^*, we can re-introduce the tuples of $r_{\mathcal{J}_i}^v$ within the original instance r, in the way $r \leftarrow r \cup \{t\}$ if and only if

1. $t \in r_{\mathcal{J}_i}^v$ and
2. $\forall C_j^*. t \models C_j^*$, with $j \in \{1, \ldots, n\}$ and $j \neq i$

i.e., a tuple t originally violating a constraint C_i and such that t participates in the construction of the itemsets for deriving C_i^*, can be safely integrated in the legal instances of the relation r, only if t does not violate any other (relaxed or not) constraint C_j^*.

Example 3. Consider, for example, the two constraints C_1 and C_2 on the Subject table. Now suppose, as shown in Example 2, that the algorithm ItemExtractor returns the set $\mathcal{J}$ composed by the itemsets $I_1 = \{\langle program, \mathtt{null}\rangle, \langle description, \mathtt{gym}\rangle\}$ and $I_3 = \{\langle program, \mathtt{null}\rangle, \langle description, \mathtt{lab}\rangle, \langle area, \mathtt{math}\rangle\}$ whereas, the output for C_2 is the empty set (i.e., no frequent violations are mined). Suppose, for example, that $t = (gymcourse, gym, \mathtt{null}, gym) \in Subject_{\mathcal{J}}^v$ cannot be inserted in the *Subject* table, because it still violates constraint C_2.   ◇

We recall that, when it is not possible to perform the relaxing process (due to insufficient support of the mined itemsets), C_i^* coincides with C_i.

The computational complexity of the integration phase described above is lower or equal to $O(|r^v| \times |\mathbb{IC}^*|)$. This worst case occurs if each tuple in r^v violates all the constraints in $\mathbb{IC}$, thus, we must check the satisfiability of r^v w.r.t. each (new) constraint.

Example 4. For constraint C_1, the output of ItemExtractor is: $\mathcal{J} = \{\{\langle program, \mathtt{null}\rangle, \langle description, gym\rangle\}, \{\langle program, \mathtt{null}\rangle, \langle description, lab\rangle \langle area, math\rangle\}\}$ where the itemsets have been mined with a support 0.90 and 0.85, respectively. This means that the table $Subject^v$ contains a high number of anomalies associating a null program either to a *gym* course, or to a *lab* course in the *math* area.

Thus, we decide that these frequent anomalies indicate real life situations and consider them as a common and acceptable behavior. The constraint is relaxed as follows:

$$C_1^* \stackrel{\text{def}}{=} \forall s, d, p. \ Subject(s, d, p, a) \rightarrow \Big((p \neq \mathtt{null}) \vee \quad (p = \mathtt{null} \wedge d = \mathtt{gym}) \vee$$
$$(p = \mathtt{null} \wedge d = \mathtt{lab} \wedge a = \mathtt{math}) \Big)$$

◇

We are now ready to prove some interesting properties. We recall that the notation r_C^v denotes the set of violating tuples for the constraint $C \in \mathbb{IC}$, while $r_{\mathcal{J}_C}^v$ denotes the set of (violating) tuples used by the algorithm ItemExtractor to produce the set of itemsets $\mathcal{J}_C$ for constraint C (i.e., those tuples that will help relax the constraint C).

Definition 3. *Given a pair $(r, \mathbb{IC})$ and $C \in \mathbb{IC}$, r is said to be C-**consistent** if r_C^v is empty, and C-**inconsistent** otherwise.*

Theorem 2. *Local correctness. Given a constraint $C \in \mathbb{IC}$ and the instances r and r_C^v, let $\mathfrak{I}_C$ be the set of itemsets returned by ItemExtractor for C, and C^* the relaxed constraint. Then, if $\mathfrak{I} \neq \emptyset$, $r_{\mathfrak{I}_C}^v$ and $r \cup r_{\mathfrak{I}_C}^v$ are C^*-consistent.*

Proof. The proof can be found in [18].

Assuming that the application of ItemExtractor returns a non empty set, we are able to state:

Corollary 1. Effectiveness of the reparation. *Suppose that $C \in \mathbb{IC}$ is the only constraint on r that has been violated. Given the instances r and r^v, if r^v has cardinality n and $r_{\mathfrak{I}}^v$ has cardinality m (with $n \geq m$) then, after the relaxing phase of the constraint C with C^*, the cardinality of r^v is exactly $n - m$.*

Lemma 1. *Let C_1 and C_2 be violated constraints of $\mathbb{IC}$ such that $r_{C_1}^v \cap r_{C_2}^v \neq \emptyset$. Then $r_{\mathfrak{I}_1}^v \cap r_{\mathfrak{I}_2}^v$ is both C_1^*- and C_2^*-consistent.*

Proof. The proof can be found in [18].

After the relaxing phase of each constraint the integration phase can be safely performed and the number of tuples that remain in r^v will be greater than or equal to $n - \sum_{i=1}^{l} m_i$, where $m_i \leq |r_{\mathfrak{I}_i}^v|$.

Theorem 3. *Global correctness. Let r be both C_1-inconsistent and C_2-consistent. Then, $r \cup r_{\mathfrak{I}_1}^v$ is C_1^*-consistent and C_2-consistent.*

Proof. The proof can be found in [18].

5   Related Work

In the database literature there have been a number of proposals that investigate the problem of constraint violations. There is a basic distinction between works that consider the problem of querying inconsistent databases [10,12,6,11] – which try to re-establish consistency in an information system by (minimally) changing the facts that violate the database constraints – and proposals for modifying constraints in order to take into account abnormal data w.r.t. the reality modeled at design-time, without changing the database instance (e.g., [19]).

In [6] the authors formalize the notion of consistent information, called *consistent query answer*, obtained from a (possibly inconsistent) database in response to a user query. A database instance r' is a *repair database instance* of r with respect to a set of integrity constraints IC if (i) r' is defined on the same schema and domain of r, (ii) r' satisfies IC, and (iii) r' differs from r by a minimal set of changes (insertions or deletions) of tuples. By using the concept of *repair database*, i.e., a database that is consistent and minimally differs from the original one, the authors define a consistent answer to a query as a set of tuples that are answers (to the same query) in every repair of the considered database. Other works consider also preferences or priorities in repairing [14,15]; the main idea is to introduce a numeric utility function assigning different reliability levels to multiple data sources involved in the query answering process.

These approaches differ from ours for one main reason: in [6,14,15,10,12,11] the authors consider the problem of inconsistent databases from a query answering point of view, thus, they discard those data that can produce inconsistency with respect to a static set of integrity constraints. By contrast, our proposal's main goal is to modify the integrity constraints with the aim of maintaining the semantics of the database as adhering as possible to the changing reality. To achieve this goal, tuples that represent constraint violations are no more considered as abnormal facts, but used as a guide to update the no longer up-to-date constraints.

More similar to our proposal is [19]: the authors introduce a framework for dealing with irregularities in a human-oriented system supporting errors and also tolerating deviations from the normal behavior. In their conceptual model, when there are facts in a database that violate integrity constraints, the human user has the possibility to evaluate whether they represent an error in the facts recorded, or an exception that must be tolerated. In this latter case, the facts are not changed, but marked as "exceptional", and the violated constraints are minimally modified in order to allow the presence of those exceptional facts. On the contrary, for us only frequent anomalies have to be taken into account, in order to better understand the underlying reality.

Other works and approaches to anomaly or outlier detection have been developed in the past years, whose aim is not to modify either abnormal facts or violated constraints. They are based on database [8] or data mining techniques [4,5,7], including knowledge discovery [1], and logic programming [3].

In [3], outliers have been formalized in the context of logic programming-based knowledge systems. The authors propose a basic framework where observations (outliers) are described by means of a set of facts encoding some aspects of the current status of the world, while the background knowledge of the system is described by means of a logic program. Outliers are identified on the basis of some disagreement with the background knowledge and supported by some evidence in the observed data, called witness sets.

In [8] the authors introduce the notion of pseudo-constraints, which are not integrity constraints specified by the designer, but predicates representing schema-level properties that have significantly few violations in the database instance. The authors use this pattern to identify rare, but interesting, events in databases. The spirit of the work is similar to our main purpose; however, our approach differs for two reasons. First, they focus on cyclic pseudo-constraints and propose an algorithm for extracting this kind of cyclic pattern, whereas we mine violations to the classical notion of integrity constraint. Moreover, we also investigate how, once significant violations have been mined, the constraints can evolve (semi-) automatically.

6   RELACS Prototype and Preliminary Results

The RELACS prototype has been developed, it uses the aforementioned strategy to store and analyze tuple constraint violations in a relational databases by using the PostgreSQL DBMS. The architecture of RELACS is sketched in Fig. 1, where we show our methodological approach.

We validated our approach by means of experiments on the TPC-H relational database [21] performed on a 3.2GHz Pentium IV system with 2GB RAM, running Kubuntu 6.10. We considered (a copy of) the LINEITEM table which has about 6.000.000 tuples and whose schema is

$$\text{LINEITEM}(\text{OrderKey}, \text{PartKey}, \text{SuppKey}, \text{LineNumber}, \text{Quantity},$$
$$\text{ExtendedPrice}, \text{Discount}, \text{Tax}, ...)$$

We have imposed the two constraints (derived by analyzing the instance of the table):

- LINEITEM.Tax > 0
- LINEITEM.Discount $\geq 0 \wedge$ LINEITEM.Discount ≤ 0.5

Then, we injected violations in the table LINEITEM^v by using a suitable Jave module: in particular 1000 violating tuples were inserted, among which 200 had the values 10000 for the Quantity attribute, 0 for Tax, and 0.7 for Discount.

In the RELACS implementation, R^v actually contains an additional attribute, which is necessary in order to be able to store in R^v also primary key violations. Indeed, commercial DBMSs do not allow to create a relation without defining its primary key, thus we add a new, ad-hoc primary key attribute (e.g., an automatic counter), so that R^v can also contain tuples that violate the primary key constraint of R.

For the itemset extraction we have used a publicly available version of Apriori [2]. The good-candidates extraction produced as output the set

$$\text{I} = \{\{\langle \text{Quantity}, 10000\rangle, \langle \text{Tax}, 0\rangle\},$$
$$\{\langle \text{Quantity}, 10000\rangle, \langle \text{Discount}, 0.7\rangle\}\}$$

Each 2-item in I has been mined with support 20%.

The execution time of the 2-itemset extraction increases almost linearly with the scale factor (i.e., with the number of violating records).

By following our methodology, the two constraints are relaxed as follows:

- LINEITEM.Tax $> 0 \vee$ (LINEITEM.Tax $= 0 \wedge$ LINEITEM.Quantity $= 10000$)
- (LINEITEM.Discount $\geq 0 \wedge$ LINEITEM.Discount ≤ 0.5) $\vee$
  (LINEITEM.Discount $= 0.7 \wedge$ LINEITEM.Quantity $= 10000$)

The two relaxed constraints cited above are quite specific (due to the set of tuples inserted as violations), but a contribution on the part of the designer, who knows the semantics of the considered scenario, can further generalize the two constraints by stating, for instance, that a value 0 for TAX (or a higher DISCOUNT) is allowed for high-quantity orders. The final relaxed constraints could be:

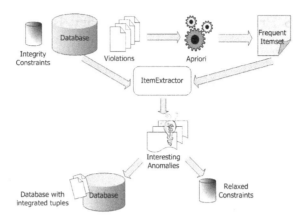

Fig. 1. The architecture of the system

- LINEITEM.Tax $> 0 \vee$ (LINEITEM.Tax $= 0 \wedge$ LINEITEM.Quantity ≥ 10000)
- (LINEITEM.Discount $\geq 0 \wedge$ LINEITEM.Discount ≤ 0.5) $\vee$
  (LINEITEM.Discount $\leq 0.7 \wedge$ LINEITEM.Quantity ≥ 10000)

7 Conclusions and Future Work

The approach we have presented in this paper proposes a mining-based support to the evolution of integrity constraints which do not reflect the reality of interest any more. The main idea of our methodology is to analyze frequent abnormal facts, which are considered as an indication of a changing reality, and learn from them the relaxing condition(s). Current research is focused on the extension of our proposal to functional and inclusion dependencies.

Currently, we are applying our methodology to peer-to-peer data integration systems; since by nature, the data coming from the various peers may be uncertain and often inconsistent with one-another.

Future work also concerns the application of this strategy to biological databases, where anomalies w.r.t. predefined constraints are very frequent. In this particular context, the problem of discovering the new "laws" of natural phenomenons plays a crucial role. By applying RELACS to this kind of data, we want to derive appropriate values for the minimal support thresholds and to analyze the performance, in term of execution time, of the itemsets extracting phase.

References

1. Abe, N., Zadrozny, B., Langford, J.: Outlier detection by active learning. In: Proceedings of Int. Conf. on Knowledge Discovery and Data Mining (KDD 2006), pp. 504–509 (2006)
2. Agrawal, R., Srikant, R.: Fast algorithms for mining association rules in large databases. In: Proceedings of the 20th Int. Conf. on Very Large Data Bases (VLDB 1994), pp. 487–499. Morgan Kaufmann Publishers Inc, San Francisco (1994)

3. Angiulli, F., Greco, G., Palopoli, L.: Outlier detection by logic programming. ACM Transaction on Computational Logic (to appear)(2006)
4. Angiulli, F., Pizzuti, C.: Outlier mining in large high dimensional datasets. IEEE Transactions on Knowledge and Data Engineering 17(2), 203–215 (2005)
5. Apiletti, D., Bruno, G., Ficarra, E., Baralis, E.: Data Cleaning and Semantic Improvement in Biological Databases. Journal of Integrative Bioinformatics, 3(2) (2006)
6. Bertossi, L.E., Chomicki, J.: Query answering in inconsistent databases. In: Chomicki, J., van der Meyden, R., Saake, G. (eds.) Logics for Emerging Applications of Databases, pp. 43–83. Springer, Heidelberg (2003)
7. Bruno, G., Garza, P., Quintarelli, E., Rossato, R.: Anomaly detection through quasi-functional dependency analysis. Special Issue of Journal of Digital Information Management Advances in Querying Non-Conventional Data Sources 5(4), 191–200 (2007)
8. Ceri, S., Giunta, F.D., Lanzi, P.: Mining Constraint Violations. ACM Transactions on Database Systems 32(1), 1–32 (2007)
9. Ceri, S., Widom, J.: Deriving production rules for constraint maintainance. In: Proceedings of the 16th Int. Conf. on Very Large Data Bases, pp. 566–577. Morgan Kaufmann, San Francisco (1990)
10. Chomicki, J.: Consistent query answering: Opportunities and limitations. In: Bressan, S., Küng, J., Wagner, R. (eds.) DEXA 2006. LNCS, vol. 4080, pp. 527–531. Springer, Heidelberg (2006)
11. Chomicki, J.: Consistent query answering: Five easy pieces. In: Schwentick, T., Suciu, D. (eds.) ICDT 2007. LNCS, vol. 4353, pp. 1–17. Springer, Heidelberg (2006)
12. Chomicki, J., Marcinkowski, J.: Minimal-change integrity maintenance using tuple deletions. Information and Computation 197(1-2), 90–121 (2005)
13. Cugola, G., Nitto, E.D., Fuggetta, A., Ghezzi, C.: A framework for formalizing inconsistencies and deviations in human-centered systems. ACM Trans. Softw. Eng. Methodol. 5(3), 191–230 (1996)
14. Flesca, S., Furfaro, F., Greco, S., Zumpano, E.: Querying and repairing inconsistent xml data. In: Ngu, A.H.H., Kitsuregawa, M., Neuhold, E.J., Chung, J.-Y., Sheng, Q.Z. (eds.) WISE 2005. LNCS, vol. 3806, pp. 175–188. Springer, Heidelberg (2005)
15. Flesca, S., Furfaro, F., Parisi, F.: Consistent query answers on numerical databases under aggregate constraints. In: Bierman, G., Koch, C. (eds.) DBPL 2005. LNCS, vol. 3774, pp. 279–294. Springer, Heidelberg (2005)
16. Greco, S., Sirangelo, C., Trubitsyna, I., Zumpano, E.: Preferred repairs for inconsistent databases. In: Proceedings of Int. Conf. on Database and Expert System Application, pp. 44–55 (2004)
17. Kanellakis, P.C.: Elements of relational database theory. In: Handbook of Theoretical Computer Science, Volume B: Formal Models and Sematics (B), pp. 1073–1156. Brown University (1990)
18. Mazuran, M., Quintarelli, E., Rossato, R., Tanca, L.: Mining violations to relax relational database constraints. Technical report, Politecnico di Milano (2009), http://home.dei.polimi.it/quintare/Papers/MQT09Constraint-RR.pdf
19. Murata, T., Borgida, A.: Handling of irregularities in human centered systems: A unified framework for data and processes. IEEE Transaction on Software Engineering 26(10) (2000)
20. Staworko, S., Chomicki, J., Marcinkowski, J.: Priority-based conflict resolution in inconsistent relational databases. In: EDBT Workshops (IIDB). Springer, Heidelberg (2006)
21. TPC-H. The TPC benchmark H. Transaction Processing Performance Council (2005), http://www.tpc.org/tpch/default.asp

Arguing from Experience to Classifying Noisy Data

Maya Wardeh, Frans Coenen, and Trevor Bench-Capon

Department of Computer Science
University of Liverpool, L60 3BX, UK
maya@csc.liv.ac.uk, frans@csc.liv.ac.uk, tbc@csc.iv.ac.uk

Abstract. A process, based on argumentation theory, is described for classifying very noisy data. More specifically a process founded on a concept called "arguing from experience" is described where by several software agents "argue" about the classification of a new example given individual "case bases" containing previously classified examples. Two "arguing from experience" protocols are described: PADUA which has been applied to binary classification problems and PISA which has been applied to multi-class problems. Evaluation of both PADUA and PISA indicates that they operate with equal effectiveness to other classification systems in the absence of noise. However, the systems out-perform comparable systems given very noisy data. Keywords: Classification, Argumentation, Noisy data.

1 Introduction

Argumentation is concerned with the logical reasoning processes required to arrive at a conclusion given two or more alternative view points. The process of argumentation can be conceptualised as a discussion about some issue that requires a solution, between a group of individuals with different points of view; where each member of the group attempts to persuade the others that his/her point of view, and the consequent solution, is the correct one. The discussion is conducted using a set of logical reasoning rules linking antecedents to consequents. Computer automation and modelling of the argumentation process has applications in legal reasoning, online auctions and so on. There is much reported work on automated argumentation, especially in the "two player" setting. In automated argumentation (persuasion dialogue games) each "player" typically has access to their own Knowledge Base (KB) which is used to propose arguments founded on the rules and facts contained in the KB [19]. Arguments can be advanced to either promote a player's own desired outcome or to attack arguments advanced by other players. However, the use of a KB to support argumentation has several disadvantages. Firstly the construction of the KB requires domain experts and entails the well established knowledge acquisition bottle neck reported in the Knowledge Based System and Expert System literature. Secondly the KB is never up to date.

An alternative to the KB approach to argumentation, and that promoted in this paper, is for each player to use data mining techniques to "mine" the desired rules from a live database. The authors refer to this process as "arguing from experience" in the

T.B. Pedersen, M.K. Mohania, and A M. Tjoa (Eds.): DaWaK 2009, LNCS 5691, pp. 354–365, 2009.

sense that each player's database can be considered to encapsulate that player's "experience". The arguing from experience idea was first explored by the authors in [25], where the PADUA two player argumentation system was introduced; and further developed in [26] where the PISA multi-player argumentation system was proposed. Both systems provided a mechanism for two (PADUA) or more (PISA) software agents to conduct dialogues to resolve disputes concerning the correct categorisation of particular examples. Both operate using an Association Rule Mining (ARM) technique to extract rules from their database repository of experience. The evidence presented in [25] and [26] indicated that the arguing from experience approach provides a natural representation of the participant's experience as a set of records, and the arguments as Association Rules (ARs).

In this paper the authors explore the application of the "arguing from experience" paradigm, advocated by both PADUA and PISA, to resolve classification (categorisation) problems, especially with respect to noisy data. The ability to handle noisy data is seen as important because it must be recognised that classification data will often contain wrongly classified examples, representing misconceptions and mistakes. In certain domains, such as welfare benefits, it is estimated that 30% or more of previous examples may have been wrongly classified [18]. Any classifier relying on such data must therefore be robust in the face of quite high levels of noise. Conceptually example cases are presented for classification, to either PADUA or PISA, and in each case the (PADUA or PISA) agents will argue for a particular classification through a persuasion process. The investigation, reported here, establishes that arguing from experience in this manner provides a classification mechanism that can produce similar accuracies to those produced by other classification systems in the absence of noise, but can cope more readily given noisy input data (noise levels of up to 50%). The rest of the paper is organised as follows. Section 2 provides some background information about the problem of classifying noisy data. Then in section 3 we give a summary of the argumentation from experience process and an overview of both PADUA and PISA. In section 4 an evaluation of PADUA, in the context of the binary classification of noisy data is presented. This is followed up in Section 5 by an evaluation of PADUA in a multi-class classification setting. Some final conclusions are presented in Section 6.

2 Background

The data classification (categorization) problem is well established in the Knowledge Discovery in Data (KDD) and data mining community. A substantial number of mechanisms have been developed to generate classifiers, including Neural Networks and Support Vector Machine, Decision Tree algorithms and Rule Induction approaches, various mechanisms influenced by ideas take from genetic programming and bio-computation, and Classification Association Rule Mining (CARM). Both PISA and PADUA operate using CARM [17]. The basic idea of CARM is to generate a set of Classification Association Rules (CAR) (a subset of the complete set of ARs) using ARM technology [1]. CARM offers a number of advantages including computational efficiency and, unlike many other classifier generators, easy understandability of the resulting classifier. One of the challenges of the classification problem is how to deal with very noisy data (and data with many missing values). Of course in an ideal domain the training data will contain no noise, no errors and no missing attributes; but unfortunately, in most real world domains, this is not the case. Tolerating noise is particularly important when designing classifiers, as the accuracy of classification depends on the quality of the input dataset. Noise can also be artificially introduced to the datasets for different purposes such as preserving privacy [2, 5].

Coping with noise can be addressed in different ways. One approach is to develop robust systems that allow for noise by avoiding over-fitting the model to the data ([2, 7]). Another approach is to pre-processing the input data before learning [6, 24] so as to eliminate tainted records [6, 24], but entails some major drawbacks: 1) Eliminating whole records of "bad data" eradicates "potentially" useful information such as the associations between uncorrupted attributes. 2) When there is a large amount of noise in the dataset, the amount of information in the remaining clean dataset may not be sufficient for building the classifier. And 3) In some cases eliminating "bad data" records is not possible because identifying these records can be an exhausting task, and may even require consulting expert opinion. This can be the case in datasets representing legal scenarios where the legislation can be misinterpreted. A number of preprocessing techniques have been developed to correct corrupted (noisy) data such as: 1) Deleting the corrupted fields and using the remaining, non-corrupted, fields for subsequent modeling and analysis [15]. 2) Cleaning of the dataset to remove noise (for example using Bayesian methods to clean corrupted data that have dependencies among features as described in [23]). Or 3) Correcting the misclassified data to improve classification accuracy based on the other predicted feature values as well as the corrected feature values [13].

However, such preprocessing is not always feasible as it often requires expert consultation (for example to provide the model for the Bayesian network in [23]), or because the noise level is so high that the correction of the corrupted data is neither easy nor effective. The following sections provide an overview of how the proposed arguing from experience framework can cope with noisy data, without any need for (i) pre-processing or (ii) initial removal of corrupted data by providing a moderation mechanism; whereby several agents engage in an argumentation dialogue, each using their own database of cases (representing their former experience). The idea is that this will allow the agents to correct each others "misconceptions".

3   Classification through Argumentation Using PISA and PADUA

The objective of both PADUA and PISA is to allow a number of agents, each with their own "private" database of examples, to debate the correct classification of a new case. The classification can be binary (PADUA) or non-binary (PISA). In PADUA (Persuasive Argumentation Using Association Rules), a protocol to enable two agents to argue about the classification of a case was established. PISA (Pooling Information from Several Agents) extended the PADUA protocol to allow any number of software agents to engage in a dialogue. This was found to be particularly useful for multi-class classification (i.e. non-binary classification), since each possible classification can then have its own champion. As noted above the distinguishing feature of both PADUA and PISA was that the arguments used by the agents were derived directly from a database of previous examples using ARM [25]. In PADUA the background dataset of each agent was represented by the means of a T-tree (Total tree) data structure, a reverse set enumeration tree structure with fast look up properties [9]. Both PADUA and PISA operate using a basic set of speech acts for argument from experience dialogues between two or n parties respectively. These speech acts are supported by three different forms of dynamic ARM request: 1) Find a subset of the possible set of ARs that conform to a given set of constraints. 2) Distinguishing a given AR by adding additional attributes. And 3) Generalising a given AR by removing attributes. Using their distinct databases PADUA and PISA agents produce reasons for and against classifications.

ARs [1] are probabilistic relationships which can be viewed as rules of the form X → Y (read as if X is true then Y is likely to be true, or X is a reason to think Y is true) where X and Y are disjoint subsets of some global set of attributes. Likelihood is usually represented in terms of a *confidence* value expressed as a percentage. This is calculated as *support*(XY)×100/*support*(X) where the *support* of an itemset is the number of records in the data set in which the itemset occurs. To limit the number of ARs generated only itemsets whose support is above a user specified support threshold, referred to as frequent itemsets, are used to generate associations. To further limit the number of ARs only those rules whose confidence exceeds a user specified confidence threshold are accepted. In the context of this paper the antecedent of an AR represents a set of reasons for believing the example should be classified as expressed in the consequent. Neither PADUA nor PISA use a specialized CARM algorithm, instead they are found on the Apriori-T ARM algorithm described in [9] and then classify the test cases by the means of the dialogue.

There are six speech acts (moves) used in PADUA [25] and PISA [26] dialogues which form three categories of "move" as follows: 1) *Propose Rule*: Move that allows generalizations of experience to be cited, by which a rule (AR) with a confidence higher than a certain threshold is proposed. 2) *Attacking Moves*: These moves argue that the reasons given in a rule proposed by another agent are not decisive in this case. This can be achieved using one of the following three speech acts: a) *Distinguish*: Add one or more premises (antecedent items) to a previously proposed rule, so that the confidence of the new rule is decreased. b) *Counter Rule*: Similar to the "propose rule" move, but used to cite a generalization leading to a different classification; and c) *Unwanted Consequence*: Move to suggest that a certain consequent (conclusion) of the proposed rule does not match the case under consideration. 3) *Refining Moves*: Moves that enable a rule to be refined to meet objections. This can be achieved using either of the following two speech acts: a) *Increase Confidence*: Replace one or more premises (antecedent items) in a previously proposed AR so as to increase the confidence of the rule; and b) *Withdraw unwanted consequences*: Exclude unwanted consequences of a rule that has been previously proposed (while maintaining a certain level of confidence). In other words, by trying to withdraw unwanted consequences, the player aims to refine a rule it previously proposed (instead of proposing a new rule). For each of the above six moves a set of legal next moves (i.e. moves that can possibly follow each move) is defined. Table 1 summarizes the rules for "next moves", and indicates where a new set of reasons is introduced to the discussion.

Table 1. Speech acts (moves) in PADUA-PISA

Move	Label	Next Move	New AR
1	Propose Rule	3, 2, 4	Yes
2	Distinguish	3, 5, 1	No
3	Unwanted Cons	6, 1	No
4	Counter Rule	3, 2, 1	Yes
5	Increase Conf	3, 2, 4	Yes
6	Withdraw Unwanted Cons	3, 2, 4	Yes

4   Evaluation Using Welfare Benefits Dataset

In this section we assess the effectiveness and robustness of PADUA as a classifier with respect to noise using a Welfare Benefits dataset. The model used to introduce noise was the same as that reported in [18]; for an N% noise level in a dataset of I instance, (N*D) instances were randomly selected and the class label changed to some other randomly selected value (with equal probability) from the set of available classes. The noise levels used in this study are: 2%, 5%, 10%, 20% and 40%. The noise was introduced to training sets only and not to the test sets. The rest of this section is organised as follows. The Welfare benefits dataset, used for the evaluation, is discussed in Sub-section 4.1. The various classifiers with which PADUA was compared are presented in Sub-Section 4.2. The ensuing results are discussed in Sub-Section 4.3.

4.1   The Welfare Benefits Dataset

The Welfare Benefits dataset was originally developed by Bench-Capon [3] and has been used in several experiments [18, 4, 14]. The data in this dataset concerns a fictional welfare benefit paid to pensioners to defray expenses for visiting a spouse in hospital. The benefit is payable if six conditions are satisfied: 1) The person is of pensionable age (60 for a woman, 65 for a man); 2) The person has paid contributions in four out of the last five relevant contribution years; 3) The person is a spouse of the patient; 4) The person is not absent from the UK; 5) The person does not have capital resources amounting to more than 3000; and 6) If the patient is an in-patient the hospital should be within a certain distance: if an out-patient, beyond that distance.

  Conditions 3 and 4 are Boolean necessary conditions, one which must be true and one which must be false. Condition 5 is a threshold on a continuous variable representing a necessary condition. Condition 2 relates five Boolean variables, only four of which need be true. Conditions 1 and 6 relate the relevance of one variable to the value of another: in 1 gender is relevant only for ages between 60 and 65, and in 6 the effect of the distance variable depends on the Boolean saying whether the patient is an in-patient or an outpatient. The wide range of conditions covered by this dataset, is one of the reasons why the dataset was selected to evaluate PADUA, as it demonstrates how well PADUA can cope with noise and how well it can cope with correlated conditions (as well as the other types of conditions used in this dataset). The dataset comprises of 2400 records such that half are classified as "entitled" (to benefit) and the other half to "not entitled". 70% of these rows were used as the training set and the rest (30%) as the test set. Noise was then applied to the training set (as defined above). The training set used for each of the noise levels, was split into two equal subsets, one given to the proponent and the other to the opponent in PADUA. The two players argued to classify the 720 cases in the testing set.

4.2 Comparator Classifiers

The operation of PADUA was compared against a variety of standard classifiers, covering a range of classification paradigms, as follows:

Decision Trees: Classification using *decision trees* was one of the earliest reported classification approaches. Quinlan's C4.5 is arguably the most referenced decision tree algorithm [22]. One of the most significant issues in decision tree generation is deciding on the *splitting criteria*. Of the approaches have been proposed in the literature, two have been used in the evaluation described here: 1) Selects most frequently occurring item; the Random Decision Tree (RDT) algorithm. And 2) Selects according to highest information gain; the Information Gain Decision Tree (IGDT) algorithm. Information gain [20] is one of the standard measures used in decision tree construction.

TFPC (Total From Partial Classification) ([10]), is a CARM algorithm founded on the TFP (Total From Partial) ARM algorithm ([11], [12]); which, in turn, is an extension of the Apriori-T (Apriori Total) ARM algorithm. TFPC is designed to produce Classification Association Rules (CARs) whereas Apriori-T and TFP are designed to generate Association Rules (ARs).

CBA (Classification Based on Associations) is another CARM algorithm developed by Liu et al [16]. CBA operates using a two stage approach to generating a classifier: (i) generate a complete set of CARs, (ii) prune the set of CARs, using the cover principle, to produce a classifier.

CMAR (Classification based on Multiple Association Rules) is a further CARM algorithm developed by Li et al [17]. CMAR also operates using a two stage approach to generating a classifier: (i) generate the complete set of CARs according to a user supplied support threshold to determine frequent (large) item sets, and a confidence threshold to confirm CRs, (ii) prune this set to produce a classifier.

FOIL – CPAR – PRM: FOIL (First Order Inductive Learner) is an inductive learning algorithm for generating Classification Association Rules (CARs) developed by Quinlan and Cameron-Jones [21]. This algorithm was later further developed by Yin and Han to produce the PRM (Predictive Rule Mining) CAR generation algorithm PRM was then further developed, by Yin and Han, to produce CPAR (Classification based on Predictive Association Rules) [27].

CN2 and ABCN2: The CN2 algorithm [7, 8] consists of a "covering" algorithm and a search procedure that finds individual rules by performing a beam search. Roughly, the covering algorithm starts by finding a rule, and then it removes from the set of learning examples those examples that are covered by this rule, and adds the rule to the set of rules. This process is repeated until all the examples are removed. **ABCN2** (Argument Based CN2) [18] is an extension of CN2. ABCN2 augmented the original CN2 algorithm to take into account arguments that explain misclassified examples: another pass uses these arguments to constrain the rules generated.

Table 2. Accuracy versus Noise (PADUA – Welfare Dataset). The CN2 and ABCN2 results are those given in [18].

Noise	PADUA	Rand DT	Info Gain DT	TFPC	CBA	CMAR	FOIL	CPAR	PRM	CN2	ABCN2
0	99.86	100	92.50	98.47	99.17	96.81	99.72	67.08	66.67	99.47	99.76
2	99.86	98.6	88.19	98.33	100	98.75	100	65.36	65.36	97.78	98.42
5	99.31	99.6	93.33	99.86	98.75	98.1	94.17	65.36	65.36	96.36	96.96
10	98.47	98.3	92.78	97.08	91.94	97.19	93.19	64.44	64.44	93.51	94.69
20	97.78	97.3	90.97	98.75	86.94	97.33	88.89	61.67	63.61	88.69	92.00
40	97.08	96.4	90.44	96.25	94.03	96.80	89.44	58.06	57.92	83.26	85.03

4.3 The Results

For the experiments the support threshold value was fixed to 1% and the confidence threshold value to 70% for all the relevant classifiers (including PADUA). Table2 shows the affect of adding noise to the Welfare dataset on the accuracy of each classifier. As expected the accuracy of all the classifiers drops as the noise level increases. When using clean data (no noise) RDT out performs all the other classifiers, with PADUA producing acceptable results. However, as the noise level increases it can be observed that PADUA is more tolerant to noise: the PADUA accuracy drops only 2.78% even when the noise level is increased to 40%, while the accuracy of RDT drops 3.61%. The other classifiers suffer more severe drops in their accuracy levels, for example the FOIL accuracy drops 10.28% between the noise levels. The results therefore indicate that PADUA is more tolerant to noise than all the other classifiers. The results for CN2 and ABCN are taken from [18], while the others were produced as part of the experiment.

5   Experimenting with Housing Benefit Dataset

In the above section PADUA, a two player argumentation protocol, was evaluated in the context of binary valued classification using an artificial welfare benefits dataset. In this section multi-class classification problems are considered using a second artificial housing benefits data set where benefits are again payable if certain conditions are satisfied. This dataset, although also originally a two class set, was selected because it is easy to modify from two classes to three, four, or five classes so as to evaluate the operation of PISA. For completeness PADUA was also applied to the dataset. The scenario that the housing benefits dataset is intended to reflect is a fictional benefit Retired Persons Housing Allowance (RPHA), which is payable to a person who is of an age appropriate to retirement, whose housing costs exceed one fifth of their available income, and whose capital is inadequate to meet their housing costs. Such persons should also be resident in the UK, or absent only by virtue of "service to the nation", and should have an established connection with the UK labour force. These conditions need to be interpreted and applied [25]. For this data set we used an interpretation very similar to the previous example, the only difference was that here we employed more flexible contribution and residency conditions. We also removed the patient-distance correlated condition. This simplified the dataset, and made modification, for the purpose of PISA, an easier task.

5.1 Evaluation Using PADUA

In this sub-section PADUA is further evaluated by applying it to the above housing benefits set configured in terms of two classes: entitled and not entitled. For the evaluation 2400 records were again generated distributed evenly over the two classes. The not entitled cases were generated such that they fail to meet one and only one condition of the five conditions listed above. Noise was then applied to this dataset in the same manner as in the previous evaluation. However, in this case an extra noise level of 50% was added to this experiment. The dataset was randomly split into a 70% training set and a 30% test set. Noise was then applied to the training set in the same manner as reported above. Again the training dataset used for each of the noise levels was split equally between two PADUA players and the two players allowed to "argue" to classify the 720 cases in the test set (using the same support\confidence level as in the previous test). This experiment was not applied to CN2 or ABCN2, which were not available to us.

Table3 shows the affect of adding noise to the housing benefit dataset on the accuracy of each classifier. Here it can be notice that FOIL is the best classifier when using correct data (unlike the previous experiment), but again it can be observed that as the accuracy of all the classifiers drops with the increase in noise level in the data, PADUA is again more tolerant of noise that the other classifiers. The accuracy of PADUA drops 5.83% as the noise level is increased from 0% to 50% whereas the accuracy of FOIL drops 21.81% and the accuracy of RDT drops 10.97%.

Table 3. Accuracy versus Noise (PADUA – Hosuing Benefit Dataset)

Noise	PADUA	RDT	IGDT	TFPC	CBA	CMAR	FOIL	CPAR	PRM
0%	99.86	99.72	77.00	98.33	97.36	99.31	100.00	64.03	66.81
2%	99.72	97.78	76.25	98.61	99.86	98.01	96.67	63.75	64.72
5%	99.58	98.89	64.31	96.53	97.50	98.61	94.44	65.28	65.14
10%	98.61	98.75	73.61	93.61	91.11	95.69	87.08	63.61	64.92
20%	96.81	98.19	73.06	93.89	96.25	96.50	86.39	62.28	64.58
40%	96.11	92.22	64.44	83.06	92.08	92.92	86.11	60.97	61.25
50%	94.03	88.75	62.22	54.72	84.17	85.31	78.19	59.58	61.81

5.2 Evaluation Using PISA

In order to use the Housing Benefits datasets to test PISA, the conditions mentioned in the previous sections were interpreted such that the final output would be increased from just two classes (entitled or not entitled). For the purpose of the example presented here a fourfold classification was used: fully entitled, entitled with priority, partially entitled and not entitled. The requirements for each class were defined as follows:

1. *Fully Entitled*: Candidates will be entitled to full housing benefit allowance if they satisfy all the above five conditions.
2. *Entitled with Priority*: candidates will entitle to housing benefit allowance with priority if they satisfy the entitling conditions and also satisfy the following: 1) Paid Contribution in four out of the last five years and either a) Have less capital than the original limit (this is interpreted as 1000£ less than the original limit). Or b) Have has less income (5%) than the original limit. Or 2) They are member of the armed forces and have paid the contribution fees in five out of the last five years.
3. *Partially Entitled*: Candidates will be entitled to a lower rate of benefit if they satisfy the age condition but they either: 1) Have slightly more capital than the original limit (e.g. +1000£ more than the original limit), but have paid contributions in 4 (or 5) years out of the last five. Or 2) they have slightly more available income (i.e. +5%) than the original limit, but have paid contributions in 4 (or 5) years out of the last five. Also 3) Merchant navy members are also partially entitled if they satisfy all the other conditions and have paid the contribution in five out of the last five years.
4. *Not Entitled*: If the candidate fails to satisfy the conditions for full or partial entitlement.

In the same manner as reported above 2400 records were generated equally distributed over the 4 possible classifications. The same noise levels used for PADUA were applied to the dataset. The training dataset used for each of the noise levels, was split into four equal subsets, each subset was given to one PISA player, and the four players in each subtest argued to classify the 720 cases in the test set. The support value was again fixed to 1% and confidence to 50% for all the CARS classifiers. Table 4 shows the affect of adding noise to the housing benefit dataset on the accuracy of each classifier. From the table it can be seen that the overall accuracy level is lower than that recorded for the binary classification. The best overall classifier is PISA with an accuracy level starting with 98.47% for clean data and dropping to 93.75% when a 50% noise level is introduced indicating that the PISA protocol copes extremely well with noisy data compared to the other classifiers used in the evaluation.

Table 4. Accuracy versus Noise (PISA)

Noise	PISA	RDT	IGDT	TFPC	CBA	CMAR	FOIL	CPAR	PRM
0%	98.47	94.44	68.19	92.56	90.28	86.75	92.25	75.83	75.83
2%	97.64	90.56	67.75	91.81	90.14	86.25	92.22	75.42	68.06
5%	97.36	93.47	62.92	89.72	90.69	85.00	91.39	73.33	73.89
10%	96.53	92.92	60.97	86.81	89.17	84.25	92.36	70.83	72.64
20%	95.69	91.94	60.56	80.83	88.89	83.75	89.31	70.78	70.61
40%	94.44	90.31	56.35	69.86	86.81	81.75	80.56	63.06	63.06
50%	93.75	88.36	61.81	45.83	62.71	80.50	70.42	63.06	65.83

6 Further Evaluation

The tests described above, use artificial datasets, mainly because we have full understanding of these datasets. But relying on just artificial datasets is not enough to demonstrate the tolerance to noise of PISA and PADUA. In this section we list some of the results obtained when testing PISA and PADUA using 7 real datasets. PADUA was used with the datasets containing 2 classes only (Mushrooms, Congress and PIMA) while PISA was applied to datasets with 3 classes (Wave Forms), 4 classes (Nursery and Car Evaluation) and 5 classes (Page Blocks). This test compares the operation of both PADUA and PISA with the same classifiers as above, but in this section we only report on the comparison with decision trees classifiers, because decision trees were found to be the closest "competitors" to PADUA and PISA. The results of this evaluation (figure 1(a) and figure 1(b)) show a similar pattern to the benefits experiments: the accuracy of almost all the classes dropped when the noise percentage was increased. The only case in which PADUA or PISA performed worse

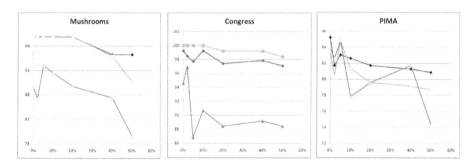

Fig. 1 (a). Real datasets study (1) (*the horizontal axe represents the noise level and the vertical represents the accuracy – white squares = RDT, dark squares =PISA/PADUA and white Triangles = GDT*)

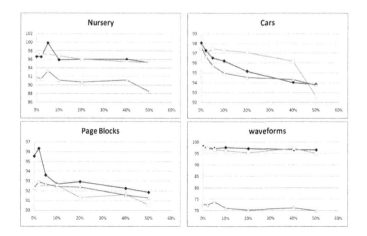

Fig. 2 (b). Real datasets study (2) (*same representation as fig 1(a)*)

than random trees with high level of noise is when the Congress dataset was used. The reason is that this dataset is very small (435 rows), therefore each player has only 152 cases from which they should mine their arguments (ARs). This is rather a small size when a high level of support or confidence is used.

7 Conclusions

In this paper we have presented an overview of PADUA and PISA, two argumentation from experience systems applicable to two and multiplayer argumentation respectively. We have described how both systems can be applied to the classification problem and illustrated this by detailed experiments using two artificial welfare scenarios/data sets, and summary results for seven real datasets. Of note, other than the operation of the two systems is that the argumentation from experience concept can successfully be applied to address classification. The results obtained indicate that the systems' performance is comparable to, or better than, other classification approaches. The particular advantage that the approach offers is that it operates very successfully in noisy environments, outperforming competitor classification systems. Ability to handle noisy data sets is of significant importance in many domains where sufficient data can only be obtained at the cost of including misclassified records. The authors are greatly encouraged by the reported results and are currently undertaking further investigation to evaluate the systems performance on a wider range of datasets.

References

[1] Agrawal, R., Imielinski, T., Swami, A.N.: Mining association rules between sets of items in large databases. In: Proc. SIGMOD Conf., pp. 207–216. ACM Press, New York (1993)

[2] Agrawal, R., Srikant, R.: Privacy-preserving data mining. In: Proc. ACM SIGMOD Conf on Management of Data (SIGMOD 2000), pp. 439–450. ACM Press, New York (2000)

[3] Bench-Capon, T.: Neural Nets and Open Texture. In: Fourth International Conference on AI and Law, pp. 292–297. ACM Press, Amsterdam (1993)

[4] Bench-Capon, T., Coenen, F.: An Experiment in Discovering Association Rules in the Legal Domain. In: DEXA 2000, pp. 1056–1060. IEEE Computer Society, Los Alamitos (2000)

[5] Bendou, M., Munteanu, P.: Learning Bayesian networks from noisy data. In: Proc. ICEIS, pp. 26–33 (2003)

[6] Brodley, C., Friedl, M.: Identifying and eliminating mislabeled training instances. In: AAAI/IAAI, vol. 1 (1996)

[7] Clark, P., Niblett, T.: The CN2 induction algorithm. In: Machine Learning, vol. 3(4), pp. 261–283 (1989)

[8] Clark, P., Boswell, R.: Rule Induction with CN2: Some Recent Improvements. In: Kodratoff, Y. (ed.) EWSL 1991. LNCS, vol. 482, pp. 51–163. Springer, Heidelberg (1991)

[9] Coenen, F., Leng, P.H., Ahmed, S.: Data structure for association rule mining: T-trees and p-trees. IEEE Trans. Knowl. Data Eng. 16(6), 774–778 (2004)

[10] Coenen, F., Leng, P.: Obtaining Best Parameter Values for Accurate Classification. In: Proc. ICDM 2005, pp. 597–600. IEEE, Los Alamitos (2005)

[11] Coenen, F., Leng, P., Ahmed, S.: Data Structures for association Rule Mining: T-trees and P-trees. IEEE Transactions on Data and Knowledge Engineering 16(6), 774–778 (2004)

[12] Coenen, F., Leng, P., Goulbourne, G.: Tree Structures for Mining Association Rules. Journal of Data Mining and Knowledge Discovery 8(1), 25–51 (2004)

[13] John, G.H.: Robust decision trees: Removing outliers from databases. In: Proc. 1st Int Conf on Knowledge Discovery and Data Mining (KDD 1995), pp. 174–179. AIII Press (1995)

[14] Johnston, B., Governatori, G.: Induction of Defeasible Logic Theories in the Legal Domain. In: Proc. 9th Int. Conf. on AI and Law, pp. 204–213. ACM Press, Edinburgh (2003)

[15] Kubica, J., Moore, A.: Probabilistic noise identification and data cleaning. Technical Report CMU-RI-TR-02-26, CMU (2002)

[16] Liu, B., Hsu, W., Ma, Y.: Integrating Classification and Association Rule Mining. In: Proc KDD 1998, pp. 80–86. AAAI, New York (1998)

[17] Li, W., Han, J., Pei, J.: CMAR: Accurate and Efficient Classification Based on Multiple Class-Association Rules. In: Proc ICDM, pp. 369–376 (2001)

[18] Mozina, M., Zabkar, J., Bench-Capon, T., Bratko, I.: Argument based machine learning applied to law. Artificial Intelligence 13(1), 53–73 (2005)

[19] Prakken, H.: Formal systems for persuasion dialogue. Knowledge Eng. Review 21(2), 163 (2006)

[20] Quinlan, J.: Simplifying decision trees. Man-Machine Studies 27(3), 221–234 (1987)

[21] Quinlan, J., Cameron-Jones, R.: FOIL: A Midterm Report. In: ECML, pp. 3–20 (1993)

[22] Quinlan, J.: C4.5: Programs for Machine Learning. Morgan Kaufmann Publishers, San Francisco (1998)

[23] Schwarm, S., Wolfman, S.: Cleaning data with Bayesian methods (2000)

[24] Teng, C.M.: Correcting Noisy Data. Machine Learning (1999)

[25] Wardeh, M., Bench-Capon, T., Coenen, F.P.: Arguments from experience: The PADUA protocol. In: Proc. COMMA, Frontiers in AI and Apps., vol. (172), pp. 405–416. IOS Press, Amsterdam (2008)

[26] Wardeh, M., Bench-Capon, T., Coenen, F.P.: PISA: Pooling Information from Several Agents: Multiplayer Argumentation from Experience. In: Proc. AI 2008, pp. 133–146. Springer, Heidelberg (2008)

[27] Yin, X., Han, J.: CPAR: Classification based on Predictive Association Rules. In: Proc. SIAM, pp. 331–335 (2003)

Dynamic Clustering-Based Estimation of Missing Values in Mixed Type Data

Vadim V. Ayuyev[1], Joseph Jupin[2], Philip W. Harris[3], and Zoran Obradovic[2,*]

[1] FN1-KF Department, Bauman Moscow State Technical University (Kaluga Branch),
Bazgenova Str. 2, Kaluga, 248600, Russian Federation
vadim.ayuyev@gmail.com
[2] Center for Information Science and Technology, Temple University, 303 Wachman Hall,
1805 N. Broad St., Philadelphia, PA 19122, USA
phone: +1.215.204.6265, fax: +1.215.204.5082
joejupin@temple.edu, zoran@ist.temple.edu
[3] Department of Criminal Justice, Temple University, 512 Glatfelter Hall, 1115 W Berks
Str., Philadelphia, PA 19122, USA
Phil.Harris@temple.edu

Abstract. The appropriate choice of a method for imputation of missing data becomes especially important when the fraction of missing values is large and the data are of mixed type. The proposed dynamic clustering imputation (DCI) algorithm relies on similarity information from shared neighbors, where mixed type variables are considered together. When evaluated on a public social science dataset of 46,043 mixed type instances with up to 33% missing values, DCI resulted in more than 20% improved imputation accuracy over Multiple Imputation, Predictive Mean Matching, Linear and Multilevel Regression, and Mean Mode Replacement methods. Data imputed by 6 methods were used for prediction tests by NB-Tree, Random Subset Selection and Neural Network-based classification models. In our experiments classification accuracy obtained using DCI-preprocessed data was much better than when relying on alternative imputation methods for data preprocessing.

Keywords: data pre-processing, data imputation, clustering, classification.

1 Introduction

A common approach to analyzing data with missing values is to remove attributes and/or instances with a large fraction of missing values. Such data preprocessing is appealing because it is simple and also reduces dimensionality. However, this is not applicable when missing values cover a lot of instances, or their presence in essential attributes is large [1].

Another common and practical way to address the problem of missing values in data is to replace them as estimates derived from the non-missing values by a linear function. The missing attribute j from an instance i, denoted as $x_{i,j}{}^{ms}$, is estimated as:

---

* Coressponding author: alternative e-mail: zoran.obradovic@temple.edu

T.B. Pedersen, M.K. Mohania, and A M. Tjoa (Eds.): DaWaK 2009, LNCS 5691, pp. 366–377, 2009.
© Springer-Verlag Berlin Heidelberg 2009

$$x_{i,j}^{ms} = f\left(x_{1,j}, x_{2,j}, \ldots, x_{p,j}, \ldots, x_{P_j,j}\right),$$ (1)

where f is a linear function of P_j variables; P_j is the number of instances in the data with non-missing values for attribute j; and $x_{p,j}$ is a non-missing attribute j from an instance p.

A special case of (1), which is simple, fast, and often provides satisfactory results when the number of missing values is relatively small and their distribution is random, is mean (or mode for categorical attributes) value based imputation:

$$x_{i,j}^{ms} = \frac{1}{P_j} \sum_{p=1}^{P_j} x_{p,j} .$$ (2)

The limitation of mean value based imputation and its variations is its focus on a specific variable without taking into account the overall similarities between instances. For example, consider the following 5 data points with 6 attributes, where a categorical attribute (fifth column) is missing one value (denoted as "ms"):

$$\begin{bmatrix} 1 & 10.2 & 1 & 1 & ms & 1 \\ 1 & 9.8 & 1 & 1 & 2 & 1 \\ 0 & 1.1 & 0 & 0 & 1 & 0 \\ 0 & 1.1 & 0 & 0 & 1 & 1 \\ 1 & 0.3 & 0 & 0 & 1 & 0 \end{bmatrix}.$$ (3)

Here, it would be reasonable to replace "ms" by "2" since the first two instances are very similar. However, mean/mode value-based imputation methods would replace "ms" by "1" as it is the most common value for this attribute in the dataset.

One of the most powerful approaches to missing values estimation is replacement by multiple imputation [1, 2]. The idea is to generate multiple simulated values for each incomplete instance, and iteratively analyze datasets with each simulated value substituted in turn. The purpose is to obtain estimates that better reflect the true variability and uncertainty in the data than are done by regression. Multiple imputation methods yield multiple imputed replicate datasets each of which is analyzed in turn. The results are combined and the average is reported as the estimate. For continuous attributes and a fairly small fraction of missing values, reliable estimates are obtained by combining only a few imputed datasets.

A clustering based approach for missing data imputation was considered as a local alternative to global estimation [3]. The premise was that instances could be grouped such that all the imputations in identified groups are independent from other groups. However, previous distance-based [4] clustering work was focused mainly on development of supervised clustering methods and mean/mode based imputations in these clusters. Also, prior studies were based on a strict separation for objects within clusters, such that it was assumed that there is no influence of instances in one cluster to an imputation process in other clusters.

In our DCI approach an independent cluster of similar instances with no missing values for a particular attribute is constructed deterministically around each instance with a missing value. In contrast to a typical clustering method, we allow cluster intersections such that the same instance may be included in many clusters. DCI relies on a distance measure that considers both categorical and continuous variables and is applicable for estimation of missing values in high dimensional mixed type data.

2 Methodology

We assume that the given data consist of M instances with N attributes where N is a mixture of tens to hundreds of categorical and continuous attributes. For the proposed Dynamic Clustering based Imputation (DCI) we use a dissimilarity measure between instances in a mixed type dataset described in Section 2.1. This measure is used in a clustering algorithm for identification of similar instances as described in Section 2.2 to perform a dynamic cluster-specific imputation of missing values as described in Section 2.3. An evaluation method and alternative imputation approaches are described in Section 2.4.

2.1 Measuring Dissimilarity between Instances in Mixed Type Data for DCI

The Minkowski distance, the Simple Matching Coefficient, the Jacquard Similarity Coefficient or other metrics could be used separately to measure the distance between instances for each type of attribute. However, such approaches are of limited applicability for mixed type data consisting of categorical and continuous attributes in the presence of many missing values [5]. In DCI, given N dimensional data, to measure the dissimilarity between two instances x_i and x_j of mixed type in the presence of missing values, we compute [6]:

$$\mathrm{dst}\left(x_i, x_j\right) = \left[\sum_{n=1}^{N} d_{i,j}^{(n)} \delta_{i,j}^{(n)}\right] \Big/ \sum_{n=1}^{N} \delta_{i,j}^{(n)}$$

$$d_{i,j}^{(n)} = \begin{cases} \left|x_{i,n} - x_{j,n}\right| \Big/ \left[\max_{p=1\ldots P_n} x_{p,n} - \min_{p=1\ldots P_n} x_{p,n}\right] & \text{for continuous attribute;} \\ 1 \text{ if } x_{i,n} \neq x_{j,n} \text{ and } 0 \text{ if } x_{i,n} = x_{j,n} & \text{for categorical attribute} \end{cases}, \quad (4)$$

$$\delta_{i,j}^{(n)} = \begin{cases} 0, \text{ if one of } x_{i,n} \text{ or } x_{j,n} \text{ is missing;} \\ 1, \text{ otherwise} \end{cases}$$

where "max" and "min" means the minimal and maximal value computed over all non-missing vales of the n-th attribute.

2.2 Clustering for Identification of Similar Instances in DCI

To identify similar instances in DCI we employ a new clustering algorithm consisting of the following steps:

1. Computing the similarity matrix (*SM*) for all instances:

$$
SM = \begin{pmatrix}
\infty & \mathrm{dst}(x_1, x_2) & \cdots & \mathrm{dst}(x_1, x_M) \\
\mathrm{dst}(x_2, x_1) & \infty & \cdots & \mathrm{dst}(x_2, x_M) \\
\vdots & \vdots & \ddots & \vdots \\
\mathrm{dst}(x_M, x_1) & \mathrm{dst}(x_M, x_2) & \cdots & \infty
\end{pmatrix}.
\tag{5}
$$

2. Computing the neighborhood matrix (*NM*):

$$
NM = \begin{pmatrix}
nm_{1,1} & \cdots & nm_{1,M} \\
\vdots & \ddots & \vdots \\
nm_{M,1} & \cdots & nm_{M,M}
\end{pmatrix},
\tag{6}
$$

where $nm_{i,j}$ is the number of common neighbors among K nearest neighbors for instances i and j, and M is the total number of instances in the dataset.

3. Constructing an ordered list $list_{i,j}$ of all neighbor instances with no missing value in j-th attribute for each missing value $x_{i,j}{}^{ms}$ by ascending sort according to the key value:

$$
\mathrm{dst}\left(x_i^{ms}, x_p\right) \Big/ nm_{i,p}, \text{ where } p = \overline{1, P_j}; nm_{i,p} > 0 ,
\tag{7}
$$

where x_i^{ms} denotes i-th instance with missing value in j-th attribute, and x_p denotes p-th instance with no missing in j-th attribute. Here, if two instances have the same dst/nm rate, the one with fewer missing attributes is listed first in the list.

4. Constructing a cluster $C_{i,j}$ for each missing value $x_{i,j}{}^{ms}$ by using first R elements of $list_{i,j}$, where R is a user-specific parameter that defines a cluster size, and $R < |list_{i,j}|$.

2.3 Cluster-Specific Imputation Methods for Mixed Type Data

In a cluster constructed as described in Section 2.2 using the similarity measure introduced in Section 2.1, a missing value could be imputed based on (a) the mean value of the corresponding attribute in other items contained in this cluster, or (b) similarity to the nearest instance with a non-missing value. Averaging in (a) and identification of the nearest instances from the same cluster in (b) could be based on various metrics. The dynamic nature of DCI derives from the ability to recalculate SM and NM for adding a newly imputed value (or all values from the certain attribute) into further imputation process. In DCI, we use the following categorical and continuous data specific metrics aimed to provide a balance in terms of imputation quality and computational complexity:

Categorical variable: A missing value is estimated by the corresponding attribute in an instance from the same cluster that has the largest number of common neighbors with the imputed instance:

$$x_{i,j}^{ms} = x_{q,j}^{(C_{i,j})} \text{ such that } nm_{i,q} = \max_{r=1...R}\{nm_{i,r}\} \ . \tag{8}$$

Continuous variable: A missing value is estimated based on all instances in the same cluster where each non-missing value is weighted by the appropriate entry of the neighborhood matrix NM:

$$x_{i,j}^{ms} = \left[\sum_{r=1}^{R} x_{r,j}^{(C_{i,j})} nm_{i,r}\right]\bigg/\sum_{r=1}^{R} nm_{i,r} \ . \tag{9}$$

2.4   Evaluation Measures and Alternative Imputation Methods

For evaluating imputation quality, different measures were used when comparing imputed categorical and imputed numerical data versus the corresponding true values.

The mean and absolute squared error measurements tend to be very sensitive to outliers. Therefore, for continuous attributes and for a given tolerance τ we measured a Relative Imputation Accuracy (*RIA*, also known as relative prediction accuracy [7]) defined as

$$RIA_\tau = [n_\tau/Q] \times 100\% \ , \tag{10}$$

where n_τ is the number of imputed elements estimated within τ percent of accuracy from the true value of the corresponding missing value and Q is the total number of imputed values in the data. In practice, *RIA* is a very useful as an approximation for an absolute precision imputed continuous values, which is often not needed. A nice property of *RIA* measure is that it is not affected by an individual incorrect imputation (e.g. large value instead of small) that could affect considerably some statistical measures (e.g., *MSE*-based [8]).

In categorical attributes we measured a fraction of Correct Imputations (*CI*) defined as

$$CI = [s/Q] \times 100\% \ , \tag{11}$$

where s is the number of correctly estimated imputed elements.

As a simple imputation alternative to DCI, we used a WEKA implementation [9] of Mean and Mode Replacement (denoted here as MMR). We also compared DCI to four statistically well-founded techniques: Multiple Imputation [1, 2], Predictive Mean Matching [10] (denoted here as PMM), Linear Regression [11], and Multilevel Regression [11] (denoted here as MLR).

The Multiple Imputation Method used for comparison and implemented in Amelia II software [12] enables the drawing of random simulations from the multivariate normal observed data posterior, and uses standard Expectation Maximization

(EM) for finding an appropriate set of starting values for data argumentation. Multiple Imputation begins with EM and adds an estimation of uncertainty for receiving draws from the correct posterior distribution followed by a resampling based on importance.

The Predictive Mean Matching comparison method implemented in WinMICE software [13] combines both parametric and nonparametric techniques. It imputes missing values by means of the nearest neighbors where the distance is computed as the expected values of the missing variables conditional on the observed covariates, instead of directly on the values of the covariates.

Linear and Multilevel Regression models, also implemented in WinMICE, are well known statistical approaches that allow variance in imputed variables to be analyzed at multiple hierarchical levels, whereas in linear regression all effects are modeled to occur at a single level.

3   Results and Discussion

We first performed experiments on a social science dataset with mixed-type attributes to compare quality of imputation by the proposed method and alternatives in presence of various fractions of missing values. In another set of experiments mixed-type data preprocessed by various imputation methods was used for classification by several algorithms to determine practical effects of an imputation method on classification accuracy (reported in Section 3.2).

A public domain *Adult* dataset [14] from the UCI Machine Learning Repository was used for comparing different data imputation methods. The dataset contained a subset of records about the US population collected by the US Census Bureau. The 48,842 individuals in this database are described by 8 categorical and 6 continuous attributes (with some missing data) related to prediction of annual income. In our experiments etalon data with 46,043 instances were constructed by removing all instances from the *Adult* dataset with missing values. To make the dataset balanced in terms of different attribute types, two categorical attributes ("education" and "native country") were also removed.

Eight test datasets with missing values ("holes") were constructed by randomly hiding 0.2%, 0.5%, 1.1%, 1.8%, 5.4%, 10.9%, 16.3% and 32.6% of data elements (which correspond to 1,000; 3,000; 6,000; 10,000; 30,000; 60,000; 90,000 and 180,000 missing values) in both categorical and continuous attributes of the etalon data. Each test database was fully independent from others, which means that places of "holes" were independent.

3.1   Evaluation of Imputation Quality on Mixed Type Data

The DCI and other imputation algorithms described in section 2.4 were compared using the eight datasets with different fractions of introduced missing values. Imputed values were compared to the true values in *Adult* dataset. To provide a comparison to

Table 1. Fraction of correct imputation (*CI*) in categorical attributes for 0.2%-32.6% imputed values

Imputation Methods	CI for different fractions of missing values							
	0.2%	0.5%	1.1%	1.8%	5.4%	10.9%	16.3%	32.6%
DCI	66.0	69.2	67.5	70.0	71.0	71.5	70.6	65.6
MMR	54.5	38.0	53.7	56.2	54.4	54.7	54.7	54.7
PMM	34.2	37.8	35.9	36.1	36.8	35.9	35.6	35.2
Linear Regression	28.1	30.3	28.8	29.0	29.0	28.4	28.4	28.1
Multiple Imputation	46.9	49.1	49.0	49.8	48.1	47.8	47.3	45.5
MLR	29.3	29.7	27.9	29.8	28.8	28.5	28.6	28.2
Random	19.1	20.8	19.3	21.1	19.9	20.3	19.7	20.2

a trivial estimate, we also report the results obtained by using the corresponding attribute value in a randomly selected instance (denoted here as Random). The imputation accuracy by DCI and alternative methods are summarized in Tables 1-4. All reported DCI results were obtained for 50 nearest neighbors and 9 the most common neighbors ($K=50$, $R=9$ defined in section 2.2). Very similar findings (within 5% of reported) were obtained for $40<K<60$ and $R=7$ or $R=11$ (stability results are omitted due to lack of space).

Imputation accuracy results for estimation of categorical attributes (Table 1) revealed that for all fractions of missing values. DCI was much more accurate than the alternative five imputation methods (1.2-1.4 times more accurate than the best of the remaining methods). Mean Mode Replacement was the second most accurate imputation method for categorical attributes. The results of the remaining imputation methods had more than 50% imputation error, but were still much better than random replacements.

The Relative Imputation Accuracy of DCI for imputation of continuous attributes (Tables 2-4) was also much better than alternative imputation methods. Here, Predictive Mean Matching was the second most accurate method. For 5% tolerance DCI provided 1.4-1.8 times better accuracy than PMM and was 6-9 times other better than the other alternatives (Table 2). Still, even the weak imputation methods were significantly more accurate than random replacements.

Table 2. Relative imputation accuracy (*RIA*) with 5% tolerance in continuous attributes for 0.2%-32.6% imputed values

Imputation Methods	RIA ($\tau = 5\%$) for different fractions of missing values							
	0.2%	0.5%	1.1%	1.8%	5.4%	10.9%	16.3%	32.6%
DCI	33.8	28.3	28.1	31.2	29.5	30.3	30.2	28.3
MMR	3.7	4.9	1.4	5.5	1.2	1.5	1.4	5.5
PMM	18.6	20.9	20.0	20.2	18.7	19.6	19.4	19.4
Linear Regression	3.7	4.5	4.4	4.5	4.2	4.3	4.4	4.2
Multiple Imputation	5.5	11.8	3.9	4.7	4.6	4.7	4.6	4.5
MLR	3.7	4.3	4.6	4.0	4.2	4.4	4.4	4.3
Random	1.8	2.1	2.9	3.3	3.0	3.2	3.0	3.0

Table 3. Relative imputation accuracy (*RIA*) with 10% tolerance in continuous attributes for 0.2%-32.6% imputed values

Imputation Methods	*RIA* (τ= 10%) for different fractions of missing values							
	0.2%	0.5%	1.1%	1.8%	5.4%	10.9%	16.3%	32.6%
DCI	38.7	35.4	35.6	37.4	36.7	37.2	37.2	35.6
MMR	10.2	11.8	12.0	11.9	11.5	11.6	11.6	12.0
PMM	25.6	30.0	29.2	28.8	27.6	28.5	28.2	28.2
Linear Regression	10.7	13.6	13.6	13.0	13.1	13.3	13.2	13.1
Multiple Imputation	13.3	20.4	13.4	13.5	13.3	13.3	13.4	13.3
MLR	10.7	12.9	13.9	13.0	13.1	13.2	13.3	13.2
Random	10.5	12.2	13.2	12.8	12.8	13.0	12.9	12.8

Table 4. Relative imputation accuracy (*RIA*) within 15% tolerance in continuous attributes for 0.2%-32.6% imputed values

Imputation Methods	*RIA* (τ= 15%) for different fractions of missing values							
	0.2%	0.5%	1.1%	1.8%	5.4%	10.9%	16.3%	32.6%
DCI	42.0	40.1	40.0	41.8	41.5	42.0	42.0	40.5
MMR	15.6	17.6	17.5	17.4	17.4	17.4	17.4	17.8
PMM	30.3	34.7	33.9	33.5	32.6	33.5	33.1	33.3
Linear Regression	15.4	18.3	18.4	17.7	18.1	18.3	18.2	18.1
Multiple Imputation	17.0	25.5	18.1	17.8	18.2	18.1	18.2	18.1
MLR	15.0	18.1	18.5	17.5	18.2	18.2	18.2	18.2
Random	15.8	17.7	18.0	17.9	18.2	18.3	18.2	18.2

Experiments with double and triple tolerance for allowed estimation errors of 10% and 15% (Tables 3 and 4) resulted in reduced differences in accuracy between imputation methods. However, even for larger tolerance DCI was still 20-50% more accurate (in relative difference) than the second best PMM method. These experiments suggest that the Mean Mode Replacement, Regression methods, and even Multiple Imputation methods are not appropriate for larger tolerance estimation in continuous variables as the corresponding results were comparable to random replacement. On the other hand, all methods outperformed the Mean Mode Replacement, which is commonly used in practice due to its simplicity.

3.2 Effect of an Imputation Method on Classification Accuracy for Mixed Type Data

The next stage of our experiments was devoted to practical comparison of how well different imputation techniques would suit for real life classification tasks. The idea was to explore a scenario where clean mixed type data provided by excluding all instances with missing values were used for training a classification model while it was applied to real data with various fractions of missing values. For this purpose we built several kinds of classifiers by training them on the first 16,043 subjects from the etalon *Adult* database, where for each instance all 12 attributes were available. For a test subject drawn from the remaining 30,000 instances the task was to predict if he/she makes over 50,000 U.S. dollars a year where a fraction of variables was missing at random. Different fractions of missing values were considered and preprocessing was achieved by 6 imputation methods described in Section 2. As a measure of accuracy, the percent of correctly classified instances was calculated.

As a classification method we applied three models implemented in WEKA: NB-Tree [15], Random Subset Selection [16] and Multilayer Perceptron [17]. NB-Tree is considered as one of the best classification methods for the *Adult* database according to [14]. Random Subset Selection and Multilayer Perceptron were used as alternative solutions that in other domains has shown good speed and classification accuracy, respectively. The classification results reported in Tables 5-7 are compared to the upper bound obtained by testing on complete data without missing values.

All imputation methods resulted in very similar accuracy for small fractions (0.2-1.8%) of missing values (Table 5). However, the difference was substantial when more than 10% of missing values were imputed. Though DCI provided the most accurate NB-Tree classifier for all fractions of missing values, its advantage was the most evident for the largest fraction of missing values (32.6%) where it had 14-22% less relative difference in error (3-7% difference in accuracy) than alternatives.

Table 5. Classification Accuracy (*CA*) of NB-Tree classification model applied to datasets with 0.2%-32.6% imputed values

Imputation Methods	CA of NB-Tree for different fractions of missing values							
	0.2%	0.5%	1.1%	1.8%	5.4%	10.9%	16.3%	32.6%
DCI	86.1	86.1	86.0	86.0	85.8	85.4	84.8	83.6
MMR	86.1	86.0	85.9	85.9	85.1	84.2	83.0	79.6
PMM	86.1	86.1	85.9	85.9	85.0	84.5	83.6	81.0
Linear Regression	86.1	86.0	85.7	85.6	84.3	82.9	81.4	76.8
Multiple Imputation	86.1	86.1	85.7	85.7	84.6	83.1	81.4	77.1
MLR	86.1	86.0	85.8	85.6	84.3	82.9	81.4	76.8
Complete data	86.1	86.1	86.1	86.1	86.1	86.1	86.1	86.1

Table 6. Classification Accuracy (*CA*) of Random Subspace Selection classification model applied to datasets with 0.2%-32.6% imputed values

Imputation Methods	CA of Random Subset for different fractions of missing values							
	0.2%	0.5%	1.1%	1.8%	5.4%	10.9%	16.3%	32.6%
DCI	84.9	84.9	84.9	84.8	84.8	85.0	84.7	84.8
MMR	84.9	84.9	84.9	84.8	84.5	84.3	83.8	81.6
PMM	84.9	84.9	84.8	84.7	84.4	84.2	84.1	83.1
Linear Regression	84.8	84.9	84.8	84.7	84.2	83.7	83.4	82.1
Multiple Imputation	84.9	84.9	84.8	84.8	84.5	84.2	83.7	82.6
MLR	84.9	84.9	84.8	84.7	84.3	83.8	83.3	81.8
Complete data	84.9	84.9	84.9	84.9	84.9	84.9	84.9	84.9

Table 7. Classification Accuracy (*CA*) of Multilayer Perceptron classification model applied to datasets with 0.2%-32.6% imputed values

Imputation Methods	CA of Multilayer Perceptron for different fractions of missing values							
	0.2%	0.5%	1.1%	1.8%	5.4%	10.9%	16.3%	32.6%
DCI	84.5	84.6	84.5	84.6	84.6	84.7	84.5	84.7
MMR	84.5	84.5	84.4	84.4	84.1	83.6	83.0	80.8
PMM	84.5	84.5	84.3	84.3	83.8	83.2	82.7	80.8
Linear Regression	84.5	84.5	84.2	84.1	83.2	82.0	81.0	77.4
Multiple Imputation	84.5	84.5	84.3	84.2	83.4	82.3	81.2	77.5
MLR	84.5	84.5	84.3	84.2	83.2	82.2	80.9	77.5
Complete data	84.5	84.5	84.5	84.5	84.5	84.5	84.5	84.5

When using the Random Subset Selection classifier, the overall results were consistent to classification by NB-Tree classifier (Table 6). However, Random Subset Selection classifier was more tolerant to an increase in fraction of missing values. Once again, DCI outperformed other approaches on the largest fractions of missing values for an 11-20% relative difference in error (2-3% difference in accuracy).

The Neural Network based classifier, represented by a 3-layer Perceptron, showed similar characteristics to NB-Tree and Random Subspace Selection (Table 7). DCI imputation resulted in more accurate classification in all datasets with a large fraction of missing values. For 0.5%, 1.8%, 5.4% 10.9%, and 32.6% imputed values a neural network achieved somewhat better accuracy than the upper bound obtained on complete data without missing values. This may be due to the Multilayer Perceptron's tolerance to noise in data.

To address class misbalance for the target variable in the *Adult* dataset (12,092 subjects in one class vs. 3,951 in another for the training subset, and 22,529 vs. 7,471 subjects for the test subset), we also measured Kappa coefficient [18] and F-score [19] for the three classification models when imputing 32.6% of missing values by the six methods (Table 8).

Table 8. Kappa coefficient and F-score of NB-Tree, Random Subspace Selection, and Multilayer Perceptron classification models applied to datasets with 32.6% of missing values imputed by 6 methods and to complete data without missing values

Imputation Methods	NB-Tree		Random Subset		Multilayer Perceptron	
	κ	F	κ	F	κ	F
DCI	0.52	0.83	0.54	0.84	0.54	0.83
MMR	0.42	0.79	0.49	0.81	0.45	0.80
PMM	0.47	0.81	0.47	0.81	0.44	0.80
Linear Regression	0.37	0.77	0.44	0.80	0.37	0.77
Multiple Imputation	0.40	0.77	0.48	0.81	0.38	0.77
MLR	0.37	0.77	0.43	0.80	0.38	0.77
Complete data	0.61	0.86	0.55	0.84	0.53	0.83

Here, Kappa coefficient is defined as:

$$\kappa = (Ra - Pa)/(1 - Pa) ,\qquad(12)$$

where Ra is the relative observed agreement, and Pa is the hypothetical probability of chance agreement, using the observed data to calculate the probabilities of each classifier randomly choosing each category.

F-score is defined as:

$$F = 2\,precision \times recall/(precision + recall) ,\qquad(13)$$

where, in a classification context, *precision* denotes the number of true positive predictions divided by the total number of items labeled as positive in the test set, while *recall* denotes the number of true positive predictions divided by the total number of items that were predicted as positive.

The obtained results clearly suggest that DCI based pre-processing results in accuracy nearest to the upper bound in terms of both Kappa coefficient and F-score statistics. We also observe that our results on imputed data confirms previous findings obtained on complete data that NB-Tree based classifier is a good choice for classification of *Adult* data. However, we also observe that the most stable results in terms of accuracy were obtained by Random Subset classifier.

4 Conclusion

Data imputation to replace missing values is often an important preprocessing step in data analysis. This study identified some limitations of a commonly used heuristic and of four known statistical methods when applied to mixed type data with a large fraction of missing values. In our approach, the main idea was to make all replacements independently for data within clusters created around each missing value. Our experiments on social science mixed type data provide evidence that the proposed data imputation method is more accurate than the evaluated alternatives and is effective when a large fraction of data is missing.

While the computational complexity of the proposed imputation method of $O(M^3 \log M)$ could be a limiting factor in large scale applications, many possibilities for improvements remain. For example, cluster-specific imputation techniques based on DCI idea could be developed. Also, specialized algorithms for defining the optimal size of specific clusters may be created. Finally, organizing data to KD-trees may improve the overall matrix processing speed.

An assumption of our research was that data were missing at random. This is not necessarily the case in real problems like the ProDES [20] data set that initiated this study. Therefore, in future research, we will explore how to tailor the proposed method to properties of missing data of a specific problem.

Acknowledgments. This study was funded in part by Award No. 2006-IJ-CX-0022 awarded by the National Institute of Justice, Office of Justice Programs, US Department of Justice. The opinions, findings and conclusions or recommendations expressed in this publication are those of the authors and do not necessarily reflect the views of the Department of Justice. Funding support from Russian Ministry of Science and Education is also acknowledged.

Authors thank the Department of Human Services (DHS), Philadelphia, and the Crime and Justice Research Center at Temple University for creating ProDES and granting us access to the dataset. We also thank Dr. Alan Izenman and Dr. Jeremy Mennis for extremely valuable comments on an early version of the methods and the results reported in this manuscript. Finally, we thank Pavel Karpukhin who rewrote all Matlab code, which made it possible to do most of our experiments.

References

1. Little, R.J.A., Rubin, D.B.: Statistical analysis with missing data. John Wiley & Sons, New York (1987)
2. Schafer, J.L.: Multiple imputation: a primer. Statistical Methods in Medical Research 8(1), 3–15 (1999)
3. Fujikawa, Y., Ho, T.-B.: Cluster-based algorithms for dealing with missing values. In: Chen, M.-S., Yu, P.S., Liu, B. (eds.) PAKDD 2002. LNCS (LNAI), vol. 2336, pp. 549–554. Springer, Heidelberg (2002)
4. Mantaras, R.L.: A distance-based attribute selection measure for decision tree induction. Machine Learning 6, 81–92 (1991)
5. Gan, G., Ma, C., Wu, J.: Data Clustering: Theory, Algorithms, and Applications. SIAM Press, Philadelphia (2007)
6. Wishart, D.: K-means clustering with outlier detection, mixed variables and missing values. In: Schwaiger, M., Opitz, O. (eds.) Exploratory Data Analysis in Empirical Research, pp. 216–226. Springer, New York (2003)
7. Nelwamondo, F.V., Mohamed, S., Marwala, T.: Missing Data: A comparison of neural network and expectation maximization techniques. Current Science 3(11), 1514–1521 (2007)
8. Bermejo, S., Cabestany, J.: Oriented principal component analysis for large margin classifiers. Neural Networks 14(10), 1447–1461 (2001)
9. Witten, I.H., Frank, E.: Data Mining: Practical machine learning tools and techniques, 2nd edn. Morgan Kaufmann, San Francisco (2005)
10. Landerman, L.R., Land, K.C., Pieper, C.F.: An Empirical Evaluation of the Predictive Mean Matching Method for Imputing Missing Values. Sociological Methods & Research 26(1), 3–33 (1997)
11. Gelman, A., Hill, J.: Data Analysis Using Regression and Multilevel/Hierarchical Models. Cambridge University Press, Cambridge (2006)
12. King, G., Honaker, J., Joseph, A., Scheve, K.: Analyzing Incomplete Political Science Data: An Alternative Algorithm for Multiple Imputation. American Political Science Review 95(1), 49–69 (2001)
13. Oudshoorn, C.G.M., Buuren, V.S., Rijckevorsel, V.: Flexible Multiple Imputation by Chained Equations of the AVO-95 Survey. In: TNO Prevention and Health (1999)
14. Asuncion, A., Newman, D.J.: UCI Machine Learning Repository. University of California, Irvine, http://archive.ics.uci.edu/ml/datasets/Adult
15. Kohavi, R.: Scaling Up the Accuracy of Naive-Bayes Classifiers: A Decision-Tree Hybrid. In: Proc. in 2-nd Int. KDDM Conf., pp. 202–207. AAAI Press, Portland (1996)
16. Ho, T.K.: The Random Subspace Method for Constructing Decision Forests. IEEE Transactions on Pattern Analysis and Machine Intelligence 20(8), 832–844 (1998)
17. Haykin, S.: Neural Networks: A Comprehensive Foundation, 2nd edn. Prentice-Hall, Englewood Cliffs (1998)
18. Gwet, K.: Statistical Tables for Inter-Rater Agreement. StatAxis, Gaithersburg (2001)
19. Van Rijsbergen, C.J.: Information Retrieval, 2nd edn. Butterworth, London (1979)
20. Crime and Justice Research Center, Temple University, http://www.temple.edu/prodes/

The PDG-Mixture Model for Clustering

M. Julia Flores, José A. Gámez, and Jens D. Nielsen

Computing Systems Dept. & SIMD Lab in I^{3}A
University of Castilla-La Mancha, Albacete, Spain
{julia,jgamez,dalgaard}@dsi.uclm.es

Abstract. Within data mining, clustering can be considered the most important unsupervised learning problem which deals with finding a structure in a collection of unlabeled data. Generally, clustering refers to the process of organizing objects into groups whose members are *similar*. Among clustering approaches, those methods based on probabilistic models have been extensively developed, such as Naïve Bayes (NB) with a latent class (cluster identifier) found via an EM algorithm.

Probabilistic Decision Graphs (PDGs) are a class of graphical models that can naturally encode some context specific independencies that cannot always be efficiently captured by other commonly used models. In this paper we propose to use a mixture of PDG models in cluster discovery, and an algorithm for automatic induction of the mixture and the models is introduced.

The proposed approach was experimentally evaluated on both synthetic and real-world databases, and the presentation of the results includes a comparison with related techniques. The comparison demonstrates competitive performance of the mixture of PDG models with respect to likelihood. Also, the mixture of PDG models have a tendency to use fewer models (clusters) to represent domains where other models use large amounts of clusters.

Keywords: Probabilistic graphical models, clustering, data mining.

1 Introduction

The increasing availability of data in our information society has led to the need for valid tools for its modeling and analysis. One core task in data mining is classification. Classification is the process of assigning labels to data instances using a function that takes a unlabeled data-instance as input and outputs a label. Unlike classification (aka supervised classification), which analyzes class-labeled data objects, clustering (aka unsupervised classification) analyzes data objects without consulting a known class label. In general, the class labels are not present in the training data simply because they are not known to begin with. Clustering can be used to generate such labels. The data instances are clustered or grouped based on the principle of maximizing the intraclass similarity and minimizing the interclass similarity. Each formed cluster can be viewed as a class of objects. Clustering can also facilitate taxonomy formation, that is, the

T.B. Pedersen, M.K. Mohania, and A M. Tjoa (Eds.): DaWaK 2009, LNCS 5691, pp. 378–389, 2009.

organisation of observations into a hierarchy of classes that group similar events together. Clustering also facilitates knowledge discovery through learning of new concepts that characterize common features or patterns, being used in many fields such as pattern recognition, image analysis and bioinformatics.

Among the different existing approaches, we will focus on the so called model-based methods [1]. Model-based clustering assumes that the data were generated by a specific model and tries to recover the original (generative) model from the data. The model that we recover from the data then defines clusters and can be used to assigns a label (or a set of possible labels) to new unlabeled data instances. EM and COBWEB belongs to this family [1].

Another possible classification on clustering methods uses as parameter the nature of the produced clusters and distinguishes *Hard, Soft, Hierarchical* and *Probabilistic*. These do not necessarily have to be disjoint sets, for example the Independency Tree [2] clustering is both hierarchical and probabilistic. The Probabilistic Decision Graph (PDG) [3] was originally proposed as an efficient representation of probabilistic transition systems. In this study, we consider the more generalized version of PDGs proposed by [4]. PDGs constitute a class of probabilistic graphical models that can represent some context specific independencies that can not efficiently be captured by other commonly used models.

The performance of the PDG model w.r.t. general probability estimation has previously been studied and results suggest that the model performs competitively when compared to state of the art models[5]. The PDG model has also been successfully applied to supervised classification problems [6] and to the problem of learning from incomplete data[7]. In this paper we extend the application area of PDGs to include also the clustering problem. The motivation for initiating this study was not only the previous successes of the PDG model to related problems such as classification and learning from incomplete data. But also the natural way in which a mixture of PDG models can take advantage of common sub-patterns in different clusters by reusing of parameters. As a result, a mixture of PDG models may provide a more compact model than conventional probabilistic clustering models. Furthermore, if context specific independencies exists within the same cluster, a PDG model may be able to capture this in a single model of this cluster, while other model that does not have this flexibility in representation may need to break the cluster into different clusters conditioning on the context.

2   Notation

We will denote random variables by uppercase letters, e.g. X, and sets with boldface uppercase letters, e.g. $\mathbf{X}$. When X_i is a discrete categorical random variable, we will by lowercase letter $x_{i,j}$ refer to the j'th state of X_i under some ordering. We will by $R(X_i)$ refer to the set of possible states of X_i, and by $R(\mathbf{X}) = \times_{X_i \in \mathbf{X}} R(X_i)$ when $\mathbf{X}$ is a set of variables. We will use r_i as a shorthand for $|R(X_i)|$. By lowercase bold letters we refer to joint states of sets of variables, e.g. $\mathbf{x} \in R(\mathbf{X})$. When $X_i \in \mathbf{X}$ and $\mathbf{x} \in R(\mathbf{X})$ we denote $\mathbf{x}[X_i]$ the projection of $\mathbf{x}$ onto coordinate X_i.

By $P(\mathbf{X})$ we will denote a joint probability distribution over $\mathbf{X}$, and by $P(\mathbf{Y}|\mathbf{Z})$ for disjoint $\mathbf{Y}$ and $\mathbf{Z}$ the conditional distribution of $\mathbf{Y}$ given $\mathbf{Z}$. To refer the probability of $\mathbf{X} = \mathbf{x}$ we use $P(\mathbf{X} = \mathbf{x})$ or simply $P(\mathbf{x})$. When computing a probability using a model M, we may indicate this by conditioning on the model $P(\mathbf{x}|M)$, however, we will use only $P(\mathbf{x})$ when M is clear from context.

Let $G = \langle \mathbf{V}, \mathbf{E} \rangle$ be a directed graph structure with set of nodes $\mathbf{V} = \{V_1, \ldots, V_n\}$ and set of directed edges $\mathbf{E} \subset \mathbf{V} \times \mathbf{V}$. We will then by $ch_G(V_i)$ and $pa_G(V_i)$ refer the set of children of V_i and parents of V_i respectively in structure G, hence $ch_G(V_i) = \{V_j \in \mathbf{V} : (V_i, V_j) \in \mathbf{E}\}$ and $pa_G(V_i) = \{V_j \in \mathbf{V} : (V_j, V_i) \in \mathbf{E}\}$. When G is clear from context we drop the subscript. A tree is a directed acyclic graph where one unique node $V_r \in \mathbf{V}$ is designated root and has no parents $pa_G(V_r) = \emptyset$ while all other nodes have exactly one parent. A forest structure is a set of such trees.

3   Techniques That Perform Probabilistic Clustering

Expectation-Maximisation and Naïve Bayes. In statistical computing, an expectation-maximisation (EM) algorithm [8] is an algorithm for finding maximum likelihood estimates of parameters in probabilistic models, where the model depends on unobserved latent variables. EM is frequently used for data clustering in machine learning and computer vision. EM alternates between performing an expectation (E) step, which computes the expected sufficient statistics by including the latent variables as if they were observed, and a maximization (M) step, which computes the maximum using expected sufficient statistics of the parameters by maximizing the expected likelihood on the observed cases found in the E step. The parameters found on the M step are then used to begin another E step, and the process is repeated.

In a probabilistic clustering task, one often includes in the model a special latent variable C that is never observed. Each states of this C then corresponds to a cluster, and inferring cluster membership is then done by answering queries like $P(C = c_i|\mathbf{X} = \mathbf{x})$. The Naïve Bayes (NB) model for clustering takes this approach, and represents a joint probability distribution that incorporate one strong independence assumptions which often have no bearing in reality, hence are (deliberately) naïve: all the variables are independent given cluster membership. The NB model is a special instance of a Bayesian Network model [9]with the structure shown in Fig. 1.(a).

The independencies that are assumed by the NB model yields the factorisation of the joint probability distribution $P(\mathbf{X}, C)$ over the domain $\mathbf{X}$ of observed variables and cluster variable C: $P(\mathbf{X}, C) = P(C) \prod_{X = \mathbf{X}} P(X|C)$. For a given observation $\mathbf{y}$ of variables $\mathbf{Y} \subseteq \mathbf{X}$, the probability of $\mathbf{y}$ being a member of cluster c_i is $P(C = c_i|\mathbf{Y} = \mathbf{y}) = \frac{1}{P(\mathbf{Y} = \mathbf{y})} P(C = c_i) \prod_{Y \in \mathbf{Y}} P(Y = \mathbf{y}[Y]|C = c_i)$ where $P(\mathbf{Y} = \mathbf{y}) = \sum_{c \in R(C)} P(C = c) \prod_{Y \in \mathbf{Y}} P(Y = \mathbf{y}[Y]|C = c)$.

For learning the parameters in the NB model one needs to reason from incomplete data as no data contains observations for C, and the standard approach

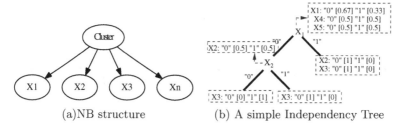

(a)NB structure   (b) A simple Independency Tree

Fig. 1. Two examples of probabilistic structures for clustering

is to use EM. For estimating the optimal number of clusters (states of C) a common approach is to use cross-validation.

Independency Trees. In [2] the Independency Tree (IT) was presented as a model able to perform clustering and also as an approximate way for factorisation. In Fig. 1(b) we show an example of an IT model. In general, the IT model can be interpreted as an extended probability tree [10] which introduces a new and very important element: a list of probabilistic marginal potentials associated to every node.

Given a leaf-node n in an IT structure, let $\mathbf{X}_n$ be the variables for which a marginal potential is associated with n, then all $\mathbf{X}_n$ are pair-wise marginally independent given the path to n. So, if the distribution for a given variable is shared by all leaves in a sub-tree, it can be stored in the root of that sub-tree for simplicity. For example, in Fig. 1(b) variables $X4$ and $X5$ are independent w.r.t. all the rest, and that is why their distributions appear in the root node.

Then, when one variable appears in a list for a node n it means that this distribution is common for all levels from here to a leaf, including intermediate nodes. On the other hand, there might be distributions that vary depending on the branch. For instance, in Fig. 1.(b) $X2$ distribution is different depending on the path (left or right) taken from the root, that is if $X1 = 0$ or $X1 = 1$.

The intuition underlying this model is based on the idea that inside each cluster the variables are independent. When we have a set of data, groups are defined by common values in certain variables, while the rest of the variables may vary independently. In an IT every cluster will be represented by a complete branch, with an associated factorisation. For the example, three clusters have been found, each one with a probability of $\frac{1}{3}$. If we look at the second branch/cluster it is characterised by $\{X1 = 0, X2 = 1\} + [X3 : 1.0/0.0, X4 : 0.5/0.5, X5 : 0.5/0.5]^1$.

3.1   The Probabilistic Decision Graph Model

A PDG encodes a joint probability distribution over a set of categorical random variables $\mathbf{X} = \{X_1, \ldots, X_n\}$ by a factorisation defined by a structure over a set of local distributions.

---

[1] $X1$ to $X5$ are binary, Xi:p1/p2 indicates that $P(Xi = 0) = p1$ and $P(Xi = 1) = p2$.

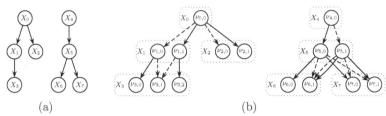

$$(a) \qquad\qquad\qquad (b)$$

$\mathbf{p}^{\nu_{0,0}} = P(X_0)$	$\mathbf{p}^{\nu_{3,0}} = P(X_3\vert X_0 = 0, X_1 = 1)$ $\mathbf{p}^{\nu_{5,1}} = P(X_5\vert X_4 = 0)$
$\mathbf{p}^{\nu_{1,0}} = P(X_1\vert X_0 = 0)$	$\mathbf{p}^{\nu_{3,1}} = P(X_3\vert X_1 = 0)$ $\qquad\qquad \mathbf{p}^{\nu_{6,0}} = P(X_6\vert X_4 = 1, X_5 = 1)$
$\mathbf{p}^{\nu_{1,1}} = P(X_1\vert X_0 = 1)$	$\mathbf{p}^{\nu_{3,2}} = P(X_3\vert X_0 = 1, X_1 = 1)$ $\mathbf{p}^{\nu_{6,1}} = P(X_6\vert X_4 = 0 \vee \{X_4 = 1, X_5 = 0\})$
$\mathbf{p}^{\nu_{2,0}} = P(X_2\vert X_0 = 0)$	$\mathbf{p}^{\nu_{4,0}} = P(X_4)$ $\qquad\qquad\qquad \mathbf{p}^{\nu_{7,0}} = P(X_7\vert X_4 = X_5)$
$\mathbf{p}^{\nu_{2,1}} = P(X_2\vert X_0 = 1)$	$\mathbf{p}^{\nu_{5,0}} = P(X_5\vert X_4 = 1)$ $\qquad\qquad \mathbf{p}^{\nu_{7,0}} = P(X_7\vert X_4 \neq X_5)$

$$(c)$$

Fig. 2. A forest F over binary variables $\mathbf{X} = \{X_0, \ldots, X_7\}$ is shown in (a), and a PDG-structure over $\mathbf{X}$ w.r.t. variable forest F is shown in (b). In the PDG-structure in (b), solid edges are labelled with value 1 and dashed edges are labelled with value 0. In (c), we have indicated the probabilistic interpretation of the parameters for each node in the PDG structure of (b).

Definition 1 (The PDG Structure). *Let F be a forest structure over $\mathbf{X} = \{X_1, \ldots, X_n\}$. A PDG-structure $G = \langle \mathbf{V}, \mathbf{E} \rangle$ for $\mathbf{X}$ w.r.t. F is a set of rooted acyclic directed graphs over nodes $\mathbf{V}$, such that:*

1. *Each node $\nu \in \mathbf{V}$ represents a unique $X_i \in \mathbf{X}$ and all $X_i \in \mathbf{X}$ are represented by at least one node $\nu \in \mathbf{V}$. We will by $\nu_{i,j}$ refer to the j'th node representing X_i under some ordering of the set of nodes representing X_i.*
2. *For each node $\nu_{i,j}$, each possible state $x_{i,h}$ of X_i and each successor $X_k \in ch_F(X_i)$ there exists exactly one edge $(\nu_{i,j}, \nu_{k,l}) \in \mathbf{E}$ with label $x_{i,h}$, where $\nu_{k,l}$ is some node representing X_k.*

Let $X_k \in ch_F(X_i)$. By $succ(\nu_{i,j}, X_k, x_{i,h})$ we refer to the unique node $\nu_{k,l}$ representing X_k that is reached from $\nu_{i,j}$ by following the edge with label $x_{i,h}$.

Example 1. A forest F over binary variables $\mathbf{X} = \{X_0, \ldots, X_7\}$ can be seen in Figure 2(a), and a PDG structure over $\mathbf{X}$ w.r.t. F in Figure 2(b). The labelling of nodes in the PDG-structure is indicated in subscripts and (redundant) by the dashed boxes, e.g., the nodes representing X_2 are $\{\nu_{2,0}, \nu_{2,1}\}$. Dashed edges correspond to edges labelled 0 and solid edges correspond to edges labelled 1, for instance $succ(\nu_{5,0}, X_6, 0) = \nu_{6,1}$.

A PDG structure is instantiated by assigning to every node a local probability distribution over the variable that it represents. By a PDG model over discrete random variables $\mathbf{X} = \{X_1, \ldots, X_n\}$ we refer to a pair $\mathcal{G} = \langle G, \boldsymbol{\Theta} \rangle$ where G is a PDG structure over $\mathbf{X}$ and $\boldsymbol{\Theta}$ is an instantiation of G. We denote by $\mathbf{p}^{\nu_{i,j}}$ the local distribution assigned to node $\nu_{i,j}$, and by $p_{x_{i,h}}^{\nu_{i,j}}$ the probability for state $x_{i,h}$ in local distribution $\mathbf{p}^{\nu_{i,j}}$. The semantics of the local distribution $\mathbf{p}^{\nu_{i,j}}$ is defined by the path(s) leading to the node $\nu_{i,j}$ from the root, that is, how $\nu_{i,j}$ can be *reached*. Let G be a PDG structure over variables $\mathbf{X}$ w.r.t. forest F. A node $\nu_{i,j}$ in G is *reached* by $\mathbf{x} \in R(\mathbf{X})$ if

- $\nu_{i,j}$ is a root in G, or
- $X_i \in ch_F(X_k)$, $\nu_{k,l}$ is reached by $\mathbf{x}$ and $\nu_{i,j} = succ(\nu_{k,l}, X_i, \mathbf{x}[X_k])$.

By $reach_G(X_i, \mathbf{x})$ we denote the unique node representing X_i reached by $\mathbf{x}$ in PDG-structure G.

A PDG model $\mathcal{G} = \langle G, \boldsymbol{\Theta} \rangle$ over variables $\mathbf{X}$ represents a joint distribution $P(\mathbf{X})$ by the following factorisation:

$$P(\mathbf{X} = \mathbf{x}) = \prod_{X_i \in \mathbf{X}} p_{\mathbf{x}[X_i]}^{reach_G(X_i, \mathbf{x})}. \tag{1}$$

Example 2. To instantiate the PDG structure in Fig. 2(b), we assign a local distribution to each node in the structure with the probabilistic interpretation given in Fig. 2(c). We can read some context specific independencies of this table, e.g. X_6 is independent of X_5 only in the context $X_4 = 0$.

4   Mixtures of PDG Models

In this section, we will describe our approach to probabilistic clustering using mixtures of PDG models.

In the previous Section 3.1 we introduced the PDG model for representing joint probability distribution over a finite set of discrete random variables. A typical approach to probabilistic clustering is to use a mixture of models. We propose a model that is a mixture of k PDG models by introducing a latent variable with one categorical state for each of the k PDGs. The marginal distribution of the latent variable defines the mixture of the k models. In Example 3 a specific mixture of 2 PDG models is presented.

Example 3. Consider 3 binary random variables X_0, X_1 and X_2. Let the distribution of X_2 be depending on X_0 and X_1, and furthermore, let the specific dependence be governed by an unobserved random variable C such that:

$$P(X_2|C = 0) = \begin{cases} P_1(X_2) & \text{if } X_0 \text{ and } X_1 \text{ have even parity,} \\ P_2(X_2) & \text{otherwise.} \end{cases} \tag{2}$$

$$P(X_2|C = 1) = \begin{cases} P_3(X_2) & \text{if } X_0 \wedge X_1 \text{ is true} \\ P_4(X_2) & \text{otherwise.} \end{cases} \tag{3}$$

The PDGs in Fig. 3(a) and (b) encodes the relations of Eq. (2) and (3) respectively when solid edges ecode value 0 and dashed encode value 1. For the numerical part of the models we have in Fig. 3(a): $\mathbf{p}^{\nu_3} = P_1(X_2)$ and $\mathbf{p}^{\nu_4} = P_2(X_2)$, while in (b): $\mathbf{p}^{\nu_3} = P_3(X_2)$ and $\mathbf{p}^{\nu_4} = P_4(X_2)$. Finally, by introducing the latent variable C in Fig. 3(c) we mix the two models to obtain a PDG model representing the full domain, and prior distribution of C is specified in $\mathbf{p}^{\nu_0}$.

In Ex. 3 we introduced an example of a mixture over two specific PDG models. Please note that the logical expressions as those used here (Eq. (2) and Eq. (3))

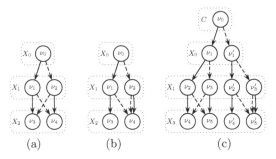

(a)   (b)   (c)

Fig. 3. In all structures, solid edges represent value 0 while dashed edges represent value 1. (a) A PDG encoding the relation in Eq. (2). (b) A PDG encoding the relation in Eq. (3). (c) A PDG that represents a mixture of the two PDG models of (a) and (b) through the latent variable C.

demonstrates some of the representation power of PDG models. While most other models has structures that grow exponentially by the number of variables included in such logical expressions, PDGs usually grow only linearly. For the toy-example presented here the difference obviously diminishes.

4.1   Learning Mixtures of PDG Models

In order to induce a PDG mixture model from data, we will have to establish a strategy for learning both k, a strategy for learning the variable structure to be shared between all component models, and both parameters and local structure of each of the k models. Lastly, the marginal distribution $P(C)$ also needs to be estimated.

Learning PDG models from complete data was addressed in [11], and for the case of incomplete data in [7]. We will combine these two approaches in a new algorithm that learns PDG mixture models.

Learning a common variable structure. The first step of our approach will induce a good structure over the variables to be shared between all k mixture component models. Here we use the approach presented in [11]. A statistical test of independence is used to decide the best organisation of variables. Initially, marginally dependent variables are grouped together. Then, incrementally the a tree is build for each group by inducing PDG models including more and more variables, placing variables that are conditionally independent in different subtrees, where the condition used in the test of independence is defined by a partition of the state space induced by the current PDG structure. The reader is referred to [11] for details.

Introducing the mixture. Once having learned an initial structure over the variables (as described above), we continue by adding a latent variable C to the model with $k = 1$ states, $R(C) = \{c_0\}$. One outgoing edge with label c_0 is connected to each of the roots of the PDG structure induced in the first step. We

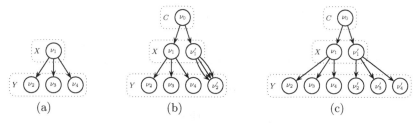

Fig. 4. (a) An inital model learned without latent variable. (b) The initialisation of the mixture by adding latent variable C to the model and extending the model with a new component by method 1. (c) Equivalent to (b) but method 2 is used for extending with a new component.

then extend C by adding one new state (incrementing k by one) and optimise structure and parameters by the structural-EM. If the likelihood of a separate hold-out dataset was increased by incrementing k, we loop and increment k once more.

Incrementing k. We consider two different strategies for introducing a new latent state.

1. We can extend the model with a new parameter node for each non-latent variable in the model. These new nodes are connected linearly without further bifurcations, and the edge labelled c_{k+1} is connected to the new root.
2. We can extend the model with a copy of the subtree(s) corresponding to an existing latent state c_i, $1 < i \leq k$. We choose the latent state with highest prior probability.

After extending the latent variable with a new variable state, splitting and merging is performed to refine the model. In Ex. 4 we give examples of the two methods listed above.

Example 4. Fig. 4(a) shows an initial PDG structure over two random variables X and Y. In Fig. 4(b) the latent variable C is added and a new latent state is initialise by method 1, that is, using a single new node for each variable. In Fig. 4(c) we show the structure resulting from initialising the new latent state by method 2, that is, using a copy of one of the existing subtrees.

When incrementing k, the new marginal probability $p^{\nu_0}_{c_{k+1}} = P(C = c_{k+1})$ is initialised to $\frac{1}{k+1}$, and the existing probabilities of the old k states $p^\nu_i : 1 < i \leq k$ are scaled accordingly by a factor $\frac{k}{k+1}$. For the new parameter nodes created for each of the variables in the domain, the initialisation depends on the method we used for creating them (the two described above). When using method 1, the new parameter node for variable X is initialised by either using the relative frequency (empirical marginal distribution) $\hat{P}(X)$. When using method 2, the parameters are copied from the relevant sub-tree. We then draw a data instance

d at random from the data set used for training, and use this as the "centre" of the new cluster, hence, we want to increase the probability assigned to d given the new cluster, $P(\mathbf{X} = d|C = c_{k+1})$. This achieved by tuning parameter $\mathbf{p}^\nu$ where ν represents X and $reach(X, \{d, c_{k+1}\}) = \nu$ as follows:

$$p_{x_i}^\nu \leftarrow \begin{cases} \frac{1.1+0.1p_{x_i}^\nu}{1.1} & \text{if } d[X] = x_i, \text{ and} \\ \frac{0.1p_{x_i}^\nu}{1.1} & \text{otherwise.} \end{cases} \qquad (4)$$

We finally arrive at framework presented as Algorithm 1. Please note that in line 3 we optimise the BIC score of the model using the score+search method presented in [11]. In this method the score of a model is optimised by iteratively splitting and merging parameternodes in a given structure. In lines 4 and 8 we use the structural-EM approach of [7] optimising the expected BIC score. Basically, this approach uses the same operators (split and merge of parameternodes) but uses expected score instead of actual score as the actual score is not tractable to compute in the precense of missing values. Finally, in lines 4 and 7 we need to choose either method 1 or 2 for extending the latent variable, yielding two different versions of the algorithm.

Algorithm 1.

1: **procedure** LEARNMIXTUREOFPDGS(**D**)
2:     Divide **D** into $\mathbf{D}_v$ being random sample of 10% of **D** and $\mathbf{D}_t = \mathbf{D} \setminus \mathbf{D}_v$.
3:     Learn PDG $\mathcal{G}$ from $\mathbf{D}_t$.
4:     Initialise mixture $\mathcal{G}^0$ from $\mathcal{G}$, and optimise $\mathcal{G}^0$ by structural-EM and $\mathbf{D}_t$.
5:     $k \leftarrow 0$.
6:     **repeat**
7:         $\mathcal{G}^{k+1} \leftarrow \{\mathcal{G}^k$ extended with 1 latent state$\}$.
8:         Optimise $\mathcal{G}^{k+1}$ by structural-EM and $\mathbf{D}_t$.
9:         $k \leftarrow k + 1$.
10:     **until** $P(\mathbf{D}_v|\mathcal{G}^k) < P(\mathbf{D}_v|\mathcal{G}^{k-1})$
11:     **return** $\mathcal{G}^{k-1}$.

---

5   Empirical Evaluation

In this section we perform a comparative analysis based on experimentation on 10 datasets. The *Exclusive* dataset is a dataset that is artificially generated from a boolean formula containing 5 boolean variables. Three of the variables are dependent such that one of them always assumes the value true while the other two assumes false. The last two variables are independent. The *TicTacToe* dataset is taken from the USI repository[12], and encodes the complete set of possible board configurations at the end of tic-tac-toe games. The *Greenhouse* datasets belong to data obtained when analysing an important economical factor in the south-east of Spain, greenhousing production at Almería. Some of these datasets were studied in [13] by using Bayesian networks. The *Sheep* datasets are historical data of sheep and has previously been used to analyse genetic merit for milk production[14]. The *PDG-mixture* data is a dataset artificially generated

Table 1. Datasets used in the empirical evaluation

Id	Name	# Vars	size train	size test
1	Exclusive	5	48	24
2	TicTacToe	9	641	317
3A	Greenhouse-A	8	830	410
3B	Greenhouse-B	17	981	484
3C	Greenhouse-C	33	883	435
3D	Greenhouse-D	6	1026	506
3E	Greenhouse-E	6	981	484
4A	Sheep-A	24	2068	1019
4B	Sheep-B	23	2068	1019
5	PDG-mixture	3	1000	500
6	NB	5	1000	500
7	IT	6	1000	500

Table 2. Results of Independency Tree (IT) learning, Naïve Bayes (NB) learning and the mixture of PDG models, the mixt-PDG-1 and mixt-PDG-2 columns refers to variation 1 and 2 of Alg. 1, respectively. The table contain log-likelihood (LL) of the learnt model measured over a separate test dataset, the number of clusters identified by the model (C), and the size of the model (S) measured in the number of independent parameters the model contains.

Id	IT			NB			mixt-PDG-1			mixt-PDG-2		
	LL	C	S	LL	C	S	LL	C	S	LL	C	S
1	-2.5481	3	8	-2.7920	3	17	-2.6942	4	16	-2.6383	4	17
2	-9.3111	5	78	-9.4090	2	37	-9.2862	7	308	-9.4228	3	396
3A	-6.3433	18	392	-6.3760	6	221	-6.3487	4	326	-6.2743	3	501
3B	-7.5005	7	214	-7.3797	4	151	-7.4285	5	332	-7.4957	3	604
3C	-17.0196	16	1270	-16.7899	10	1059	-16.8512	10	1345	-16.8505	11	6905
3D	-5.0027	6	96	-5.0174	5	124	-5.0191	3	166	-5.0609	3	262
3E	-5.6823	8	152	-5.6948	5	134	-5.7372	4	224	-5.7250	4	378
4A	-19.7172	63	3526	-19.7393	13	948	-19.3950	10	2163	-18.8837	6	7367
4B	-18.8637	53	2835	-18.4460	24	1679	-19.1973	6	1736	-18.8946	8	7852
5	-2.7918	7	22	-2.8033	7	48	-2.7976	3	32	-2.7943	3	40
6	-2.3400	4	14	-2.3440	2	11	-2.7526	4	31	-2.7528	5	46
7	-3.2364	3	12	-3.2768	2	13	-3.2541	4	32	-3.2592	3	29

from a mixture of 3 PDG models. The *NB* dataset has been sampled from a NB model with 3 latent states and 5 observable variables. Finally, the *IT* dataset was sampled from a IT model defining 3 clusters over 6 random variables. A brief description of the datasets can be found in table 1.

For learning IT models from the databases in Tab. 1, we have used the method presented in [2]. NB models was learnt using the Weka[15] system, using default settings of EM. When establishing the number of clusters, Weka uses cross-validation. The mixture of PDG models was learned using the algorithm discussed in Section 4.1.

In Tab. 2 we have listed log-likelihood (LL) for the learnt models measured over the test data which is a special separate dataset only brought in after the learning process to assess the quality of the learnt model. In Tab. 2 we also list the number of clusters (C) identified by the models and the size (S) of the models measured in the number of independent parameters defined by the respective models.

6   Discussion

We have compared the four algorithms over the ten datasets by using non-parametric Wilcoxon paired Signed-Ranks Test (α=0.05). From this statistical study we are in a position to say that the four models perform equally well in terms of log-likelihood but that some significant differences appear in the other two studied parameters (size and clusters). Clearly, IT produces a greater number of clusters than NB and both PDGs, but no significant difference is obtained when comparing pairwise NB, mixt-PDG-1 and mixt-PDG-2. However, the average number of clusters identified are 6.92 for NB, 5.33 for mixt-PDG-1 and 4.67 for mixt-PDG-2, which we find quite remarkable. With respect to size, there is no surprise, and the simplicity of NB makes it statistically superior in this parameter with respect to PDG models, and almost statistically superior (p-value = 0.0639) with respect to IT (average number of parameters is 370 (NB) vs 718 (IT)). There exist also statistically difference between PDG-2 and the other three models, indicating that PDG-2 is the model needing more parameters. Finally, no statistical difference appear between PDG-1 and IT, although PDG-1 needs 23% fewer parameters than IT.

Having done this general comparison, we continue with the analysis of the behaviour of the algorithms in some particularly interesting cases. Thus, when investigating the number of clusters identified by the different models, one interesting behaviour is evident. Both the IT and NB models uses many more clusters to model the *Sheep* domains than does the mixture of PDG models. This is interesting with respect to the practical use of the model for clustering, where usually a smaller number of clusters is preferable as it may be easier to assign meaningful semantics to each cluster. The two databases *Sheep-A* and *Sheep-B* differs only in that *Sheep-A* includes a variable that represents the breeding value of the sheep, while *Sheep-B* excludes this variable. Following domain experts (the shepherds), this variable can naturally be used as a classification of the given sheep into 4 different classes (the 4 possible values of this variable). For *Sheep-B* we see that the IT model identifies 53 clusters, NB identifies 24 while our mixt-PDG-1 and mixt-PDG-2 approaches identify only 6 and 8 clusters respectively, though with a somewhat lower score in likelihood.

Finally, investigating the datasets 5, 6 and 7, sampled from a mixture of PDGs, a NB and an IT respectively, we find one surprise: IT scores higher likelihood than both PDGs and NBs on all three datasets. We expected each model to provide the closest representation of data sampled from that exact model type. However, when investigating the number of clusters identified we see that for dataset 5 only PDG approaches identifies the correct number of clusters. For dataset 6 non of the methods identifies the correct number of clusters, and finally for dataset 7 both IT and mixt-PDG-2.

7   Conclusion

In this paper we have shown how the PDG model can be used in data clustering by extending the model with a latent variable, yielding a mixture of PDG models.

We have shown how to induce such models from data using EM and score based model-selection. Using two variations over the framework for induction of PDG mixtures, we have shown our proposal to be competitive with IT and NB model approaches, the latter being a standard technique in probabilistic clustering. On average, the PDG based approach identifies fewer numbers of clusters than both IT and NB, though non of the differences are statistically significant.

References

1. Han, J., Kamber, M.: Data Mining: Concepts and Techniques, 2nd edn. Morgan Kaufmann, San Francisco (2006)
2. Flores, M.J., Gámez, J.A., Moral, S.: The independency tree model: A new approach for clustering and factorisation. In: Proc. of the 3rd PGM workshop, pp. 83–90 (2006)
3. Bozga, M., Maler, O.: On the representation of probabilities over structured domains. In: Halbwachs, N., Peled, D.A. (eds.) CAV 1999. LNCS, vol. 1633, pp. 261–273. Springer, Heidelberg (1999)
4. Jaeger, M.: Probabilistic decision graphs - combining verification and AI techniques for probabilistic inference. International Journal of Uncertainty, Fuzziness and Knowledge-Based Systems 12, 19–42 (2004)
5. Nielsen, J.D., Jaeger, M.: An empirical study of efficiency and accuracy of probabilistic graphical models. In: Proc. of the 3rd PGM workshop, pp. 215–222 (2006)
6. Nielsen, J.D., Rumí, R., Salmerón, A.: Supervised classification using probabilistic decision graphs. Computational Statistics & Data Analysis 53(4), 1299–1311 (2009)
7. Nielsen, J.D., Rumí, R., Salmerón, A.: Structural-EM for learning PDG models from incomplete data. In: Proc. of the 4th PGM workshop, pp. 217–224 (2008)
8. Dempster, A.P., Laird, N.M., Rubin, D.: Maximum likelihood from incomplete data via the EM algorithm. Journal of the Royal Statistical Society, Series B 39(1), 1–38 (1977)
9. Jensen, F.V.: Bayesian Networks and Decision Graphs. Springer, Heidelberg (2001)
10. Salmerón, A., Cano, A., Moral, S.: Importance sampling in bayesian networks using probability trees. Computational Statistics & Data Analysis 34, 387–413 (2000)
11. Jaeger, M., Nielsen, J.D., Silander, T.: Learning probabilistic decision graphs. International Journal of Approximate Reasoning 42(1-2), 84–100 (2006)
12. Newman, D., Hettich, S., Blake, C., Merz, C.: UCI repository of machine learning databases (1998), http://www.ics.uci.edu/~mlearn/MLRepository.html
13. Céspedes, A., Rumí, R., Salmerón, A., Soler, F.: Analysis of the agricultural sector in the west-area of Almería by using bayesian networks (in spanish). In: Proc. of the 27th Spanish Conf. on Statistics & Operations Research, pp. 3438–3455 (2003)
14. Flores, M.J., Gámez, J.A., Mateo, J.L.: Mining the ESROM: A study of breeding value classification in manchego sheep by means of attribute selection and construction. Computers and Electronics in Agriculture 60(2), 167–177 (2008)
15. Witten, I.H., Frank, E.: Data Mining: Practical machine learning tools and techniques, 2nd edn. Morgan Kaufmann, San Francisco (2005)

Clustering for Video Retrieval

Petr Chmelar, Ivana Rudolfova, and Jaroslav Zendulka

Brno University of Technology, Faculty of Information Technology, Bozetechova 2,
612 66 Brno, Czech Republic
{chmelarp,rudolfa,zendulka}@fit.vutbr.cz

Abstract. The paper deals with an application of clustering we used as one of
data reduction methods included in processing huge amount of video data provided for TRECVid evaluations. The problem we solved by means of clustering
was to partition the local feature descriptors space so that thousands of
partitions represent visual words, which may be effectively employed in video
retrieval using classical information retrieval techniques. It has proved that
well-known algorithms as K-means do not work well in this task or their computational complexity is too high. Therefore we developed a simple clustering
method (referred to as MLD) that partitions the high-dimensional feature space
incrementally in one to two database scans. The paper describes the problem of
video retrieval and the role of clustering in the process, the MLD method and
experiments focused on comparison with other clustering methods in the video
retrieval application context.

Keywords: Incremental clustering, MLD, Leader, ART, video retrieval, feature
extraction, SURF, MSER, SIFT, cosine distance.

1 Introduction

Currently, the amount of multimedia data stored in repositories or processed as
streams is rapidly increasing. This data is an important source of potentially useful
information. There are two important and required operations on video data – content-
based retrieval and high-level feature extraction. The objective of content-based retrieval in video, also known as video search is to retrieve effectively and accurately
key frames and shots containing particular objects or events. High-level feature extraction in video is a classification of shots to classes of concepts that should reflect as
most as possible the human perception of presence or absence of concepts in the
frame. Examples of concepts are a person, a group of people, a hand, telephone or
a bus driver.

An activity the aim of which is to compare methods and techniques related to video
retrieval is the TREC Video Retrieval Evaluation (TRECVid). It is a series of confer-
ences sponsored by the National Institute of Standards and Technology (NIST). The
main goal of TRECVid is to promote progress in content-based analysis and retrieval
from digital video via open, metrics-based evaluation. TRECVid is a laboratory-style

T.B. Pedersen, M.K. Mohania, and A M. Tjoa (Eds.): DaWaK 2009, LNCS 5691, pp. 390–401, 2009.

evaluation that attempts to model real world situations or significant component tasks involved in such situations. It is also a forum for presentation and communication between the participants [17]. Automatic video retrieval plays the pivotal role of TRECVid evaluations. The task models a work of somebody looking for segments of video containing persons, objects or events of interest - the goal is to find shots of the video which best satisfy a given multimedia query. The query could have a form of a reference to an image (as in figure 1) or a reference to a video clip (e.g. people entering a vehicle). Both are supplemented with a brief textual clarification as "Find shots of a president entering or leaving a vehicle (e.g., car, airplane, helicopter), he and the vehicle both visible at the same time."

The video search task is considered to be hard. The procedure is based on the automatic extraction of low-level features, usually based on visual properties of video frames. The metadata representing a content of a frame or its region is referred to as feature vector. The video search can be based on finding the most similar feature vectors – similarity search task. Or it can be based on relations to high-level concepts. In such a case the low-level feature vectors classify to high-level concept classes.

The main problem of the task is enormous amount of video data and of low-level feature vectors derived from them. For example, the data sets provided to participants for purposes of methods development and testing for TRECVid 2008 evaluation contained 438 videos, which took about 120GB. There were extracted about 74 million low-level feature vectors. The metadata database size to be searched was more than 100 GB. Under these conditions it is practically unfeasible to search this data directly, even with support of indexing. Therefore, it is necessary to apply appropriate reduction methods - clustering of local feature vectors to find a set of classes (visual words), referred to as a visual vocabulary. Then low-level feature vectors are replaced with vectors representing the occurrence of the visual words in video frames, similarly as words in text documents [16]. This allows us to use text retrieval techniques [2] and to perform the video search in tenths of seconds.

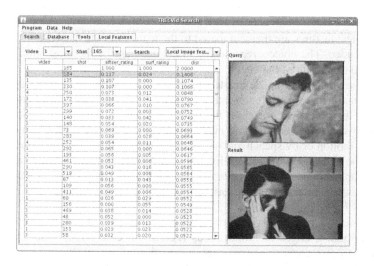

Fig. 1. An example of a successful retrieval of an image query (the top one)

Though, we had to cope with several problems related to the clustering itself. First, the data set is very large and also the data dimensionality is high. Second, this clustering task is specific in that the data objects we need to cluster are similar to each other and they cover the feature space almost continuously. As a result, it is difficult to find a division of the elements to very high number of clusters, which would correspond to the classical idea of a sufficiently large aggregations separated by an empty (or low density) space [19]. We tried several common clustering methods but they failed either because of their computational complexity or their inability to create an appropriate number of clusters. Thus we developed a simplified method we named later MLD (Modified LeaDer), according to Hartigan's Leader principle published in 1975 [5] and later used in BIRCH, DBSCAN or ART [19].

The reminder of the paper is organized as follows: Section 2 describes in more details the process of video retrieval, including a brief introduction to techniques and data formats that we used. Section 3 summarizes the state of the art in clustering large data sets of high-dimensional data. The MLD method is presented in Section 4. Experiments are described and results discussed in Section 5. Finally, Section 6 presents conclusions and future research issues.

2   Video Retrieval Based on Local Features

The goal of the video retrieval is to retrieve information contained in the video in a way as close as possible to the human perception. The content-based retrieval based on high-level semantic concepts suffers both from subjectivity of the human perception and the absence of effective methods. For example, the best automatic search system on TRECVid 2008 returned on average 1 or 2 of the top 10 shots containing the desired information [17]. This even has not been improved by a human-assisted query reformulation, but only by intensive and interactive assistance.

In general, the video retrieval is based on similarity search in a database containing the video metadata. A video query has a form of an example – a multimedia object (image, image video, sound) from which visual or high-level conceptual features are extracted. The features are then compared to the features in the database that represent objects in video. The objects represented in the database by the most similar features are selected as a result of the query.

Feature extraction is an automated process of extracting structured information from the unstructured, both representing the same object. Its output is a feature vector describing the object. In our case, the object is a multimedia shot – a sequence of frames between two cuts. The shot may be represented by several keyframes, camera and object motion or by speech and sound. The paper deals only with visual features. They can be either global - concerning the whole video frame (e.g. color histograms, layout or an amount of motion), or local – concerning regions in a frame. Video search based on local feature descriptors currently seems to outperform other approaches [17].

Local features represent visual objects as a compound of statistically interesting regions as points, edges or homogenous regions (in color) [12]. The local feature extraction process consists of two steps – detection of remarkable objects in images and their description. The main property of these steps is the repeatability – an ability to detect and describe the same object uniformly under various fotometric, geometric conditions and noise. We have employed two types of detectors and descriptors.

Maximaly Stable Extremal Regions (MSER, [10]) is used to find connected components of a tresholded image to be maximally stable. Here the 'extremal' means that pixels inside are lighter or darker than the surroundings. For description of these regions we use *Scale Invariant Feature Transform* (SIFT, [10]). SIFT captures information about (an eliptical) region using histogram of localy oriented gradients [18] (8 orientations, 4x4 locations), thus the feature vector is 128-dimensional. The second technique we use is the *Speeded Up Robust Features* (SURF, [1]) for region detection, based on the computation of determinant of Hessian matrix (second partial derivations) of an integral (summarized) image. Moreover it describes the region using Haar Wavelet Transform [18], similarly to SIFT. These regions are illustrated in figure 3.

If a multimedia object is represented by many (thousands) local feature vectors, their processing, storing and similarity search are time and space consuming. This problem solved as the first Josef Sivic in 2003 [16] by clustering high-dimensional local vectors into a large number (thousands to hundreds of thousands) classes and treated them as visual words to be searched in multimedia documents using information retrieval (IR) methods [2]. We applied this approach too.

The process consists of four basic steps, as illustrated in figure 2. Local visual features are extracted from objects (video frames) first. Each local feature is represented by a 128-bit feature vector. Then the clustering is applied to discover clusters representing visual words. The set of visual words forms a visual vocabulary and the more words the better [14]. Each shot is represented by a weighted vector, which we call document vector (in analogy to IR terminonogy [2]). We use *Term Frequency–Inverse Document Frequency* (*TF – IDF*) weighting:

$$tf\text{-}idf(w) = tf(w)idf(w), \text{ where } tf(w) = \frac{|d(w)|}{|d|}, \quad idf(w) = \log\left(\frac{|D|}{|D(w)|}\right), \quad (1)$$

where $|d|$ is the number of words in a document d, $|d(w)|$ is number of occurences of a word w in d, D is a document database and $D(w)$ means all documents containg word w [2]. Weighted document vectors are indexed using the Generalized Inverted Index (GIN, [15]), which is an effective implementation of a structure holding pairs of keys (words) and pointers to the documents containing them.

Same procedure is applied to the visual query – local visual features are extracted, classified by means of the visual vocabulary and the weighted document vector representing the query is created. Finally, the database si searched and the most similar document vectors are selected. They identify the visual documents most similar to the query. We use the Cosine distance r as a similarity measure of the query document vector d_q to any document d_d in the database:

$$r(d_q, d_d) = \frac{d_q \cdot d_d}{|d_q||d_d|}, \quad (2)$$

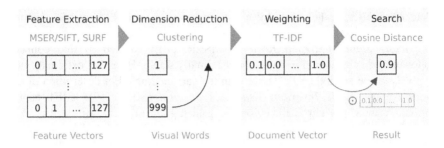

Fig. 2. The visual document retrieval process

3   Clustering of Large Data Sets of High-Dimensional Data

We have already mentioned in the Introduction that there are four main problems related to the clustering of local feature vectors:

- The large number of vectors (order of 10^7)
- The high dimensionality (more than 100)
- The large number of clusters required to represent visual words (10^4 to 10^6)
- Nearly uniform distribution of the vectors in some subspace contra outliers.

Most of clustering algorithms which are often used in other application domains are not suitable for these conditions. Table 8.1 on page 214 in [19] summarizes computational complexity of clustering algorithms. Classical hierarchical algorithms are not appropriate for clustering large-scale data because of their quadratic time and space complexity. Well-known and often used the K-means algorithm has time complexity $O(NKdT)$ where N is number of vectors, K is number of clusters, d is dimensionality and T is the number of iterations. Its time and space complexity is linear to the number of vectors but it also grows linearly with dimensionality and with the number of clusters. In our case the product Kd gives the value of the order close to the order of N. There are modifications of the K-means algorithm that employ a kd-tree to speed up the access to centroids [8], [13] but the problems of high dimensionality and large number of clusters remain.

There are several approaches to cope with large scale data, which include [19]:

- Random sampling – a random sample of original data is used instead of the entire data set. The minimum size of the sample can be estimated by using Chernoff bounds. We can apply this principle in MLD method by interrupting its first step as described in chapter 4.
- Data condensation – the clustering algorithm works with calculated summary statistics instead of the entire data set. A representative of this approach is the algorithm BIRCH [20] that we used in experimental evaluation described in section 5. Our algorithm is based on this principle.
- Density-based approach – relies on the density of data points in the data set. The representative of this and the following approach is DBSCAN [4]. However, a relatively uniform distribution of the data is a problem when joining clusters. MLD separates clusters based on the density (outliers).

- Grid-based approach – uses a grid to partition a data space at different resolution levels. The clustering algorithm then usually applies data condensation and density-based approaches to find clusters.
- Divide and conquer – approach that divides a data set into several subsets that can fit to the main memory and does clustering separately on the subsets. After that it merges these partial results to get the final result. The reverse (top-down) principle is employed in [14] for visual object retrieval. MLD is capable of data division.
- Incremental learning – in this approach clusters are built incrementally. In addition, an algorithm based on Adaptive Resonance Theory (ART) [Hudík] is another algorithm used in the experimental evaluation. MLD is of this type.
- Common approach that allows coping with large data is parallelization. A parallel data-clustering tool, P-CLUSTER, designed to execute on a network of workstations, is described in [7]. MLD is capable of parallelization.

Because of unsatisfactory results of several mentioned methods that we tried, we developed a simplified clustering method. The MLD method is similar to the Leader algorithm [5] (as we found out later), but this partitioning strategy is similar to [20], [4] and [6]. It is described in the next section.

4 The MLD Clustering Method

The method requires one (if summarized information is sufficient) or two passes over data set containing feature vectors extracted from video frames. Our goal was to develop a method able to cope with problems mentioned in the previous section. In addition, we aimed to meet some of requirements specified by Bradley et al. for their scalable framework for large-scale data clustering [3]. We have focused on minimizing the number of database scans and the possibility of early termination if appropriate; ability to incrementally incorporate additional data; utilization of variety of possible scan modes and on ability to operate on forward-only cursor over a view of the database.

The method has three basic steps (third is optional). Representatives of candidate clusters are selected during the first step. Let *mindistance* be a minimum distance of two representatives of two clusters. Feature vectors are processed sequentially. Let fv_i be a feature vector being processed, *Clusters* be a set of candidate clusters cc_j represented by representative feature vectors fv_j ($j = 1, 2, ..., \|Clusters\|$) selected so far. In addition, $distance(fv_k, fv_l)$ be a distance of two feature vectors fv_k and fv_l in the feature space. Then fv_i becomes a representative feature vector of a new candidate cluster cc_i if

$$\min\{distance(fv_i - fv_j), j = 1, 2, ..., \|Clusters\|, i \neq j\} > mindistance \qquad (3)$$

We will use a SQL-like pseudocode to describe the three steps more formally. Let *fvectors*(*vector_id*, *vector*, *cluster*), *clusters*(*cluster_id*, *vector*, *n_vectors*) be two database tables containing data of feature vectors and clusters, respectively. In the *fvector* relation, the attribute *vector_id* is an identificator of a feature vector, *vector* is the feature vector itself and *cluster* identifies a cluster which the feature vector belongs to. In the *clusters* relation, the attribute *cluster_id* is an identificator of a cluster, *rvector* is a representative feature vector of the cluster and *n_vectors* is the number of feature vectors assigned to the cluster.

The pseudocode for the first step of the MLD method could be the following:

```
for each ( SELECT f.vector_id, f.vector
           FROM fvectors as f ) {
   SELECT c.cluster_id,
           distance( f.vector, c.vector ) AS dist
   FROM clusters AS c
   ORDER BY dist LIMIT 1;
   if ( dist > mindistance )
       INSERT INTO clusters
       VALUES ( next(cluster_id), f.vector, 1 );
   else
       UPDATE clusters SET n_vectors += 1
       WHERE cluster_id = c.cluster_id
};
```

The first step requires the first database scan (the table *fvectors*). The number of clusters is much lower; therefore we ignore the time of the scan of the *clusters* table. Thus, it can be well parallelized except the situation, when a new cluster is detected (minor database load). If running in parallel, there is additionally a summarization of feature vectors count (from the local copy) at the end.

This step scan can be terminated and the threshold *mindistance* changed if the number candidate clusters seems to be too high or too low with respect to the number of required classes, i.e. visual words. Other (recommended) possibility is to count (in parallel) more minimum distance thresholds during the only database scan.

The objective of the second step, which is performed after the first step is completed, is to exclude candidate clusters the density of which is low (represent outliers or visual words that appear very rarely). Let *minvectors* be a threshold for the minimum number of feature vectors in a cluster, *maxclusters* be a maximum number of clusters and N_{ccj} be the number of feature vectors in a candidate cluster cc_j. Then a candidate cluster cc_j becomes a final cluster c_j if it contains at least *minvectors* feature vectors and belongs to the *maxclusters* number of clusters with the highest density:

$$N_{ccj} \geq minvectors \text{ and } \|\{cc_i \mid N_{cci} > N_{ccj}\}\| < maxclusters \tag{4}$$

The pseudocode for the second step could be:

```
DELETE FROM clusters
WHERE n_vectors < minvectors
   OR
   cluster_id NOT IN (
       SELECT cluster_id FROM clusters
   ORDER BY n_vectors DESC LIMIT maxclusters );
```

The *minvectors* and *maxclusters* thresholds allow controlling the number of clusters during the second step. However, in comparison to other methods, these parameters can be determined easily upon the process using pre-counted aggregate candidate clusters information - using the distribution of items in candidate clusters.

For example, a graph of candidate cluster count dependency on *minvectors* or *maxclusters* or both can help to determine these values intuitively.

The third step assigns feature vectors that have not yet been assigned to clusters (if needed). The feature is assigned to a cluster whose representative feature vector is the closest – the third step requires another database scan, but it can be well parallelized. As a result, the feature space is partitioned as a Voronoi tessellation. The pseudocode of the step could be the following:

```
for each ( SELECT f.vector_id, f.vector
              FROM fvectors as f ) {
    SELECT c.cluster_id,
              distance(f.vector, c.vector) AS dist
    FROM clusters AS c
    ORDER BY dist LIMIT 1;
    UPDATE fvectors SET cluster=c.cluster_id
    WHERE vector_id = f.vector_id;
};
```

The number of feature vectors N_{cj} in a cluster c_j represented by a feature vector fv_j and some other summary statistics (e.g. centroid and standard deviation) of the cluster c_j are also calculated and stored during the third step. They can be used later to find medoids, to apply other clustering algorithms on condensed clusters etc.

5 Evaluation

For the experimental evaluation of the MLD method and its comparison, we chose K-means, K-medoids, DBSCAN, BIRCH and ART algorithms [19]. The input datasets contained the local feature vectors extracted from the frames of the first five videos from the TRECVid 2008 developmental video keyframes sets. We had two data sets named here SURF and SIFT/MSER with respect to the applied feature extraction technique.

The dimensionality of feature vectors was 128 for both techniques. The domain of each dimension was quantized to an interval 0 .. 255. The standard deviation of values in the dimensions was between 40 and 60 for both techniques, independently on the video being processed. The number of shots in the videos was 1,765. For the sake of simplicity, each shot was represented by one (middle) frame in our experiments. There was detected up to 1 000 regions of interest in each frame. As a result we got 179,000 SIFT/MSER and 302,000 SURF feature vectors (more than 62 and 109 MB in plain text; 223 and 367 MB in the database).

We used post-relational database PostgreSQL [15], feature vectors were stored as integer fields (Int32 []). K-means and DBSCAN implementation come from an excellent machine learning tool RapidMiner [11]. There are implemented all common techniques of machine learning – both supervised and unsupervised. The ART method has been implemented by Tomas Hurdik [6] and we have our own implementation of BIRCH. All the methods run at 2x AMD Opteron (2 cores 2.8GHz), 8GB RAM 4TB RAID-5 machine with single PostgreSQL database.

Table 1. Methods comparison - the number of created classes and the execution time [hours]

Method	Setup	MSER/SIFT		SURF	
		Classes	Duration	Classes	Duration
	100 classes	100	18:36	100	15:20
	1000 classes	0	>> 500	0	>> 320
BIRCH	100 classes	100	550	100	296
	1000 classes	0	>> 500	0	>> 500
DBSCAN	E 1-512	1	>> 320	1	>> 320
ART	V 0.0001-0.8	74 000	17:38	302 000	138
	D 128	11 257	8:05	50 594	6:11
MLD	D 256	21 776	5:45	44	0:03
	D 512	24	0:02	1	0:02

We have performed three kinds of experiments. The first one was focused on time complexity of the methods and the ability to find number of classes large enough to represent a visual word vocabulary. The results are summarized in table 1. The Setup column shows the values of main method parameters, values in the column Classes denote the number of classes created and values in Duration the time consumed, respectively. The value 0 in Classes indicates that the method hasn't finished in time mentioned in column Duration.

Table 1 also shows the scalability of compared methods for the required amount of data while maintaining a reasonable performance. For comparison, in [16] are references to innovated K-means methods, which both referred to as working with up to one million vectors or 5 000 classes in [14] as their maximum (except the divide and conquer strategy, which results doesn't fit in our servers memory). In our case, even with the reduced (1%) data set, the RapidMiner implementation of K-means (and K-medoids) consumed 1.9GB of memory. This makes these methods definitely unusable on large data sets, which is concluded in chapter 6. In addition, K-means do not converge well (there is no K-medoids in table 1, because it has similar results).

The DBSCAN method has two compound disadvantages in comparison to "an ideal method". The first is setting the Epsilon (or maximum distance, E in table 1) too high, resulting in the only class created. Second is setting it lower, then the calculation lasted a very long time that, which was caused by many neighborhood searches.

BIRCH seemed to satisfy us first; however the weakest link is the final hierarchical clustering of the produced CF-tree. The problem is that the tree was not shaped correctly using these datasets and the hierarchical clustering method had not enough information about hierarchies - there were just about 2% of generated classes in different levels and thus it degraded to simple K-means.

As the best alternative method, we found ART [6]. However, setting up the three parameters: Vigilance (V in the table 1), Alpha and Beta and using the trial and error method to acquaint the serious number of classes is a burden task. We tried it and with the MSER/SIFT dataset we were lucky in contrast to the SIFT dataset. However even for MSER/SIFT, the ART method didn't converge - at each pass (60-100th) it was replaced about 40% of objects in different classes, we expected 5% only.

The MLD method gave us good results. In the table 1 D next to the MLD method refers to the *mindistance* measure. Moreover, we have proved that out method can cluster 25 and 38 million vectors in less than 100 hours and it can create thousands of clusters consuming about 100MB of memory (both the client and server side except the caches) without any parallelization. However, there might be objections to the quality of created classes.

Thus we have also performed tests according to the Quality measure [9]:

$$\sum_{i \in 1..K} \sum_{j \in 1..N_{ci}} distance(centroid_i, fv_{ij}) \tag{5}$$

where $centroid_i$ is the centroid of a cluster or the medoid, depending on the method. We found out, that the Quality measure depends most on the number of classes. For instance the "best" (0) Quality was achieved clustering SURF dataset using ART to 302 000 classes (which is the count of all objects). Moreover, all methods have very similar Quality ($\sim 10^6$) while analyzing thousands of clusters, so we don't present an extensive overview table.

Thus, the only experiment, how to validate the clusters quality is human assessment of the retrieval process, similarly to TRECVid [17], which is summarized in the following figure 3. The three graphs there show the ratio of first 50 retrieved documents (Positive) to the relevant documents (True Positive). A retrieved document was

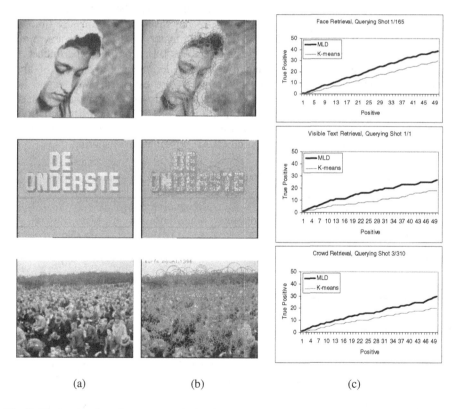

(a)   (b)   (c)

Fig. 3. The retrieval performance analysis of three queries (a) – Face, Visible Text and Crowd, extracted local features using SURF (b) and retrieval performance (c) of MLD and K-means

assessed relevant if it contained the high-level concept present in the query image (selected randomly, but having a human-recognizable concept): Face – at least one face recognizable; Text visible in a shot; and Crowd – four or more people together.

6 Conclusions

In brief, we have developed an incremental clustering technique suitable for clustering large datasets of high dimensional data when a large number of clusters is expected, later called MLD.

We have successfully tested properties of the method on local feature descriptors reduction into visual words used for video retrieval. The advantage of MLD lies in the stream-like data processing, a simple parallelization and effective outliner elimination. In addition, it is possible to stop execution and restart it with changed parameter (*mindistance*) in an early stage if the values of parameters seem to be inappropriate. In fact, several instances of the method can be executed simultaneously with different minimal cluster distances. Moreover, the number of clusters and the minimal cluster cardinality can be discovered easily using simple aggregate queries.

Although the method includes only one optimization step, the result seems to be good in the context of the visual-word-based video retrieval. We have tested several other methods, but and none of them provided better results if we consider time and space complexity, ability to create large number of clusters and subjective assessment of the result of video retrieval on the tested data set. Based on our experiments, we summarize in table 2 our recommendations for clustering large data sets of high-dimensional data when a large number of clusters are expected. The second best method we consider ART, but its result requires a lot of try-error parameter settings.

Our future research will be focused on more in-depth evaluation of the quality of the clusters measured in the context of this specific domain, fuzzy clustering and adaptation of semi-supervised learning and OLAP principles in clustering.

Table 2. Methods recommendations based on the theory and our experiments

Method	Complexity (theoretical)	Recommendations		
		Large dataset	High dimension	Many clusters
k-Means	$O(NKd)$	No	Yes	No
DBSCAN	$O(NlogNd)$	No	No	Yes
BIRCH	$O(Nd)$	Yes	Yes	No
ART	$O(Nd)$	Yes	Yes	Hard
MLD	$O(Nd)$	Yes	Yes	Yes

References

1. Bay, H., Tuytelaars, T., Van Gool, L.: SURF: Speeded up robust features. In: Leonardis, A., Bischof, H., Pinz, A. (eds.) ECCV 2006. LNCS, vol. 3951, pp. 404–417. Springer, Heidelberg (2006)
2. Baeza-Yates, R.: Modern information retrieval, p. 513. ACM Press, New York (1999)

3. Bradley, P., Fayyad, U., Reina, C.: Scaling Clustering Algorithms to Large Databases. In: Proceedings of 4th International Conference on Knowledge Discovery and Data Mining (KDD 1998), pp. 9–15. AAAI Press, Menlo Park (1998)
4. Ester, M., et al.: A density-based algorithm for discovering clusters in large spatial databases with noise. In: Proceedings of 2nd International Conference on Knowledge Discovery and Data Mining (KDD 1996), pp. 226–231. AAAI Press, New York (1996)
5. Hartigan, J.: Clustering Algorithms. John Wiley and Sons, New York (1975)
6. Hudik, T.: The machine-learning methods in the environmental risk assessment spatial modelling. In: Proc. of the 2nd Int. Summer School on Comp. Biology, pp. 52–57 (2006)
7. Judd, D., McKinley, P., Jain, A.: Large-Scale Parallel Data Clustering. IEEE Trans. on Pattern Analysis and Machine Intelligence 20, 8 (1998)
8. Kanungo, T., et al.: An Efficient k-Means Clustering Algorithm: Analysis and Implementation. IEEE Trans. on Pattern Analysis and Machine Intelligence. 24(7), 881–892 (2002)
9. Kogan, J.: Introduction to clustering large and high-dimensional data, p. 205. Cambridge University Press, Cambridge (2007)
10. Lowe, D.G.: Distinctive Image Features from Scale-Invariant Keypoints. Int. Journal of Computer Vision 60(2), 91–110 (2004)
11. Mierswa, I., et al.: YALE: Rapid Prototyping for Complex Data Mining Tasks. In: Proc. of the 12th ACM SIGKDD Int. Conf. on Knowledge Discovery and Data Mining (2006)
12. Mikolajczyk, K., et al.: A Comparison of Affine Region Detectors. Int. Journal of Computer Vision 65(1), 43–72 (2005)
13. Pelleg, X., Moore, A.: Accelerating exact k-means algorithms with geometric reasoning. In: Proc. of the fifth ACM SIGKDD Int. Conf. on Knowledge discovery and data mining, pp. 277–281 (1999)
14. Philbin, J., et al.: Object retrieval with large vocabularies and fast spatial matching. In: Video proc. CVPR, pp. 1–8 (2007)
15. PostgreSQL Global Development Group: PostgreSQL 8.3 Documentation: GIN Indexes, http://www.postgresql.org/docs/8.3/static/gin.html
16. Sivic, J., Zisserman, A.: Efficient Visual Search for Objects in Videos. Proc. of IEEE 96, 4 (2008)
17. Smeaton, A.F., Over, P., Kraaij, W.: Evaluation campaigns and TRECVid. In: Proc. of the 8th ACM Int. Workshop on Multimedia Information Retrieval (2006), http://www-nlpir.nist.gov/projects/trecvid/
18. Sonka, M., Hlavac, V., Boyle, R.: Image Processing, Analysis, and Machine Vision, 3rd edn., Thomson Engineering, Toronto, p. 800 (2007)
19. Xu, R.: Clustering, p. 358. Wiley, IEEE Press, Hoboken (2009)
20. Zhang, T., Ramakrishnan, R., Livny, M.: BIRCH: An efficient data clustering method for very large databases. In: Proc. of the ACM SIGMOD Conf. on Management of data, pp. 103–114 (1996)

Trends Analysis of Topics Based on Temporal Segmentation

Wei Chen and Parvathi Chundi*

University of Nebraska at Omaha,
NE, 68182 Omaha, NE, US
{wchen,pchundi}@mail.unomaha.edu

Abstract. Extracting interesting information from large unstructured document sets is a time consuming task. In this paper, we describe an approach to analyze the temporal trends of a given topic in a time-stamped document set based on time series segmentation. We consider topics containing multiple keywords and use a fuzzy set based method to compute a numeric value to measure the relevance of a document set to the given topic. The measure of relevance is then used to assign a discrepancy score to a segmentation of the time period associated with the document set. The discrepancy score of a segmentation represents the likelihood of the topic across all segments in a segmentation. Given a user specified value k, we then define a min different k segmentation to capture the k-segmentation with the maximum possible discrepancy score and describe a dynamic-programming based algorithm to compute it. The proposed approach is illustrated by several experiments using a subset of the TDT-Pilot Corpus data set. Our experiments show that the min difference k segmentation successfully highlights the temporal trends of a topic using k segments.

Keywords: Temporal Text Mining, Temporal Segmentation, Topics, Fuzzy sets.

1   Introduction

Unstructured data is pervasive in the information age. The time stamp associated with each document in a stream of text, such as news articles, blogs, emails and so on, can be used to extract information of different topics and their temporal progression. Temporal text segmentation is aimed to find such temporal patterns and is important for many natural language processing tasks, including information retrieval [6], summarization [12], and other analysis [11].

The activity related to a topic/event may be different in different time intervals. Identifying these different periods of activity related to a topic may provide

---

* This work was partially supported by NSF Grant IIS-0534616 and by Grant Number P20 RR16469 from the National Center for Research Resources (NCRR), a component of the National Institutes of Health (NIH).

T.B. Pedersen, M.K. Mohania, and A M. Tjoa (Eds.): DaWaK 2009, LNCS 5691, pp. 402–414, 2009.

rich and useful information [3,4,8,16]. To analyze the temporal trends of the activity surrounding a topic, a typical process measures the information related to a topic in each hour/day/month in the time period and construct a time series of these numeric values. Visualizing or processing this time series further can be then be used to identify temporal trends. It has been well-recognized that a time series of values recorded at fine-grained intervals such as an hour or a day includes too many values, much of which could be noise. Therefore, analyzing for trends at the fine-grained level may be unnecessary or time consuming. Time series segmentation is a well-known technique for removing noise from a time series so that temporal information such as trends can be more readily identified. Time series segmentation combines consecutive time points into segments while replacing the measured values at each of these time points with a single value. The single value at the segment is typically computed as a function of the values at each of the time points.

In this paper, we discuss an approach for identifying the temporal trends of a given topic from a set of time stamped documents based on temporal segmentation. We assume that we are given a time stamped document set published over a time period, which is a list of time points. A time point is a unit of time such as an hour, day, month, etc. We define a **topic** as a list of one or keywords and associated weights. We use the fuzzy set theory to compute the relevance of a set of documents to the topic as a numeric value. We use a variation of the temporal scan statistic to capture the likelihood of a topic in a time point/interval. In our earlier work [3,4], we used the temporal scan statistic (http://www.satscan.org) to identify *hot spots* of a topic; i.e., the intervals where there is a burst of activity surrounding the topic. Scan statistics are well established for identifying clusters in temporal, spatial, as well as spatio-temporal data in fields such as epidemiology and astronomy.

A segmentation of a time period simply divides the time period into a desired number of segments specified by the user. The number k of segments in a segmentation is typically much smaller compared to the number of time points in the time period. We assign a value called the **discrepancy score** to each segmentation which measures the likelihood of a topic across all segments in a segmentation. We prove that the segmentation where each segment contains a single time point has the highest discrepancy score and hence captures the likelihood of the topic entirely. However, a user may desire a segmentation containing only a few segments, say k. Then, we define the **min-difference** k segmentation problem which constructs a segmentation containing k segments such that the difference between the discrepancy score of the k segmentation and that of the finest segmentation is minimum. We describe a dynamic programming based algorithm to compute the k segmentation with minimum difference.

We used two data sets – the titles of Reuters and CNN news articles from the TDT-Pilot Corpus published during 1994-1995, to conduct a preliminary set of experiments. We constructed 5 topics and the preliminary results of experiments show that the k segmentation constructed by our method preserves the temporal trends of the topic in each segment.

The rest of the paper is organized as follows. After discussing some preliminaries in Section 2, we introduce the definition of a topic and its measure in Section 3. In Section 4, we mainly discuss the discrepancy score of a segmentation, definition of a min-difference k segmentation, and describe an algorithm to compute it. Section 5 describes experimental results. Section 6 discusses some related work. Section 7 concludes the paper.

2   Preliminaries

2.1   Concepts Related to Segmentation

A *time point* is an instance of time with a given *base granularity*, such as a second, minute, day, month, year, etc. A time point can be represented by a single numerical value, specifying a given second, minute, day, etc. Let T be a time period. We use $|T|$ to denote the number of time points in T.

An *interval* of T is a list of two or more consecutive time points of T. We use the notation $[i, j]$ $(i \leq j)$ to denote the interval of T containing time points t_i, $t_{i+1}, ...t_j$. We also use T_i, T_j etc., to denote arbitrary intervals.

A **segmentation** Γ of a time interval T, is a sequence of subintervals T_1, T_2, ... T_k, such that T_{i+1} immediately follows T_i for $1 \leq i < k$, and T is equal to the concatenation of the k time intervals, which we can write as $T = T_1 * T_2 * ... * T_k$. Each T_i is called a **segment** of Γ. The **size** of segmentation Γ is the number of segments k in Γ and is denoted by $|\Gamma|$. If The time interval associated with a segmentation Γ is denoted as $\Gamma(T)$. Let Γ be a segmentation of a time period T. If $|\Gamma| = |T|$, then Γ is called the **finest** segmentation of T and is denoted by Γ_f. The finest segmentation of a segment T_i is denoted by $\Gamma_f(T_i)$.

Let D be a document set spanning over the time period T. Suppose T is a list of time points $t_1, \ldots, t_n$. Let $b_i \subseteq D$ denote the document set published/created at time point t_i. The document set of an interval $[i, j]$, denoted by b_{ij}, is simply the union of the document sets at each time point in $[i, j]$.

2.2   Fuzzy Set Theory

Fuzzy set theory has been established by L.A.Zadeh in 1965 [18]. Fuzzy set extends traditional set such that every element has degree of membership. A fuzzy set F is a pair (A, f), where A is a traditional set and f is a membership function $f : A \rightarrow [0, 1]$. For each $x \in A$, $f(x)$ is the grade of membership of x belongs to the fuzzy set. If $A = \{x_1, ..., x_n\}$ the fuzzy set (A, f) can be denoted as $\{f(x_1)/x_1, ..., f(x_n)/x_n\}$. For element x_i, if $f(x_i) = 1$, x_i certainly belongs to F, and if $f(x_i) = 0$, then x_i certainly does not belong to F.

3   Measure of a Topic

In this section, we define a topic and describe how to compute the measure of a multi-keyword topic using fuzzy set theory in a document set. The definition of a topic and its measure were first defined in [4].

3.1   Definition of a Topic

We define a **topic** p as a list of pairs, $\{ (kw_i, w_i)|1 \leq i \leq s\}$, where kw_i's are topic keywords and w_i's denote the keyword weights. The value of each $w_i \in (0,1]$. We require that the sum of all w_i in a topic p add up to 1. If p contains a single keyword, then the weight of that keyword is 1.

Given a topic p, $S_{kw}(p)$ denotes the set of keywords in topic p. Function $Ex(kw_i, d_j)$ denotes whether keyword kw_i appear in the document d_j; if it exists, the function returns a 1, otherwise it returns 0.

Given a topic $p= \{ (kw_i, w_i)|1 \leq i \leq s\}$ and a document d_j, we define a **membership function**, denoted by mb, as follows.

$$mb(p, d_j) = \sum_{i=1}^{s} w_i \times Ex(kw_i, d_j) \tag{1}$$

3.2   Measure of a Topic

Given the document set D and a topic p, a **related topic set** $Rt_{p,D}$ is a fuzzy set (D, mb) where mb is the membership function. For each document $d \in D$, $mb(p, d)$ reflects how d related to p.

We define the **measure** of a topic p over a document set D as the sum of the membership of each document in D to topic p.

$$m(p, D) = |Rt_{p,D}| = \sum_{d_i \in D} mb(p, d_i) = \sum_{d_i \in D} \sum_{kw_j \in Kw(p)} w_j \times Ex(kw_j, d_i) \tag{2}$$

Measure of a topic p in a time point t_i, denoted by m_i, is simply $m(p, D_{ii})$ and that measure in an interval of $[i, j]$, $m(p, D_{ij})$, denoted by m_{ij}. If p has a single keyword, then w_i is 1 and $m(p, D)$ is the number of documents containing the keyword. If document d_i contains all keywords in $S_{kw}(p)$, its $mb(p, d_i) = 1$.

The measures of a topic over different document sets qualitatively describes the relevance of the document set to the specific topic. This value denotes the amount of presence of the topic in each document set.

4   Discrepancy Score of a Segmentation

Discrepancy score of a segmentation is designed to evaluate the likelihood of the distribution of the measure (or useful information) of topic p across all the segments in the segmentation. We define discrepancy score of a segmentation based on the temporal scan statistic [1]. Scan statistics have been are used to detect and evaluate clusters of case in temporal, spatial or space-time setting. The scan statistic considers not only interested information, but also adjusts to the background knowledge.

The likelihood function for the distribution of the measure of a topic p in a segmentation is $\prod_{i=1}^{k}(\frac{m_i}{b_i})^{m_i 1}$, where m_i is the measure of p and b_i is the

---

[1] This function measures the presence of p as $\frac{m_i}{b_i}$ as well as boost the presence of consecutive signals.

number of documents (background knowledge) in the ith segment. This must be contrasted with $(\frac{M}{B})^M$, where M is sum of all m_is, and B is the sum of all b_is, i.e., the whole time period is treated a single segment. If the value of the likelihood function is high compared to the ratio of the whole period, then there is a value in partitioning the time period into k segments. We use logarithms to convert the above formula into summations for ease of computation. Therefore, we define the discrepancy score of a k size segmentation as follows.

Let Γ_k be a segmentation containing k segments. Let p be a given topic. Given a segment $[i, j]$, we first define the **segment score** $d(p, i, j)$ as follows.

$$d(p, i, j) = m_{ij}(p, D) \log \left(\frac{m_{ij}(p, D)}{b_{ij}}\right) \tag{3}$$

We define the **discrepancy score** of Γ_k as follows.

$$ds(\Gamma_k, p) = \sum_{[i,j] \in \Gamma} d(p, i, j) \tag{4}$$

The following theorem proves that the finest segmentation Γ_f has the highest discrepancy score of any segmentation.

Theorem 1. *Let T be a time period containing n time points. Let Γ be a segmentation of size k where $k < n$. Then, $ds(\Gamma_k, p) \leq ds(\Gamma_f, p)$.*

Proof. For function $F(m_i, ..., m_j, b_i, ..., b_j) = \sum_{l=i}^{j} m_l \log \frac{m_l}{b_l}$ with constraints that $\sum_{l=i}^{j} m_l = m_{ij}$ and $\sum_{l=i}^{j} b_l = b_{ij}$. $d(i, j) = m_{ij} \log \frac{m_{ij}}{b_{ij}}$, Using Lagrange multipliers method, The minimum value of $F(m_i, ..., m_j, b_i, ..., b_j)$ is achieved when $\frac{m_i}{b_i} = ... = \frac{m_j}{b_j} = \frac{m_{ij}}{b_{ij}}$. So $\sum_{l=i}^{j} m_l \log \frac{m_l}{b_l} \geq m_{ij} \log \frac{m_{ij}}{b_{ij}}$. We can use this relationship to expand Equation 4

$$ds(\Gamma_k, p) = \sum_{[i,j] \in T(\Gamma)} d(p, i, j) \leq \sum_{l=1}^{n} m_{l,l} \log \frac{m_{ll}}{b_{ll}} = ds(\Gamma_f, p) \tag{5}$$

Where Γ is any partition with K segments.

We can also prove that the discrepancy score of a segmentation may increase with its size.

Lemma 1. *Let Γ and Γ' be two segmentations where $|\Gamma| = k - 1$ and $|\Gamma'| = k$. Then, $ds(\Gamma, p) \leq ds(\Gamma', p)$.*

4.1   Min-Difference Segmentation

The finest segmentation of the measure of topic p shows the actual progression of the topic over time. Therefore, the sum of the segment scores of the finest segmentation gives the maximum likelihood of the measure of the topic over the entire time period. However, the finest segmentation may contain noise which may hide the real temporal change in the measure of a topic. Therefore, we can find a segmentation that contains fewer segments, say k, than the finest segmentation where the changes in the measure of p is very similar to that in the finest segmentation. We call such a segmentation as a **min-difference k segmentation**.

The desired number k of segments in a segmentation is specified by the user. The min-difference segmentation problem can be defined as follows. Given the desired number k of segments, find a segmentation Γ containing at most k segments and $ds(\Gamma_f, p) - ds(\Gamma, p)$ is a minimized.

We solve this problem using a dynamic programming based algorithm. The algorithm works in a bottom-up manner, recording the minimum difference between the segment score of segment $[i, j]$'s of finest segmentation and that of the k segmentation of $[i, j]$.

The biggest discrepancy score for time interval $[i, j]$ is $ds(\Gamma_f([i, j]), p)$, which is $\sum_{l=i}^{j} d(p, l, l)$. Our goal is to find a k segmentation, $1 \leq k < (j - i + 1)$, such that the difference between the finest and k segmentation is minimized. Therefore, k is must result in $\min_{\Gamma([i,j])} (ds(\Gamma_k([i, j]), p) - ds(\Gamma_f([i, j]), p)$.

For each value of k $(1 \leq k < (j - i + 1))$, the minimum difference incurred for partitioning interval $[i, j]$ into k intervals is recorded in a table S as entry $S(k, i, j)$. The values are computed as follows.

$$S(k, i, j) = \begin{cases} ds(\Gamma_f([i, j]), p) - d(p, i, j) & k = 1 \\ \min_{l \in [k-1, j]} (S(k-1, i, l) + ds(\Gamma_f([l+1, j]), p) - d(p, l+1, j)) & k > 1 \end{cases}$$
(6)

In Equation 6, if $i > j$ then $d(p, i, j)$ is undefined. In addition, if segmentations have the same minimum difference, we select the segmentation with the smaller size.

4.2 Preparing the Measure

In this section, we describe an efficient method for computing the min-difference segmentation. To compute $S(1, n, k)$, we need values m_{ij} and b_{ij} efficiently. Let us assume that the time period of the given documents T contains n time points. The input two vectors are V_m and V_b of size n are constructed from the input document set. Vector $V_m[i]$ contains the values of m_i and vector $V_b[i]$ contains the values of b_i for each of the n time points in T. We preprocess the input vectors V_m and V_b so that values m_{ij} and b_{ij} in all segments $[i, j]$ can be computed in constant time. We construct cumulative two vectors CV_m and CV_b of size n as follows. It is easy to observe that both vectors CV_m and CV_b can be constructed in $O(n)$ time.

$CV_m[1] = m_1$;
$CV_m[i] = CV_m[i - 1] + m_i; 2 \leq i \leq n$;

Vector CV_b is similarly computed from b_i values. We can access m_{ij} and b_{ij} in constant time as follows.

$$m_{ij} = CV_m[j] - CV_m[i-1];$$
$$b_{ij} = CV_b[j] - CV_b[i-1];$$

4.3 Computing Discrepancy Score of Min-Difference k Segmentation

According Equation 3, given time interval $[i,j]$ we can compute $d(p,i,j)$ in constant time.

We can compute $ds(\Gamma_f[i,j],p)$ for each $[i,j]$ using cumulative vectors to in constant time.

$$CDs[1] = d(p,i,i);$$
$$CDs[i] = CDs[i-1] + d(p,i,i); \; 2 \le i \le n;$$

And $ds(\Gamma_f([i,j]),p)$ can be computed as following

$$ds(\Gamma_f([i,j]),p) = CDs[j] - CDs[i-1];$$

Let k be the desired number of segments specified by the user and recall that $|T| = n$. The algorithm uses an array s to store the minimum difference values. We always set the initial time point i as 1, so we only keep record of $s[g][j]$ which denote to $S(g,1,j)$ in Equation 6. Entry $s[g][j]$ stores that minimum difference value incurred in partitioning the interval $[1,j]$ into g ($g \le k$) segments. In addition to array s, the algorithm maintains another array called *path* in which it stores the information to construct the min-difference segmentation. Entry $path[g][j]$ records the value l at which the interval $[1,j]$ is partitioned to achieve the minimum difference value. The min difference value for splitting the time period T into at most k segments will be stored in $s[k,n]$ and the starting point of the kth segment will be in $path[g][n]$.

We assume that vectors CV_m, CV_b and CD_s are available to the algorithm so that segment score $d(p,i,j)$ and $ds(\Gamma_f([i,j]),p)$ can be computed in constant time. We note that the running time of this algorithm is $O(n^2 k)$.

Algorithm to construct the k segmentation of the time period T

```
MinScore (s[][],path[][])

  begin
    Initiate every unit in s[][] as +infinity;
    Initiate every unit in path[][] as 0;
    For i=1 to n s[1][i]=Ds(1,i)-d(p,1,i);
    For k=2 to K
      For i=k to n
        s[k][i]=s[k-1][i];
        path[k][i]=i;
        For l=k-1 to i-1
          sd=s[k-1][l]+Ds(l+1,i)-d(p,l+1,i);
          if sd<s[k][i] then
```

```
        s[k][i]=sd;
        path[k][i]=1; // The minScore are stored in s[][]
    \\ use path[][] to trace back the partition result
  Initiate rp[K] as all 0s;
  use path[][] trace back and store partition series into rp[];
  return rp[];
end.
```

5   Experiments

5.1   Data Sets

We applied our approach to document set from the TDT-Pilot Corpus. We extracted the titles of news articles from the TDT-pilot corpus (http://projects.ldc .upenn.edu/TDT-Pilot/). The TDT-Pilot corpus covers about 16,000 stories, in which half is collected from Reuters newswire and half from CNN broadcast news transcripts from July 1, 1994 to June 30, 1995. This data set contains 365 time points, one for each day.

5.2   Results

We used a Red Hat Linux System with a dual core Intel Xeon 3.73GHz processor with a 3GB of RAM to conduct experiments. We implemented the algorithms described in previous section using the R Project (http://www.r-project.org).

To study the performance of the min-difference k segmentation algorithm described in the paper, we randomly constructed the following five topics: **Simpson, Simpson case, Iraq Kuwait, Haiti invasion, North Korea**. All keywords in a topic are weighed equally. We then computed the measure m_i of each topic in each of the time points t_i and collected the b_i values required to the segment score and the discrepancy score. We then constructed the cumulative vectors – CV_m, CV_b, and CDs and then applied the min-difference k segmentation algorithm to construct a k segmentation for each of the topics.

In our experiments, we set k to different values to observe the difference (or $S(k, 1, n)$ in Equation 6) between discrepancy score of the segmentation and the finest one. In fact, the reduce of difference become smaller or more stable when k is beyond 20.

In the Figure 1, we show how these difference change with increasing k . In Figures 1, the X-axis is the value of k, the Y-axis is the difference of discrepancy score between min-difference k segmentation and finest segmentation, which we denote as S. To show the trend of S, we pick S of five topics when number of segmentation is set as 10, 20, 30 and 40.

For example, in Figure 1, for topic "Simpson", the value of S at 10 segmentation is 208, and this means that the discrepancy score of min-difference 10 segmentation on topic "Simpson" is 208 less than finest segmentation and value of S decreases to 123 when k sets to 40. It is clear that S reduces as k increases in all 5 topics. This trend confirms to the Lemma 1.

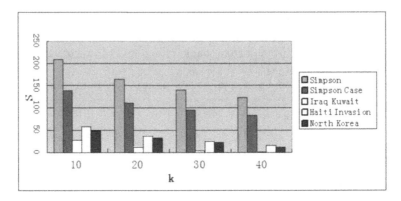

Fig. 1. The Trends of Min-Difference k Segmentation for Five Topics from CNN

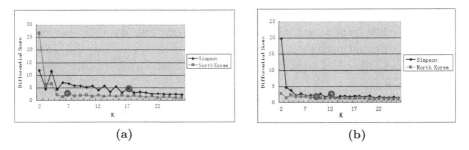

(a)   (b)

Fig. 2. (a)The Differential Score in Min-Difference k Segmentation from CNN. (b)The Differential Score in Min-Difference k Segmentation from Reuters.

Our next experiment analyzed the differential score between min-difference k and $k+1$ segmentations to observe changes in discrepancy score of different size segmentations. The algorithm outputs the minimum difference between the discrepancy score of a min-difference k segmentation and that of the finest segmentation of T. As proved in Lemma 1, k segmentation has a discrepancy score that is as much as or less than the $k+1$ segmentation. Therefore, $s[k, n] \geq s[k+1, n]$. Differential score analysis studies $s[k + 1][n] - s[k][n]$ $(1 \leq k < n)$. If the differential score is of a high value, then $k + 1$ segments capture the information from the finest segmentation more closely than k segments. On the other hand, if the differential score is a small value, one can conclude that both k and $k+1$ segmentations represent the finest segmentation closely. Therefore, the differential score analysis can be used to decide the value for k that best represents the finest segmentation without losing significant trends.

In Charts 2a and 2b, X-axis plots k values and the Y-axis plots differential score. It can be observed that the curve of differential score changes sharply after the segmentation capture a big trend among data. For example, in Chart 2a, the differential score between the size 17 and size 18 segmentations of topic "Simpson" is 4.88, after $k = 17$, the curve of differential score appears much

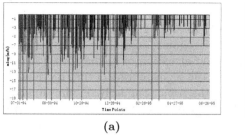
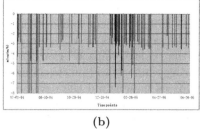

(a)   (b)

Fig. 3. (a)17 Segmentation of Topic "Simpson" from CNN. (b)12 Segmentation of Topic "Simpson" from Reuter

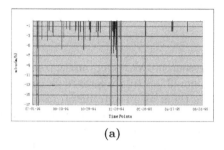
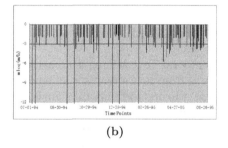

(a)   (b)

Fig. 4. (a)Min-Difference 7 Segmentation of Topic "North Korea" from CNN. (b)Min-Difference 9 Segmentation of Topic "North Korea" from Reuter.

smoother than it was for values of k less than that, therefore, we choose 17 as the size of min-difference segmentation for the topic "Simpson" in CNN data. In the same reason, we choose 7 as the size of min-difference segmentation for topic "North Korea" from CNN. Similarly, in Figure 2b we select 12 as size of segmentation for the topic "North Korea" in CNN data and 9 as size of segmentations Reuters data. The results of segmentations for these topics are showed in Figures 3 and 4.

We use Figures 3 and 4 to display the final segmentation itself. The horizontal axis in these figures lists all 365 time points, starting from July 1, 1994 and ending at June 30, 1995. The vertical axis plots the segment score of each of the time points. Higher difference between the segment scores of t_i and t_j implies less likelihood the t_i and t_j are in the same segment of some min-difference k segmentation. Therefore, if segment $[i, j]$ is in the min-difference segmentation of size k, then, values $m_g log(m_g/b_g)$ $g \in [i, j]$ should be more similar to each than to those in consecutive segments.

We use parallel vertical lines to show the segments resulting from a min-difference segmentation. In Chart 3a, there are 16 vertical lines separating the time period into 17 segments. For example, it can be observed that the pattern of $m_g log(m_g/b_g)$ value in the 15th segment is very different than that in the 14th or 16th segment. We selected the 1st segment from 07-01-94 to 07-04-94, and the

2nd segment from 07-05-94 to 07-07-94. We checked the CNN document titles containing the topic "Simpson" in these two segments. The measure of this topic is 11 in 1st segment and increases sharply to 33 in 2nd segment. The contents of the documents in 1st segment and 2nd segment also show the progression of the topic "Simpson" from capturing the attention of media to a hot topic during segments . From this point of view, the segmentation of time on the topic helps us understand the development of the topic. Similarly, Figure 3b shows the min-difference 12 segmentation of the same topic "Simpson" in the Reuters data set. It can be observed from these two figures that the topic may have had different progression patterns in different data sets. We conjecture that it is due to the fact that different media outlets may have different reporting interests. The difference in progression patterns can also be observed for topic "North Korea" as shown in Figure 4.

6   Related Work

Temporal Text Mining is an active area of research. Detecting bursty topic patterns from text streams is an emerging problem [3,4,8,16]. Reference [8] converts text streams to temporal signals and models them as an automaton to identify bursty features in their structure. Reference [16] mines multiple coordinated text streams to identify correlated bursty topic patterns. In our earlier work [3,4], we used the temporal scan statistic to identify the interval where a topic has an increased presence.

Time series segmentation is well established field, more information can be seen in [5], [7], [9] and [15]. An interesting related work is [11] where authors used a general probabilistic method to capture theme features, construct an evolution graph of themes and analyze life cycles of themes. The current paper is different from [11] in that we are focused on capturing the trends of multi-word topics by segmenting the time period associated with the document set.

Scan statistics such as the temporal statistics scan were proposed by Kulldorff [10].This approach is an important statistical tool to detect an abnormal cluster and is applied in the public health community for disease trend detection. Temporal scan statistic has been well studied in [1,13,14].

Fuzzy logic systems was applied to the area of information systems in Europe in the early 1990s (see [2]). In paper [17] authors provide a fuzzy clustering algorithm for the analysis of document collections. The method illustrated in the current paper to compute a numeric value of relevance of a document set to a multi-keyword topic based on the fuzzy set theory was first discussed in [4].

7   Conclusion

This paper describes an approach based on the segmentation technique to extract temporal trends of topics from a time stamped document set. We assume that the time period associated with the document set is a list of time points, where each time point is an hour, a day, etc. A segmentation of the time period of the document partitions the time period into segments where each segment contains one or more consecutive time points. We define topics with multiple keywords and give a method based on the fuzzy set theory to compute the relevance of a document set to the topic. The measure of relevance for a given topic in a document set is used to assign a discrepancy score to a segmentation. We define the notion of a min difference k-segmentation and give a dynamic-programming based algorithm to compute it. A min difference k-segmentation maximizes the likelihood of the topic in each of its segments. We then conducted a preliminary set of experiments using the titles of Reuters and CNN news articles from the TDT-Pilot Corpus. The experiments confirm that the discrepancy score of segmentation increases with its size. Our experiments also show that the change in the discrepancy score for segmentations of different sizes can be used to determine the number of segments appropriate for capturing all of the temporal trends of the topic.

References

1. Agarwal, D., McGregor, A., Phillips, J.M., Venkatasubramanian, S., Zhu, Z.: Spatial scan statistics: approximations and performance study. In: KDD 2006: Proceedings of the 12th ACM SIGKDD international conference on Knowledge discovery and data mining, pp. 24–33. ACM Press, New York (2006)
2. Baeza-Yates, R.A., Ribeiro-Neto, B.A.: Modern Information Retrieval. ACM Press/Addison-Wesley (1999)
3. Chen, W., Chundi, P.: An approach for discovering hot spots of topics from time stamped documents. In: 2008 SDM Text Mining Workshop (2008)
4. Chen, W., Chundi, P.: Extracting hot spots of basic and complex topics from time stamped documents. In: IEEE Conference on Computational Intelligence and Data Mining 2009 (2009)
5. Gionis, A., Mannila, H.: Segmentation algorithms for time series and sequence data. In: A Tutorial in the SIAM International Conference on Data Mining (2005)
6. Marti, A.: Hearst and Christian Plaunt. In: Subtopic structuring for full-length document access, pp. 59–68 (1993)
7. Keogh, E., Kasetty, S.: On the need for time series data mining benchmarks: a survey and empirical demonstration. In: KDD 2002: Proceedings of the eighth ACM SIGKDD international conference on Knowledge discovery and data mining, pp. 102–111. ACM Press, New York (2002)
8. Kleinberg, J.: Bursty and hierarchical structure in streams. Data Min. Knowl. Discov. 7(4), 373–397 (2003)
9. Leung, C.K.-S., Ng, R.T., Mannila, H.: Ossm: A segmentation approach to optimize frequency counting. In: International Conference on Data Engineering, p. 0583 (2002)
10. Martin, K.: A spatial scan statistic. Communications in Statistics: Theory and Methods 26, 1481–1496 (1997)
11. Qiaozhu, M., ChengXiang, Z.: Discovering evolutionary theme patterns from text: an exploration of temporal text mining. In: KDD 2005: Proceedings of the eleventh ACM SIGKDD international conference on Knowledge discovery in data mining, pp. 198–207. ACM Press, New York (2005)

12. Nakao, Y.: An algorithm for one-page summarization of a long text based on thematic hierarchy detection. In: Proceeding of ACL 2000, pp. 302–309 (2000)
13. Neill, D.B., Moore, A.W.: A fast multi-resolution method for detection of significant spatial disease clusters. In: Advances in Neural Information Processing Systems, vol. 16, pp. 651–658. MIT Press, Cambridge (2004)
14. Neill, D.B., Moore, A.W., Pereira, F., Mitchell, T.: Detecting significant multidimensional spatial clusters. In: Advances in Neural Information Processing Systems, vol. 17, pp. 969–976. MIT Press, Cambridge (2005)
15. Siy, H., Rosenkrantz, P., Chundi, D.J., Subramaniam, M.: Discovering dynamic developer relationships from software version repositories using time series segmentation. In: 23rd IEEE International Conference on Software Maintenance, pp. 437–445. ACM Press, New York (2007)
16. Xuanhui, W., ChengXiang, Z., Xiao, H., Richard, S.: Mining correlated bursty topic patterns from coordinated text streams. In: KDD 2007: Proceedings of the 13th ACM SIGKDD international conference on Knowledge discovery and data mining, pp. 784–793. ACM Press, New York (2007)
17. Witte, R., Bergler, S.: Fuzzy clustering for topic analysis and summarization of document collections. In: Kobti, Z., Wu, D. (eds.) Canadian AI 2007. LNCS, vol. 4509, pp. 476–488. Springer, Heidelberg (2007)
18. Asker Zadeh, L.: Fuzzy sets. Information Control 8, 338–353 (1965)

Finding N-Most Prevalent Colocated Event Sets

Jin Soung Yoo and Mark Bow

Department of Computer Science, Indiana University-Purdue University,
Fort Wayne, Indiana, USA
{yooj,bowmg01}@ipfw.edu

Abstract. Recently, there has been considerable interest in mining spatial colocation patterns from large spatial datasets. Spatial colocations represent the subsets of spatial events whose instances are frequently located together in nearby geographic area. Most studies of spatial colocation mining require the specification of a minimum prevalent threshold to find the interesting patterns. However, it is difficult for users to provide appropriate thresholds without prior knowledge about the task-specific spatial data. We propose a different framework for spatial colocation pattern mining: finding N-most prevalent colocated event sets, where N is the desired number of event sets with the highest interest measure values per each pattern size. We developed an algorithm for mining N-most prevalent colocation patterns. Experimental results with real data show that our algorithmic design is computationally effective.

1   Introduction

The evolution of location sensing and mobile computing is generating lots of rich spatial datasets. Examples of such data include geographic search logs, GPS logs, environmental observation data, climate measurements, disease occurrence records, and so on. As one of important spatial data mining tasks, spatial colocation pattern mining has been popularly studied in spatial data mining literature [16,21,22,5,14,11,4,18]. A spatial colocation pattern represents a subset of spatial events whose instances are frequently located together in a neighborhood area. For example, a mobile service provider may be interested in service request types frequently queried by geographically neighboring users. A {'shopping mall', 'parking lot', 'restaurant'} might be a colocation pattern discovered from logs of service search engines. The mining result can be used for providing location-sensitive advertisements, recommendations, and so on.

Let E be a set of event types, S be a set of their objects, and R be a spatial neighbor relationship over S. A set of event $C \subseteq E$ becomes a *colocation* pattern if its instance objects $I \subset S$ frequently form cliques under the neighbor relationship R, that is, the prevalence strength of the colocated event set C on space, $Pi(C)$, is greater than a given threshold min_prev. A common framework of mining spatial colocation patterns, therefore, requires a user specified minimum prevalence threshold min_prev to find the interesting patterns. However, without prior knowledge about the task-specific spatial data, users may have

T.B. Pedersen, M.K. Mohania, and A M. Tjoa (Eds.): DaWaK 2009, LNCS 5691, pp. 415–427, 2009.

difficulties in setting proper prevalence thresholds to obtain desired results. If the prevalence threshold is set too high, there may be only a small number of results or even no result. If the threshold is too low, too many results can be generated with an exceedingly long computational time. Users also need extra efforts to screen interesting patterns. Another argument against the use of a uniform prevalence threshold for all colocation patterns is that the probability of occurrence of larger size colocations is inherently much smaller than that of smaller size colocations. To solve these problems, we propose a different framework for colocation pattern mining, in which users specify a threshold on the amount of results instead of a prevalence threshold. In particular, we explore the task of **mining N-most prevalent colocated event sets** from a spatial dataset, which finds N k-colocated event sets with the highest prevalence values for each size k up to a certain k_{max} value.

Similar frameworks such as Top-K frequent association pattern mining were proposed in classical data mining literature [6,2,1,17,8]. However, it is nontrivial to reuse the association pattern mining algorithms [6,2,1] for discovering our N-most prevalent colocation patterns due to the complexity of spatial data types, spatial relationships and spatial autocorrelation [15]. State-of-the-art algorithms for spatial colocation discovery [14,11,16,21,22] use generation-and-test methods like *Apriori*, and reduce the number of candidate sets for colocation patterns using the *downward closure property* of their interest measure [16]. However, the property does not hold in our problem. Although a colocated event set is in the result set of the N-most patterns, its subsets cannot be included in the result set. In the worst case, therefore, we have to consider all possible subsets of events. It is crucial to effectively reduce the search space of colocated event sets, and efficiently find their colocation instances from a large spatial dataset.

The main contributions of this work are first to propose a different framework for spatial colocation pattern mining, and then develop an efficient algorithm for mining the N-most prevalent colocation patterns. We proved the algorithm is correct and complete in finding the N-most patterns. Experimental results with real datasets show that our algorithmic design is effective in reducing the number of candidates, and the proposed algorithm is scalable in various parameter settings. The remainder of the paper is organized as follows. Section 2 introduces the basic concepts of colocation pattern mining, and describes our problem statement and the related work. Section 3 presents our algorithmic design concepts, and the proposed algorithm. Section 4 discusses the experimental results. Section 5 ends with the conclusion and future work.

2 Problem Statement and Related Work

2.1 Basic Concepts and Problem Statement

Given a set of event types $E = \{e_1, \ldots, e_m\}$, a set of their instance objects S, and a spatial neighbor relationship R over S, a **colocation** C is a subset of spatial events, $C \subseteq E$, whose instance objects frequently form cliques using R. When the Euclidean metric is used for the neighbor relationship R, two

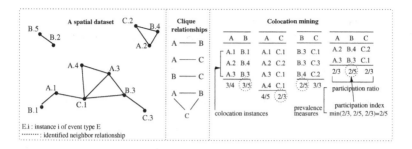

Fig. 1. Spatial colocation pattern

spatial objects are neighbors if the distance between them is not greater than a neighborhood distance bound, e.g., 0.5 mile. Fig. 1 shows an example dataset with three event types, A, B and C. Each object is represented by its event type and unique instance id, e.g., A.1. Identified neighbor objects are connected by a line in the figure. A **colocation instance** I of a colocation C is a set of objects, $I \subset S$, which includes all event types of C, and forms a clique relationship. For example, {A.2, B.4, C.2} is a colocation instance of {A, B, C}. The prevalence of a colocation is often measured with participation index defined in [16].

Definition 1. *The* **participation index** $Pi(C)$ *of a colocation C is defined as* $Pi(C) = \min_{e_i \in C}\{Pr(C, e_i)\}$, *where $Pr(C, e_i)$ is the* **participation ratio** *of event type e_i in a colocation $C = \{e_1, \dots, e_k\}$ that is the fraction of objects of event e_i in the neighborhood of instances of colocation $C - \{e_i\}$, i.e.,*
$$Pr(C, e_i) = \frac{Number\ of\ distinct\ objects\ of\ e_i\ in\ instances\ of\ C}{Number\ of\ objects\ of\ e_i}.$$

Consider the prevalence values of colocation c={A, B, C} in Fig. 1. The instances of colocation c are {A.2, B.4, C.2} and {A.3, B.3, C.1}. The participation ratio of event A in the colocation c, $Pr(c, A)$ is $\frac{2}{4}$ since only A.2 and A.3 among four objects of event A are involved in the colocation instances. In the same way, $Pr(c, B)$ is $\frac{2}{5}$ and $Pr(c, C)$ is $\frac{2}{3}$. Thus the participation index of colocation c, $Pi(c)$, is $min\{Pr(c, A), Pr(c, B), Pr(c, C)\} = \frac{2}{5}$. In this paper, we use 'colocation' and 'colocated event set' terms interchangely. A k-**colocated event set** means a colocation containing k spatial event types.

Definition 2. *The N-most prevalent k-colocated event sets: Let L be a list of all k-colocated event sets by descending their participation index values, and let p be the participation index of the Nth k-colocated event set in the list L. The N-most prevalent k-colocated event sets are a set of k-colocated event sets having participation index $\geq p$.*

The N-**most prevalent colocated event sets** are the union of the N-most prevalent k-colocated event sets for $2 \leq k \leq k_{max}$, where k_{max} is the upper bound of the size of colocation patterns we would like to find. The problem statement of finding N-most prevalent colocated event sets is:

Given:
1) A set of spatial event types $E = \{e_1, \ldots, e_m\}$
2) A dataset of spatial point objects $S = S_1 \cup \ldots \cup S_m$ where $S_i (1 \leq i \leq m)$ is a set of objects of event type e_i. Each object $o \in S_i$ has a vector information of $<$ event type e_i, object id j, location $x, y >$ where $1 \leq j \leq |S_i|$.
3) A spatial neighbor relationship R (a distance function and a distance bound)
4) A maximum size of colocated event sets (max_k), and a number of colocated event patterns per each size (N)
Objective:
Find N-most prevalent colocated event sets per each size k where $2 \leq k \leq max_k$ while reducing the computation cost.
Constraints:
R is a distance based neighbor relationship, and is symmetric and reflexive.

2.2 Related Work

The problem of mining association rules based on spatial relationships (e.g., proximity, adjacency) was first discussed in [11]. After that, spatial colocation pattern mining has been popularly studied in [16,21,22,5,14,11,4,18]. Most works present different approaches in identifying colocation instances and choosing the interest measures for colocation patterns. Their methods focus on finding colocation patterns that satisfy a given minimum prevalence threshold. To the best of our knowledge, there is no previous work in mining colocation patterns without a prevalence threshold. In classical association pattern mining, Fu, et al. [6] proposed the first algorithm for discovering N-most interesting itemsets. Cheung, et al. [2] extended their previous work [6] and developed a different method based on a frequent pattern-growth algorithm. Arshad, et al. [1] solved the problem using a support-ordered trie structure. Hirate, et al. [8] proposed a TF^2P-growth algorithm for mining frequent patterns without any thresholds. Wang, et al. [17] proposed the problem of mining Top-K frequent closed patterns without minimum support. It is nontrivial to use these Top-k methods for our problem due to the complexity of spatial data types and spatial relationships. A similar framework for graph mining was proposed in [23]. The work locates top-k graphs matching a given query graph in a large data graph.

3 Algorithmic Design

In this section, we describe our algorithmic design concept for mining N-most prevalent colocation patterns, and present the proposed algorithm.

3.1 Preprocess

An input spatial dataset can be represented as a neighbor graph with the spatial objects being its vertex set, and an undirected edge between two objects where they are neighbors each other. For discovering colocation patterns, we need to find all colocation instances forming cliques from the neighbor graph, and then compute a participation index per event set. However, it is computationally

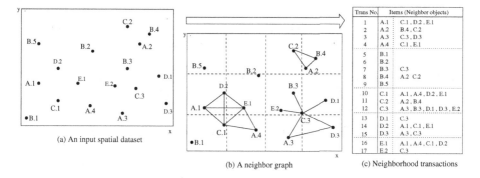

Fig. 2. Preprocess: neighborhood transaction

expensive to find all cliques from a graph [3]. On the other hand, if we can convert a spatial dataset into a graph transaction set, we may consider frequent subgraph mining algorithms [12,19,10] for our problem. However, it is not easy to represent a spatial dataset to a set of distinct graphs. An explicit partitioning of the data may lose some neighbor relationships across different partitions. Another difference from a graph data for frequent subgraph mining is that our object is represented with its event type and instance id. There are many instance objects per event type. Most frequent subgraph mining algorithms do not consider multiple instances of a subgraph in a transaction. In this paper, we represent an input spatial data to a set of neighborhood transactions.

Definition 3. *Given a spatial object $o_i \in S$, the* **neighborhood transaction** *of o_i is defined as a set of spatial objects $\{o_i, o_j \in S | R(o_i, o_j) = true \land o'_j s$ event type $\neq o'_i s$ event type$\}$, where R is a neighbor relationship.*

In Fig. 2, the neighborhood transaction of A.1 is {A.1, C.1, D.2, E.1}. All objects of the set have neighbor relationships with the first object, A.1, which is called a reference object. This neighborhood transaction approach gives several advantages for colocation pattern mining. First, the neighborhood transaction set does not loose any object and neighbor relationship of the original data. Second, the neighborhood transactions can be easily constructed from the neighbor pair objects of the input data. Third, they can give the information of upper bound of participation index of a colocated event set. Finally, we can also use the neighborhood transactions for filtering candidate event sets.

3.2 Candidate Generation

The total number of sub event sets examined for the N-most patterns is $2^m - m - 1$ since there is no prevalence threshold given, where m is the number of event types, and single events and empty set are not considered. Rather than considering all possible sets, it would be desirable to focus on potential event sets having at least one colocation instance. We generate colocation candidates using a project based pattern mining paradigm [7]. The idea was also used in [9].

1) Event trees

An *event tree* is similar with an FP-tree which is one of popular data structures for classical association rule mining [7]. The FP-tree is a prefix-tree structure for storing compressed information about frequent patterns without generating candidates. However, it is hard to represent an input spatial data to one FP-tree with preserving all neighbor relationships among the objects. Instead, we construct one tree per each event type for storing its neighbor information.

Definition 4. *A* **reference event pattern tree** *(or* **event tree** *in short) is a tree structure defined: (1) It consists of one root labeled as a reference event type, a set of event prefix subtrees as the children of the root, (2) Each node consists of three fields: event-type, count, and node-link, where event-type denotes a event this node represents, and count registers the number of neighborhood transactions represented by the portion of the path reaching this node. The transactions start with an object whose event type is the same with the event type of the root note. Node-link links to the next node in the tree carrying the same event-type.*

For example, in Fig. 2 (c), the transactions which start with 'A' event type are used for building 'A' reference event pattern tree. Fig. 3 (a) shows all event trees constructed from the transactions in Fig. 2 (c).

2) Colocation candidates

We use FP-growth algorithm [7] for generating event sets from each event tree. The difference from the output of the original FP-growth is that each set has the event type of root node as the first element. Fig. 3 (b) shows the event sets generated from each event tree. We call the result sets **star candidate** sets since all elements of a set have a neighbor relationship with its first element. The output also gives the frequency information of an event set, i.e., support. The support value of a set presents the frequency that its first item event has a neighbor relationship with the other events in the set. For example, the support value of {B,A,C} represents how many objects of event B has a neighbor relationship with the objects of both A and C.

After generating star candidate sets, we combine them for filtering candidates which can have at least one colocation instance. Fig. 3 (c) shows the combined candidate set which is called a **clique candidate** set or a **colocation candidate** set. The colocation candidate inherits the frequency value of each event from its star candidates. For example, {A,B,C} has the frequency values of A, B, and C events, i.e., $\frac{1}{4}$, $\frac{1}{5}$ and $\frac{2}{3}$ from each star candidate {A,B,C}:$\frac{1}{4}$, {B,A,C}:$\frac{1}{5}$ and {C,A,B}:$\frac{2}{3}$. The values represent the upper bound of the chance (i.e., participation ratio) that each event has a clique relationship with the other events in the set. The minimum value of the upper bound participation ratios becomes the **upper bound of participation index** of the colocation.

3) Candidate pruning

We present additional scheme to reduce the candidates further. Since our problem is to discover n most prevalent patterns, there is no prevalence threshold

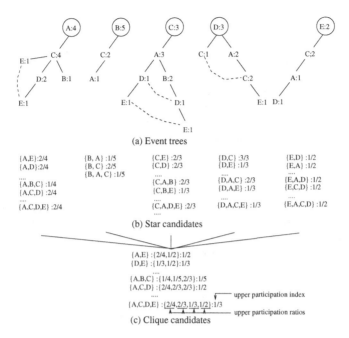

(a) Event trees

{A,E}:2/4	{B, A} :1/5	{C,E} :2/3	{D,C} :3/3	{E,D} :1/2
{A,D}:2/4	{B, C} :2/5	{C,D} :2/3	{D,E} :1/3	{E,A} :1/2
	{B, A, C} :1/5	....		
....			....	....
{A,B,C} :1/4		{C,A,B} :2/3	{D,A,C} :2/3	{E,A,D} :1/2
{A,C,D} :2/4		{C,B,E} :1/3	{D,A,E} :1/3	{E,C,D} :1/2
....		....	....	....
{A,C,D,E} :2/4		{C,A,D,E} :2/3	{D,A,C,E} :1/3	{E,A,C,D} :1/2
		....		

(b) Star candidates

{A,E} :{2/4,1/2}:1/2
{D,E} :{1/3,1/2}:1/3
....
{A,B,C} :{1/4,1/5,2/3}:1/5
{A,C,D} :{2/4,2/3,2/3}:1/2
....
{A,C,D,E} :{2/4,2/3,1/3,1/2}:1/3
(c) Clique candidates

upper participation index
upper participation ratios

Fig. 3. Candidate generation

initially given. However, during the pattern mining phase, we can maintain a minimum prevalence threshold θ_k per size k to determine if an event set can be included in the result set $result_k$. We first make a sorted list of candidates with their upper bound participation index per each size. Any top candidate with its true participation index is blindly included in the result set. Once we have encountered n event sets, we are only interested in event sets whose participation index is greater than the smallest participation index in the result. θ_k is set to the participation index of the nth event set of $result_k$. If the upper bound participation index of a candidate is less than θ_k, all candidates after the candidate in the sorted list are pruned without examining their colocation instances. In addition, we can estimate θ_k in the beginning stage of size k mining.

Lemma 1. *Let θ_l be the smallest participation index in the result set of the N-most prevalent l-colocated event sets, $result_l$, and θ be the participation index of a N-most prevalent k-colocated event set where $k < l$. If there are at least N different event types in $result_l$, then $\theta \geq \theta_l$ where $k < l$.*

Proof. The participation index is monotonically non increasing with increases in the size of colocation by the proof of [16]. That is, for a colocation $C_k \subseteq C_l$, $Pi(C_k) \geq Pi(C_l)$. The participation ratio of an event has the monotonicity property too, i.e., $Pr(C_k, e_i) \geq Pr(C_l, e_i)$ where $C_k \subseteq C_l$, $e_i \in C_k$ and $e_i \in C_l$. Suppose $result_l$ contains N different events, $\{e_1, e_2, ..., e_N\}$. If θ_l is the smallest participation index in $result_l$, the smallest participation ratio of $e_1, e_2, ..., e_N$ is θ_l. The number of size k sets (T_k) generated from the different N events is $\binom{N}{k}$

$\geq N$, where $k < l$ and $k \leq N$. The participation index of any set in T_k is not less than θ_l by the monotonicity property of Pi. Therefore, in order that a candidate from a k-candidate set (C_k), $T_k \subseteq C_k$, becomes the N-most pattern, the set's participation index θ should be at least θ_l.

Therefore, for mining N-most prevalent size k patterns, θ_k can be initialized with the largest θ_l if the result set of size l has at least N different events, where $k < l \leq k_{max}$ and $k \leq N$. If the upper bound participation index of any candidate of size k is less than θ_k, the candidate is eliminated.

3.3   Instance Filtering

After generating colocation candidates, the next steps are to gather their colocation instances, compute their true participation index values, and find the N most prevalent patterns in each size. To find the colocation instances efficiently, we use the neighborhood transactions generated in the preprocess step.

Definition 5. *Let $I = \{o_1, \ldots, o_k\} \subseteq S$ be a set of spatial objects whose event types $\{e_1, \ldots, e_k\}$ are different each other. If all objects in I are neighbors to the first object o_1, I is called the* **star instance** *of colocation $C = \{e_1, \ldots, e_k\}$.*

The star instances of a colocation can be gathered from the neighborhood transactions whose first item's event type is the same as the first event of the colocation. For example, the star instances of {A, C, D} are gathered from the transactions whose first item's event type is 'A'. Therefore the number of candidates examined in each transaction is much smaller than the number of actual candidates. The true colocation instances can be filtered from the star instances. If a star instance {A.1, C.1, D.2} has an additional neighbor relationship between C.1 and D.2, it becomes a colocation instance of {A, C, D}.

3.4   Algorithm

We developed an efficient N-Most Colocation mining algorithm (NMColoc). Algorithm 1 shows the pseudo code of NMColoc algorithm. After generating the neighborhood transactions and colocation candidates, we first find size 2 patterns to reuse the neighbor pair information. We then discover the patterns from size k_{max} to size 3 to use the property of Lemma 1. Due to the page limit, we omit the detail explanation of the algorithm. You may refer to [20] for it. The correctness and completeness of the algorithm are proved in [20].

4   Experimental Evaluation

We compare the efficiency of our proposed algorithm with a general colocation mining algorithm [16]. For the latter, we used a tuned minimum prevalence threshold that is the participation index of the nth pattern of size K_{max}. We use

Inputs
$E = \{e_1, \ldots, e_m\}$:a set of spatial event types
S:a spatial dataset, R:a spatial neighbor relationship
k_{max}:a maximum size of colocated events of interest
N: a number of the patterns of interest per each size

Variables
ST: a set of all neighborhood transactions.
$Tree_i$:an event tree of type e_i, k:interest colocation size
θ_k:a minimum prevalence threshold for N-most k colocated patterns
θ:a minimum prevalence threshold for N-most colocated patterns
C:a set of all candidate sets, C_k:a set of size k candidates
$upper\_pi$: an approximate PI of a candidate, pi: true PI
CI_c: a set of clique instances of a candidate c
SI_c: a set of star instances of a candidate c
SI_k: a set of star instances of size k colocated event sets $SI_c \in SI_k$
R_k:a set of size k patterns, each record has <size, event set, pi, rank>
$R_k.last$:a Nth prevalent k-colocated event set

Preprocess
1) ST=gen_neighbor_transactions(S, R);//*Generate neighborhood transactions*

Candidate generation
2) for i=1 to m do
3) $Tree_i$=build_event_tree(e_i, ST); //*Build each reference event tree*
4) end do
5) C=gen_candidates($Tree_1, \ldots, Tree_m$); //*Generate candidate event sets*
6) calculate_upper_pi(C); //*Compute the upper bound of PI of candidates*

Pattern finding
7) k=2;
8) C_k=get_size_k_candidates(C, k); //*Filter size 2 from candidate pool*
9) SI_k = gather_instances(C_k, ST); //*Find size 2 colocation instances*
10) R_2 =find_size2_N-most_patterns(SI_K); //*Sort and find size2 results*
11) θ=0;
12) for k=k_{max} to 3 do
13) R_k=∅; θ_k=θ; // *θ_k is initialized with previous θ*
14) C_k=get_size_k_candidates(C, k);
15)   sort_candidates_by_upper_pi(C_k);
16)   for each candidate set $c \in C_k$ do //*Prune a candidates by Lemma 1*
17)   if $k \leq N$ and c's $upper\_pi < \theta_k$ then C_k = C_k - c;
18) SI_k=gather_star_instances(C_k, ST);// *Find candidate instances*
19)   for each candidate set $c \in C_k$ do
20)   if $|R_k|$ == N and c's $upper\_pi \leq \theta_k$ then //*All N k-patterns found*
21)   if(R_k_has_at_least_ different_N_events) then θ=$R_k.last.pi$;
22)   exit;
23) CI_c=find_clique_instances(SI_c);// *Otherwise, filter true instances*
24) pi=calculate_true_pi(CI_c); // *Compute true participation index*
25)   if $|R_k|$ < N then // *Update the result set*
26)   insert(c, pi, R_k);
27)   else if $|R_k|$==N and $pi > \theta_k$ then
28)   remove($R_k.last$); insert(c, pi, R_k); θ_k=$R_k.last.pi$;
29)   end do
30) end do
31) return $\bigcup(R_2, \ldots, R_k)$;

Algorithm 1. NMColoc Algorithm

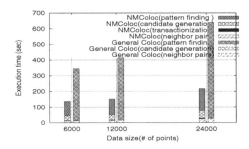

(a) Computation costs

		N-Most	N-Most(N=20)	N-Most(N=40)
size	Number of possible candidates	Number of generated candidates	Number of *final candidates	Number of *final candidates
2	1225	476	20	40
3	19600	1273	23	45
4	230300	1718	30	56
5	2118760	1390	37	57
6	15890700	706	38	64
7	99884400	227	35	77

* : candidates which examine their colocation instances

(b) Number of candidates

Fig. 4. Effect of pruning

GeneralColoc to denote the latter method. A real data about points of interest in California [13] was used. The total number of points is 104,770. The number of distinct events is 63. We prepared several test datasets from this base dataset with selecting a part of data or modifying the data to increase the number of event types. All the experiments were performed on a Sun SunBlade with 1GB main memory. The following describes the experimental results.

1) Comparison of the computation performances of subtasks: We compared the computation times that NMColoc and GeneralColoc perform each subtask: (1) finding neighbor pairs, (2) generating neighborhood transactions, (3) generating candidates, and (4) finding the interesting patterns. Fig. 4 (a) shows the results with three datasets with different size of data points. The number of distinct event types is 40. We used 1000 for the neighbor distance, 20 for the N parameter, and 10 for the k_{max}. Note that GeneralColoc also has a preprocess step for finding neighbor pairs. The preprocess time for neighborhood transactions is relatively smaller than other computation times of NMColoc. The candidate generation cost of NMColoc is much larger than GeneralColoc. However, the performance of finding the N-most patterns in NMColoc is speeded up. The overall performance difference is increased with increase of the data size.

2) Comparison of the number of candidates: Next, we compared the number of candidates generated from NMColoc with different N values. The dataset has 50 event types. Fig. 4 (b) shows the numbers of candidates generated from NMColoc with the number of possible candidates per each size. The values of the second column represent the numbers of colocated candidates generated as combining star candidates from event trees. The values of the third and fourth columns represent the numbers of candidates left after the theta pruning. Only these candidates will be examined for finding their colocation instances. We can notice NMColoc dramatically reduced the numbers of candidates.

3) Effect of the number of data points: In the third experiment, we compared the effect of the number of data points. We prepared five different datasets in size. As shown in Fig. 5 (a), the execution time of NMColoc increased much slower than GeneralColoc.

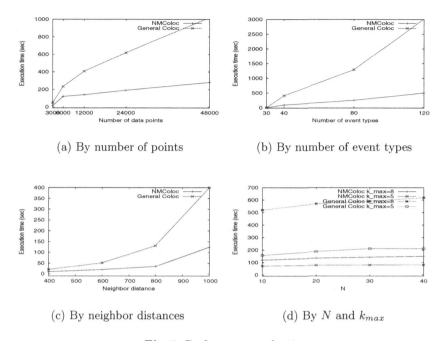

(a) By number of points

(b) By number of event types

(c) By neighbor distances

(d) By N and k_{max}

Fig. 5. Performance evaluation

4) Effect of the number of event types: We compared the performance with a function of the number of different event types. We used four different datasets. The number of data points in each dataset is around 9000. The neighbor distance is 1000. Fig. 5 (b) shows the results. The overall execution time of GeneralColoc increased faster than that of NMColoc. We can see that the speedup with the theta pruning in NMColoc increases with the number of events.

5) Effect of neighbor distance: Next, we examined the effect of different neighbor distances, 400, 600, 800, and 1000. We used a dataset with 12000 data points. The increase of neighbor distance makes a neighborhood area larger and increases the number of colocation instances. As shown in Fig. 5 (c), the overall execution time increased with increase of the neighbor distance.

6) Effect of N and k_{max} parameters: Finally, we examined the performance effect with different Ns and k_{max}s. With small increases of N, NMColoc showed similar performance. The execution time of GeneralColoc increased a little due to the decrease of the tuned minimum prevalence threshold with increase of N. On the other hand, when k_{max} is increased, the time of GeneralColoc is dramatically increased.

5   Conclusion and Future Work

In this paper, we proposed a different problem setting for colocation pattern mining. Instead of a minimum prevalence threshold, users can control their interesting patterns with the number of desired patterns. We developed an efficient algorithm for finding N-most prevalent colocation patterns. The proposed framework still needs a user specified neighborhood distance threshold. In the future work, we plan to explore methods to estimate a distance threshold to help users choose their interesting neighborhood distance bound.

References

1. Arshad, M.U., Ayyaz, M.N.: Mining N-most Iteresting Itemsets using Support-Ordered Tries. In: IEEE Int'l Conf. on Computer Systems and Applications (2006)
2. Cheung, Y., Fu, A.W.: Mining Frequent Itemsets without Support Threshold: With and Without Item Constraints. IEEE Transactions on Knowledge and Data Engineering 16(9) (2004)
3. Cormen, T., Leiserson, C., Rivest, R., Stein, C.: Introduction to Algorithms. McGraw-Hill Science, New York (2003)
4. Ding, W., Jiamthapthaksin, R., Parmar, R., Jiang, D., Stepinski, T.F., Eick, C.F.: Towards Region Discovery in Spatial Datasets. In: Washio, T., Suzuki, E., Ting, K.M., Inokuchi, A. (eds.) PAKDD 2008. LNCS, vol. 5012, pp. 88–99. Springer, Heidelberg (2008)
5. Eick, C.F., Parmar, R., Ding, W., Stepinski, T.F., Nicot, J.: Finding Regional Colocation Patterns for Sets of Continuous Variables in Spatial Datasets. In: Proc. of ACM SIGSPATIAL International Conference on Advances in Geographic Information Systems(ACM-GIS) (2008)
6. Fu, A.W., Kwong, R.W., Tang, J.: Mining N-most Interesting Itemsets. In: International Syposium on Methodologies for Intelligent Systems (2000)
7. Han, J., Pei, J., Yin, Y.: Mining Frequent Patterns Without Candidate Generation. In: Proc. of ACM SIGMOD Conference on Management of Data (2000)
8. Hirate, Y., Iwahashi, E., Yamana, H.: TF^2P-growth: An Efficient Algorithm for Mining Frequent Patterns without any Thresholds. In: Proc. of Workshop on Alternative Techniques for Data Mining and Knowledge Discovery (2004)
9. Huang, Y., Zhang, L., Yu, P.: Can we apply projection based frequent pattern mining paradigm to spatial co-location mining? In: Ho, T.-B., Cheung, D., Liu, H. (eds.) PAKDD 2005. LNCS, vol. 3518, pp. 719–725. Springer, Heidelberg (2005)
10. Inokuchi, A., Washio, T., Motoda, H.: An Apriori-based Algorithm for Mining Frequent Substructures from Graph Data. In: Proc. of European Conference on Principles of Data Mining and Knowledge Discovery (2000)
11. Koperski, K., Han, J.: Discovery of Spatial Association Rules in Geographic Information Databases. In: Proc. of International Symposium on Large Spatial Data bases, Maine, pp. 47–66 (1995)
12. Kuramochi, M., Karypis, G.: Fequent Subgraph Discovery. In: IEEE International Conference on Data Mining (2001)
13. Li, F., Cheng, D., Hadjieleftheriou, M., Kollios, G., Teng, S.: On Trip Planning Queries in Spatial Databases. In: Proc. of Interational Symposium on Advances in Spatial and Temporal Databases(SSTD) (2005)
14. Morimoto, Y.: Mining Frequent Neighboring Class Sets in Spatial Databases. In: Proc. ACM SIGKDD Int'l Conf. on Knowledge Discovery and Data Mining (2001)
15. Shekhar, S., Chawla, S.: Spatial Databases: A Tour. Prentice-Hall, Englewood Cliffs (2003)

16. Shekhar, S., Huang, Y.: Co-location Rules Mining: A Summary of Results. In: Proc. of International Symposium on Spatial and Temporal Database, SSTD (2001)

17. Wang, J., Han, J., Lu, Y., Tzvetkov, P.: TFP: An Efficient Algorithm for Mining Top-K Frequent Closed Itemsets. IEEE Transactions on Knowledge and Data Engineering 17(5) (2005)

18. Xiao, X., Xie, X., Luo, Q., Ma, W.: Density based co-location pattern discovery. In: Proc. of ACM SIGSPATIAL International Conference on Advances in Geographic Information Systems (ACM-GIS) (2008)

19. Yan, X., Han, J.: gSpan: Graph-Baseed Substructure Pattern Mining. In: IEEE International Conference on Data Mining (2001)

20. Yoo, J.S., Bow, M.: Finding N-Most Prevalent Colocated Event Sets: A Summary of Results. Technical Report (2009),
http://users.ipfw.edu/yooj/publications/NMost-Summary.pdf

21. Yoo, J.S., Shekhar, S.: A Partial Join Approach for Mining Co-location Patterns. In: Proc. of ACM ACM SIGSPATIAL International Conference on Advances in Geographic Information Systems(ACM-GIS) (2004)

22. Yoo, J.S., Shekhar, S.: A Join-less Approach for Mining Spatial Co-location Patterns. IEEE Transactions on Knowledge and Data Engineering 18(10) (2006)

23. Zou, L., Chen, L., Lu, Y.: Top-k Subgraph Matching Query in a Large Graph. In: Proc. of Conference on Information and Knowledge Management (2000)

Rule Learning with Probabilistic Smoothing

Gianni Costa, Massimo Guarascio, Giuseppe Manco, Riccardo Ortale,
and Ettore Ritacco

ICAR-CNR
Via P. Bucci 41c
87036 Rende (CS) - Italy

Abstract. A hierarchical classification framework is proposed for discriminating rare classes in imprecise domains, characterized by rarity (of both classes and cases), noise and low class separability. The devised framework couples the rules of a rule-based classifier with as many local probabilistic generative models. These are trained over the coverage of the corresponding rules to better catch those globally rare cases/classes that become less rare in the coverage. Two novel schemes for tightly integrating rule-based and probabilistic classification are introduced, that classify unlabeled cases by considering multiple classifier rules as well as their local probabilistic counterparts. An intensive evaluation shows that the proposed framework is competitive and often superior in accuracy w.r.t. established competitors, while overcoming them in dealing with rare classes.

1 Introduction

Rule learning is a mainstay of research in the field of concept learning, because of various desirable properties such as, e.g., its high expressiveness and immediate intelligibility to humans. In particular, associative classification [16] is an advance in rule learning, that relies upon the associations in the available data between the co-occurrence of certain combinations of attribute values and their observed class labels. The resulting classification models, referred to as associative classifiers, consist of association rules, whose consequents are restricted to predict the values of the target class attribute. Associative classification retains the advantages of conventional rule learning and also tends to achieve a higher predictive accuracy [19]. Indeed, rule induction does not operate on the whole training data. Rather, it is generally performed as a heuristic separate-and-conquer process, that progressively excludes subsets of the training data from further consideration as soon as covered by locally-optimal, biased rules. Instead, associative classification yields rules with an appropriate degree of generality/specificity, that summarize the co-occurrence patterns across the whole training data.

Several approaches to associative classification are available from the literature, with differences in three major aspects, i.e discovery of classification rules, the extraction of a compact classifier and the classification of unlabeled cases [16]. Classification rules are mined through search strategies based on Apriori [1] in [2,14,15], whereas a variant of the FP-growth algorithm [11] is used in [13]. A row enumeration method is leveraged in [3]. Often, the huge number of resulting classification rules, that may overfit the training data, is pruned to distil a compact associative classifier. A variety of methods is used for this purpose, such as χ^2 testing [13], minimum class support [14], complement

T.B. Pedersen, M.K. Mohania, and A M. Tjoa (Eds.): DaWaK 2009, LNCS 5691, pp. 428–440, 2009.
© Springer-Verlag Berlin Heidelberg 2009

class support [15] as well as database coverage [2,14,15]. As to the classification of an unlabeled case, some methods exploit the top-quality rule covering the case [14,15]. Other approaches take into account multiple rules applicable to the case [13,19] and resort to suitable scoring mechanisms as well as voting.

Unfortunately, like most classification models, associative classifiers exhibit a poor predictive accuracy in highly imprecise learning settings, such as fraud and intrusion detection, manufacturing line monitoring, risk management as well as medical diagnosis, where primary aspects (i.e. cases and classes) of the concept to learn are rare and noisy. Additionally, cases of distinct classes may be hardly separable, which conceptually calls for classification rules with possibly (very) limited coverage and still high predictive accuracy, especially on the minority classes.

As it is pointed out in [18], rare classes originate several accurate rules targeting the predominant classes, supplemented by very few (if any) error-prone rules predicting minority classes, which are of primary interest in practical applications. Rare cases, instead, tend to materialize within the resulting classifier as strongly inaccurate rules, referred to as *small disjuncts* [12]. These difficulties are exacerbated by noise, that may further skew class imbalance and be nearly indistinguishable from rare cases.

Yet, the decision regions induced by a rule-based classifier and the true distribution of the classes in the space of data do not match. Indeed, classes form regions with irregular and interleaved shapes, whereas the induced decision regions are neatly separated by boundaries parallel to the features of the data space. As a consequence, those cases falling within and close to the boundary of a decision region may be misleadingly predicted as belonging to the class associated with that decision region, even if the true class membership in the surroundings of the boundary is different. This is problematic when there is a low separability between classes, i.e. when these form true overlapping (or embedded) regions. In such cases, indeed, the true regions formed by rare classes may be overlapped by the decision regions associated to the predominant classes.

In this paper, we combine associative classification with probabilistic learning [5] to improve classification performance on the rare classes. In imprecise environments, this is preferable with respect to simply increasing classification accuracy, since the latter is strongly biased against rare classes, which as anticipated may also be hardly discriminated from predominant classes. The idea is to use the individual rules of an associative classifier to segment the training data. Segments are used to build as many local probabilistic generative models, that refine the predictions from the corresponding classifier rules. This is particularly useful both in the surroundings of the rule boundaries as well as inside the associated decision regions, wherein local probabilistic generative models act so that classes other than the ones associated to the whole regions influence the classification of nearby unlabeled cases. In practice, local probabilistic models are involved into the classification of unlabeled cases for more effectively dealing with those globally rare cases/classes, that become less rare in the corresponding segments. Two new schemes for tightly combining associative classification and probabilistic learning are proposed, wherein the class of an unlabeled case is decided by considering multiple class association rules as well as their relative probabilistic generative models. An intensive empirical evaluation shows that, although many possible lines of research for further improvements exist, the hierarchical framework is competitive and often superior in accuracy w.r.t. established competitors, while overcoming them in the ability to deal with rare classes. The paper proceeds as follows. Section 2 presents the hierarchical classification framework. Section 3 reports on the empirical evaluation of our approach. Finally, section 4 concludes and highlights future research.

2   The Hierarchical Predictive Framework

In this section, we discuss our approach to learning a hierarchical framework. We start with some preliminary notions. Let $\mathcal{D}$ be a relation storing the labeled training cases. Also, let the schema of $\mathcal{D}$ be a set $\mathcal{A} = \{A_1 : Dom(A_1), \ldots, A_n : Dom(A_n), L : \mathcal{L}\}$ of descriptive attributes. Features $A_1, \ldots, A_n$ are defined over as many categorical or numerical domains, whereas the target class attribute L is a categorical feature. The generic labeled training case $t \in \mathcal{D}$ is a structured tuple, i.e. $t \in Dom(A_1) \times \ldots Dom(A_n) \times Dom(L)$. t can also be equivalently represented in a transactional form. Therein, assume that $\mathcal{M} = \{i_1, \ldots, i_m\}$ is a finite set of items denoting relationships between any attribute of $\mathcal{A}$ but L and a corresponding value. Precisely, the generic item i has the form $A \, [rel] \, v$ where $A \in \mathcal{A} - L, v \in Dom(A)$ and $[rel] \in \{=, \leq, \geq\}$ denotes a relationship between A and v. In our formulation, $A = v$ is admissible iff A is a categorical attribute. The remaining relationships $A \leq \tau$ and $A \geq \tau$ are instead allowed iff A is a numeric attribute and, in such a case, τ indicates a generic split point. Split points reflect the discretization of numeric attributes. Any (un)labeled case defined over $\mathcal{A}$ can be modeled as a suitable subset of items in $\mathcal{M}$. Let $\mathcal{L}$ be a finite domain of class labels, the original dataset $\mathcal{D}$ can thus be redefined over $\mathcal{M}$ as a collection $\mathcal{D} = \{t_1, \ldots, t_n\}$ of labeled cases, such that the generic case $t \in 2^{\mathcal{M}} \times \mathcal{L}$. The class label of t is denoted as $class(t)$. Henceforth, we shall adopt the transactional notation.

A class association rule (CAR) $r : I \rightarrow c$ catches an association that occurs in $\mathcal{D}$ between any subset of items $I \subseteq \mathcal{M}$ and a class label $c \in \mathcal{L}$. Notation $class(r)$ represents the class c targeted by r.

The notions of support, coverage and confidence are employed to define the interestingness of a rule r. In particular, A training case $t \in \mathcal{D}$ is said to *support* rule $r : I \rightarrow c$ if it holds that $(I \cup c) \subseteq t$. The support of r is the fraction of training cases supporting r, i.e., $supp(r) = \frac{|\{t \in \mathcal{D}|(I \cup c) \subseteq t\}|}{|\mathcal{D}|}$, where $|\mathcal{D}|$ indicates the cardinality of $\mathcal{D}$.

Rule $r : I \rightarrow c$ is said to *cover* a training case $t \in \mathcal{D}$ (and, dually, t is said to trigger or fire r) if the condition $I \subseteq t$ holds. The set of all training cases covered by r is denoted by $\mathcal{D}_r = \{t \in \mathcal{D}|I \subseteq t\}$. The coverage of r can is the fraction of cases in $\mathcal{D}$ covered by r, i.e. $coverage(r) = \frac{|\mathcal{D}_r|}{|\mathcal{D}|}$. The foresaid rule $r : I \rightarrow c$ is said to *cover* an unlabeled case $I' \subseteq \mathcal{M}$ if $I \subseteq I'$ holds. The confidence of a rule r, denoted by $conf(r)$, is the ratio of support to coverage, i.e. $conf(r) = \frac{supp(r)}{coverage(r)}$.

An associative classifier $\mathcal{C}$ is a suitable disjunction of propositional if-then CARs, that predicts the class of an unlabeled case I, i.e. $\mathcal{C}(I) = c \in \mathcal{L}$.

Our goal is to learn a hierarchical framework from $\mathcal{D}$, that consists of two classification levels. At the higher level, an associative classifier is built such that its component CARs meet some requirements on the minimum support and confidence. For each CAR $r \in \mathcal{C}$, the lower level of the framework includes a local probabilistic generative model $P^{(r)}$ that allows to confirm or rectify r in the classification of an unlabeled case.

The idea is to build, at the higher level, an associative classifier whose CARs are coupled with local probabilistic generative models, sited at the lower level, that confirm or rectify the predictions from the corresponding CARs. The overall learning process is shown in fig. 1. Given a database $\mathcal{D}$ of training cases (defined over a set $\mathcal{M}$ of items and a set $\mathcal{L}$ of class labels), the algorithm begins (at line 1) by discovering a set $\mathcal{R}$ of association rules from $\mathcal{D}$ via the MINECARs search strategy. The latter is essentially an enhancement of the Apriori algorithm [1] that integrates multiple minimum class support [15] and complement class support [3] to uncover, within each class, an appropriate number of interesting association rules, whose antecedents and consequents are positively correlated. In particular, within the generic class, multiple minimum class support automatically adjusts the global minimum support threshold σ, provided by the user, to a minimum support threshold specific for that class. Instead, an important property of complement class support is used to retain in $\mathcal{R}$ positively correlated CARs. These are CARs for which the ratio of the observed confidence to the confidence expected by chance (i.e. if the CAR antecedent and consequent were independent) exceeds a class-specific threshold, that is selected without any additional parameter. The exploitation of positively correlated rules allows to overcome a flaw with the support and confidence framework, that produces CARs with poor implicative strength when class distribution is imbalanced, since antecedents and consequents can be negatively correlated [3].

The rule set $\mathcal{R}$ is then sorted (at line 2) according to the total order $\prec$, which is a refinement of the one introduced in [14]. Precisely, given any two rules $r_i, r_j \in \mathcal{R}$, r_i precedes r_j, which is denoted by $r_i \prec r_j$, if (i) the confidence of r_i is greater than that of r_j, or (ii) their confidences are the same, but the support of r_i is greater than that of r_j, or (iii) both confidences and supports are the same, but r_i is shorter than r_j.

The learning process proceeds (at line 3) to distil a classifier $\mathcal{C}$ by pruning $\mathcal{R}$, which is likely to include a very large number of CARs, that may overfit the training cases. The adopted strategy for overfitting avoidance involves item and rule pruning. Briefly, rule items and/or whole rules are removed from $\mathcal{R}$ whenever this does not worsen the accuracy of the classifier being distilled. The effects of item and rule pruning on the accuracy of the resulting classifier are evaluated using statistical arguments, omitted due to space restrictions. The interested reader is referred to [6] for further details.

The resulting classifier $\mathcal{C}$ may leave some training cases uncovered. Therefore, a default rule $r_d : \emptyset \rightarrow c^*$ is appended to $\mathcal{C}$ (at line 5), such that its antecedent is empty and the targeted class c^* is the majority class among the uncovered training cases.

Finally, for each CAR $r \in \mathcal{C}$ other than the default rule r_d, a local probabilistic model $\mathcal{P}^{(r)}$ is built (lines 7-9) over $\mathcal{D}_r$ to catch a better generalization of those globally rare cases/classes that become less rare within $\mathcal{D}_r$. This allows to refine the prediction from r with a local generative model that is better suited to deal with the local facets of rarity. The TRAINLOCALCLASSIFIER step is treated in the following subsection 2.1, that covers the classification of unlabeled cases (not reported in fig. 1) in the context of two schemes for a tight integration between associative and probabilistic classification.

As a concluding remark, notice that, due to the total order $\prec$ enforced over $\mathcal{R}$, the associative classifier $\mathcal{C}$ is actually a decision list: each training case is classified by the first CAR in $\mathcal{C}$ that covers it. In other words, the CARs in $\mathcal{C}$ are mutually exclusive, i.e. a training case is covered by at most one rule of the classifier. Formally, the definition

---

HIERARCHICALLEARNING($\mathcal{M},\mathcal{D},\mathcal{L},\sigma$)
   Input: a finite set $\mathcal{M}$ of boolean attributes;
        a training dataset $\mathcal{D}$;
        a set $\mathcal{L}$ of class labels in $\mathcal{D}$;
        and a support threshold σ;
   Output: An associative classifier $\mathcal{C} = \{r_1 \vee \ldots \vee r_k\}$ and a set of local classifier $\mathcal{P}_{r_i}$;
   1: $\mathcal{R} \leftarrow$ MINECARS$(\mathcal{M}, \mathcal{D}, \sigma)$;
   2: $\mathcal{R} \leftarrow$ ORDER$(\mathcal{R})$;
   3: $\mathcal{C} \leftarrow$ PRUNE$(\mathcal{R})$;
   4: **if** there are cases in $\mathcal{D}$ that are not covered by any rule within $\mathcal{C}$ **then**
   5:     $\mathcal{C} \leftarrow \mathcal{C} \cup \{r_d\}$;
   6: **end if**
   7: **for** each rule $r \in \mathcal{C}$, such that $r \neq r_d$ **do**
   8:     $\mathcal{P}^{(r)} \leftarrow$ TRAINLOCALCLASSIFIER(r);
   9: **end for**
  10: RETURN $\mathcal{C}$ and $\mathcal{P}^{(r)}$ for each $r \in \mathcal{C}$

Fig. 1. The hierarchical learning framework

of the set of training cases covered by the generic CAR $r \in \mathcal{C}$ hereafter becomes $\mathcal{D}_r = \{t \in \mathcal{D} | r \subseteq t \wedge \not\exists r' \in \mathcal{C} : r' \prec r, r' \subseteq t\}$. Moreover, the addition to $\mathcal{C}$ (at line 5) of the default rule r_d ensures that the classifier is also exhaustive, i.e. that every training case of $\mathcal{D}$ is covered by at least one CAR of $\mathcal{C}$.

2.1 Training Local Classifiers

To improve the predictive accuracy both in the surroundings of decision boundaries as well as within the inner areas of decision regions (wherein classes other than the ones associated to the whole regions may influence the classification of nearby unlabeled cases), each CAR $r \in \mathcal{C}$ is associated with a local probabilistic generative model $\mathcal{P}^{(r)}$, trained over the regularities across the training cases local to $\mathcal{D}_r$. In principle, such regularities are likely to be more descriptive of those globally rare cases/classes that become less rare within $\mathcal{D}_r$. Hence, the individual $\mathcal{P}^{(r)}$ can be involved into the classification process for more accurately dealing with the corresponding forms of rarity.

In the following, we adopt two different probabilistic generative models based, respectively, on the naïve Bayes and nearest neighbor classification models. Precisely, naïve Bayes naturally allows to incorporate the effects of locality on classes and cases in terms of, respectively, class priors and item posteriors. To elucidate, an unlabeled case $I \subseteq \mathcal{M}$ is assigned by the generic generative model $\mathcal{P}^{(r)}$ to the class $c \in \mathcal{L}$ with highest posterior probability

$$\mathcal{P}^{(r)}(c|I) \triangleq p(c|I,r) = \frac{p(I|c,r)p(c|r)}{\sum_{\overline{c} \in \mathcal{L}} p(I|\overline{c},r)p(\overline{c}|r)} = \frac{\prod_{i \in I} p(i|c,r)p(c|r)}{\sum_{\overline{c} \in \mathcal{L}} \prod_{i \in I} p(i|\overline{c},r)p(\overline{c}|r)}$$

Locality influences factors $p(c|r)$'s and $p(i|c,r)$'s, whose values are estimated by computing $p(c)$ and $p(i|c)$ over $\mathcal{D}_r$, and allows to better value rare cases/classes. Indeed, if a significant extent of some form of rarity falls within $\mathcal{D}_r$, the corresponding cases/classes are obviously less rare than in $\mathcal{D}$ and, hence, factors $p(c)$'s and $p(i|c)$'s are accordingly higher (w.r.t. their values in $\mathcal{D}$). Dually, $p(c)$'s and $p(i|c)$'s are sensibly lower, if the density of that form of rarity within $\mathcal{D}_r$ is much lower than in $\mathcal{D}$. However, this is acceptable, since most of that form of rarity is still captured within some other region(s). An inconvenient behind the adoption of naïve Bayes as the underlying model for local probabilistic classifiers is their performance degrade (e.g. accuracy loss) due to the violation of the attribute independence assumption. To alleviate such an issue, the weaker attribute independence assumption postulated in AODE [17] can be plugged into the above formulation, that simply refines naïve Bayes by considering each attribute dependent upon at most n other attributes in addition to the class. This is more realistic in practice and is empirically shown in section 3 to yield a better performance.

Another difficulty behind naïve Bayes is that the estimates of some class priors and item posteriors may not be reliable when data is too rare within $\mathcal{D}_r$. In such cases, the nearest neighbor model can be alternatively used to compute probabilities $\mathcal{P}^{(r)}(c|I)$ from the distribution of classes within $\mathcal{D}_r$ through the generative approach below

$$\mathcal{P}^{(r)}(c|I) \triangleq \frac{\sum_{I' \in \mathcal{D}_r} w_{I'} p(c|I')}{\sum_{\bar{c} \in \mathcal{L}} \sum_{I' \in \mathcal{D}_r} w_{I'} p(\bar{c}|I')}$$

The above is essentially a probabilistic re-formulation of a distance-weighted voting scheme, in which each neighbor I' votes for the class that should be assigned to I. The vote from the generic neighbor I' is suitably weighted by a corresponding factor $w_{I'}^{(r)}$, which takes into account the actual distance between I' and I. Formally,

$$w_{I'} = \frac{e^{-d^2(I,I')}}{\sum_{I' \in \mathcal{D}_r} e^{-d^2(I,I')}}$$

where $d(I, I')$ is any suitable function that defines a notion of distance between I and I'. Notice that, whatever the distance between cases, the chosen weight-definition attributes higher influences to those neighbors in $\mathcal{D}_r$ that are actually closest to I.

Two alternative approaches for refining the predictions from the associative classifier $\mathcal{C}$ through the local probabilistic generative models $\mathcal{P}^{(r)}$'s are discussed next.

Local priors and local instance posteriors. The idea is to reformulate a generative approach to classification which spans into local generative models. Starting from the observation that the exhaustive and exclusive rules within $\mathcal{C}$ partition the space of covering events relative to a tuple, it is possible to define the joint probability over unlabeled cases and a class labels as shown below

$$p(c, I) = \sum_{r \in \mathcal{C}} p(c, I, r) = \sum_{r \in \mathcal{C}} p(c, I|r)p(r) = \sum_{r \in \mathcal{C}} \mathcal{P}^{(r)}(c|I)p(I|r)p(r)$$

Within the above formula, $p(I|r)$ represents the compatibility of I with the rule r. We choose to model $p(I|r)$ as the relative number of items that I shares with r: intuitively, the number of (mis)matches represents the closeness of I to the region bounded by r. $\mathcal{P}^{(r)}(c|I)$ denotes the probability associated with c by the local naïve Bayes classifier $\mathcal{P}^{(r)}$ trained over $\mathcal{D}_r$. $p(r)$ indicates the support $supp(r)$ of CAR r and weights its contributions to $p(c, I)$ by the relative degree of rarity of its antecedent and consequent.

Finally, the probability of class c given the unlabeled case I can be formalized as the following generative model

$$p(c|I) = \frac{p(c,I)}{\sum_{\bar{c} \in \mathcal{L}} p(\bar{c}, I)}$$

Cumulative rule effect. A stronger type of interaction between global and local effects can be injected into the classification process, if the predictions from a CAR r and unrelated local generative model $\mathcal{P}^{(r')}$ (with $r \neq r'$) are compared for selecting the most confident one. The overall approach sketched in fig. 2. Precisely, the generic unlabeled case $I \subseteq \mathcal{M}$ is presented to the associative classifier $\mathcal{C}$ and the first CAR $r : I \to c$ (in the precedence order $\prec$ enforced over $\mathcal{C}$) is chosen (at line N1). If r does not cover I, it is skipped and the next rule is recursively taken into account (at line N20). Otherwise, r is used for prediction. However, its target class c is not directly assigned to I. Rather, the local probabilistic generative model $\mathcal{P}^{(r)}$ corresponding to r is exploited to produce a possibly more accurate prediction (at line N4). Some tests are performed to identify the more confident prediction (lines N9- N15). If both counterparts agree or one is deemed to be more reliable than the other one, the better prediction (in terms of class-membership probability distribution) is returned (lines N10 and N12). Otherwise, in the absence of strong evidence to reject the prediction from $\mathcal{P}^{(r)}$ (which is in principle preferable to r, being more representative of the local regularities that may come from globally rare cases/classes that fall within $\mathcal{D}_r$), r is skipped in favor of the next CAR $r' \in \mathcal{C}$ covering I (at line N14). To this point, if $\mathcal{P}^{(r')}$ predicts I more confidently than $\mathcal{P}^{(r)}$ (at line N5), the probability distribution from $\mathcal{P}^{(r')}$ replaces the current best distribution yielded by $\mathcal{P}^{(r)}$ (at line N6) and the choice of a better prediction is hence made between r' and $\mathcal{P}^{(r')}$. In the opposite case, the choice involves r' and the current best distribution $\mathcal{P}^{(r)}$. If no prediction is clearly eligible as the most confident throughout the search, the process halts when the default rule is met and the current best distribution is returned (at line N17). Notice that the sofar best class-membership probability distribution is remembered throughout the consecutive stages of the search process via the input arguments $p_1, \ldots, p_k$ (such arguments are individually set to 0 at the beginning of the search process). A key aspect of the overall search process is represented by the criteria adopted to choose the more confident prediction between the ones from a CAR r_h and a local probabilistic generative model $\mathcal{P}^{(r_i)}$. Accuracy is used as a discriminant between the alternatives. In particular, the accuracy $acc^{(c)}\left(\mathcal{P}^{(r_i)}\right)$ is the percent of cases in $\mathcal{D}^{(r)}$ correctly predicted by $\mathcal{P}^{(r_i)}$ as belonging to class c.

The accuracy $acc^{(c)}(r_h)$ of a CAR r_h predicting class c is its confidence $conf(r_h)$. When comparing the accuracies of a CAR r_h and a local probabilistic generative model $\mathcal{P}^{(r_i)}$ there are four possible outcomes.

1. $\mathcal{P}^{(r_i)}$ is clearly deemed more reliable than r_h (at line N9), if the weighted accuracy of the former, p^*, is greater than the accuracy of the latter.
2. r_h is preferred to $\mathcal{P}^{(r_i)}$ (at line N11) if the accuracy of the former is greater than or equal to the weighted accuracy of the latter and both agree anyhow.
3. r_h is preferred to $\mathcal{P}^{(r_i)}$ (again at line N11) if its accuracy is much greater than the weighted accuracy of $\mathcal{P}^{(r_i)}$. Therein, $\frac{p^*}{p} > p*$ is a prudential threshold, that

---

PREDICTION($\mathcal{C}, I, p_1, \ldots, p_k$)
   Input: An associative classifier $\mathcal{C}$;
          an unlabeled case $I \subseteq \mathcal{M}$;
   Output: the class distribution for I;
   N1:   select the first rule $r : I' \to c_h$ in sequence within $\mathcal{C}$;
   N2:   **if** r covers I (i.e. $I' \subseteq I$) **then**
   N3:   **if** $|\mathcal{C}| > 1$ (i.e. r is not the default rule) **then**
   N4:   let $\overline{p}_i = \mathcal{P}^{(r)}(c_i|I) \cdot acc^{(c_i)}\left(\mathcal{P}^{(r)}\right), \forall i = 1, \ldots, k$;
   N5:   **if** $max_i(\overline{p}_i) > max_i(p_i)$ **then**
   N6:   let $p_i = \overline{p}_i, \forall i = 1, \ldots, k$;
   N7:   **end if**
   N8:   let $p^* = max_i(p_i)$ and $i^* = argmax_i(p_i)$ and $p = \sum_i p_i$;
   N9:   **if** $acc^{(c_h)}(r) < p^*$ **then**
   N10:   RETURN the distribution $(p_1/p, \ldots, p_k/p)$;
   N11:   **else if** $i^* = h$ **or** $acc^{(c_h)}(r) > \frac{p^*}{p}$ **then**
   N12:   RETURN the distribution $(acc^{(c_1)}(r), \ldots, acc^{(c_k)}(r))$;
   N13:   **else**
   N14:   PREDICTION($\mathcal{C} - \{r\}, I, p_1, \ldots, p_k$);
   N15:   **end if**
   N16:   **else**
   N17:   RETURN the distribution $(p_1/p, \ldots, p_k/p)$;
   N18:   **end if**
   N19: **else**
   N20:   PREDICTION($\mathcal{C} - \{r\}, I, p_1, \ldots, p_k$);
   N21: **end if**

Fig. 2. The scheme for classifying an unlabeled case under the cumulative rule effect

represents the normalized weighted accuracy from $\mathcal{P}^{(r_i)}$. In practice, r_h is actually preferable to $\mathcal{P}^{(r_i)}$ iff its accuracy exceeds $\frac{p^*}{p}$.

4. There is no strong evidence (at line N16) to reject either r_h or $\mathcal{P}^{(r_i)}$ when the accuracy of r_h lies in the interval $(p*, \frac{p^*}{p})$. In such a case, r is skipped and the search proceeds to considering the next CAR in the associative classifier $\mathcal{C}$ that covers I (through the recursive call at line N14).

3   Evaluation

We experimentally evaluate the behavior of the hierarchical classification framework to understand whether it exhibits improvements in classification performance with respect to established competitors. For the comparative evaluation, we use some standard datasets from the UCI KDD repository [4] with high class imbalance. Tests are performed over two further datasets. kdd99 is the KDD99 intrusion detection dataset, wherein class distribution is strongly skewed and low-frequency classes are affected by noise. fraud is a (non-publicly available) real-life fraud detection dataset, with a very low class separability.

We remark that, as pointed out in [18], the effectiveness of a classification strategy on rare cases cannot be directly evaluated, since these are usually unknown. Notwithstanding, both rare classes and rare cases are argued to be two strongly related facets of rarity, whose issues can be addressed with the same methods. Hence, we expect that if an approach is effective with rare classes, it is also useful for dealing with rare cases. Experiments consists in comparisons against several established rule-based and associative classifiers. The selected rule-based competitors are Ripper [8] and PART [10], while the associative ones include CBA [14] and CMAR [19]. In particular, we exploited the implementations of CBA and CMAR in [7]. All tests are conducted on an Intel Itanium processor with 4Gb of memory and 2Ghz of clock speed running Windows XP. Numeric attributes in the chosen datasets are discretized for all schemes but Ripper, through equal-frequency binning. Moreover, the test involving CBA and CMAR are re-iterated several times, under different settings for the minimum support and confidence parameters: we next report the results corresponding to the best parameter configuration allowed by the implementations at [7]. Overall, the results from the individual classifiers were averaged over ten-fold cross-validation.

Our schemes simply require the specification of a global minimum support. Due to the adoption of minimum class support [14], such threshold is automatically adjusted to become a class specific threshold. In particular, we fixed the global support threshold to 20%, which is transparently adjusted to be, within the individual class in the data at hand, the 20% of the frequency of that class. The exploitation of complement class support [15] permits to avoid specifying a minimum confidence threshold.

We compare the approaches using accuracy, some meaningful ROC curves and the Area Under the Curve (AUC) relative to the minority class. Tables 1 and 2 display the results. Within the tables, (1) indicates Ripper, (2) corresponds to PART, while (3) and (4) stand for CBA and CMAR, respectively. Our schemes are instead numbered from (5) to (10). More specifically, (5) and (6) indicate naive Bayesian smoothing (respectively through local priors or cumulative effect). (7) and (8) stand for nearest-neighbor

Table 1. Classification accuracy

Dataset	(1)	(2)	(3)	(4)	(5)	(6)	(7)	(8)	(9)	(10)
anneal	96.63	96.66	92.81	96.33	96.76	96.76	96.70	96.63	96.76	96.76
balance-scale	77.29	77.27	68.81	68.49	77.84	77.53	74.04	77.60	**79.67**	78.01
breast-cancer	71.46	68.54	69.20	67.67	70.52	70.52	72.65	71.92	72.44	72.44
horse-colic	84.26	81.95	81.62	83.96	82.42	82.42	84.12	83.85	82.11	82.11
credit-rating	86.28	85.07	81.74	83.76	85.78	85.78	**86.57**	86.43	86.06	86.06
german-credit	71.74	72.24	73.10	73.34	74.21	74.21	72.48	71.85	**74.53**	74.53
pima-diabetes	77.41	76.84	77.87	73.03	78.06	78.06	77.97	76.86	77.63	77.63
glass	71.95	74.94	72.69	74.23	75.63	74.93	73.72	72.94	**76.66**	75.65
cleveland-14-heart-disease	82.24	80.46	82.12	75.12	81.78	81.97	82.43	82.20	82.34	82.33
hungarian-14-heart-disease	80.48	81.29	82.06	79.69	82.87	82.90	81.60	81.67	**83.00**	82.97
heart-statlog	82.89	83.33	82.59	84.19	83.34	84.19	82.74	81.93	84.52	84.52
hepatitis	80.58	78.20	79.89	81.08	**82.17**	81.08	80.38	80.19	80.85	80.85
ionosphere	91.68	90.03	87.89	89.74	**93.72**	89.74	92.28	92.28	92.85	92.85
labor	83.33	84.63	86.67	88.77	87.17	88.77	83.33	83.33	88.23	88.23
lymphography	79.14	80.20	81.18	80.59	**84.16**	80.45	79.68	79.54	80.21	80.21
sick	97.60	97.87	97.51	97.64	97.64	97.64	97.51	97.57	97.65	97.65
sonar	79.00	81.26	80.00	82.78	63.36	82.78	80.10	79.67	82.64	82.64
fraud	93.07	93.02	80.82	90.52	91.79	91.79	**93.05**	92.96	92.61	92.61
kdd99	96.61	96.98	94.65	94.63	95.98	95.98	96.78	96.73	96.65	96.65

Table 2. Area Under the Curve

Dataset	(1)	(2)	(5)	(6)	(7)	(8)	(9)	(10)
anneal	0.79	0.91	0.94	0.94	0.93	0.93	0.95	0.95
balance-scale	0.82	0.90	0.94	0.92	0.93	0.92	0.94	0.92
breast-cancer	0.60	0.59	0.68	0.68	0.63	0.63	0.69	0.69
horse-colic	0.83	0.83	0.85	0.85	0.87	0.87	0.87	0.87
credit-rating	0.87	0.91	0.92	0.92	0.91	0.91	0.93	0.93
german-credit	0.63	0.71	0.77	0.77	0.72	0.72	0.78	0.78
pima-diabetes	0.75	0.81	0.84	0.84	0.83	0.82	0.84	0.84
glass	0.87	0.89	0.87	0.85	0.87	0.85	0.87	0.86
cleveland-14-heart-disease	0.83	0.84	0.90	0.90	0.89	0.88	0.90	0.90
hungarian-14-heart-disease	0.78	0.86	0.90	0.90	0.89	0.89	0.90	0.90
heart-statlog	0.83	0.85	0.90	0.90	0.88	0.87	0.90	0.90
hepatitis	0.70	0.69	**0.84**	**0.84**	0.78	0.77	**0.84**	**0.84**
ionosphere	0.92	0.92	0.95	0.95	0.94	0.94	0.98	0.98
labor	0.81	0.83	0.96	0.96	0.90	0.90	0.96	0.96
lymphography	0.46	0.56	**0.99**	0.79	0.81	0.68	0.97	0.92
sick	0.91	0.93	0.96	0.96	0.96	0.96	0.96	0.96
sonar	0.81	0.86	0.92	0.92	0.90	0.89	0.92	0.92
fraud	0.68	0.77	0.81	0.81	0.78	0.78	**0.92**	0.90
kdd99	0.98	0.99	0.99	0.99	0.99	0.99	0.99	0.99

Predicted ->	good	bad
good	607	93
bad	155	145

AODE local priors (9)

Predicted ->	good	bad
good	611	89
bad	194	106

Ripper (1)

Fig. 3. The confusion matrices yielded by AODE local priors (9) and Ripper (1)

smoothing (respectively, through local priors or cumulative effect). (9) and (10) are AODE smoothing (respectively, through local priors or cumulative effect).

The results clearly state that the combination of associative classification and probabilistic smoothing is at least as accurate as the seminal rule-based classifiers chosen for the comparison. In many cases, however, (5) and (11) achieve improvements in accuracy, reported in bold within table 1, that are statistically significant according to the t-test. In addition, a deeper analysis reveals that the response versus the classes of interest is strongly improved. Such an improvement can be appreciated by looking at the details of the individual datasets. We report in fig. 3 the confusion matrices originated by (1) and (9) over the german-credit dataset: the probabilistic smoothing here recovers 39 tuples to the minority class, thus allowing to achieve a higher precision.

A further analysis of the results obtained over the fraud and the kdd99 datasets provides an in-depth into the effects of smoothing. Fig. 4 shows the ROC curves relative to (1), (2), (5), (7) and (9). There is an evident improvement in the underlying area with respect to the competitors (1) and (2), whose trends are plotted in red. Results with the kdd99 dataset are even more surprising, and in particular with the u2r class, as shown in fig. 5, that represents the curves relative to the schemes (1), (2) and (9). The u2r class is made of 56 tuples (out of 150K), and still the probabilistic adjustment is capable of recovering some problematic cases.

Finally, the ability of the approaches at dealing with the classes is compared in table 2, which reports the AUC values across the selected datasets. The AUC is a measure

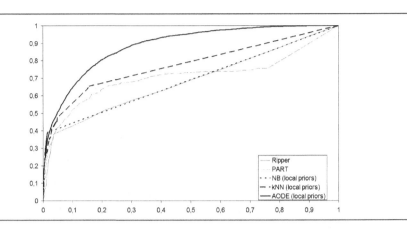

Fig. 4. ROC curve for the minority class in the `fraud` dataset

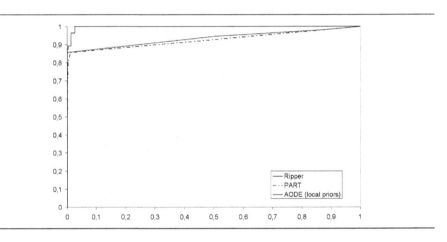

Fig. 5. ROC curve for the minority `u2r` class in the `kdd99` dataset

of the separability of two classes. Therefore, for the two-class datasets, table 2 simply reports a scalar value, which indicates the capability of the classification schemes at discriminating between the positive (i.e. rare) and negative (i.e. predominant) classes. For the multi-class datasets, table 2 combines multiple pairwise separability values by following the class reference approach in [9] and reports the weighted sum of the resulting pairwise AUC values (weights are the occurrence frequencies of each reference class). The devised schemes exhibit an improved performance across all classes within the distinct datasets and, in particular, with `hepatitis`, `lymphography` and `fraud`, where the improvement is over 10%. As witnessed by the graphs in figg. 4 and 5, such an overall improvement is primarily obtained on the minority classes.

4   Conclusions and Future Work

This paper proposed two novel smoothing approaches that tightly integrate rule-based (associative) classification and probabilistic learning [5], to improve the classification performance on the rare classes in multi-class imprecise learning environments.

We planned to investigate the overhead due to the enrichment of the base associative classifier with one generative probabilistic model for each rule. The analysis of the amount of additional time required in the learning and classification steps will be useful to better balance the accuracy benefits with the computational time requirements.

Also, we intend to improve the accuracy of the local probabilistic generative models through ROC analysis. The classification threshold used in our framework assigns a class label when the associated probability is higher than 0.5. However, the latter may not necessarily be the best threshold, especially if we consider the bias introduced by the CAR associated with the probabilistic classifier. In general, higher thresholds produce improvements in recall, by contemporarily degrading precision. However, by automatically selecting the best class-specific threshold, probabilistic smoothing can still allow to remove some locality effects within the CAR and maintain high precision as well.

References

1. Agrawal, R., Srikant, R.: Fast algorithms for mining association rules. In: Proc. of Int. Conf. on Very Large Data Bases, pp. 487–499 (1994)
2. Antonie, M.-L., Zaïane, O.R.: Text document categorization by term association. In: Proc. on IEEE Int. Conf. on Data Mining, pp. 19–26 (2002)
3. Arunasalam, B., Chawla, S.: CCCS: A top-down association classifier for imbalanced class distribution. In: Proc. of ACM SIGKDD KDD, pp. 517–522 (2006)
4. Asuncion, A., Newman, D.J.: UCI machine learning repository (2007)
5. Bishop, C.M.: Pattern Recognition and Machine Learning. Springer, Heidelberg (2006)
6. Cesario, E., Folino, F., Locane, A., Manco, G., Ortale, R.: Boosting text segmentation via progressive classification. Knowledge and Information Systems 15(3), 285–320 (2008)
7. Coenen, F.: LUCS KDD implementations of CBA and CMAR (2004)
8. Cohen, W.W.: Fast effective rule induction. In: Proc. of Int. Conf. on Machine Learning, pp. 115–123 (1995)
9. Fawcett, T.: An introduction to ROC analysis. Pattern Recognition Letters 27(8), 861–874 (2006)
10. Frank, E., Witten, I.H.: Generating accurate rule sets without global optimization. In: Proc. of Int. Conf. on Machine Learning, pp. 144–151 (1998)
11. Han, J., Yin, Y.: Mining frequent patterns without candidate generation. In: Proc. of ACM SIGMOD Int. Conf. on Management of data, pp. 1–12 (2000)
12. Holte, R.C., Acker, L., Porter, B.: Concept learning and the problem of small disjuncts. In: Proc. of Int. Joint Conf. on Artificial Intelligence, pp. 813–818 (1989)
13. Li, W., Han, J., Pei, J.: CMAR: Accurate and efficient classification based on multiple class-association rules. In: Proc. of IEEE Int. Conf. on Data Mining, pp. 369–376 (2001)
14. Liu, B., Hsu, W., Ma, Y.: Integrating classification and association rule mining. In: Proc. of ACM SIGKDD Int. Conf. on Kwnoledge Discovery and Data Mining, pp. 80–86 (1998)

15. Liu, B., Ma, Y., Wong, C.K.: Improving an association rule based classifier. In: Proc. of Principles of Data Mining and Knowledge Discovery, pp. 504–509 (2000)
16. Thabtah, F.: A review of associative classification mining. The Knowledge Engineering Review 22(1), 37–65 (2007)
17. Webb, G., Boughton, J., Wang, Z.: Not so naive bayes: Aggregating one-dependence estimators. Machine Learning 58(1), 5–24 (2005)
18. Weiss, G.M.: Mining with rarity: A unifying framework. ACM SIGKDD Explorations 6(1), 7–19 (2004)
19. Xin, X., Han, J.: CPAR: Classification based on predictive association rules. In: Proc. of SIAM Int. Conf. on Data Mining, pp. 331–335 (2003)

Missing Values: Proposition of a Typology and Characterization with an Association Rule-Based Model

Leila Ben Othman[1,2], François Rioult[2], Sadok Ben Yahia[1], and Bruno Crémilleux[2]

[1] Department of Computer Science,
Faculty of Sciences of Tunis. Tunisia
[2] University of Caen Basse-Normandie, France
GREYC - CNRS UMR 6072
{lbenothm,francois.Rioult,Bruno.Cremilleux}@info.unicaen.fr,
sadok.benyahia@fst.rnu.tn

Abstract. Handling missing values when tackling real-world datasets is a great challenge arousing the interest of many scientific communities. Many works propose completion methods or implement new data mining techniques tolerating the presence of missing values. It turns out that these tasks are very hard. In this paper, we propose a new typology characterizing missing values according to relationships within the data. These relationships are automatically discovered by data mining techniques using generic bases of association rules. We define four types of missing values from these relationships. The characterization is made for each missing value. It differs from the well-known statistical methods which apply a same treatment for all missing values coming from a same attribute. We claim that such a local characterization enables us perceptive techniques to deal with missing values according to their origins: the way in which we deal with the missing values should depend on their origins (e.g., attribute meaningless w.r.t. other attributes, missing values depending on other data, missing values by accident). Experiments on a real-world medical dataset highlight the interests of such a characterization.

Keywords: Data mining, missing values, association rules.

1 Introduction

Many data sets are incomplete and handling missing values is a major challenge in data analysis. There are two main approaches to analyze incomplete data: using a data mining method which is adjusted to cope with missing values or completing the data by imputation. The first approach is clearly expensive. Indeed, even if a technique can be updated to handle missing values, the process has to be repeated for each technique. The second is appealing: once a database is completed, it enables us the running of any method. It may explain why many works deal with imputation. Obviously, this approach requires accurate imputations in order to provide proper and unbiased data sets. Basic methods such as the mean, the most common value, a default value, are not satisfactory because they exaggerate correlations [9]. Several techniques are based on the Expectation Maximization (EM). It has been shown that EM provides accurate

T.B. Pedersen, M.K. Mohania, and A M. Tjoa (Eds.): DaWaK 2009, LNCS 5691, pp. 441–452, 2009.

probability estimations but it requires a model which has to be chosen according to the data [7]. By using the minimum description length and the idea that the best completion is that allowing for the best compression, a method taking into account how specific values co-occur locally has been proposed in [19]. A rule-based system designed for symbolic data is presented in [5]. More generally, there are several techniques based on the exploitation of local regularities, e.g., association rules [2,4,14,17,20], concise representations of patterns [15] or rough sets [8,11].

Results from the literature show that completion is a very hard task. In these works, completion methods are evaluated by missing values which are artificially introduced in a *reference* dataset. Then, a completion process is applied and the completed dataset is compared to the reference one. However, missing values are introduced according to statistical hypotheses, typically by removing values fully randomly or removing values randomly dependent on a present value. Unfortunately, as recently highlighted in [13], these assumptions may be wrong in real-world datasets and current approaches do not avoid the pitfall of non-random missing values. This issue of different types of missing data mechanisms is addressed in the well-known work of Little and Rubin [10] which distinguishes three types of missing values (cf. section 2.3). This work is interesting because it brings out a better understanding of the origin of missing values and the scope of the completion methods. Not surprisingly, the type *NMAR* [10] (i.e., Not Missing At Random) is generally not tackled by the current completion methods [19]. We also think that these types suffer from limitations in practical uses of the data mining. First, it relies on a global characterization of the missing values. We claim that completion methods can benefit from local characterization. For instance, we will see in Section 4 that the explanation of missing values suggests several completion strategies and it should be illusory to settle for a unique characterization for all the missing values coming from one attribute. Second, the types proposed by Little and Rubin are based on both the true (or complete) data and the available data. But, in real-world data mining tasks, the true data remain unknown.

We propose in this paper a new typology characterizing missing values according to relationships within the data. These relationships are automatically discovered by an association rule-based model. We define four types of missing values leading to a precise characterization of each missing value. Contrary to [10], these types rely on local regularities so that each missing value has its own type independently of the other missing values. Moreover, relationships are only based on the available data. We claim that such a characterization brings out a twofold advantage. First, it provides a better understanding of the underlying reasons of the missing values thus contributing to better control data quality. For instance, in Section 4, we will see that this characterization suggests attributes meaningless w.r.t. other attributes and consequently the data set should be restructured. Second, this characterization enables us perceptive techniques to deal with missing values according to their origins. Experiments in Section 4 show that the type *indirect* of this characterization highlights missing values which are not tackled by the current methods [19]. About completion, previous works have already shown the importance of a preliminary analysis of missing values to propose relevant methods [6].

More generally, we address the following questions that we feel crucial to handle missing values in data mining tasks: what kind of missing values models can be recognized from the available data? Is it possible to explain the presence of the missing values? How can we characterize these missing values? Could we have different characterizations for the missing values coming from one attribute?

The remainder of the paper is organised as follows. Section 2 presents the terminology used throughout the paper and summarizes missing values models from the literature. Section 3 proposes a new typology of missing values corresponding to our method characterizing missing values. Experiments (Section 4) on a real-world medical dataset highlight the interests of this typology.

2   Preliminaries

This section introduces the technical concepts (real and measured contexts, itemsets and association rules) and the missing values models proposed by Little and Rubin [10].

2.1   Definitions and Notations

Let us consider a database in an "attribute/value" format. Figure 1 provides a toy example. Each object is described by four attributes A_1, A_2, A_3 and A_4. A domain of values is associated to each attribute, e.g., $dom(A_1) = \{a, b\}$, $dom(A_2) = \{c, d\}$, $dom(A_3) = \{e, f, g\}$ and $dom(A_4) = \{h, i\}$. An attribute A_i may have an unknown value, called a *missing value*, noted by "?". We give now the definition of a *real context* (an example of such a context is given by the left part of Figure 1).

Definition 1 (Real context). *A **Real Context** is a triplet $\mathcal{K} = (\mathcal{O}, \mathcal{I}, \mathcal{R})$, where $\mathcal{O}$ is the set of objects or transactions, $\mathcal{I}$ the set of items and $\mathcal{R}$ is a function over $\mathcal{O} \times \mathcal{I}$ which takes its values in $\{present, absent\}$. $\mathcal{R}(o, i) = present$ indicates that the item $i \in \mathcal{I}$ is present in the object $o \in \mathcal{O}$. $\mathcal{R}(o, i) = absent$ means that i is not in o.*

When missing values occur, the real context is converted in a measured context.

Definition 2 (Measured Context)
*A missing value modelling operator, noted mv, maps a **real context** $\mathcal{K} = (\mathcal{O}, \mathcal{I}, \mathcal{R})$ into a **measured context** noted by $mv(\mathcal{K}) = (\mathcal{O}, \mathcal{I}, mv(\mathcal{R}))$. The new function $mv(\mathcal{R})$*

	A_1		A_2		A_3			A_4	
	a	b	c	d	e	f	g	h	i
o_1	×		×		×			×	
o_2		×	×	×	×				×
o_3	×		×			×		×	
o_4	×			×	×				×
o_5	×		×			×			×
o_6		×	×	×		×		×	
o_7	×			×			×		×
o_8		×		×			×		×

	A_1		A_2		A_3			A_4	
	a	b	c	d	e	f	g	h	i
o_1	×		×		?	?	?	×	
o_2	?	?	×		×			?	?
o_3	×		×		?	?	?	×	
o_4	×			×	×			?	?
o_5	?	?	×		×			?	?
o_6		×	?	?	×			×	
o_7	×		?	?		×		?	?
o_8	?	?		×		×		?	?

Fig. 1. Boolean context. Left : *real* context. Right : *measured* context.

takes its values in $\{present, absent, missing\}$ *and fulfills the following properties for value* $\in \{present, absent\}$:

1. $mv(\mathcal{R})(o, i) = value \Rightarrow \mathcal{R}(o, i) = value$
2. $\mathcal{R}(o, i) = value \Rightarrow mv(\mathcal{R})(o, i) \in \{value, missing\}$

A real context $\mathcal{K}$ corresponds to the complete dataset (which stays unknown in real-world applications) whereas a measured context $mv(\mathcal{K})$ refers to the available data, i.e., the data that we have to tackle in practice. The modelling operator $mv()$ models a data erasing, *i.e.,* some values were deleted (moved to $missing$). When a value is missing in $mv(\mathcal{K})$, it becomes impossible to guess its real value in $\mathcal{K}$. However, when the value is known in $mv(\mathcal{K})$, it corresponds to the same value in the real context $\mathcal{K}$ (first property of Definition 2). The second property ensures that a value (either present or absent) in $\mathcal{K}$ keeps its value or will be missing in $mv(\mathcal{K})$.

Figure 1 (Right) is the measured context associated to the left part of this figure. As this context comes from an attribute/value format, a missing value affects all the possible values of an attribute. For example, the missing values on the items a and b in the object o_8 in $mv(\mathcal{K})$ (Figure 1 - Right) hide actually the presence of the item b in $\mathcal{K}$ (Figure 1 - Left).

2.2 Association Rules

An *itemset* (or *pattern*) $X \subset \mathcal{I}$ is a set of items. An object $o \in \mathcal{O}$ contains the itemset X and we note $X \subset o$ if $\forall i \in X$, $\mathcal{R}(o, i) = present$. The absolute support of X, noted by $Supp(X)$, is defined as follows : $Supp(X) = |\{o \in \mathcal{O}|X \subset o\}|$. An association rule R, based on a pattern Z, is an expression $R : X \to Y$ where $X \subsetneq Z$ and $Y = Z \backslash X$. The itemsets X and Y are respectively called *premise* and *conclusion* of R. An association rule is quantified by its *support* and *confidence*: the *support* is equal to the one of Z and the confidence is defined as $Conf(R) = \frac{Supp(Z)}{Supp(X)}$. Valid association rules are those whose support and confidence are greater than or equal to minimal thresholds, respectively noted by $minsup$ and $minconf$. If $Conf(R) = 1$, then R is *exact*, otherwise it is *approximative*. For example, $fh \to c$ is an exact association rule (cf. Figure1 - Left).

Presenting all the steps of the association rule mining is out the scope of this paper. However, we use techniques for non-redundant association rules computation, specially the basis of proper implications [18]. These rules are composed in their premise part by a free-set[1][3] (*aka* key [12] or minimal generator), and in the conclusion part, by an attribute of its closure[2], which does not appear in the closure of one of the subsets of the premise part. This computational technique considerably restricts the number of redundant exact association rules, while faithfully preserving all the induced knowledge.

2.3 Classical Models of the Missing Values Appearance

Little and Rubbin [10] distinguish the following three types of missing values, which are also called models for missing values appearance.

---

[1] Since exact association rules cannot be built on part of these patterns.
[2] The closure of a pattern is composed by attributes which are systematically present with it.

- **MCAR** *(Missing Completely at Random)*: the probability that an attribute A_i is missing is unrelated to the value of A_i nor to the value of any other attributes. It may affect any record and any object. For example, the missing values in the attributes A_1 and A_2 of Figure 1 (Right) are *MCAR*. These missing values have *a priori* no particular explanation.
- **MAR** *(Missing at Random)*: such a missing value depends on the values of other attributes but it does not depend on the true value of any of the missing values. In Figure 1 (Right), the missing values which affect the attribute A_3 are examples of *MAR*. Note that all missing values on A_3 are related to the presence of the itemset ac.
- **NMAR** *(Not Missing at Random)*: if the missing value is related to the true value itself, then the missing value is said to be *NMAR*. The missing values affecting the attribute A_4 in Figure 1 (Right) are *NMAR*. If the item i cannot be recorded, then a missing value arises each time that i should be observed.

2.4 Discussion and Position Statement

The use of the Little and Rubbin models is not easy [16]. A first difficulty is that these models require knowledge or assumptions on the real context (i.e., the true values) whereas, in practice, only the measured context is known. With the *MAR* model, the missing values coming from an attribute have to satisfy a property of randomness. The characterization of the *NMAR* model needs assumptions on the real context. For instance, we can assume that a missing weight is more likely linked to an obese people than a healthy one. Such an assumption can be formulated by experts but an expert is not always available. Moreover, there are relationships inside the data which are not taken into account by these models. For instance, in our experiments (cf. Section 4), we note that the size of an invasion of a location is missing when there is no ganglion. In practice, it should be useful to integrate such relationships to deal with missing values (in this example, this relationship suggests that the attribute on the size is meaningless when there is no ganglion and thus the data set should be restructured). Finally, note that these models are based on a global characterization of the missing values: they give the same explanation for all the missing values coming from one attribute. We think that a local characterization is more powerful to deal with missing values, especially to propose completion techniques.

3   A New Missing Values Typology

In this section, we propose a new typology of missing values based on a local characterization of the missing values computed only from the available data.

- **Direct missing value**: a missing value is said to be *direct*, whenever it has relations with other measured values.
- **Indirect missing value**: a missing value is said to be *indirect*, whenever it has relations with other missing values.
- **Hybrid missing value**: a missing value is said to be *hybrid*, whenever it has relations with both measured and missing values.

– **Random missing value**: a missing value is said to be *random*, whenever it does not have any relation with other measured values or missing ones.

The next section formally defines this typology. It uses association rules for characterizing the missing values.

3.1 Association Rule Based Model for Missing Values Characterization

The definition of association rules characterizing missing values beforehand requires to quantify the degree of the presence/absence of an itemset in a measured context $mv(\mathcal{K})$:

Definition 3 (Present itemset). *An itemset $X \subset \mathcal{I}$ is said to be Present, in $o \in \mathcal{O}$ if and only if $\forall x \in X, mv(\mathcal{R})(o, x) = present$, and is noted by $Present(X, o)$.*

Definition 4 (Missing itemset). *An itemset $X \subset \mathcal{I}$ is said to be missing, in $o \in \mathcal{O}$ if and only if $\forall x \in X, mv(\mathcal{R})(o, x) = missing$, and is noted by $Missing(X, o)$.*

Definition 5 (Partially present itemset). *An itemset $X \subset \mathcal{I}$ is said to be Partially present, in $o \in \mathcal{O}$ if and only if $\forall x \in X, mv(\mathcal{R})(o, x) \neq absent$ and $\exists x_1 \in X, mv(\mathcal{R})(o, x_1) = present$ and $\exists x_2 \in X, mv(\mathcal{R})(o, x_2) = missing$, and is noted by $PartPresent(X, o)$.*

Example 1. In the measured context depicted by Figure 1 (Right), we have $Present(adf, o_4)$, $Missing(ah, o_8)$ and $PartPresent(bdg, o_8)$.

The regularities allowing the characterization of missing values can be straightforwardly detected by association rules. In practice, these rules are discovered by using a minimal support value, $minsup$ to only care about regularities that appear frequently. We propose here a formalization of this new missing value typology as follows:

Definition 6 (*direct* missing value). *A missing value i is said to be* direct *in $\mathcal{T} \subset \mathcal{O}$ ($|\mathcal{T}| \geq minsup$) if and only if $\exists X \subset \mathcal{I} \setminus \{i\} \,|\, \forall o \in \mathcal{T}, Present(X, o) \Rightarrow Missing(i, o)$.*

Definition 7 (*indirect* missing value). *A missing value i is said to be* indirect *in $\mathcal{T} \subset \mathcal{O}$ ($|\mathcal{T}| \geq minsup$) if and only if $\exists X \subset \mathcal{I} \setminus \{i\} \,|\, \forall o \in \mathcal{T}, Missing(X, o) \Rightarrow Missing(i, o)$.*

Definition 8 (*hybrid* missing value). *A missing value i is said to be* hybrid *in $\mathcal{T} \subset \mathcal{O}$ ($|\mathcal{T}| \geq minsup$) if and only if $\exists X \subset \mathcal{I} \setminus \{i\} \,|\, \forall o \in \mathcal{T}, PartPresent(X, o) \Rightarrow Missing(i, o)$.*

Definition 9 (*random* missing value). *A missing value i is said to be* random *in $\mathcal{T} \subset \mathcal{O}$ ($|\mathcal{T}| \geq minsup$) if and only if $\forall X \subset \mathcal{I} \setminus \{i\}, \exists o \in \mathcal{T} \,|\, Missing(i, o) \wedge \neg Present(X, o)$.*

Example 2. Let us follow the running example in Figure 1. The rules used for the characterization associated to the context $mv(\mathcal{K})$ (Figure 1 - Right) are given by the left

	Rule	Support
R_1	$a \wedge c \rightarrow MV(A_3)$	2
R_2	$MV(A_1) \rightarrow MV(A_4)$	3
R_3	$a \wedge h \rightarrow MV(A_3)$	2
R_4	$c \wedge MV(A_4) \rightarrow MV(A_1)$	2
R_5	$c \wedge h \rightarrow MV(A_3)$	2
R_6	$d \rightarrow MV(A_4)$	2
R_7	$g \rightarrow MV(A_4)$	2

	A_1	A_2	A_3	A_4
o_1	-		{direct}	-
o_2	{hybrid}	-	-	{indirect}
o_3	-	-	{direct}	-
o_4	-	-	-	{direct}
o_5	{hybrid}	-	-	{indirect}
o_6	-	{random}	-	-
o_7	-	{random}	-	{direct}
o_8	{random}	-	-	{direct, indirect}

Fig. 2. Left: Rules concluding on missing values with *minsup*=2 from the measured context $mv(\mathcal{K})$ (cf. Figure 1). Right: Typology of the missing values associated to $mv(\mathcal{K})$.

part of the Figure 2 (Left) with $minsup = 2$. The notation $MV(A_i)$ indicates a missing value on the attribute A_i (*i.e.*, on all items of the A_i domain). The column *Support* indicates the value of the absolute support of the rule. The characterization of the missing values is given by the Right part of Figure 2. For example, the rule R_4 shows that when c is present and a missing value occurs on the A_4 attribute, then, a missing value is observed on the A_1 attribute. This rule characterizes *hybrid* missing values on the A_1 attribute over the objects o_2 and o_5 (Figure 2 - Right).

3.2 Impact of the Basis of Proper Implications for the Missing Values Characterization

As said in Section 2.2, we use the basis of proper implications for building the rules characterizing the missing values. We now show the interest of this rule basis.

The basis of proper implications can be seen as a nicety of the well-known Bastide's basis [1] which provides a cover of the exact association rules. With the Bastide's basis, every free pattern is the premise of a rule whose the conclusion is the *closure* of its premise (see Section 2.2). The basis of proper implications is a finer cover: a rule is kept only if its conclusion cannot be inferred from the closures of any subset of its premise. The rules of the basis of proper implications satisfy the following suitable properties to characterize missing values:

1. a rule has a minimum premise for a given conclusion. The redundancy, which may lead to conflicts, is limited.
2. the number of rules of the basis of proper implications is very small compared to the size of the Bastide's basis. For example, Table 1 compares the size of these basis under our experimental conditions on the HODGKIN dataset (see Section 4). The number of rules is drastically reduced with the basis of proper implications.

From our example (Figure 1), we illustrate the interest of the basis of proper implication for characterizing missing values. Figure 3 presents the rules concluding on the missing values of the A_4 attribute (noted $MV(A_4)$) which are generated by the Bastide's basis and the basis of proper implications. We note that the rule R_4' is not generated by the basis of the proper implications since that $MV(A_4)$ has already appeared in the closure of one subset of the premise of R_4' *i.e.*, $MV(A_4)$ is already in the

Table 1. Number of rules with the Bastide's basis and the basis of proper implications on the HODGKIN dataset

	Rules	Rules concluding on a missing value
Bastide's basis	2 923 070	2 681 045
Proper implications basis	49	15

	Rule	Support
R_1'	$MV(A_1) \rightarrow MV(A_4)$	3
R_2'	$d \rightarrow MV(A_4)$	2
R_3'	$g \rightarrow MV(A_4)$	2
R_4'	$c \wedge MV(A_1) \rightarrow MV(A_4)$	2

	Rule	Support
R_1''	$MV(A_1) \rightarrow MV(A_4)$	3
R_2''	$d \rightarrow MV(A_4)$	2
R_3''	$g \rightarrow MV(A_4)$	2

Fig. 3. Rules concluding on $MV(A_4)$ Left: Bastide's basis. Right: basis of proper implications

conclusion of the rule R_1''. Consequently, R_1'' is considered as more interesting than R_4' when characterizing the missing values on the A_4 attribute, since it has a non-redundant and minimum premise. As expected, the non-redundancy limits conflicts between types: objects o_2 and o_5 are characterized only by the type *indirect* with the basis of proper implications whereas the Bastide's basis proposes two types (*indirect* and *hybrid*). For the other objects, the characterization is the same with the two bases. The formalization of this intuition appears in our perspective, *i.e.,* we are particularly focusing on defining properties of this characterization.

3.3 Characteristics of the Missing Values Typologies

Table 2 summarizes the differences between the Little and Rubin models and our typology. Obviously, we get back the main features that we have introduced. The Little and Rubbin models need knowledge (or assumptions) on the true data so that their practical use is difficult. That it is why we qualify these models as "theoretical". On the contrary, our typology is only based on the available data. The Little and Rubin models perform a global characterization: they allocate the same type for all missing values coming from an attribute. With our typology, the types are evaluated by using local regularities and

Table 2. Characteristics of the missing values typologies

	Little and Rubin Typology	Our new typology
Data	Available + unavailable	Available
Framework	Theoretical	Theoretical + Operational
Characterization	Global	Local
Types	MCAR	Random
	MAR	Direct
	NMAR	-
	-	Indirect
	-	Hybrid

each missing value has its own type independently of the other missing values. The last row compares the types of the two typologies. The *NMAR* model does not appear in our typology since it is based on the unavailable data. *Indirect* and *hybrid* types are not present in the Little and Rubin typology. Besides, even if there is a matching between *MCAR/Random* and *MAR/Direct*, practical results can be different because our typology stems from local subsets of objects.

4   Experimental Results

This section describes our experiments carried out on a medical dataset about the Hodgkin disease. We have chosen this dataset because it addresses a real-world medical application with many missing values. These missing values are natural (*i.e.,* no simulation was made for artificially introducing them). Furthermore, this database is used by physicians and they can provide advice and feedback on the data and results.

The HODGKIN *dataset.* The Hodgkin disease is a cancer of the lymphatic system. The HODGKIN dataset contains 3904 patients split in three therapeutic trials (H7, H8 and H9) realized over successive temporal periods. Each patient is described by 36 attributes and 29 contain missing values. The percentage of missing values for an attribute varies between 2% and 88%. The attributes include blood and histological characteristics and several information on the locations (cervical, hilum, mediastinum, auxiliaries) and the sizes of the invasions. An invasion is a symptom of a cancer.

Results. The rules were mined with an absolute support equals 700. Only 15 rules concluding on missing values are discovered (recalling that the properties of the rule cover drastically reduce the number of rules as indicated in Table 1). We have also performed experiments with rules allowing few exceptions (i.e., non exact rules with very high confidence) and we found similar results. The 15 rules are reported in Figure 4 (Left). For example, the rule R_4 indicates that all objects containing the item $plaq <= 600$ and a missing value on the attribute ctr (right top cervical ganglion) also contain a missing value on the attribute ctl. It corresponds to an *hybrid* missing value characterization.

The rules R_1 and R_2 conclude on a missing value of the invasions of the left or right top cervical ganglion (*ctl* or *ctr*). These rules contain in their premise only the *trial H7* attribute. Therefore, the type of these missing values is *direct* and the trial H7 explains these missing values. This type highlights a characteristic situation suggesting to investigate the running of the trial H7: actually this trial did not distinguish the top and bottom cervical ganglions and that it is why these values are missing. This case of missing values reveals a classical problem of data merging. Our method enables us to be aware of such issue and therefore it allows to better control the data quality. Note also that some missing values on *ctl* attributes were characterized by other rules as *indirect* (R_3) and as *hybrid* (R_4). It states a multiple missing value characterization.

Rules R_5 until R_{10} (left part of the Figure 4) characterize missing values having the type *direct*. They mean that when a ganglion is not invaded, its invasion size is not measured. It is interesting to check that such a knowledge is automatically discovered. Rules R_{11} and R_{12} characterize *indirect* missing values on the sizes of the ganglions

	Premise	Conclusion	support
R_1	trial H7	MV(ctr)	816
R_2	trial H7	MV(ctl)	816
R_3	MV(axlsiz)∧ MV(ctr)	MV(ctl)	811
R_4	plaq<=600 ∧ MV(ctr)	MV(ctl)	778
R_5	ctr not invaded	MV(ctrsiz)	2449
R_6	ctl not invaded	MV(ctlsiz)	2407
R_7	cbr not invaded	MV(cbrsiz)	1969
R_8	cbl not invaded	MV(cblsiz)	1690
R_9	axl not invaded	MV(axlsiz)	3295
R_{10}	axl not invaded	MV(axlsiz)	3185
R_{11}	MV(ctr)	MV(ctrsiz)	908
R_{12}	MV(ctl)	MV(ctlsiz)	910
R_{13}	med not invaded ∧ vs <=30	MV(mtr)	920
R_{14}	med not invaded ∧ relapse= no	MV(mtr)	1042
R_{15}	med not invaded ∧ MV(cblsiz)	MV(mtr)	717

attribute	missing values	direct	indirect	hybrid	random
ctr	908	90%	0	0	10%
ctl	910	10.7%	10%	79%	0.3%
ctrsiz	3435	71%	3%	24	2%
ctlsiz	3398	71%	3%	24%	2%
cbrsiz	2274	87%	0	0	13%
cblsiz	2027	83%	0	0	17%
axlsiz	3444	96%	0	0	4%
axlsiz	3360	95%	0	0	5%
mtr	1512	32%	0	47%	21%

Fig. 4. Left: Exact association rules discovered from the HODGKIN dataset with $misnusp$=700. Right: Missing values characterization in the HODGKIN dataset.

$ctlsiz$ and $ctrsiz$. Missing values on these size attributes are explained by missing values on other attributes. When we do not know if a ganglion is invaded, then a missing value always occurs on its size. Rules R_{13} until R_{15} characterize missing values on the attribute *ratio mediastinum ganglion width / thorax*. The first two rules characterize *direct* missing values and the third one *hybrid* missing values.

The right part of Figure 4 summarizes the different types of the missing values according to the attributes. An important result is that most of the missing values are not due to randomness but they belong to the *direct*, *indirect* or *hybrid* types. As these types express relationships in the data, it means that our characterization is able to suggest explanations for most of the missing values.

In these experiments, we check that the missing values coming from one attribute may be characterized according to different types. It is the case for the attributes *ctl*, *ctrsiz*, *ctlsiz* and *mtr*. It illustrates the power of the local characterization of our approach, which does not force to consider only a single type for all the missing values coming from one attribute.

The database includes other attributes with a low rate of missing values (between 2% and 9%). As we used an absolute support threshold of 700 corresponding to 18% of the data, we have not discovered rules characterizing these missing values. The characterization depends on the minimal support threshold. Decreasing the minimal support threshold may lead to discover rules characterizing missing values on these attributes but it may provide multiple conflicts of characterization.

5   Conclusions and Perspectives

In this paper, we have proposed a new typology of missing values according to the relationships within the data. These relationships are automatically discovered by an association rule-based model. Contrary to models from the literature, our approach is only based on the available data and it relies on local regularities so that each missing value has its own type independently of the other missing values. This characterization enables us a better understanding of the underlying reasons of the missing values (e.g., attribute meaningless w.r.t. other attributes, missing values depending on other data, missing values by accident). We claim that it is precious because it suggests explanations about the quality of the data and also more powerful techniques to deal with missing values, especially to propose completion methods. Experiments on a real-world medical dataset highlight the interests of this typology. Among others, they show that many missing values do not stem from randomness. Further work is to show the impact of the variation of the *minsup* over the characterization and on the conflict as well as to investigate the use of this typology for the completion issue. Association rules have been shown to be efficient to complete missing values coming from random processes [2]. Our intuition is that missing values characterized by *direct*, *indirect* or *hybrid* types require the help of background knowledge to be completed since these types express specific behaviors.

Acknowledgments. The authors are grateful to the "Centre Anti-Cancéreux François Baclesse de Caen" and to the Doctor Michel HENRY-AMAR for providing the HODGKIN data. This work is partially supported by the French-Tunisian project *CMCU 05G1412*.

References

1. Bastide, Y., Pasquier, N., Taouil, R., Lakhal, L., Stumme, G.: Mining minimal non-redundant association rules using frequent closed itemsets. In: Palamidessi, C., Moniz Pereira, L., Lloyd, J.W., Dahl, V., Furbach, U., Kerber, M., Lau, K.-K., Sagiv, Y., Stuckey, P.J. (eds.) CL 2000. LNCS (LNAI), vol. 1861, pp. 972–986. Springer, Heidelberg (2000)
2. Ben Othman, L., Ben Yahia, S.: $GBAR_{MVC}$: Generic Basis of Association Rules based approach for Missing Values Completion. The International Journal of Computing and Information Sciences (to appear)
3. Boulicaut, J.-F., Bykowski, A., Rigotti, C.: Approximation of frequency queries by means of free-sets. In: Zighed, D.A., Komorowski, J., Żytkow, J.M. (eds.) PKDD 2000. LNCS, vol. 1910, pp. 75–85. Springer, Heidelberg (2000)
4. Calders, T., Goethals, B., Mampaey, M.: Mining itemsets in the presence of missing values. In: Proceedings of the ACM Symposium on Applied Computing, Seoul, Korea, pp. 404–408. ACM Press, New York (2007)
5. Dardzinska, A., Ras, Z.W.: CHASE-2: Rule based chase algorithm for information systems of type lambda. In: Tsumoto, S., Yamaguchi, T., Numao, M., Motoda, H. (eds.) AM 2003. LNCS (LNAI), vol. 3430, pp. 258–270. Springer, Heidelberg (2005)
6. Delavallade, T., Dang, T.: Using entropy to impute missing data in a classification task. In: Proceedings of the International Conference of Fuzzy Systems (FUZZ-IEEE 2007), London, UK, July 2007, pp. 23–26 (2007)
7. Dempster, A., Laird, N., Rubin, D.: Maximum likelihood from incomplete data via the EM algorithm. Journal of the Royal Statistical Society 39(1), 1–38 (1977)
8. Grzymala-Busse, J.W.: Three approaches to missing attribute values - a rough set perspective. In: Workshop on Foundations of Data Mining, associated with the fourth IEEE International Conference on Data Mining (2004)

9. Grzymała-Busse, J.W., Hu, M.: A comparison of several approaches to missing attribute values in data mining. In: Ziarko, W.P., Yao, Y. (eds.) RSCTC 2000. LNCS, vol. 2005, pp. 378–385. Springer, Heidelberg (2001)
10. Little, R., Rubin, D.: Statistical Analysis with Missing Data. John Wiley, New York (1987)
11. Nelwamondo, F., Marwala, T.: Rough set theory for the treatment of incompltete data. In: Proceedings of the IEEE International Conference of Fuzzy Systems (FUZZ-IEEE 2007), London, UK, July 2007, pp. 23–26 (2007)
12. Pasquier, N., Taouil, R., Bastide, Y., Stumme, G., Lakhal, L.: Generating a condensed representation for association rules. Journal of Intelligent Information Systems 24, 29–60 (2005)
13. Pearson, R.K.: The problem of disguised missing data. SIGKDD Explorations 8(1), 83–92 (2006)
14. Ragel, A., Crémilleux, B.: Treatment of missing values for association rules. In: Wu, X., Kotagiri, R., Korb, K.B. (eds.) PAKDD 1998. LNCS, vol. 1394, pp. 258–270. Springer, Heidelberg (1998)
15. Rioult, F., Crémilleux, B.: Mining Correct Properties in Incomplete Databases. In: Džeroski, S., Struyf, J. (eds.) KDID 2006. LNCS, vol. 4747, pp. 208–222. Springer, Heidelberg (2007)
16. Shafer, J.L., Graham, J.W.: Mising data: Our view of the state of the art. Psychological Methods 7(2), 147–177 (2002)
17. Shen, J.J., Chang, C.C., Li, Y.C.: Combined association rules for dealing with missing values. Journal of Information Science 33(4), 468–480 (2007)
18. Taouil, R., Bastide, Y.: Computing proper implications. In: Proceedings of the 9th International Conference on Conceptual Structures (ICCS 2001), Stanford, CA, pp. 49–61 (2001)
19. Vreeken, J., Siebes, A.: Filling in the blanks - krimp minimisation for missing data. In: Perner, P. (ed.) ICDM 2008. LNCS, vol. 5077, pp. 1067–1072. Springer, Heidelberg (2008)
20. Wu, C., Wun, C., Chou, H.: Using association rules for completing missing data. In: Proceedings of 4th International Conference on Hybrid Intelligent Systems (HIS 2004), Kitakyushu, Japan, December 5-8, 2004, pp. 236–241 (2004)

Recommending Multidimensional Queries

Arnaud Giacometti, Patrick Marcel, and Elsa Negre

Université François Rabelais Tours
Laboratoire d'Informatique
France
{arnaud.giacometti,patrick.marcel,elsa.negre}@univ-tours.fr

Abstract. Interactive analysis of datacube, in which a user navigates a cube by launching a sequence of queries is often tedious since the user may have no idea of what the forthcoming query should be in his current analysis. To better support this process we propose in this paper to apply a Collaborative Work approach that leverages former explorations of the cube to recommend OLAP queries. The system that we have developed adapts Approximate String Matching, a technique popular in Information Retrieval, to match the current analysis with the former explorations and help suggesting a query to the user. Our approach has been implemented with the open source Mondrian OLAP server to recommend MDX queries and we have carried out some preliminary experiments that show its efficiency for generating effective query recommendations.

1 Introduction

Traditional OLAP users interactively navigate a cube by launching a sequence of queries over a datawarehouse, which we call an analysis session (or session for short) in the following. This process is often tedious since the user may have no idea of what the forthcoming query should be [1]. This difficulty might be related to the decline of interactive analysis pointed out in [2].

To better support this process, we proposed in [3] a framework for recommending OLAP queries. The idea is to leverage what the other users did during their former navigations on the cube, and to use this information as a basis for recommending to the user what his forthcoming query could be.

In this paper, we present a significant extension of this work that results in a system for recommending multidimensional queries expressed with MDX [4], the de facto standard. Namely we have changed the core of the framework, that is the distance between queries and the distance between sessions, to better handle the peculiarities of OLAP data. We have adapted our system to deal with real-case cubes and MDX queries. More precisely, our contribution include:

- A measurement of the distance of two MDX queries that leverages the peculiarities of OLAP data,
- A measurement of the distance of two sequences of MDX queries by using Approximate String Matching [5], a technique popular in Information Retrieval,

T.B. Pedersen, M.K. Mohania, and A M. Tjoa (Eds.): DaWaK 2009, LNCS 5691, pp. 453–466, 2009.

- A framework for using these measures to search the log of an OLAP server to find a set of sessions matching the current session and generate recommendations,
- An implementation of this approach into a recommender system that fully integrates with the open source Mondrian OLAP engine [6] to recommend MDX queries on the fly during an interactive analysis session,
- Experiments conducted to assess the efficiency and effectiveness of our approach.

The paper is organized as follows: Section 2 briefly reviews related work. A motivating simple example is given in Section 3. Section 4 introduces the distance for comparing two MDX queries, and Section 5 introduces the distance for comparing two analysis sessions. Finally Section 6 completes the description of the recommender system by detailing the algorithm for computing recommendations. Section 7 presents our experimental results. We conclude and discuss future work in Section 8. The proofs of the properties are omitted due to lack of space.

2   Related Work

The only other work we know that proposes to recommend queries for supporting database exploration is that of [7]. Although this work shares some common features with ours, it differs on two important aspects: First it deals only with SQL Select-Project-Join queries and second, the fact that a session is a sequence of queries is not taken into account. To the best of our knowledge, our work is the first work dealing with the problem of recommending multidimensional, especially MDX, queries.

The only work that proposed a framework for anticipating an OLAP query is the work of [8,9]. However the main concern of this work is to prefetch data, not to guide the user based on what other users did. In addition, [8,9] does not deal with MDX queries, and the similarity between queries only relies on the schema of the query (i.e., dimensions and levels) whereas the distance that we use takes the members into account. Finally, a Markov Model is used to predict the forthcoming query, whereas our approach is based on Approximate String Matching [5], a technique popular in Information Retrieval.

To support interactive analysis of multidimensional data, Sarawagi et al. introduced discovery driven analysis of OLAP cube in [10]. This and subsequent work [11,12,1] resulted in the definition of various OLAP operators to guide the user towards unexpected data in the cube or to propose to explain an unexpected result. The main difference with our work is that these operators are applied only on query results and they do not take into account what other users might have discovered.

Computing distances between queries logged by a database server has already been investigated by [13]. In this work, language modeling is used to detect sessions within OLTP query logs. With a different goal (recommending query instead of detecting sessions) our work also proposes a way of calculating a distance between queries where the distance computation takes advantage of the particularities of OLAP queries, like the possibility of navigating multidimensional data by changing the level of detail.

Our work can be seen as a way to integrate OLAP and Information Retrieval (IR) a domain where it is very popular to leverage what the other users did to generate recommendations [14]. Note that there is a recent interest for trying to combine IR and OLAP. For instance, in [15] the authors propose to query a datacube with only a set of keywords. Among the potential answers to the query, only the subcubes that are the most surprising are presented to the user.

3 Example

In this section we illustrate with a simple example the basic idea under our recommender system. Consider an OLAP server used by several users navigating a datacube. In what follows, this cube is a simplified version of the FoodMart datacube (the demo example coming with the open source Mondrian OLAP engine [6]) that is composed of four dimension tables and the Sales fact table, having respectively the following schemas:

- $sch(\text{PRODUCT}) = \{p\_id, Name, Brand, SubCateg, Category, Family,$
$$AllProducts\},$$
- $sch(\text{TIME}) = \{t\_id, Day, Month, Quarter, Year, AllYears\},$
- $sch(\text{CUSTOMER}) = \{c\_id, Name, City, State, Country, AllCustomers\},$
- $sch(\text{STORE}) = \{s\_id, Name, City, State, Country, AllStores\},$
- $sch(\text{SALES}) = \{p\_id, t\_id, c\_id, s\_id, Unit\ Sales\}$

Each user can open a session on the server to navigate the cube by launching a sequence of queries. The server logs these sessions, i.e., the sequences of queries launched during each analysis session. Suppose the log contains the three sessions detailed in the appendix. Session s_1 analyzes the sales of alcoholic beverages in the USA, Session s_2 analyzes the sales of milk of the brand "Gorilla" in California, and Session s_3 analyzes the sales of milk and cereals in San Francisco.

Imagine now a new session, called the *current session* (or s_c), is performed by a user. Suppose the user issues the three following queries on the cube, named respectively q_1, q_2 and q_3, to analyze the sales of milk in San Francisco:

```
SELECT {[Store].[All Stores].Children} ON COLUMNS,
       {[Product].[All Products].[Food], [Product].[All Products].[Drink]} ON ROWS
FROM   [Sales]
WHERE  {[Measures].[Unit Sales]}
```

```
SELECT {[Store].[All Stores].[USA].[CA].[San Francisco]} ON COLUMNS,
       {[Product].[All Products].[Food], [Product].[All Products].[Drink]} ON ROWS
FROM   [Sales]
WHERE  {[Measures].[Unit Sales]}
```

```
SELECT {[Store].[All Stores].[USA].[CA].[San Francisco]} ON COLUMNS,
       {[Product].[All Products].[Drink].[Dairy].[Milk]} ON ROWS
FROM   [Sales]
WHERE  {[Measures].[Unit Sales]}
```

The recommender system computes the distance between the current session and each session of the log in order to find those candidate sessions that resemble the current session the most. In our example, suppose that sessions s_2 and s_3 are found the closest to the current session. Intuitively this is because each ith query of s_c is close to the ith of the session s_2 (resp. s_3) and, in the case of s_3, having one more query does not increase the distance a lot.

Among the queries composing these candidate sessions, one must be recommended to the user. Considering that the outcome of a session is very often the result of the last query of this session, the recommender system will compute the distance between the last query of the current session and each last query of the candidate sessions. It will then select as the first recommendation the query that is the closest to the last query of the current session. In our example, this query is q_6 since it is closer to q_3 than q_5.

4 Comparing MDX Queries

In this section, we present our approach for computing a distance between MDX queries. We first begin by giving basic definitions.

4.1 Basic Definitions (Cube, References, Queries)

An n-dimensional *cube* $C = \langle D_1, \ldots, D_n, F \rangle$ is defined as the classical $n + 1$ relation instances of a star schema, with one relation instance for each of the n dimensions and one relation instance for the fact table. Given a particular dimension table D_i, the members of the dimension are the values in this table[1]. These members are arranged into a graph H_i (traditionally a hierarchy)[2].

Given an n-dimensional cube $C = \langle D_1, \ldots, D_n, F \rangle$, a cell is a tuple of the fact table F. A cell *reference* (or reference for short) is an n-tuple $\langle r_1, \ldots, r_n \rangle$ where r_i is a member of dimension D_i for all $i \in [1, n]$.

MDX queries are modeled in the following way: Considering that the SELECT and WHERE clauses of an MDX expression define the set of references that the user wants to extract from the cube, we propose to see MDX queries as sets of references, for a given instance of a cube.

Formally, let $C = \langle D_1, \ldots, D_n, F \rangle$ be an n-dimensional cube, M be an MDX expression and for all $i \in [1, n]$, let R_i be the set of members of dimension D_i that is deduced from the SELECT and WHERE clause. The *query* over C that corresponds to M is the set of references $R_1 \times \ldots \times R_n$. In what follows, if q is a query we note $r \in q$ to denote that r is a reference of q.

Example 1. The query q_2 of section 3 corresponds to the following set of references: $\{\langle Drink, \quad alltime, allcustomer, San \quad Francisco \rangle, \langle Food, alltime, allcustomer, San Francisco \rangle\}$

---

[1] Note that this definition is done without loss of generality w.r.t the calculated members defined in MDX by the optional WITH MEMBER clause. Indeed, a calculated member is associated with a particular dimension, at a particular level of a hierarchy, and thus it is treated in the following as a regular member.

[2] Flat dimensions (like e.g., a measure dimension) are considered as arranged in a hierarchy as well, where all the members have as common ancestor the root of the hierarchy.

4.2   Distance between References

Given a dimension D with its hierarchy H, the distance between two members m, m' in this dimension is the shortest path [16] from m to m' in H. It is noted: $d_{members}(m, m')$. The distance between references is then defined in the following way from $d_{members}$.

Definition 1. *(Distance between references) Given two references* $r_1 = \langle r_1^1, ..., r_1^n \rangle$ *and* $r_2 = \langle r_2^1, ..., r_2^n \rangle$ *of an n-dimensional cube, the distance between* r_1 *and* r_2 *is:* $d_{references}(r_1, r_2) = \sum_{i=1}^{n} d_{members}(r_1^i, r_2^i)$

Example 2. As an example, consider the two references of query q_2 given in the previous example. These references only differ on the *Product* dimension. As members *Drink* and *Food* have the same parent in the hierarchy of the dimension *Product*, then $d_{members}(Drink, Food) = 2$. Thus the distance between these two references is $2+0+0+0=2$.

4.3   Distance between Queries

As MDX queries are modeled as sets of references, comparing two MDX queries boils down to comparing two sets of references. In our approach we use the classical Hausdorff distance [17] for comparing two sets based on a distance between the elements of the sets. Informally, two sets are close if every element of either set is close to some element of the other set.

Definition 2. *(Hausdorff distance) Given two queries* q_1, q_2*, the distance between* q_1 *and* q_2 *is:*
$$d_h(q_1, q_2) = \max\{ \max_{r_1 \in q_1} \min_{r_2 \in q_2} d_{references}(r_1, r_2),$$
$$\max_{r_2 \in q_2} \min_{r_1 \in q_1} d_{references}(r_1, r_2) \}$$

This distance d_h is combined with the distance $d_{dim}(q_1, q_2)$ that gives the number of dimensions where q_1 and q_2 differ (if $q_1 = R_1^1 \times ... \times R_n^1$ and $q_2 = R_1^2 \times ... \times R_n^2$, D_i is a dimension where q_1 and q_2 differ if $R_i^1 \neq R_i^2$). Thus the distance between queries is defined as the following function of these two distances.

Definition 3. *(Distance between queries) Given two queries* q_1, q_2*, the distance between* q_1 *and* q_2 *is :* $d_{queries}^{\gamma}(q_1, q_2) = \gamma \times d_{dim}(q_1, q_2) + (1 - \gamma) \times d_h(q_1, q_2)$ *where* $\gamma \in [0, 1]$.

Example 3. Consider query q_2 described above and the queries given in the appendix. $q_2 = \{r_1 = \langle Drink, alltime, allcustomer, San Francisco \rangle, r_2 = \langle Food, alltime, allcustomer, San Francisco \rangle\}$ and $q_2^2 = \{r_3 = \langle Drink, alltime, allcustomer, USA \rangle, r_4 = \langle Food, alltime, allcustomer, USA \rangle\}$. Note that q_2^2 rolls up q_2 from the city level to the country level. Their distance is computed as follows. $d_{references}$ is used to compare r_1 to r_3 and r_4. We have $d_{references}(r_1, r_3) = 2$ and $d_{references}(r_1, r_4) = 4$. The minimum is 2. r_2 is also compared to r_3 and r_4, the minimum being also 2. Thus the maximum of these two rounds of comparison is 2. Now r_3 is compared to r_1 and r_2 and so is r_4. In both cases the minimum is 2. Therefore $d_{queries}^0(q_2, q_2^2) = 2$.

The following property indicates the range of possible values for the distance $d_{queries}^{\gamma}$. The maximal value for this distance is denoted $d_{queries}^{max}$.

Property 1. Given an n-dimensional cube C, the distance $d_{queries}^{\gamma}$ ranges from 0 to $d_{queries}^{max} = \gamma \times n + (1 - \gamma) \times 2 \times \sum_{i=1}^{n} h_i$ where h_i is the height of the hierarchy of dimension i.

5   Comparing Analysis Sessions

In this section, we present our approach for comparing two sessions. The basic idea stems from *Approximate String Matching* [5], which we introduce briefly in the following definitions.

5.1   Definitions (Edit Distance, Session, Log)

Given two sequences s_1, s_2, Approximate String Matching is the problem of matching the sequences allowing errors. The matching relies on the computation of a distance between the sequences, which is the minimal cost of the sequences of operations transforming s_1 into s_2. The classical Levenshtein (or edit) distance [18] is commonly used. It allows the following operations: insertions, deletions, substitutions. If the cost associated with each of these operations is 1, this distance can be thought of as the minimal number of insertions, deletions or substitutions to make the two sequences equal.

In our approach, the sequences we consider are sequences of MDX queries which we call analysis sessions (or sessions for short). A log is a set of sessions.

Example 4. Session s_c of Section 3 is the sequence $\langle q_1, q_2, q_3 \rangle$. The log given in appendix is the set $\{s_1, s_2, s_3\}$ and session $s_3 = \langle q_1, q_2^2, q_3, q_6 \rangle$. If insertions, deletions and substitutions are operations allowed on sessions, a sequence of operations that transforms s_c into s_3 is: substitute q_2 by q_2^2 and insert q_6 at the end. If all operations have the same cost 1, then this sequence costs 2. Another sequence that transforms s_c into s_3 is: delete all queries from s_c and insert respectively queries q_1, q_2^2, q_3 and q_6. Obviously the cost of this sequence is not minimal.

5.2   Distance between Sessions

we compute a distance that is the minimal cost of a sequence of operations (called an edit sequence) to transform s_1 into s_2. As in the edit distance the operations permitted are:

- The substitution of a query q_1 by a query q_2. The cost of this operation is the distance between q_1 and q_2 as defined in Definition 2, that is $d_{queries}^{\gamma}(q_1, q_2)$.
- The insertion (resp. deletion) of a query in a sequence. The cost of these operation is a constant α.

An intuitive reason for a fixed cost for insertion (or deletion) is the following. Suppose we want to compute a distance between session $\langle a \rangle$ and session $\langle a, b \rangle$ on the one hand and session $\langle a \rangle$ and session $\langle a, b' \rangle$ on the other hand. There is no reason for distinguishing or favoring the adding of b from the adding of b'. In both cases, a user found these two particular queries of interest, and the sessions are distant from $\langle a \rangle$ only in that a query has been added.

Now, the value for α can range from 0 to $d_{queries}^{max}$. Low values for e.g., insertion allow not to discriminate longer sessions too much. On the other hand, the value should be high enough since it should be more expensive to delete and then insert instead of substituting. Adjusting the value for this cost is part of the experiments described section 7.

Definition 4. *(Distance between sessions) The distance between two sessions s and s' is the minimal cost of all edit sequences transforming s into s'. It is noted $d_{sessions}$.*

The following property states that $d_{sessions}$ is a metric in the mathematical sense.

Property 2. $d_{sessions}$ is a metric in that it satisfies the following properties: non-negativity, symmetry, triangle inequality.

Example 5. Consider the sessions s_c presented in Section 3, and the sessions given in the appendix. Suppose $\gamma = 0$ and α (the cost for inserting or deleting) is $d_{queries}^{max}/2 = 14$. The sequence having minimal cost for transforming s_c into s_1 is: substitute q_2 by q_2^2 and then substitute q_3 by q_4. Substituting q_2 by q_2^2 costs $d_{queries}^0(q_2, q_2^2) = 2$ and substituting q_3 by q_4 costs $d_{queries}^0(q_3, q_4) = 6$ (cf. Example 3). Thus $d_{sessions}(s_c, s_1) = 8$. The sequence having minimal cost for transforming s_c into s_2 is: substitute q_2 by q_3^2 and then substitute q_3 by q_5. Its cost is: $d_{sessions}(s_c, s_2) = 4$ (for substituting q_2 by q_3^2) $+3$ (for substituting q_3 by q_5). The sequence having minimal cost for transforming s_c into s_3 is the first one given in Example 4. Its cost is: $d_{sessions}(s_c, s_3) = 2$ (for substituting q_2 by q_2^2) $+14$ (for inserting q_6).

6   The Recommender System

In this section, we present how we use the distances defined above to recommend MDX queries. The principle is the following: The log is searched for candidate sessions matching the current session. From these candidate sessions a set of recommended queries is obtained. These recommended queries are ranked and presented to the user as recommendations in the resulting order. The best ranked queries are called the best recommendations.

Before detailing the algorithm we introduce the following definitions. The candidate sessions are the closest to the current session in the sense of the distance between sessions.

Definition 5. *(Candidate sessions) Given a set L of sessions and a session s_c, the set of candidate sessions is defined by $Cand_{sessions}(s_c, L) = \{s \in L | \nexists s' \in S, d_{sessions}(s', s_c) < d_{sessions}(s, s_c)\}$*

To define the recommended queries, we use an analogy with Web search, where it has been shown that what is seen at the end of a session can be used to enhance further searches [19]. Indeed, even in our case, it makes sense to consider that if the session ended on this particular query, it is because the user found something of interest. We adopt this point of view and simply define a recommended query to be the last query of a candidate session. The best recommendations are the recommended queries that are the closest to the last query of the current session, in the sense of the distance between queries.

Definition 6. *(Recommended queries and best recommendations) Given a set L of sessions and a session s_c, the set of recommended queries is defined by $Reco_{queries}(s_c, L) = \{last(s) | s \in Cand_{sessions}(s_c, L)\}$ where $last(s)$ is the last query of session s. Given a set C of recommended queries and a session s_c, the best recommendations are: $best(s_c, C) = \{q \in C | \nexists q' \in C, d^{\gamma}_{queries}(q', last(s_c)) < d^{\gamma}_{queries}(q, last(s_c))\}$*

Note that changing these definitions can have an important impact on the subjective quality of the recommendations. Assessing this is part of our long-term goal as discussed in conclusion.

Finally the algorithm for recommending MDX queries is:

> **Input:** A current session s_c and a log L
> **Output:** A sequence of recommendations
> 1. Generate the set C of recommended queries $C = Reco_{queries}(s_c, L)$
> 2. **Let** *Output* be the empty sequence
> 3. **Repeat until** C empty
>   (a) Generate the best recommendations $best(s_c, C)$
>   (b) Append $best(s_c, C)$ to *Output*
>   (c) Remove $best(s_c, C)$ from C

Example 6. Consider the distances computed in Example 5. There is only one candidate session which is s_2 since it is the closest to s_c. Thus there is only one candidate query q_5 which is then the recommendation. Suppose now the cost of the insertion operation used to compute the distance between sessions is 5. This means that there are now two candidate sessions s_2 and s_3. The candidate queries are q_5 and q_6. The query recommended first is q_6 since it is closer to q_3 than q_5.

7   Experiments

In this section, we present the results of the experiments we have conducted to assess the capabilities of our framework. We used synthetic data produced with our own data generator. Both our prototype for recommending queries and our generator are developed in Java using JRE 1.6.0_13. All tests are conducted with a Core 2 Duo - E4600 with 4GB of RAM using Linux CentOS5.

7.1 Data Set

We generated a set of sessions over the test database FoodMart supplied with the Mondrian OLAP engine [6]. Each session is generated in the following way: The first query of the session is selected by random among the 15 example queries supplied by Mondrian. Each subsequent queries is generated by choosing randomly one dimension and applying on the preceding query an OLAP operation (rollup, drilldown, changing the set of members) on this dimension. Our generator uses the following parameters: A number (X) of sessions in the log, a maximum number (Y) of queries per session. In our tests, we fixed the maximum number of references at 100 since it is reasonable to consider that users will very seldom produce a cross table larger than 10×10 as the answer to an MDX query.

7.2 Results

Note that, due to lack of space we have not included all the results of the experiments we have conducted.

7.2.1 Performance Analysis

Our first experiment assesses the time taken to generate the best recommendation for various log sizes. The performance is presented in Figure 1 according to various log sizes. These log sizes are obtained by multiplying parameters X (number of sessions) and Y (maximum number of queries per session). X ranges from 25 to 500 and Y ranges from 20 to 50. We thus obtain logs of size varying between 150 and 25000 queries. The best recommendation is computed for each of these logs, for current sessions of various sizes, generated with the session generator.

Figure 1 shows that the time taken to generate one recommendation increases linearly with the log size but remains highly acceptable and is slightly influenced by the current session size. Indeed, to recommend a query for a session s, the

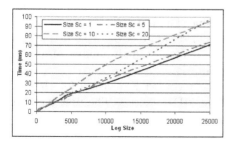

Fig. 1. Performance analysis

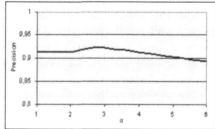

Fig. 2. Precision for various α (cost of insertion or deletion)

system only compares $last(s)$ to each query of the log and uses the distances previously computed for $s \setminus last(s)$.

7.2.2 Precision/Recall Analysis

We use a 10-fold cross validation to assess our framework in the spirit of the experimental validation done in [7]. The generated set of sessions is partitioned in 10 equally sized subsets and in each run 9 subsets are used as log and each session of the remaining subset is used as a basis for the current session. More precisely for each such session s_c of size n, we use the sequence of the first $n-1$ queries as the current session, and we compute the recommendations for the n-th query. The n-th query of s_c is called the expected query and is noted q_{ex}.

We evaluate the precision and recall [20] of the recommendations using the following metrics: precision$=|members(q_{ex}) \cap members(q_{rec})|/|members(q_{rec})|$ and recall$=|members(q_{ex}) \cap members(q_{rec})|/|members(q_{ex})|$, where $members(q)$ is the set of members of query q, q_{rec} is a recommended query and q_{ex} is the expected query. For each session, we report the maximum recall over all the recommended queries and the precision for the query achieving this maximum recall. The log generated for these tests has size 5877 queries (750 sessions).

Figures 3 and 4 show the inverse cumulative frequency distribution (inverse CFD) of the recorded precision, recall and/or F-measure[3] for the sessions. A point (x, y) in these graphes signifie that $x\%$ of sessions had precision or recall or F-measure $\geq y$.

The first experiments allow us to tune our system in order to choose for α and γ the values that achieve best precision and recall. Precision is computed for α which is the cost of the insertion (or deletion) operation (see Section 5). Figure 2 shows that a precision above 0.9 is obtain for $\alpha \in [1, 5]$. In the subsequent experiment, the value for α is 2. Precision and recall are computed for $\gamma = 0, 0.5$ or 1. Figure 3(a) and Figure 3(b) show that the worse results are obtained for $\gamma = 1$ i.e., when the distance between queries only counts the number of dimensions that differ (see Section 4). For 0 and 0.5 the curves are confounded. This shows that for $\alpha = 0.5$ d_{dim}, ranging only from 0 to n (see Property 1), contributes for nothing to the distance between queries. Thus in what follows, $\gamma = 0$.

Figure 4 shows the inverse CFD of precision, recall and F-measure of the recommendations computed with our system for $\alpha = 2$ and $\gamma = 0$. The results demonstrate the effectiveness of our method since for around 80 % of the sessions, precision and recall are above 0.8. These good results can be explained by the density of the log generated, considering the relatively small number of queries (15) in the pool we used for seeding the generation.

Figure 5 displays the inverse CFD of the recorded F-measure for the sessions for various methods for recommending MDX queries. The first method, called *ClusterH*, is the one proposed in [3] that uses a k-medoid clustering

---

[3] The F-measure, $F = 2.(\text{precision} \cdot \text{recall})/(\text{precision} + \text{recall})$, is a measure of a test's accuracy.

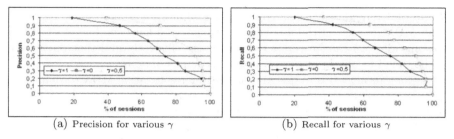

(a) Precision for various γ   (b) Recall for various γ

Fig. 3. Precision and Recall of the recommendations (for various γ)

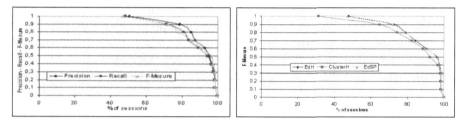

Fig. 4. Precision, Recall and F-Mesure of **Fig. 5.** F-Mesure of the recommendations
the recommendations (α = 2, γ = 0)   (α = 2, γ = 0) for the 3 possible methodes

algorithm with a simple Hamming distance to compare references. The second method, called *EdSP* (Edit Distance with Shortest Path), is the one proposed in the present paper for α = 2 and γ = 0. Finally the last method called *EdH* combines the Edit distance with the simple Hamming distance for comparing references. First we note that all methods achieve good results for our dense log. The method using a clustering algorithm performs slightly bad compared to the two others. It can also be seen that *EdSP* and *EdH* perform similarily, which may seem surprising at first since the Hamming distance for comparing references is coarse compared to the Shortest Path. However, it is to be noted that the way we compute precision and recall (inspired by [7]) favors *EdH*. Indeed, if *EdH* recommends a query close (in the sense of the Hamming distance) to the expected query it will have good precision and recall. But if *EdSP* recommends a query close (in the sense of the Shortest Path) to the expected query it can have bad precision and recall. Therefore it turns out that *EdSP* performs as well as *EdH* even though it is not favored by the computation of precision and recall.

8   Conclusion and Future Work

In this paper, we present a system for recommending MDX queries that is an evolution of the framework presented in [3]. Our framework leverages former navigations on a datacube and is based on two distances that we propose to compare MDX queries and analysis sessions. Our approach is implemented in a system that integrates with the open source Mondrian OLAP engine to recommend MDX queries on the fly. The experiments we have conducted show that recommendations can be computed on the fly efficiently and that our system can be tuned to obtain objectively good recommendations.

Our long term goal is to design a platform for generating MDX recommender systems by giving the user the possibility to adapt the approach to his/her needs. This can be done by proposing to the user various methods for computing candidate sessions and/or candidate queries. We are working on the definition of a new method that takes into account the measures' values and not only the references of the cells. A combination of the recommender system with techniques for OLAP query personalization [21] is also under consideration.

To fulfill this goal, we need to undertake experiments on real data sets with feedback from users. This will allow not only to improve the overall quality of the recommended queries but also to determine to which context a particular approach for computing candidate recommendations is adapted.

On the technical side, we need to propose an indexing method for organizing the log in order to make the search in the log even more efficient, and thus making it possible to search very large log files on the fly.

References

1. Sarawagi, S.: User-adaptive exploration of multidimensional data. In: VLDB, pp. 307–316 (2000)
2. Pedersen, T.B.: How is BI used in industry?: Report from a knowledge exchange network. In: Kambayashi, Y., Mohania, M., Wöß, W. (eds.) DaWaK 2004. LNCS, vol. 3181, pp. 179–188. Springer, Heidelberg (2004)
3. Giacometti, A., Marcel, P., Negre, E.: A framework for recommending olap queries. In: DOLAP, pp. 73–80 (2008)
4. Microsoft Corporation: Multidimensional expressions (MDX) reference (2008), http://msdn.microsoft.com/en-us/library/ms145506.aspx
5. Navarro, G.: A guided tour to approximate string matching. ACM Comput. Surv. 33(1), 31–88 (2001)
6. Pentaho Corporation: Mondrian open source OLAP engine (2009), http://mondrian.pentaho.org/
7. Chatzopoulou, G., Eirinaki, M., Polyzotis, N.: Query recommendations for interactive database exploration. In: SSDBM, pp. 3–18 (2009)
8. Sapia, C.: On modeling and predicting query behavior in OLAP systems. In: DMDW, pp. 2.1–2.10 (1999)
9. Sapia, C.: PROMISE: Predicting query behavior to enable predictive caching strategies for OLAP systems. In: Kambayashi, Y., Mohania, M., Tjoa, A.M. (eds.) DaWaK 2000. LNCS, vol. 1874, pp. 224–233. Springer, Heidelberg (2000)
10. Sarawagi, S., Agrawal, R., Megiddo, N.: Discovery-driven exploration of OLAP data cubes. In: Schek, H.-J., Saltor, F., Ramos, I., Alonso, G. (eds.) EDBT 1998. LNCS, vol. 1377, pp. 168–182. Springer, Heidelberg (1998)
11. Sarawagi, S.: Explaining differences in multidimensional aggregates. In: VLDB, pp. 42–53 (1999)

12. Sathe, G., Sarawagi, S.: Intelligent rollups in multidimensional OLAP data. In: VLDB, pp. 531–540 (2001)
13. Huang, X., Yao, Q., An, A.: Applying language modeling to session identification from database trace logs. Knowl. Inf. Syst. 10(4), 473–504 (2006)
14. Adomavicius, G., Tuzhilin, A.: Toward the next generation of recommender systems: A survey of the state-of-the-art and possible extensions. IEEE Trans. Knowl. Data Eng. 17(6), 734–749 (2005)
15. Wu, P., Sismanis, Y., Reinwald, B.: Towards keyword-driven analytical processing. In: SIGMOD Conference, pp. 617–628 (2007)
16. Dijkstra, E.W.: A note on two problems in connexion with graphs. Numerische Mathematik 1, 269–271 (1959)
17. Hausdorff, F.: Grundzüge der Mengenlehre. von Veit (1914)
18. Levenshtein, V.I.: Binary codes capable of correcting deletions, insertions, and reversals. Technical Report 8 (1966)
19. White, R.W., Bilenko, M., Cucerzan, S.: Studying the use of popular destinations to enhance web search interaction. In: SIGIR, pp. 159–166 (2007)
20. Baeza-Yates, R.A., Ribeiro-Neto, B.A.: Modern Information Retrieval. ACM Press/Addison-Wesley (1999)
21. Bellatreche, L., Giacometti, A., Marcel, P., Mouloudi, H., Laurent, D.: A personalization framework for olap queries. In: DOLAP, pp. 9–18 (2005)

A Appendix: A Toy Query Log

Session $s_1 = \langle q_1, q_2^2, q_4 \rangle$: Sales of alcoholic beverages in the USA

```
SELECT {[Store].[All Stores].Children} ON COLUMNS,
        {[Product].[All Products].[Food], [Product].[All Products].[Drink]} ON ROWS
FROM    [Sales]
WHERE   {[Measures].[Unit Sales]}
SELECT {[Store].[All Stores].[USA]} ON COLUMNS,
        {[Product].[All Products].[Food], [Product].[All Products].[Drink]} ON ROWS
FROM    [Sales]
WHERE   {[Measures].[Unit Sales]}
SELECT {[Store].[All Stores].[USA].Children} ON COLUMNS,
        {[Product].[All Products].[Drink].[Alcoholic Beverages]} ON ROWS
FROM    [Sales]
WHERE   {[Measures].[Unit Sales]}
```

Session $s_2 = \langle q_1, q_3^2, q_5 \rangle$: Sales of milk of the brand "Gorilla" in California

```
SELECT {[Store].[All Stores].Children} ON COLUMNS,
        {[Product].[All Products].[Food], [Product].[All Products].[Drink]} ON ROWS
FROM    [Sales]
WHERE   {[Measures].[Unit Sales]}
SELECT {[Store].[All Stores].[USA].[CA].[San Francisco]} ON COLUMNS,
        {[Product].[All Products].[Drink].[Dairy].[Milk].[Gorilla].Children,
        [Product].[All Products].[Drink].[Dairy].[Milk].[Gorilla]} ON ROWS
FROM    [Sales]
WHERE   {[Measures].[Unit Sales]}
SELECT {[Store].[All Stores].[USA].[CA],
        [Store].[All Stores].[USA].[CA].Children} ON COLUMNS,
        {[Product].[All Products].[Drink].[Dairy].[Milk].[Gorilla].Children,
        [Product].[All Products].[Drink].[Dairy].[Milk].[Gorilla]} ON ROWS
FROM    [Sales]
WHERE   {[Measures].[Unit Sales]}
```

Session $s_3 = \langle q_1, q_2^2, q_3, q_6 \rangle$: Sales of milk and cereals in San Francisco

SELECT {[Store].[All Stores].Children} ON COLUMNS,
       {[Product].[All Products].[Food], [Product].[All Products].[Drink]} ON ROWS
FROM   [Sales]
WHERE {[Measures].[Unit Sales]}
SELECT {[Store].[All Stores].[USA]} ON COLUMNS,
       {[Product].[All Products].[Food], [Product].[All Products].[Drink]} ON ROWS
FROM   [Sales]
WHERE {[Measures].[Unit Sales]}
SELECT {[Store].[All Stores].[USA].[CA].[San Francisco]} ON COLUMNS,
       {[Product].[All Products].[Drink].[Dairy].[Milk]} ON ROWS
FROM   [Sales]
WHERE {[Measures].[Unit Sales]}
SELECT {[Store].[All Stores].[USA].[CA].[San Francisco]} ON COLUMNS,
       {[Product].[All Products].[Drink].[Dairy].[Milk],
       [Product].[All Products].[Food].[Breakfast Foods].[Breakfast Foods].[Cereal]} ON ROWS
FROM   [Sales]
WHERE {[Measures].[Unit Sales]}

Preference-Based Recommendations for OLAP Analysis

Houssem Jerbi, Franck Ravat, Olivier Teste, and Gilles Zurfluh

IRIT, Institut de Recherche en Informatique de Toulouse
118 route de Narbonne, F-31062 Toulouse, France
{jerbi,ravat,teste,zurfluh}@irit.fr

Abstract. This paper presents a framework for integrating OLAP and recommendations. We focus on the anticipatory recommendation process that assists the user during his OLAP analysis by proposing to him the forthcoming analysis step. We present a context-aware preference model that matches decision-makers intuition, and we discuss a preference-based approach for generating personalized recommendations.

Keywords: OLAP analysis, Recommendations, Analysis context, Preferences.

1 Introduction

OLAP (On-Line Analytical Processing) systems aim to ease the decision-making process with a multidimensional data presentation. Data are organised according to subjects of analysis, called facts, and axes of analysis, called dimensions [10]. Dimensions are usually organized as hierarchies, supporting different levels of data aggregation. OLAP analyses are performed through interactive exploration of Multi-dimensional DataBases (MDB). It has been recognized that the workload of an OLAP application can be characterized by the user's navigational analysis task [4,5]: the user defines a first query then successively manipulates the results applying OLAP operations, such as drill-down, roll-up, slice and dice [7,13].

OLAP systems offering multidimensional and large data space cannot solely rely on standard exploration of MDB but need to apply recommendations to make the analysis process easy and to help users quickly find relevant data for decision-making [9]. In this paper, we focus on the conceptual framework and the implementation issues for integrating anticipatory recommendations in OLAP: the system guides the user navigation through the multidimensional data by proposing to him the forthcoming analysis step. In summary, we make the following contributions:

- We provide an efficient preference model that is intimately related to the multidimensional data model. Our model is conformed to the context-awareness nature of the analyst's interests [8].
- We present a preference-based approach to generate and rank recommendations.
- We present a prototype that implements the proposed approach and we show how progressively build recommendations based on user preferences.

T.B. Pedersen, M.K. Mohania, and A M. Tjoa (Eds.): DaWaK 2009, LNCS 5691, pp. 467–478, 2009.

The rest of the paper is organized as follows: Section 2 presents the OLAP analysis; section 3 introduces our framework, while section 4 discusses its implementation. Section 5 presents related work and section 6 concludes the paper.

2   From Multidimensional Modeling to OLAP Analysis

The analytical power of OLAP technology comes from its underlying multidimensional data model, called constellation [10,13].

2.1   Multidimensional Data Model

A *constellation* regroups several facts, which are studied according to several dimensions. It is defined as $(N^{CS}, F^{CS}, D^{CS}, Star^{CS})$ where N^{CS} is the constellation name, F^{CS} is a set of facts, D^{CS} is a set of dimensions, $Star^{CS}: F^{CS} \rightarrow 2^{D^{CS}}$ associates each fact to its linked dimensions.

A fact reflects information that has to be analysed through indicators, called measures. A *fact*, noted $F_i \in F^{CS}$, is defined as (N^{Fi}, M^{Fi}) where N^{Fi} is the fact name, $M^{Fi}=\{f_1(m_1),\ldots,f_w(m_w)\}$ is a set of *measures* associated to aggregation functions f_i.

A *dimension,* noted $D_i \in D^{CS}$, is defined as (N^{Di}, A^{Di}, H^{Di}) where N^{Di} is the dimension name, $A^{Di} = \{a^{Di}_1,\ldots, a^{Di}_u\}$ is a set of dimension attributes, $H^{Di} = \{H^{Di}_1,\ldots, H^{Di}_v\}$ is a set of hierarchies. Within a dimension, attribute values represent several data granularities according to which measures could be analysed. In a same dimension, attributes may be organised according to one or several hierarchies

A *hierarchy*, noted $H^{Di}_j \in H^{Di}$, is defined as (N^{Hj}, P^{Hj}) where N^{Hj} is the hierarchy name, $P^{Hj}=<id^{Di}, p^{Hj}_1,\ldots, p^{Hj}_{vj}, All>$ is an ordered set of attributes, called *parameters*, which represent useful graduations along the dimension, $\forall k \in [1..v_j]$, $p^{Hj}_k \in A^{Di}$.

Fig. 1 shows an example of a constellation that allows analysing online sales as well as the purchase activity of a worldwide distributor.

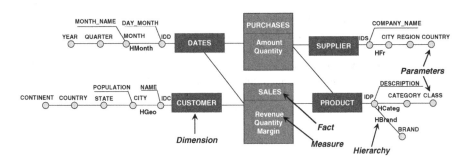

Fig. 1. Example of constellation schema

2.2 OLAP Analysis

An OLAP analysis consists in exploring interactively the MDB. It is considered as navigation: the user performs a set of OLAP operations in order to find relevant data for decision making. Thus, we define an OLAP analysis by a sequence of OLAP operations and intermediate results [9]. Each intermediate result represents an analysis context [8]. More precisely, a user analysis is described by a graph, where nodes represent the analysis contexts and the graph edges represent the user operations to move from one context to another [9].

Analysis Context Modeling. An analysis context represents a given state of the OLAP analysis. It displays analysis subject data that are aggregated according to dimension attributes. We assume two disjoint sets of elements that are displayed within an analysis context: a set of constellation structures *i.e.*, fact, measures, dimensions, and attributes, and a set of values of attributes and aggregated measures. We call *context prototype* the set of the constellation structures of an analysis context.

Definition. An OLAP analysis context is defined as $(C^F, \mathscr{C}^D, \mathscr{C}^R)$ where:

- $C^F = F \ (/ \ f_j \ (m_j) \in \{val_{mj}\})^+$ represents the analysed subject through a fact F, a set of displayed measures $m_j \in M^F$ associated with aggregate functions f_j (AVG, SUM, ...), and their underlying aggregated values $val_{mj} \in Type(m_j)^1$.
- $\mathscr{C}^D = \{C^{D1}, ..., C^{Du}\}$ where $\forall i \in [1..u]$, $C^{Di} = Di \ (/ \ p_k \in \{val_p\})^+$ represents one displayed analysis axis, where $p_k \in A^{Di}$ and $val_p \in Dom(p_k)^2$.
- $\mathscr{C}^R = \{pred^F_1, ..., pred^F_x, pred^D_1 ..., pred^D_y\}$ where $pred^F_i = f_j \ (m_j)$ *operator value* is a restriction predicate on fact data, and $pred^D_j = p_j$ *operator value* is a restriction predicate on dimension data where $p_j \in A^{Dj}, D_j \in D^{CS}$.

An analysis context is expressed by means of a tree *T(V, E)* (where *V* is the set of nodes and *E* is the set of edges) that reflects the nature of the relationship between the components of an OLAP analysis [8]. This tree structure is an internal view which is completely independent of the visualization structures of data.

Example 1. Consider the analysis context of sales revenue over 10K Euro by year (year $\geq$ 2008) according to the countries and cities of customers: $C = (C^F, \{C^{D1}, C^{D2}\}, C^R)$ where

- $C^F = SALES \ / \ Sum \ (Revenue) \in \{14, 13, 11, 20, 24, 18, 16\}$,
- $C^{D1} = CUSTOMER/Country \in \{France, USA\}/City \in \{Paris, Lyon, N-Y, Washington\}$,
- $C^{D2} = DATES \ / \ Year \in \{2008, 2009\}$, and
- $C^R = \{ 'Year \geq 2008', 'SUM(REVENUE) > 10' \}$.

The internal view of C is shown in Fig. 2 (a). Within the tree structure, the restriction predicates are integrated into the nodes of their measures or parameters.

This analysis context is displayed to the user according to a Multidimensional Table (MT) (see Fig. 2 (b)).

---

[1] Type(A) gives the set of all possible values of the attribute A.

[2] Dom(A) represents the set of values of A, *i.e.*, all $a_i \in A$ (Dom(A) $\subset$ Type(A)).

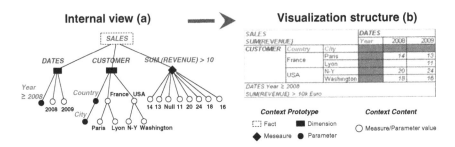

Fig. 2. Example of analysis context

3   The Framework

3.1   OLAP Recommendation Scenarios

In the following, we discuss how to apply recommendations based on the analysis pattern graph. From a conceptual standpoint, we argue that OLAP calls for the following three kinds of recommendations:

1. A part of an analysis node, *e.g.* the data granularity level: the system helps the user in building the decision-support reports. This consists in an *interactive assistance* in the query composition mechanism (see Fig. 3, step (1)).
2. An analysis node that the user would ask for by the next query (see Fig. 3, step (2)). This leads to expedite matters by anticipating the user navigation strategy and proposing the forthcoming analysis context. Such recommendations are called *anticipatory recommendations*.
3. Analysis nodes that are provided in addition to the classic result for the user operation in order to guide users toward relevant alternative results (see Fig. 3, step (3)). Alternative nodes do not necessarily belong to the classic analysis graph.

By recommendation of an analysis context throughout this paper we mean to recommend an analysis node within the user analysis graph.

In the following, we show how the recommended analysis contexts are generated. This requires a carefully dealing with the user preferences [9].

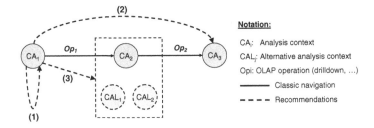

Fig. 3. Recommendation scenarios upon an OLAP analysis graph

3.2 OLAP Preferences Modeling

Problem Statement. A user preference in OLAP is associated with a specific analysis context [8]. The model in [8] captures only preferences on the constellation *structures* (dimensions and parameters). However preferences for *values* in OLAP are quite involved, as the following example shows: the left-hand side of a preference specifies the *content choice*, while the right-hand side represents the *context*.

Example 2. Decision-maker preferences include the following:
- (P_1) He is extremely interested in *export data* when analysing *sales of computers.*
- (P_2) He likes *amounts over 100k Euro* for the analysis of *sales before 5 years.*
- (P_3) He prefers data about the *biggest cities, i.e.,* cities with a population over 200 000, when analysing *sales in Africa.*

We argue that an accurate preference model for OLAP should be at both structures and content levels. Such model should depend on the analysis context. In this work, a relevance score θ_i associated with the analysis element e defines the degree of trust being the ideal choice for the user in a context c_i.

We now formalize the informal description given above, in the definitions of the preference context, the schema-level and the content-level preferences.

Preference Context. The analyst may have preferences that depend on more or less general contexts, *e.g.* a user preference can be associated with the context of analysis of sales or with a more detailed context such as the analysis of sales of a given product category. Hence, a preference context cp is a fragment of the analysis context tree. The idea behind such assumption is that the context of a user preference does not necessarily contain all the analysis context components.

Schema-level preferences. Preferences associated with a fact define relevant dimensions for the fact analysis in a specific analysis context, whereas preferences related to a dimension provide priority parameters for analysing data in a given context.

Definition. Given a constellation CS, a *schema-level preference* is a triple (E; θ; cp), where E is a dimension ($D_i \in D^{CS}$), or a dimension attribute ($p_i \in A^{Di}$), θ is a real number in the range [0, 1], and cp is a preference context.

We say that θ represents the *relevance degree* of the mapping of E to the context cp. The value 0 indicates lack of any trust in E from the user part, while value 1 indicates extreme trust.

Content-level preferences. The structure of content-level preferences is related to the features of MDB content, *i.e,* we consider preferences over fact data and preferences over dimension data. In particular, we assume that user preferences are stored at the level of atomic conditions, which are therefore called *restriction predicates.*

Definition. Given an attribute A associated with a data type Type(A), a *content-level preference* P^A is defined as (*pred*; θ; cp), where *pred* is a restriction predicate of the form A *op* a_i that specifies condition on the values $a_i \in$ Type(A), θ is a real number between 0 and 1, and cp is a preference context.

According to A, the predicate *pred* may be a restriction of fact data, *i.e.*, A is a measure associated with an aggregate function $f_i(m^F_i) \in M^F$, or a condition on dimension data, *i.e.*, A is a dimension attribute $p_i \in A^{Di}$. We assume op$\in \{=,<,>,\leq,\geq,\neq\}$ for numerical attributes and op$\in \{=,\neq\}$ for the other data types.

We call the set of user preferences that hold for a MDB, *profile P*. P consists of a set of scored mappings.

Example 3. Now let's revisit example 2. P_1 and P_2 are expressed as follows:

- $P_1^{Country}$ = ('Country $\neq$ *France*'; 0.9; c_1), where c_1 = Sales, Product/Category = *Computers*.
- $P_2^{Revenue}$ = ('Sum(Revenue) > *100* ; 0.6; c_2), where c_2 = Sales, Dates/Year $\leq$ *2004*.

The meaning of $P_1^{Country}$ is that the degree of trust to include the predicate 'Country $\neq$ France' into the qualification of the computers sales analysis is very high.

Notation. In the following, an analysis element denotes a dimension, a parameter or a restriction predicate.

3.3 Preference-Based Recommendations

Our approach supports three recommendation scenarios. In this paper, we focus on the anticipatory recommendation process. The aim of this process can be stated as follows: suppose C_i is an analysis context, which represents a node within the OLAP analysis graph. The recommendation problem consists in finding an anticipatory node C_{i+m} that follows C_i according to the user classic navigation graph.

The recommendation process transforms the analysis context C_i which results from the user query Q_i basing on his preferences. This consists of two stages: 1) recommendations building, where the system generates useful analysis nodes, and 2) recommendations ranking, where the candidate nodes are ranked with regard to preferences scores, so, only the best scored one is delivered to the user.

3.3.1 Generating Recommendations

This stage aims at generating candidate recommendations with regard to the current analysis context of the user, *i.e.*, the user query result. To generate recommendations, our framework goes through two steps: preference selection, and preference integration.

Preference Selection. The preference selection step takes as input the current analysis context C and a user profile. The output is a set of candidate mappings for C enhancement. We formulate this step as contexts matching problem: candidate preferences under a context C are those associated with contexts that match C. Following a total covering approach [8] leads to consider those preferences whose contexts are equal to or included in C. A preference context c_i (represented by the context tree t_i) matches C (represented by T) if $t_i \subseteq T$, *i.e.*, all the edges of t_i belong to T. More specifically, the system checks, for each edge E_{ij} (v_{ik}, v_{kj}) into t_i, if both nodes v_{ik} and v_{kj} belong to T and if they are laid in the same order. This leads to process with node-to-node matching. Depending on the node v of the preference context tree, we distinguish two cases of matching with the node V of T:

- v is a composite node (*i.e.*, v includes a restriction predicate): If v.predicate is not conflicting with V.predicate, then v matches V; otherwise v does not belong to T.
- v contains a simple value: If v.value = V.value, then v matches V.

Definition. We say that two restriction predicates are conflicting, if they target the same measure or parameter, and their conjunction returns no results.

If there are several preference contexts that match C, we consider the preference whose context covers more C. Actually, the more a preference context is detailed, the more specific the user interest represented by that preference is. Formally:

Definition (Covering context). Let C be an analysis context, c_i and c_j two contexts that match C, associated respectively with trees t_i and t_j. We say that c_i covers more C than c_j, if and only if $| t_i \cap T |^3 > | t_j \cap T |$ where $t_k \cap T = \{v_1,...,v_n\}$ is the common set of nodes within t_k and T.

Preference Integration. The preference integration is the process of enhancing an analysis context with user related preferences. It proceeds in two steps.

The first step considers the selected schema-level mappings. The basic idea is to gradually construct the analysis context prototype through preferences integration in decreasing order of their degrees of hierarchy: relevant dimensions are stated, then dimension attributes are specified.

The second step consists in integrating into the resulting analysis context C restriction predicates from selected content-level mappings. Recall that the current context C may include restriction predicates that arise from the user query. These predicates are preserved for the recommended node. Hence, a content-level mapping m is extracted providing that it is related to C, *i.e.*, if its restriction predicate is *not conflicting* with a predicate already there.

3.3.2 Ranking Recommendations

As there may be numerous candidate recommendations, we study in the following how to rank possible recommendations in order to provide the user with the best one. In this scenario, we intend to predict a *score S* assigned to a generated recommendation. This score can be seen as the user preference value of an analysis context (*i.e.*, an analysis node). Highest relevance is achieved when the recommendation is computed through most relevant elements. Thereby, a global preference value of an analysis context reflects the mutual relevance between the components of the analysis context. This leads to determine the user interest degree in each analysis component when integrated with the other ones. This semantic is ensured through the concept of contextual relevance degree of an analysis element.

Recall that an analysis node N is comprised of items that derive from the user query, noted N_Q, as well as items that are inferred from user preferences, called N_P. We assume that the score of an analysis node is a real-valued function that satisfies the following assumptions:

---

[3] $|S|$ is the number of elements of the set S.

- A1: Items that are settled by the user query make a maximal contribution to the global score since they traduce certainly the user exact needs as he asked for them.
- A2: Each analysis element that is inferred from user preferences contributes to the score with its contextual relevance degree.
- A3: A fact or a measure has no contribution to the global score since they are not recommended items.

In our approach, we have decided to place equal score on each member of N_Q. We define the following scoring function that associates a numeric score between 0 and 1 with every analysis node N basing on the relevance degrees of its members.

$$F(N) = \frac{\sum_{i=1}^{n} f(ei)}{\left|N_Q\right| + \left|N_P\right|} \text{ where } f(ei) = \begin{cases} 1 \text{ if } ei \in N_Q \\ \theta_i \text{ if } ei \in N_P, \text{ where } (ei, \theta_i, cp_i) \text{ is} \\ \text{a scored mapping} \end{cases} \quad (1)$$

This score is used to rank candidate recommendations. Then, the best ranked one is rendered to the user.

4 Implementation

We implement the proposed framework by extending a prototype [13] for OLAP manipulations to apply recommendations during OLAP analyses. This prototype allows the visualization of OLAP queries results through MT [7, 13] (see Fig. 2). Our prototype is summarized by the architectural view of Fig. 4. It consists of three layers:

1. User Interaction Layer. This layer allows users to express their queries through *the query formulation interface*, and to visualize the results as well as recommendations within the *analysis display interface*.

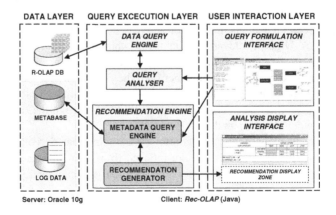

Fig. 4. System architecture

2. Query execution layer. This consists of the OLAP query engine that is extended with a *recommendation engine*. The query processing can be summarized as follows: The *query analyser* receives the user query and checks out the correctness of its expressions. A valid query is sent to the *data query engine* and the result is sent back to the recommendation engine which searches for related recommendations.

3. Data layer. The data layer addresses the representation of the constellation data as well as user preferences. We implement the constellation schema and the contextual preference model presented above into an R-OLAP database.

4.1 Recommendation Engine

The recommendation engine aims at building recommendations with respect to the user query result using the stored preferences. The basic idea is to gradually construct analysis contexts by integrating analysis elements. In this way, the *recommendation generator* asks at each step for the relevant element in order to enhance the present-built context. Preference selection is ensured by the *metadata query engine* that selects the most relevant preference under the present-built context. The recommendation generator updates progressively the context by selected elements.

Preference Extraction Algorithm. The Preference extraction algorithm generates a preference mapping *m*, that is related to the context *C* and associated with the preference target P_T. Recall that a preference associated with a fact F (resp. a dimension D) expresses the user interest to include the associated dimension (resp. parameter) into the qualification of an analysis context, whereas a preference related to a parameter (resp. to a measure) indicates the interest in an analysis context that *exactly* satisfy the associated restriction predicate.

The algorithm, presented in Fig. 5, is based on a total matching of the analysis context *C* with the set of contexts appearing in the preferences mappings *M*. Actually, the function *Match (M, C, P_T)* is invoked to generate the set of contexts of preferences whose target is P_T, that are included in *C*.

Only the preference context *cm* which covers more *C* is considered. If we are interested in a schema-level preference, then the algorithm outputs only the mapping related to *cm*, since there may be a single schema-level preference by context. For content-level preferences, the system returns the conjunction of restriction predicates that are related to but not conflicting with *C*. The global score of the conjunction of selected predicates is the average of the participating mappings' scores.

Recommendation Generation. Preferences that depend on the resulting analysis context are used to transform it in order to produce a recommended one. Then, the transformed analysis context is enhanced using the related preferences.

The integration of content-level preferences refines the resulting analysis context by keeping only relevant aggregated data. Such preferences allow predicting the user focus on a relevant restriction of the dimension or measure data.

Our prototype implements a scoring function that computes the global score of an analysis context (see formula (1)). A recommended analysis context is generated if its overall score is over a user threshold. Such threshold represents the average of the scores of recommendations the user selected in the past. If there are several generated recommendations, the system ranks them and generates the best scored one.

Input: Set of preference mappings $M = \{(e_i, \theta_i, cp_i)\}$
Current context C, Preference target P_T
Output: Selected preference mapping m

$CM = \{\}, cm = \{\}$
$CM \leftarrow Match\ (M, C, P_T)$
$cm \leftarrow cp_k \in CM$ such that cp_k covers more C
If (P_T is fact or dimension) **Then**
$\quad m \leftarrow (e_k, \theta_k, cm)$ where e_k is a dimension (resp. parameter) associated with P_T
Else if (P_T is parameter or measure) **Then**
$\quad pred = \{\}, \theta = \{\}$
$\quad$**Foreach** $m_j = (pred_j, \theta_j, cm) \in M$
$\quad\quad$**If** ($pred_j$ is conflicting with C) **Then**
$\quad\quad\quad$ discard m_j
$\quad\quad$**Else**
$\quad\quad\quad pred \leftarrow pred \wedge pred_j\ ;\ \theta \leftarrow \theta \cup \theta_j$
$\quad\quad$**End if**
$\quad\quad\quad m \leftarrow (pred, AVG(\theta_x, ..., \theta_y), cm)$
$\quad$**End for**
End if

Fig. 5. Preference extraction algorithm

4.2 Recommendation Display Interface

The recommended analysis context is displayed within *the recommendation display zone*. The system returns only the context prototype in order to not impact the query response time negatively: parameters and measure values are loaded when the user selects the recommended prototype. As the prototype of a recommended analysis context may be not intuitive, a simple explanation of the recommendation choices is displayed in the form of "context value that leads to anticipation → displayed item".

Example 4. The user has the following preferences:

– $P^{Dates}_3 =$ (Month; 0.5; c_3), and $P^{Revenue}_4 =$ ('Sum(Revenue)>100k Euro'; 0.6; c_3), where c_3= Sales/Sum(Revenue), Product/Description='Toshiba U300'; and
– $P^{Category}_5 =$ (Category='Telephony'; 0.6; c_4), where $c_4 =$ Sales/Sum(Revenue), Dates/Month).

Consider the analysis of sales revenue by year by city, where data are analysed for all products. Suppose the user focuses his analysis on the sales of the product *'Toshiba U300'*. The system displays the classic result within a MT as well as a recommendation of the next analysis step (see Fig. 6).

The classic query result is considered as the current analysis context: $CC =$ (Sales/Sum(Revenue); Dates/Year, Customer/City; Product/Description = 'Toshiba U300'). The system includes the preference P^{Dates}_3 whose context matches CC, which leads to change the dimension parameter in *Month*. This anticipates the user drilling up to the parameter month. Then, the system integrates the predicate of $P^{Revenue}_4$. Note

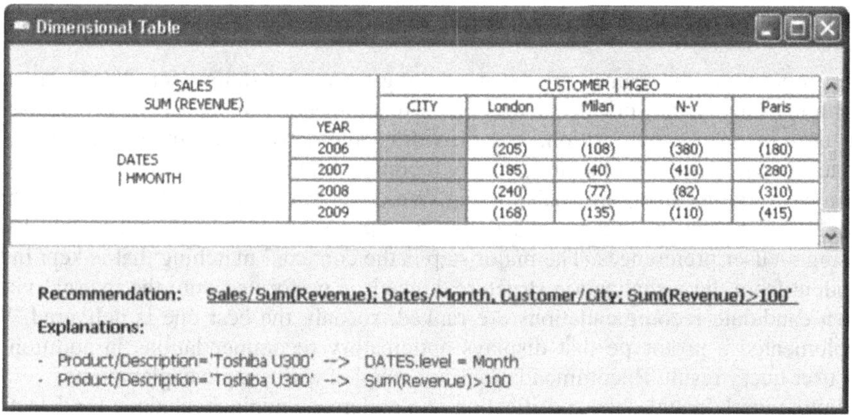

Dimensional Table						
SALES			CUSTOMER	HGEO			
SUM (REVENUE)		CITY	London	Milan	N-Y	Paris
DATES	HMONTH	YEAR					
		2006		(205)	(108)	(380)	(180)
		2007		(185)	(40)	(410)	(280)
		2008		(240)	(77)	(82)	(310)
		2009		(168)	(135)	(110)	(415)

Recommendation: Sales/Sum(Revenue); Dates/Month, Customer/City; Sum(Revenue)>100'

Explanations:

- Product/Description='Toshiba U300' --> DATES.level = Month
- Product/Description='Toshiba U300' --> Sum(Revenue)>100

Fig. 6. The result of user operation enhanced with a personalized recommendation

that $P^{Category}_5$ is discarded since its predicate is conflicting with the user query predicate: the category of 'Toshiba U300' is different from 'Telephony'. Therefore, the system recommends to the user to analyse sales revenue over 100k Euro by city by month. Explanations of the recommendation choices are displayed (see Fig. 6).

5 Related Work

We discuss related work regarding recommendation systems and OLAP preferences.

Recommendation systems are designed either based on content-based filtering or collaborative filtering. Content-based methods [12] recommend to the user items similar to the ones he preferred in the past, while collaborative filtering [11,15] recommends to the user items that people with similar preferences liked in the past. Recommendation approaches have been studied in many research communities, such as information retrieval [1], World Wide Web [2], and databases [15]. In OLAP systems, [6] proposes to recommend to the user the next query based on the log of the sequences of queries launched in the past. Recommendations are provided irrespective of user preferences, while in our approach the recommendations generation is a preference-driven process.

OLAP preferences are rather under-researched and should get more attention by OLAP community [14]. In [3] a preference is defined by a total order over the values of dimensions that consist of one attribute. User preferences are used to personalize the queries visualisation. Our work, however, targets the recommendation of helpful data for decision-making. Moreover, user preferences in [3] are independent from the user analysis context. In our earlier work [8], preferences are context-aware. However, the essence of context-awareness is that items have different relevance depending on the context of the user. For this reason, we look in this paper, how to capture different relevance degrees of the same element under different contexts. We capture such variations by mapping a preference to a context with a relevance score.

6 Conclusions and Future Work

We proposed a new framework for generating recommendations for OLAP data exploration. Our framework deals with three recommendation scenarios, *i.e.*, assisting the user in query composition, and providing alternative and anticipatory analysis contexts. We focused on the anticipatory recommendation process, which proposes to the analyst the forthcoming analysis step. We defined a preference-based approach to generate anticipatory recommendations. Recommendations are built progressively basing on user preferences. The major step is the contexts' matching that is kept independent from the visualization structure, since it is performed onto the internal view. Then candidate recommendations are ranked, so only the best one is delivered. We implemented a prototype that displays anticipatory recommendations in addition to the user query result. Recommendations are coupled with simple explanations.

Future work includes the specification of a preference mining technique for detecting preferences in the user log data. This technique must elicit the user preferences and discover the scored mappings that associate user preferences to their related contexts.

References

1. Baeza-Yates, R., Ribeiro-Neto, B.: Modern Information Retrieval. Addison-Wesley, Reading (1999)
2. Balabanovic, M., Shoham, Y.: Fab: Content-based, collaborative recommendation. Communications of the ACM 40(3), 66–72 (1997)
3. Bellatreche, L., Giacometti, A., Marcel, P., Mouloudi, H., Laurent, D.: A personalization framework for OLAP queries. In: DOLAP, pp. 9–18. ACM, New York (2005)
4. Choong, Y.W., Laurent, D., Marcel, P.: Computing Appropriate Representations for Multidimensional Data. Data & knowledge Engineering Journal 45(2), 181–203 (2003)
5. Dittrich, J.P., Kossmann, D., Kreutz, A.: Bridging the gap between OLAP and SQL. In: International Conference on Very Large Data Bases, pp. 1031–1042 (2005)
6. Giacometti, A., Marcel, P., Negre, E.: A Framework for Recommending OLAP Queries. In: DOLAP, pp. 73–80. ACM, New York (2008)
7. Gyssen, M., Lakshmanan, L.: A foundation for multi-dimensional databases. In: International Conference on Very Large Data Bases, pp. 106–115 (1997)
8. Jerbi, H., Ravat, F., Teste, O., Zurfluh, G.: Management of Context-aware Preferences in Multidimensional Databases. In: IEEE International Conference on Digital Information Management, pp. 669–675 (2008)
9. Jerbi, H., Ravat, F., Teste, O., Zurfluh, G.: Applying Recommendation Technology in OLAP Systems. In: Filipe, J., Cordeiro, J. (eds.) ICEIS 2009. LNBIP, vol. 24, pp. 220–233. Springer, Heidelberg (2009)
10. Kimball, R.: The Data Warehouse Toolkit, 1996, 2nd edn. John Wiley and Sons, Chichester (2003)
11. Konstan, J.A., Miller, B.N., Maltz, D., Herlocker, J.L., Gordon, L.R.: GroupLens: Applying Collaborative Filtering to Usenet News. Communications of the ACM 40(3), 77–87 (1997)
12. Maes, P.: Agents That Reduce Work and Information Overload. ACM 37(7), 31–40 (1994)
13. Ravat, F., Teste, O., Tournier, R., Zurfluh, G.: Algebraic and graphic languages for OLAP manipulations. International Journal of Data Warehousing and Mining 4(1), 17–46 (2008)
14. Rizzi, S.: OLAP preferences: a research agenda. In: DOLAP, pp. 99–100. ACM Press, New York (2007)
15. Satzger, B., Endres, M., Kießling, W.: A Preference-Based Recommender System. In: Bauknecht, K., Pröll, B., Werthner, H. (eds.) EC-Web 2006. LNCS, vol. 4082, pp. 31–40. Springer, Heidelberg (2006)

Author Index